Physical and Chemical Properties of Carbon Nanotubes

Physical and Chemical Properties of Carbon Nanotubes

Edited by **Lindy Bowman**

NY RESEARCH PRESS

New York

Published by NY Research Press,
23 West, 55th Street, Suite 816,
New York, NY 10019, USA
www.nyresearchpress.com

Physical and Chemical Properties of Carbon Nanotubes
Edited by Lindy Bowman

International Standard Book Number: 978-1-63238-356-3 (Hardback)

Printed in China.

Contents

Preface

This book presents a detailed account on the various physical and chemical properties of carbon nanotubes. These are tube like quasi-one-dimensional structures of nanometer-scale diameter made up of rolled graphene sheets. Over the past two decades, carbon nanotubes have drawn noteworthy attention of chemists, electronic device engineers, material scientists and physicists due to their outstanding electronic, chemical, optical, structural and mechanical properties. Carbon nanotube researches, particularly the ones focusing on industrial applications, are fast becoming more significant. The latest research topics on the structural, electric, physical and chemical properties on carbon nanotubes have been encompassed in this all-inclusive book. Reputed researchers from across the globe have contributed in this book. The aim of this book is to serve as a valuable source of reference for engineers, researchers as well as students engaged in the field of carbon nanotubes.

The information contained in this book is the result of intensive hard work done by researchers in this field. All due efforts have been made to make this book serve as a complete guiding source for students and researchers. The topics in this book have been comprehensively explained to help readers understand the growing trends in the field.

I would like to thank the entire group of writers who made sincere efforts in this book and my family who supported me in my efforts of working on this book. I take this opportunity to thank all those who have been a guiding force throughout my life.

Editor

Section 1

Physical Properties

Chapter 1

Carbon Nanotubes in a Fluidic Medium: Critical Analysis

Maria Alexandra Fonseca, Sylvio Freitas,
Bruno Lamas, Bruno Abreu, Hugo Calisto,
Nelson Martins and Mónica Oliveira

Additional information is available at the end of the chapter

1. Introduction

Under the current energy and climate change scenario, the implementation of measures that simultaneously aid resource preservation and environmental protection have gained a significant priority at a global scale. The continuous growth of fossil fuel price and the intensification of natural events severity are different but concurrent indicators of mankind's dangerous proximity to a point of no return, as far as long-term sustainability and environmental conservation are concerned. Over the past few decades, scientists and engineers have proposed and successfully employed all kinds of improvements in most industrial areas, however, progress still needs to be made.

Heat exchangers are widely used as a part of most mechanical systems. They are a basic component in industrial plants, transportation equipment and even buildings. The last 20-30 years have witnessed a continuous rise in heat transfer exchange necessities together with a sustained tendency to the equipment miniaturization, leading to notorious heat transfer intensification needs. Although heat transfer intensification has been mostly a core engineering need, it may be regarded as an indirect way to improve systems sustainability. Intensified heat transfer processes require smaller heat exchangers, less heat transfer fluids, lower pumping energy requirements and consequently more sustainable systems. With a perspective of passively increase heat exchanging thermal performance through its intensification, the use of colloidal dispersions of solid nanoparticles in common base fluids (nanofluids) has recently been proposed. Experimental studies have demonstrated that, compared to the base fluids alone, specifically tailored carbon nanotubes (CNT) based nanofluids may show significantly increased heat transfer capabilities, making them very promising from an intensification point of view and consequently from the system sustainability point of view.

Nanofluids can be categorized as a function of the nanoparticle material and shape, which are typically dispersed in an engineering base fluid such as water, ethylene glycol, oil or even a mixture of these. The thermo physical properties of a nanofluid, besides the expected variation with pressure and temperature, will also depend of the considered nanoparticle, base fluid and nanoparticle concentration. In order to predict the performance of their use in industrial applications, a precise understanding of nanofluid thermophysical properties is essential. To that effect, a large amount of studies have been reported in the literature, most of them focus on spherical particle based nanofluids. The majority of the research work in this area has been centred on the quantification of the thermal conductivity enhancement via two approaches: experimental measurement and/or theoretical formulation. The most consensual feature of these studies is that the addition of nanoparticles does increase the effective thermal conductivity of common fluids and that tubular particle nanofluids exhibit the highest enhancements. However, the mechanisms responsible for such anomalous behaviour remain unknown and theoretical models can only match experimental data for specific situations, at short margined parameter values. Among the proposed mechanisms of enhancement, the most accepted are the thermal conductivities of the solid particles and base fluids, particle volume fractions, nano-layer formation, particle clustering and Brownian motion. In the case of nanotube suspensions, the percolation effect is an additional widely accepted heat conductivity enhancement mechanism. Despite the current understanding of the possible mechanism behind the observed behaviour, the exact influence of each parameter is yet to be quantified.

Significantly less attention has been conceded to investigating the convective heat transfer involving nanofluids and possible enhancement factors, an essential requirement in enabling the practical application of nanofluids, e.g., in heat exchangers. The limited available studies report the enhancement of the convective heat transfer coefficient which is explained with the increase of the thermal conductivity expected for the respective nanofluids. Virtually all experimental studies of convective heat transfer, typically studying nanofluid flow through circular tubes, agree that the particle volume fraction is influential to the thermal performance increases.

However, higher particle concentrations contribute to higher pressure drop of the fluid flow, which bares the inconvenience of increased pumping requirements. Another noted trend, though not as well agreed upon, is the dependence on the fluid flow state, most authors reporting convective heat transfer increases as function of growing Reynolds numbers. Several authors also found that the thermal enhancement occurred in the tube entry region, proposing the low boundary layer thickness as a possible culprit.

Most of the experiment model validations, as well as theoretical studies, apply the single-phase approach in nanofluid behaviour predictions. This approach treats the solid particle suspension in a liquid as a single-phase fluid, enabling the employment of classic correlations for convective heat transfer. Despite its reduced application namely due to a higher complexity, authors using the alternative two-phase approach have boasted better agreement with the experimental data. Still, no consensual model has been defined the responsible mechanisms, as well as their levels of influence are yet to be known. Research attempts to test the thermal performance of alternative heat exchanger types have been found, but these are still quite unique and lack proper validation.

Numerical studies of nanofluid heat transfer performances are the scarcest amongst the available literature, most conducted to evaluate forced convection in circular tubes, as the majority of experimental investigations. Here, the influence of the definition of boundary conditions and the associated theoretical assumptions selected are essential. Most studies employ the constant wall heat flux condition and are theoretically based on modifications of classical correlations for laminar and turbulent flow, whenever applicable. With regard to the mentioned anomalous nanofluid behaviour, exhibited in the tube entrances, some authors have employed finer meshes in the region, creating a non-uniformity of overall tube elements. In order to predict and understand the flow and thermal behaviour of nanofluids, further numerical investigation is still required.

2. Heat transfer fluids

The main limitation of the heat transfer enhancement techniques effectiveness is the poor thermal performance of the employed fluids, obstructing increases in performance and compactness of heat exchangers [1]. In what regards heat transfer performance, the fundamental thermo-physical properties of fluids are convective heat transfer, thermal conductivity, viscosity, and specific heat at constant pressure [2].

Table 1 presents the thermophysical properties of the conventional heat transfer fluids commonly used in cooling processes. As shown, water is the most efficient, having a higher thermal performance than glycols or engine oil and being more inexpensive. When freeze conditions or the need to increase the fluid boiling point exist, the addition of ethylene or propylene glycol is frequent, providing freeze and burst protection. Glycols have inferior thermal transfer properties than water and superior densities, resulting in higher flow-rates or heat transfer surface areas, leading to increased pressure drop, energetic requirements and equipment wear [3]. Engine oils can accumulate various functions specific to individual parts of engines, including heat dissipation, friction reduction, detergency and area sealing [4]. Oils also have inferior thermal transfer properties than water, being most suitable for heat transfer duties in which the fluid has increased boiling point requirements. The success rate of engine oils depends on the complex additives that are blended into these, being categorized as chemically active, with the capacity of interaction with metals at low oxidation and degradation costs, and chemically inert, which improve physical properties and effective performance [4, 5].

Fluid Property	Water	Ethylene Glycol	Engine Oil
Thermal conductivity @ 300K (W/mK)	0.613	0.258	0.145
Dynamic viscosity @ 300K (N.s/m^2)	0.798x10^{-3}	4.8x10^{-3}	319x10^{-3}
Kinematic viscosity @ 294K /(m^2/s)	0.801x10^{-6}	17.8x10^{-6}	-
Density @ 298K (Kg/m^3)	995.7	1096.78	1114.62
Specific heat capacity @ 295K (KJ/(Kg.K))	4.179	2.36	2.3927

Table 1. Thermophysical properties of the conventional heat transfer fluids [6].

3. Nanofluids

For heat transfer enhancement, the alternative to the conventional approach of increasing device size or altering component geometry is improving fluid performance. As was observed in Table 1, conventional fluids have poor thermal properties that constitute a paramount limitation in heat transfer enhancement. Following a general trend for system miniaturization with improved heat transfer requirements, experimental studies into the application of particles, with higher thermal performance, to these fluids have taken place. This has led to the development of nanofluids, nano-sized solid particles suspended in conventional fluids (base fluids) with the purpose of increasing the heat transfer performance of the fluids.

It is well known that solid materials tend to have much higher thermal conductivities than fluids [7], as it can be observed in Figure 1. Therefore, it is expected that fluids containing suspended solid particles, millimetre and micrometer sized, display significantly enhanced thermal conductivities, as compared with conventional fluids, due to their large surface areas. However, its usage has not been reported on practical applications due to sedimentations, erosion, fouling and increasing pressure drops for pumping the fluid. Modern material processing technologies provided the possibility of nano-scaled material production, which has originated the emergence of nanofluids, at a time in which heat transfer requirements are summiting. Due to the smaller size, nanoparticles exhibit higher mobility and less particle interaction, allowing for improved stability and heat transfer. Such factors are significant in reducing pumping power, fluid inventory and eliminating clogging issues, thus increasing global interest in nanofluid research.

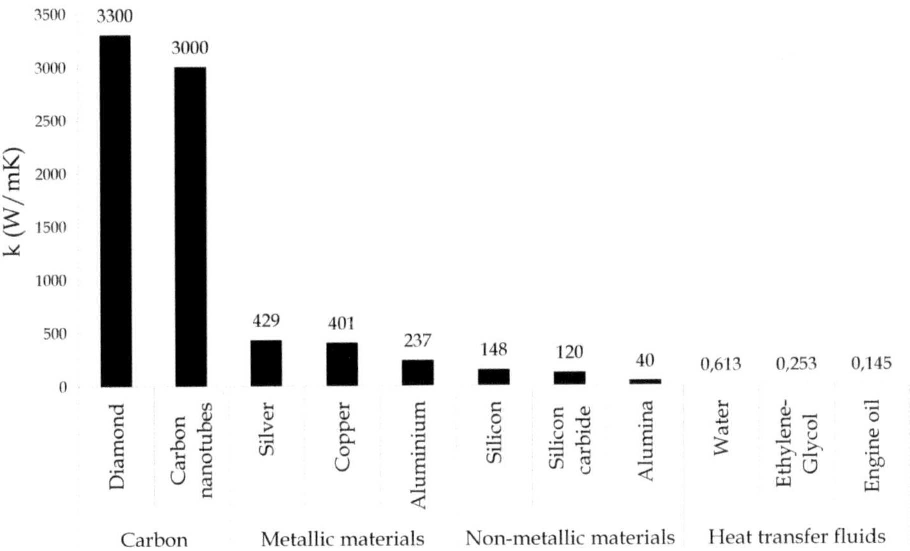

Figure 1. Thermal conductivities of common solids and liquids, at room temperature [7].

The application of different types of metallic and oxide nanoparticles has been investigated by several researchers. The most widely studied nanoparticles are the Al_2O_3 and CuO. In the last decade, carbon nanotubes (CNT) have attained a great interest from researchers, due to their exceptional physical properties, namely their enormous thermal and electrical conductivity [8]. The possibility to incorporate a small amount of CNTs into base engineering fluids in order to enhance its thermal conductivity has been studied.

However, the major drawback of using CNTs lies with its poor dispersability and homogeneity into the base fluid [9] (pristine CNTs tend to aggregate). The aggregation of carbon nanotubes may cause their settlement, the clogging of the flow channels and the decay of the overall effective properties.

3.1. Nanofluid tailoring

Nanofluid preparation is a complex process due to a number of factors with a high degree of difficulty in maintaining under control. Masuda et al. [10] found that particle instabilities resulted in agglomerations while Grimm [11] had to conduct experiments in an accelerated manner due to imminent particle settling. Nowadays, nanofluid development is based on two methods: the one-step method and the two-step method, the latter being the one employed in earlier investigations. The most important features when considering a massive nanofluid production are the nanoparticle materials and the host liquids. It is vital that nanoparticles show an high dispersability and stability and chemical compatibility with the selected host fluid.

Particle agglomeration can be prevented by balancing attractive forces between the nanoparticles within the fluid, obtainable through the employment of two common mechanisms: electrostatic stabilization (also known as mechanical or physical) or steric stabilization (chemical) [12, 13]. Mechanic stabilization generally includes ultrasonication, i.e., different pressure wave generation which induces cavitation and consequent particle deagglomeration. It consists of the placement of an electric charge on the particle surfaces with the purpose of to force kinetic stability. The absorption of ions to the electrophilic metal surface during the process creates an electrical multi-layer, resulting in repulsive electric forces between nanoclusters [14, 15]. Electrostatic stabilization is pH sensitive and of limited use, therefore steric stabilization is being most frequently applied [12]. This chemical methodology consists in the addition of a bulky material layer, such as a polymer or surfactant, to the nano particles, which provides a steric barrier enabling cluster prevention [16-18]. An alternative mechanism of stabilization, involving the addition of charged nanoparticles to microspheres, has recently been reported. The stabilization is obtained through the addition of small charged nanoparticle concentrations which repel Van der Waals attraction forces [19].

CNT nanofluids have been found to have a greater instability than spherical particle nanofluids, a consequence of the tendency to assemble into bundles or ropes which results from the stronger Van der Waal attractive forces between carbon surfaces, intensified by the higher nanotube specific areas [13, 20]. Nanotube morphology and attractive forces between tubes are highly influential in successful dispersions [21]. CNTs are usually suspended with the assistance of steric stabilization though surfactant employment or other functionalization techniques, which generally involves nanotube treatment with acids at high temperatures [13, 22]. Functionalization also aids in effectively preventing nanotube aggregation [23]. As with the spherical nanoparticles, electrostatic (or physical) stabilization is typically avoided, as it has also been found to cause the CNTs destruction [24]. According to Hilding et al. [21] the most important challenges in CNT nanofluid production are the chemical and morphological purification of the nanotubes, the uniform and reproducible dispersion and the orientation of the nanotubes in liquid and melt phases. The use of the mentioned surfactants is advantageous in interface absorption and accumulation in supra-molecular structures, thus aiding a uniform dispersion [20]. Nasiri et al. [13] experimentally concluded that functionalized suspensions present better stability, dispersion and thermal conductivity than suspensions obtained through ultrasonication, resulting in a lower propensity for agglomeration and precipitation. The same authors also found that the thermal conductivity of all suspensions decrease with time, the reduction rate varying with the preparation method. However, an experimental study using plasma coating on MWCNT nanoparticles to improve stability was successfully conducted by Kim et al. [25]. A desired nanotube orientation can be obtained through shear flows, elongation flows or by electric and magnetic field application [21].

4. Conductive heat transfer

Being below the critical length scale, the physical properties of nanoparticles differ from conventional bulk solids, serving as motivation for a large number of studies on nanofluid behaviour. Most of these studies echo higher heat transfer performances than that of the base fluids alone, though some contradictory results have been reported [26]. Experiments have shown that the thermal conductivity of nanofluids depends on a large number of parameters, but an accurate, consensual prediction of its variation with these parameters is yet to be established [27].

Measurement of the thermal conductivity of liquids can be a difficult task, this due to the necessity of establishing a steady one-dimensional temperature field. The measurements should be taken in a reduced timeframe, so that convection currents cannot develop, while liquid heating should take place from above to facilitate heat conduction in a layer wise manner. The most common measurement methods are the transient hot-wire method, cylindrical cell method, temperature oscillation method and *3-omega* method.

The transient hot-wire technique is the most employed method, consisting in the measurement of temperature and time response of a platinum wire, acting as a probe, subjected to an abrupt electrical pulse [27, 28]. The wire, heated resistively, usually employing a *Wheatstone bridge* resistor setup, is suspended in the nanofluid with the purpose of increasing its temperature [27, 29]. The temperature increase of the fluid, measured by the wire, depends on its thermal conductivity, which is calculated from the temperature-time profile of the wire [28-32]. This method has several advantages, the most significant being the capacity to eliminate experimental errors associated to natural convection, as well as the relatively accelerated measurement process [27]. The main drawback is the need for a chemical wire coating for measurements in electrically conducting fluids [28, 29].

Over the last decade, a large number of experimental studies investigating the enhancement in thermal conductivity of nanofluids have been conducted. Most have shown increases in thermal conductivity of nanofluids, even at diminutive particle volume fractions, when compared to base fluids, in most cases exceeding the predictions of theoretical models developed for suspensions of larger particles. The studies have also revealed that the thermal conductivity depends on several parameters and that these dependencies cannot be discarded when investigating heat transfer mechanisms of nanofluids [33-36].

Eastman et al [37] has shown an increase in thermal conductivity of about 60% for the nanofluids consisting of water as base fluid with 5%vol of CuO nanoparticles. Wang *et al* [9] measured the thermal conductivity of Al_2O_3 and CuO nanoparticles dispersed in distilled water (DW), ethylene-glycol (EG) and engine oil (EO). The increase of thermal conductivity was different for each base fluid, suggesting an effective thermal conductivity dependent on the base fluid thermophysical properties. Xuan *et al.* [38] studied the thermal conductivity enhancement for Cu-DW nanofluid. The result was the increase of 56% for the thermal conductivity with a 5% of volume fraction. Xie *et al.* [39] investigated the effects of pH value of the suspensions, the specific surface area of the nanoparticles and the thermal conductivity of the base fluid. The enhancement observed seemed to increase with the decrease of pH values, and with the increase of nanoparticles specific surface area. Das *et al.* [40] investigated the influence of the temperature in the enhancement of thermal conductivity of nanofluids based on Al_2O_3 and CuO. The experimental results have shown that the thermal conductivity rise with an increase in temperature. Murshed, et al [31] reported a maximum of 33% enhancement on thermal conductivity for 5% volumetric loading of TiO_2 nanoparticles in water based nanofluids.

In the last decade, carbon nanotubes (CNT) have attained a great interest from researchers, due to their exceptional physical properties, such as its enormous thermal and electrical conductivity [41]. The possibility to incorporate a small amount of CNTs into base fluids, in order to enhance its thermal conductivity has been studied. Effective thermal conductivity measurements of dispersed CNTs in synthetic poly-oil were carried out [42] and it was found out a 160% increase in the thermal conductivity of oil at 1% volume fraction of CNTs. Thermal conductivity of Cu-ethylene-glycol nanofluids was assessed [43], and it was observed a 40% enhancement containing approximately 0.3% vol. of Cu nanoparticles with a mean diameter of 10 nm. A thermal conductivity increase of 35% to 79% for 0.5 to 1.0% vol. of carbon nanoparticles in water was also observed [44]. The viscosity of CNT nanofluids as a function of shear rate was also measured at different temperatures and concentrations, at pH=6. It was observed a shear thinning behaviour, at low shear rates, but a slightly shear thickening is seen at shear rates greater than $200s^{-1}$. It was observed that the thermal conductivity enhancement reaches up to 17.8% at the volume fraction of 0.01 (1%vol.) for CNTs – ethylene glycol base nanofluids [45]. At low volume fractions (<0.4 %vol.), nanofluids have lower viscosities than the corresponding basefluid, due to the lubricative effect of the nanoparticles. Various nanoparticles were studied [46], such as multiwalled carbon nanotubes (MWCNT), fullerene, copper oxide, silicon dioxide (SiO_2) and silver, and used it to produce nanofluids with enhancing thermal conductivity and lubrification. As basefluids, distilled water (DW), ethylene glycol (EG), silicon oil and poly-α-olefin oil (PAO) were used. These authors observed the highest thermal conductivity enhancement in MWCNT water based nanofluids, whereas the lowest one was observed for silicon dioxide (SiO2). They have also concluded that the higher thermal conductivity enhancement can be obtained for basefluids with the lower thermal conductivity.

At present there are no general and precise analytic models able to predict the thermal conductivity of nanofluids. Most of the tentative proposals in this field underestimate the nanofluid thermal conductivity when compared to experimental results [1, 47].

Earlier theoretical models are derived from the Maxwell model (1881), a semi-empiric correlation used to describe the effective thermal conductivity (k_{eff}) of larger scale spherical solid suspensions in liquids, with emphasis on the conductivity of the solid particles (k_p) and base liquids (k_b) used, as well as the volume fraction (ϕ) of solid particles [1, 48, 49]. However, despite working best for low concentrations of solid particles, the Maxwell model was rapidly found to under-predict effective thermal conductivities of smaller nano-scaled suspensions, also failing for non-spherical particle suspensions [1, 49].

$$k_{eff} = \frac{k_p + 2k_b + 2\left(k_p - k_b\right)\varphi}{k_p + 2k_b - \left(k_p k_b\right)\varphi} k_b \tag{1}$$

In 1935, Bruggeman proposed a similar model that didn't present the low volume fraction constraint of its predecessor, satisfactorily agreeing with early experimental data [49, 50].

$$\varphi\left(\frac{k_p - k_{eff}}{k_p - 2k_{eff}}\right)(1-\varphi)\left(\frac{k_b - k_{eff}}{k_b - 2k_{eff}}\right) = 0 \tag{2}$$

Later, the Hamilton and Crosser model (1962) introduced an empirical particle shape factor (n), dependent on sphericity, to account for its effect, obtaining the same result as the Maxwell model for spherical particles [1, 49].

$$k_{eff} = \frac{k_p + (n-1)k_b - (n-1)\left(k_p - k_b\right)\varphi}{k_p + (n-1)k_b - \left(k_p - k_b\right)\varphi} k_b \tag{3}$$

This model was the first to reveal the importance of the geometrical shape of the dispersed particles. However, since it is based on Maxwell model, this is only valid for heterogeneous systems where the interaction between the particles is negligible. Such heterogeneous mixtures are known as dispersed phase distributions. Although, for specifics volume fractions and shape of the particles, the mixture phase distribution may modify for aggregate structure or percolation-like structure.

The percolation theory predicts the existence of a critical particle concentration threshold, characterized by the formation of a continuous solid path, formed by highly conducting nanotubes coming into contact with each other, thus assisting the increase in thermal conductivity of several orders of magnitude [23, 51-53] consider heat percolation as one of the main culprits for the higher thermal conductivity exhibited by CNTs.

Xie and Chen [54], producing CNT nanofluids via ball milling, found that longer milling times lead to higher aggregation levels, which promote percolation, having concluded that the positive influence of aggregation surpasses the negative effect of aspect ratio deterioration that results from excessive milling. The percolation threshold of CNTs depends on nanotube dispersion, alignment, aspect ratio and surface modifications [23].

Munson-McGee [55] found that a reduction of the threshold can be obtained with an increase in aspect ratio, demonstrating that the critical volume of cylindrical particles can vary from less than 1% to just over 20%, whereas Biercuk, et al [56] indicated that the threshold for SWCNTs is approximately equal to the inverse of the aspect ratio, which was roughly 1000 in value, and that the percolation network formation occurs at inferior loadings, even for randomly oriented nanotubes.

Martin et al. [52] established that the threshold for MWCNTs could be controlled by diffusion during particle dispersion. The same authors also consider that reported inconsistencies between experimental observation and the statistical percolation theory are owed to the lack of considerations with regard to inter-particle interactions, neglecting the effects of *Van der Waals* forces and *Coulomb* interactions due to static particle charging, and kinetic effects, such as particle Brownian motion. They also predicted that aging high aspect ratio nanotube dispersions would lead to lowered percolation thresholds.

Lamas et al. [51] backed by theoretical studies of Biercuk et al. [56] and Nan et al. [57], indicate thermal conductivity discontinuities, attributed to *Kapitza* resistances and the anisotropy thermal properties of the CNTs, credited to play a significant role in defining the percolation threshold. These interfacial thermal resistances were recognized to depend on the bonding strength between CNTs and the surrounding medium, as well as low functionalization levels [58, 59]. The dependence of the *Kapitza* resistance on the strength of liquid-solid interactions was found to exhibit two distinct regimes: an exponential dependence for weak bonding and a power law dependence for strong bonding, in which thermal resistance is inversely proportional to the solid-liquid connection strength [59]. Shenogin et al. [58] found that the functionalization of SWCNTs leads to significant decreases of the referred thermal resistances, but they also witnessed drops in thermal conductivity with increases in functionalized atom fractions. They used the effective medium theory to predict that this dependence could be eliminated for highly functionalized CNTs.

The previously mentioned Hamilton and Crosser model (1962) was the first to enable the prediction of the effective thermal conductivity of non-spherical particles. However, this classic model was formulated for larger sized particles than the nano-sized ones employed in nanofluids, being found to under-predict the effective thermal conductivity of CNT nanofluids.

Nan et al. [57] presented a simple formula based on Maxwell's effective medium model, accounting for high particle aspect ratios and volume fractions. However, the model was later found to over-predict the thermal conductivity, explained by not accounting for the influence of the interfacial thermal resistances [51]. These conclusions led the authors to propose a modified model the following year, formulated to include, to some extent, the interface thermal resistance effect on the thermal conductivity [60]. According to the revised model, the effective thermal conductivity enhancement is a function of the volume fraction (ϕ) and β coefficients along the transverse direction (β_x) and the longitudinal direction (β_z). These coefficients depend on the thermal conductivities in each direction (k^c_{11} and k^c_{22}), in turn influenced by the particle thermal conductivity (k_p), nanotube diameter (d) and length (L), as well as the radius in which Kapitza resistance is influential (a_k). The Kapitza radius is a function of interface thermal resistance (R_k) and base fluid thermal conductivity (k_b). Although the increasing complexity, the revised model was found to lack precision for increasing volume fractions and also doesn't account for percolation effects, CNTs assumed to be isolated from each other [61].

$$k_{eff}=\frac{3+\varphi(\beta_x+\beta_z)}{3-\varphi(\beta_x)}k_b$$
$$\beta_x=\frac{2\left(k_{11}^c-k_b\right)}{k_{11}^c+k_b};\ \beta_z=\frac{k_{33}^c}{k_b}-1 \tag{4}$$
$$k_{11}^c=\frac{k_p}{1+\frac{2a_k}{d}\frac{k_p}{k_d}};\ k_{33}^c=\frac{k_p}{1+\frac{2a_k}{L}\frac{k_p}{k_b}};\ a_k=R_kk_b=8x10^{-8}\frac{m^2K}{W}xk_b$$

Xue [61] proposed a theoretical model which incorporates interfacial thermal resistances using an average polarization theory, as well as simultaneously considering the effects of nanotube dimensions and concentrations. The deduced expression leads the author to state that increases in thermal conductivity can be obtained via an increase in nanotube length, regardless of the corresponding diameter, which indicates that thermal variations along the transversal direction can be neglected. Same author also proposed a Maxwell-based model to account for the effect of the nanotube orientation distribution, founded on the discontinuity theory of dispersions in a continuous phase [62]. Once again, the effect of percolation was overlooked and the model only predicts increases in effective thermal conductivity for increases in particle volume fraction [51],

Sastry et al. [63] presented a model based on percolation and the contact resistance in the consequent thermal resistance network. A dimensionless parameter was introduced to represent the effect of percolation, it being a function of conductance between CNTs (G), CNT length (L_f) and particle volume fraction (ϕ). According to their model, the effective thermal conductivity depends on the number of parallel CNT chains (M), the number of connected segments (N) over a distance (X), CNT diameter (d) and the heat transfer area (A),

$$k_{eff}=\frac{X}{A}\left(\sum_{i=1}^{N}\frac{1}{\frac{k_bA}{dx_i}+\frac{M}{\frac{L_i}{\pi k_pd^2}+\frac{2}{Gd^2}}}\right)^{-1};\ M=\frac{4\varphi X^3}{\pi d^2LN} \tag{5}$$

Based on the Sastry et al. model [63], Koo et al. [64] proposed a revised model which takes the non-linear conductivity enhancement with particle concentration increases into account using the excluded volume concept, where the excluded volume is the volume surrounding an object in which the centre of an identical object should be missing in order to avoid object inter-penetration. According to their model, the role of percolation is represented not only by the Sastry et al. dimensionless parameter, but also by the number of contacts per cylinder (N_c) of randomly oriented cylinders, quantified by the product of the excluded volume and the CNT volume fraction, as follows:

$$N_c = \frac{\pi}{2} L^2 d \frac{\varphi}{\frac{\pi}{4} d^2 L} = 2\varphi \frac{L}{d} \tag{6}$$

Patel et al. [53] derived a model for CNT nanofluids based on a spherical particle conductivity model previously proposed by Kumar et al. [65], announcing a reasonable enhancement trend prediction for both oil and water based CNT nanofluids. Additionally, two paths for heat flow are assumed: one through the base liquid and another through the CNTs; both considered to be acting in parallel to each other. According to the model, the effective thermal conductivity is a function of the base liquid molecular size (r_b), the average CNT diameter (d) and the CNT volume fraction (ϕ).

$$k_{eff} = \left(1 + \frac{k_p \varphi r_b}{k_b \left[1-\varphi\right] d}\right) k_b \tag{7}$$

With a view to predict the effects of anisotropy, aspect ratio, non-straightness, CNT volume fraction and interfacial thermal resistance on the effective thermal conductivity, Deng et al. [66] and Deng and Zheng [67] proposed several analytical formulas. The most significant of these models takes into account a non-straightness of CNTs (η), a high thermal anisotropy of CNTs ($k_{11}^c / k_{33}^c << 1$), a random CNT orientation and a tube-end thermal resistance. The model also assumes the formation of CNT thermal cables, while the role played by the aspect ratio ($p = L \ / d$) is reflected by parameter H, as follows,

$$k_{eff} = 1 + \frac{\eta \frac{\varphi}{3}}{\frac{\frac{k_b}{\eta}}{k_{33}^c / \left(1 + 2R_k k_{33}^c / L\right)} + H(\eta p)} k_b$$
$$H = \frac{1}{p_2 - 1}\left[\frac{p}{\sqrt{p^2-1}} \ln\left(p + \sqrt{p^2-1}\right) - 1\right] \tag{8}$$

5. Convective heat transfer

In order to employ nanofluids in concrete applications, a full understanding of their convective heat transfer features is essential. When compared to the reported studies of thermal conductivity, convective heat transfer research is scarce, little attention having been given to determining the convective heat transfer characteristics of nanofluids [34, 68]. However, in recent years this mode of heat exchange has gained more awareness, result of a necessary comprehension for practical application. Convective heat transfer can be divided into two categories: closed flows forced convection, with induced fluid flow through confined regions such as pipes and channels, and natural convection, with fluid motion due to buoyancy both in closed and opened flow conditions [33].

5.1. Forced convection

Of the limited available literature, most studies are focused on the forced convective heat transfer of nanofluids in circular tubes. The convective heat transfer coefficient depends on thermal conductivity, specific heat capacity, viscosity, flow rate and density of fluids [29]. The fluid flow nature is of up most importance, studies being conducted for both laminar and turbulent flows. The reported investigations characterize the heat transfer with emphasis on the convective transfer coefficient (h) or more frequently, in a non dimensional approach, by the Nusselt number (Nu), as a function of the Reynolds number (Re) and the Prandtl number (Pr), also non dimensional numbers. The Nusselt number presents the ration between convective and conductive heat transfer and is given by

$$Nu = hL/k_f \tag{9}$$

where L is the characteristic length, k_f is the thermal conductivity of the fluid and h is the convective heat transfer coefficient.

A Nusselt number close to one, namely convective and conduction of similar magnitude, is characteristic of "slug flow" or laminar flow. A larger Nusselt number corresponds to more active convection, with turbulent flow typically in the 100-1000 range.

These parameters are dependent of the nanofluid thermal transport properties, most significantly the viscosity (μ) and the thermal conductivity (k), that are a function of the temperature. As a result, most experimental procedures involve the measurement of the fluid temperature in different regions along the tube. In experimental procedures, laminar flow convective heat transfer studies are frequently validated using the predictions of the Shah equation, whereas turbulent flows are typically validated by the Gnielinski or the Dittus-Boelter equations, defined in terms of the Nusselt number [69].

Shah equation for laminar flows (D, tube diameter and x, axial position along tube axis):

$$Nu = \begin{cases} 1.953\left(\mathrm{RePr}\frac{D}{x}\right)^{1/3}, & \left(\mathrm{RePr}\frac{D}{x}\right) \geq 33.3 \\ 4.364 + 0.0722\,\mathrm{RePr}\frac{D}{x}, & \left(\mathrm{RePr}\frac{D}{x}\right) < 33.3 \end{cases}$$

$$f \approx 0.078\,\mathrm{Re}^{-1/4} \tag{10}$$

$$Nu = \frac{\frac{f}{2}\left(\mathrm{Re}^{-10^3}\right)\mathrm{Pr}}{1+12.7\left(\frac{f}{2}\right)^{\frac{1}{2}}\left(\mathrm{Pr}^{\frac{2}{3}}-1\right)} \tag{11}$$

Dittus-Boelter equation for turbulent flows:

$$Nu = 0.023\,\mathrm{Re}^{0.8}\,\mathrm{Pr}^{0.4} \tag{12}$$

5.1.1. Laminar flow

Ding et al. [44] studied the heat transport properties of MWCNT/water nanofluids along a uniformly heated copper tube. The experimental data demonstrated good agreement with the Shah equation for laminar flows under the constant heat flux boundary condition. The convection coefficient was found to increase significantly with nanoparticle concentration and the Reynolds number, both for spherical and elongated particles (MWCNT). The highest values of the convective coefficient were witnessed in the tube entrance region, leading the authors to indicate the creation of multiple artificial entrances along tubes, to maximize the heat transfer via boundary layer degradation, for future studies. However, the most significant enhancement occurred in the vicinity of the tube midway point (110 times the tube diameter), about 375% at Re=800, for 0.5wt% CNTs.

Amrollahi et al. [70] conducted a similar experiment for both laminar and turbulent flows, using functionalized MWCNT nanofluids. Their experiment resulted in less significant values for the convective heat transfer enhancement (33-40% at Re=15920) when compared to the previous study.

Chen et al. [45] investigated the heat transfer behaviour of TiO_2 nanotube suspensions in water, through a tube, under the constant wall heat flux boundary condition. The convective heat transfer enhancement varied minimally for different nanotube loadings, decreasing along the tube length.

Garg et al. [71] prepared an experimental study to evaluate the influence of the ultrasonication extent, during MWCNT/water nanofluid preparation, on the heat transfer performance. To that effect, four distinct samples, with ultrasonication times ranging from 20 to 80 minutes, were prepared. The convective heat transfer was analysed along a copper tube, under a constant heat flux condition, with Reynolds numbers varying between 600 and 1200. As with previous studies, the convection coefficient was highest in the tube entry region but its maximum enhancement was found to occur in the developed boundary layer region. Surprisingly, the increase in Reynolds numbers resulted in a decrease of the convective heat transfer coefficient. The authors found that the optimum ultrasonication time was 40 minutes, above which the tube breakage rate increased, and that the viscosity of nanofluids increases with ultrasonication times.

Silva and Abreu et al. [72, 73] conducted similar experiments to evaluate the convective heat transfer of low particle concentration MWCNT/water nanofluids along a stainless steel tube, under the constant wall heat flux boundary condition. In both investigations the convection coefficient enhancement was greatly superior to that observed for the thermal conductivity, reported to be maximum in the tube entry region. Table 2 resumes the main experimental studies of forced convection heat transfer, under laminar flow, of tubular particle nanofluids. Greater attention will be paid to these studies further on, as these will act as an experimental basis for numerical model development and validation.

Author	Nanofluid	Nanotube Loadings	Test Tube Dimensions	Heat Transfer Enhancement
[44]	MWCNT/Water	0.1-0.5wt%	L=0.96m d=4.5mm	375% for h at 0.5wt% and Re=800
[71]	MWCNT/Water	0.25wt%	L=0.914m d=1.55mm	32% for h at Re=600; 29% at Re=900
[74]	MWCNT/Water	0.2vol%	L=1.8m d=5mm	10% for k at Re=1200
[70]	MWCNT/Water	0.1-0.25wt%	L=1m d=11.4mm	12% for h at 0.12wt%; 40% at 0.25wt%; Re=1592
[72]	MWCNT/Water	0.25vol%	L=1.2m d=6mm	105% at Re=2060
[73]	MWCNT/Water	0.25-0.5vol%	L=1.2m d=6mm	94% for h at 0.5vol% and Re=2061

Table 2. Summary of experimental studies of forced convective heat transfer, under laminar flow, of tubular particle nanofluids.

5.1.2. *Turbulent flow*

Despite extensive research, limited experimental investigations of tubular shaped particle based nanofluids were found. Amrohalli et al. [70] evaluated the convective heat transfer enhancement of functionalized MWCNT/water nanofluids for both laminar and turbulent flow modes. The convection coefficient in turbulent flows displayed a greater increase than the values obtained for laminar flows and the enhancement decreased with increasing temperatures. The coefficient was found to become constant with increasing the Reynolds number in the tube entry region.

Liu and Liao [75] studied the heat transfer behaviour of CNT dispersions in an aqueous solution of cetyltrimethyl ammonium chloride (CTAC), intentionally used to reduce drag. The heat transfer enhancement was found to be greater with higher temperatures, even when the drag reduction is insignificant, and higher particle loadings. The dependence of the convection coefficient on the Reynolds number was noticed to be minimal. Table 3 resumes the main experimental studies of forced convective heat transfer, under turbulent flow, of tubular particle nanofluids.

Author	Nanofluid	Nanotube Loadings	Test Tube Dimensions	Heat Transfer Enhancement
[70]	MWCNT/Water	0.1-0.25wt%	L=1m d=11.4mm	25% at 0.12wt% and Re=4778
[75]	CNT/CTAC	0.5-4wt%	L=1.08m d=25.6mm	70% for h at 2wt%; 40% increase at 4wt%; Re~45000

Table 3. Summary of experimental studies of forced convective heat transfer, under turbulent flow, of tubular particle nanofluids.

5.1.3. Theoretical studies

Despite a recent interest boom regarding convective heat transfer intensification by the use of nanofluids, theoretic models of convective heat transfer remain scarce, most are derived from classical correlations, such as the aforementioned Shah or Dittus-Boelter equations for laminar and turbulent flow, respectively. This bares the consequence of only being valid for specific nanofluids over small parameter variations [49].

Pak and Cho [76], following their experimental study of the heat transfer performance of Al_2O_3/water and TiO_2/water nanofluids under turbulent flow conditions, proposed the following correlation, a modified version of the Dittus-Boelter equation.

$$Nu = 0.021 Re_{nf}^{0.8} Pt_{nf}^{0.5} \tag{13}$$

Xuan and Roetzel [77] derived correlations for the convective heat transfer of nanofluids proposing two different approaches: the first treating nanofluids as single-phase fluids, the second assuming nanofluids as solid-liquid mixtures. The first method assumes that classical correlations for pure fluids, such as, can be applied to convective heat transfer predictions. The second approach continues to treat the nanofluid as a single-phase fluid but also takes into account the heat transfer enhancement due to the thermal dispersion that results from random particle motion. Both approaches indicated that the heat transfer enhancement depends on thermal conductivity increase and chaotic particle motion, which accelerates energy exchanges. For the second approach, the effective thermal conductivity of the nanofluid is given by the sum of the contributions of the single-phase thermal conductivity of the nanofluid (k_{nf}) and the thermal conductivity of the dispersion (k_d), as follows,

$$k_{eff} = k_{nf} + k_d;\ k_d = C(\rho Cp)_{nf}\, \mu_X \varphi d_p r_0 \tag{14}$$

Based on the thermal dispersion and their experimental results for Cu/water nanofluids, Li and Xuan [78] proposed a model for predicting forced convective heat transfer of nanofluids inside circular tubes. According to this correlation, the thermal dispersion promoted by micro-convection and micro-diffusion is quantified by the Peclet number (Pe). Based on their experimental data for both laminar and turbulent flows, the following correlations were proposed.

Laminar flows: $Nu_{nf} = 0.4328\left(1.0 + 110285\varphi^{0.754} Pe_d^{0.218}\right)\mathrm{Re}_{nf}^{0.333}\mathrm{Pr}_{nf}^{0.4}$

Turbulent flows: $Nu_{nf} = 0.0059\left(1.0 + 7.628\varphi^{0.6886} Pe_d^{0.01}\right)\mathrm{Re}_{nf}^{0.9238}\mathrm{Pr}_{nf}^{0.4}$

Ding and Wen [79] focused on particle migration derived from Brownian motion, as well as shear stress and viscosity gradients, indicating a non-uniform property distribution that causes radial variations in thermophysical properties, most notably of temperature and flow velocity, thus being proposed as a possible enhancement mechanism in heat transfer of nanofluids.

A comprehensive study of different analytical approaches was conducted by Mansour et al. [80], in which laminar and turbulent flows, for an Al_2O_3/water nanofluid, were considered and common correlations used for nanofluids were evaluated for fully developed streams in a tube subjected to a constant heat flux boundary condition. Their analysis was focused on the determination of the specific heat, the viscosity and the thermal conductivity, followed by pressure drop and heat transfer correlation studies for singular tube conditions using distinct particle volume fractions. The most significant discrepancies were found for the laminar flow cases, some contradictory.

5.2. Natural convection

The natural convection of nanofluids having been found to be influenced by liquid unstable density distributions, which result from temperature and particle distribution differences due to particle sedimentation [81, 82].

Few experimental and analytical studies to ascertain the natural convective heat transfer behaviour of nanofluids have been performed. Kang et al. [81] and Putra et al. [83] conducted similar experiments using cylindrical and rectangular vessels, respectively, in which the nanofluid was heated from one side (or wall) and cooled from the other. Kang et al. [81] found that the formation and deterioration of multiple layers around SiO_2 nanoparticles occurred, these influenced by the increase of the temperature gradient between opposite sides. Putra et al. [83] found that the enhancement of the thermal conductivity of CuO/water nanofluids was higher than that of Al_2O_3/water nanofluids, both displaying improved convection than that of common slurries but inferior to that of the base fluid. Additionally, both nanofluids' natural convective heat transfer properties were found to deteriorate with an increase in particle concentration and density, characterized by decreasing Nusselt numbers.

Analytic studies conducted by Khanafer et al. [84] and Kim et al. [85] led the authors to conclude that natural convective heat transfer increased with the particle volume fraction (at any Grashof number), contradicting the experimental conclusions of Putra et al. [83].

6. Numerical studies

The computational analysis of fluid mechanics and heat transfer phenomena (CFD) of nanofluids requires strong and validated models regarding their thermo physical properties dependency with temperature and pressure. It is also fundamental to establish the physical nature of a nanofluid, namely if it should be considered a single-phase or a multi-phase material, since the numerical modelling approach may depend on that. The lack of knowledge regarding the above mentioned aspects is reflected on the limitations showed by the results of exiting studies focus on nanofluids CFD analysis. Although the referred limitations, CFD analysis of nanofluid systems may be quite important, namely as a support tool to the more fundamental experimental work.

A pioneer study in this area is due to Xuan and Roetzel,[77] the simpler single-phase assumption for nanofluids has been found to be numerically more efficient, result of its reduced computational workloads. The advantage of this approach resides in the assumption that the base fluid and the nanoparticles are in thermal equilibrium as well as equal velocities [86].

Maiga et al. [87]developed a numeric model to simulate the flow of Al_2O_3/water and Al_2O_3/ethylene glycol nanofluids, under both laminar and turbulent flows, applying single-phase and constant wall heat flux conditions to a circular tube (L=1m; d=10mm). Flow symmetry was assumed and, for turbulent flow, the semi-empirical K-ε model was used to describe the heat flux of the nanofluids. Results demonstrated a higher heat transfer enhancement of the ethylene glycol based nanofluid, increasing with growing particle loads.

Applying the same theoretical considerations, Roy et al. [6] numerically evaluated the thermal performance and wall shear stress of nanofluids in a radial cooling system, subjected to laminar, uniform velocity flows. The obtained data indicates convection coefficient increases for growing particle volume fraction and Reynolds number. More recently, a similar numerical study was conducted by Saeedi [86] to assess the thermal performance of a CuO/water nanofluid in a tube.

Xuan et al. [88] proposed a thermal Lattice Boltzmann model for flow and energy transport simulation of a Cu/water nanofluid. The distinguishing feature of this model is the hypothesis of particle location at a series of lattices, presenting a Boltzmann distribution within these. Another relevant feature is the temperature independence with regard to particle density distribution. The model predicted a 27% enhancement of the nanofluid Nusselt number over that of water alone.

Following their previously cited experimental investigations with regard to Al_2O_3/water nanofluids under laminar flow, Heris et al. [89] established a model which employed the thermal dispersion theoretical hypothesis, proposed by Xuan and Roetzel [77]. Simulations indicated that maximal heat transfer enhancements could be gained by the simultaneous effect of volume fraction increase and nanoparticle size decrease.

Employing the two-phase approach, Behzadmehr et al. [90] numerically investigated the turbulent flow of a nanofluid through a tube. The model takes into account both nanoparticle and base fluid molecule velocity gradients and uses a numerical solution to enable the application of the constant wall heat flux boundary condition. The authors performed a simulation using the experimental data gathered by Li and Xuan [78], claiming good agreement for the Cu/water nanofluids, contrary to the single-phase assumption for the same experimental data.

Pfautsch [68] conducted a numerical analysis of the thermal transfer behaviour of Al_2O_3/water and Al_2O_3/ethylene glycol nanofluids in a flat PHE assuming fluid and nanoparticle continuity, as well as momentum conservation via the Navier-Stokes equation. Due to a high non-linearity of the governing equations, simulations were performed employing the finite difference method. Results demonstrated that the laminar flow convection coefficient increases dramatically with particle size reduction and volume fraction increase. For well dispersed particles in water the maximum enhancement was predicted to be 130%, whereas for ethylene glycol the enhancement was significantly higher, 275%).

More recently, Mohammed et al. [91] conceived a model of an aluminium square micro-channel heat exchanger (25 channels) with the intent to evaluate the thermal performance of four nanofluids (Al_2O_3, SiO_2, Ag and TiO_2) under laminar flow. They assumed single-phase fluids and steady-state flow, while simulations followed the finite volume methodology. Simulations proved the better convective heat transfer performance of the Al_2O_3 nanofluid and the disadvantageous increase in pumping requirements with increasing Reynolds numbers.

Kalteh et al. [92] proposed a two-phase model to study the behaviour of a Cu/water nanofluid in an isothermally heated parallel plate micro-channel. Once again, the governing mass, momentum and energy equations were solved via the finite volume method using a non-uniform mesh. Perhaps anticipating the more anomalous behaviour reported in experimental studies for singular tube exchangers, the mesh was most refined in the micro-channel entry region. Consequent simulations demonstrated independence between particle viscosity and Nusselt number at Re=100. As with the Behzadmehr et al. [90] model, a comparison with the homogeneous single-phase assumption indicated the higher precision of the two-phase approach.

A unique numerical study was established by Manca et al. [93] in assessing the heat transfer enhancement resultant of an Al_2O_3/water nanofluid employed in confined slot jet impingement on a heated wall. The single-phase approach was used and the impingement temperature was considered constant. Other relevant considerations included steady-state, turbulent and constant property flow conditions, as well as nanofluid incompressibility.

Heris et al. [94] performed numerical simulations to evaluate the thermal behaviour of Al_2O_3, CuO and Cu nanoparticle suspensions in water under constant wall temperature boundary conditions when transiting through a square duct, which is less penalizing in pressure drop but limited in heat transfer when compared to the circular profiles. In an analogous approach to that taken by Kalteh et al. [92], the model was composed of a non-uniform mesh, finer elements packed in the duct entrance region. Of the tested nanofluids, the Cu/water presented the best thermal characteristics, Al_2O_3 displaying the worst.

7. Conclusions

Despite the extensive research on nanofluids, especially spherical particle types, a precise description of the anomalous heat transfer enhancements experimentally displayed is still a paramount limitation. The influential factors, as well as the precise quantification of their contribution to the observed results, remain a source of speculation amongst the scientific community. This is the major issue that currently forbids the employment of nanofluids in practical applications, which could do wonders in reducing fluid inventories and heat exchanger sizes. Obviously, both of these features converge in assisting the achievement of a worldwide goal: sustainable development. Therefore, a full understanding of how nanofluids will perform under real operating conditions is a important goal and a new frontier concerning the sustainable development promotion.

Of the reviewed literature, a limited amount of investigations regarding CNT nanofluids is currently available. A consensual position of the respective authors is that CNT nanofluids exhibit significantly higher thermal properties enhancements than other nanofluids, which indicates that these will promote top efficiencies when used in heat exchanging devices. To this extent, any enlightenment on the possible enhancement mechanisms, as well as the level at which these contribute to such promising behaviour, would be welcome. It can easily be concluded that the lack of experimental and theoretical studies regarding the convective heat transfer behaviour of nanofluids is currently their application in real industrial processes.

In the past few years, with the growing interest concerning nanofluids, some researcher groups have dedicated their attention to the analysis of the thermal properties of CNT based nanofluids, extended to the study of the influence of nanofluid preparation, stabilization and testing conditions, which are time-consuming and expensive tasks. However, with a view to attempting an adequate prediction of the thermal performance of these nanofluids, consistent experimental data is being gathered. This systematic approach is expected to fulfil the imperative requirement of behaviour prediction and allow further progress of these CNT nanofluid studies and some innovative and validated models are expected to be proposed in a near future.

Author details

Maria Alexandra Fonseca[1,2], Sylvio Freitas[2], Bruno Lamas[1,2], Bruno Abreu[1,2], Hugo Calisto[1,2], Nelson Martins[2] and Mónica Oliveira[2]

1 TEMA – Centre for Mechanical Technology and Automation, Portugal

2 Department of Mechanical Engineering, University of Aveiro, Portugal

References

[1] Trisaksri V, Wongwises S. Critical review of heat transfer characteristics of nanofluids. Renewable and Sustainable Energy Reviews. 2007;11:512-23.

[2] Incropera F, DeWitt D. Fundamentals of Heat and Mass Transfer, 5th Edition: Wiley; 2001.

[3] Kurt H, Kayfeci M. Prediction of thermal conductivity of ethylene glycol–water solutions by using artificial neural networks. Applied Energy. 2009;86:2244-8.

[4] Abou-Ziyan HZ. Heat transfer characteristics of some oils used for engine cooling. Energy Conversion and Management. 2004;45:2553-69.

[5] Korcek S, Jensen RK, Johnson MD, Sorab J. Maximising the fuel efficiency of engine oils: The role of tribology. Tribotest. 2001;7:187-201.

[6] Roy G, Nguyen CT, Lajoie P-R. Numerical investigation of laminar flow and heat transfer in a radial flow cooling system with the use of nanofluids. Superlattices and Microstructures. 2004;35:497-511.

[7] Eastman JA, Phillpot SR, Choi SUS, Keblinski P. Thermal Transport in Nanofluids. Annual Review of Materials Research. 2004;34:219-46.

[8] Eastman JA, Choi US, Li S, Thompson LJ, Lee S. Enhanced thermal conductivity through the development of nanofluids. Materials Research Society. 1997.

[9] Wang X, Xu X, Choi SUS. Thermal conductivity of nanoparticle–fluid mixture. Journal of Thermophysics and Heat Transfer. 1999;13.

[10] Masuda H, Ebata A, Teramae K, Hishinuma N. Alteration of Thermal Conductivity and Viscosity of Liquid by Dispersing Ultra-Fine Particles. Netsu Bussei. 1993;7:227-33.

[11] Grimm A. Powdered Aluminium-containing Heat Transfer Fluids. German Patent. Germany1993.

[12] Botha SS. Sythesis and Characterization of Nanofluids for Cooling Applications [Doctor of Phylosophy]. South Africa: University of the Western Cape; 2007.

[13] Nasiri A, Shariaty-Niasar M, Rashidi A, Amrollahi A, Khodafarin R. Effect of dispersion method on thermal conductivity and stability of nanofluid. Experimental Thermal and Fluid Science. 2011;35:717-23.

[14] Frederic J, Harun HS, Paul B, Heinrich H. Fabrication of large-area ordered arrays of nanoparticles on patterned substrates. Nanotechnology. 2005;16:1311.

[15] Warad HC, Ghosh SC, Hemtanon B, Thanachayanont C, Dutta J. Luminescent nanoparticles of Mn doped ZnS passivated with sodium hexametaphosphate. Science and Technology of Advanced Materials. 2005;6:296.

[16] Aiken Iii JD, Finke RG. A review of modern transition-metal nanoclusters: their synthesis, characterization, and applications in catalysis. Journal of Molecular Catalysis A: Chemical. 1999;145:1-44.

[17] Egorova EM, Revina AA. Optical Properties and Sizes of Silver Nanoparticles in Micellar Solutions. Colloid Journal. 2002;64:301-11.

[18] Rensmo H, Ongaro A, Ryan D, Fitzmaurice D. Self-assembly of alkane capped silver and silica nanoparticles. Journal of Materials Chemistry. 2002;12:2762-8.

[19] Karanikas S, Louis AA. Dynamic Colloidal Stabilization by Nanoparticle Halos. Physical Review Letters. 2004;93:248303.

[20] Vaisman L, Wagner HD, Marom G. The role of surfactants in dispersion of carbon nanotubes. Advances in Colloid and Interface Science. 2006;128–130:37-46.

[21] Hilding J, Grulke EA, George Zhang Z, Lockwood F. Dispersion of Carbon Nanotubes in Liquids. Journal of Dispersion Science and Technology. 2003;24:1-41.

[22] Keblinski P, Eastman JA, Cahill DG. Nanofluids for thermal transport. Materials Today. 2005;8:36-44.

[23] Sahoo NG, Rana S, Cho JW, Li L, Chan SH. Polymer nanocomposites based on functionalized carbon nanotubes. Progress in Polymer Science. 2010;35:837-67.

[24] Lamas B, Abreu B, Fonseca A, Martins N, Oliveira M. Assessing colloidal stability of long term MWCNT based nanofluids. Journal of Colloid and Interface Science.

[25] Young Jo K, Hongbin M, Qingsong Y. Plasma nanocoated carbon nanotubes for heat transfer nanofluids. Nanotechnology. 2010;21:295703.

[26] Khanafer K, Vafai K. A critical synthesis of thermophysical characteristics of nanofluids. International Journal of Heat and Mass Transfer. 2011;54:4410-28.

[27] Paul G, Chopkar M, Manna I, Das PK. Techniques for measuring the thermal conductivity of nanofluids: A review. Renewable and Sustainable Energy Reviews. 2010;14:1913-24.

[28] Penas JRV, de Zarate JMO, Khayet M. Measurement of the thermal conductivity of nanofluids by the multicurrent hot-wire method. Journal of Applied Physics. 2008;104:044314-8.

[29] Eastman JA. Mechanisms of enhanced heat transfer in nanofluids. Fluctuations and Noise in Out of Equilibrium Systems2005.

[30] Hong KS, Hong T-K, Yang H-S. Thermal conductivity of Fe nanofluids depending on the cluster size of nanoparticles. Applied Physics Letters. 2006;88:031901-3.

[31] Murshed SMS, Leong KC, Yang C. Enhanced thermal conductivity of TiO2—water based nanofluids. International Journal of Thermal Sciences. 2005;44:367-73.

[32] Zhang X, Gu H, Fujii M. Effective thermal conductivity and thermal diffusivity of nanofluids containing spherical and cylindrical nanoparticles. Experimental Thermal and Fluid Science. 2007;31:593-9.

[33] Yulong Ding HC, Liang Wang, Chane-Yuan Yang,Yurong He, Wei Yang, Wai Peng Lee, Lingling Zhang and Ran Huo. Heat Transfer Intensification Using Nanofluids. KONA. 2007;25.

[34] Murshed SMS, Leong KC, Yang C. Thermophysical and electrokinetic properties of nanofluids – A critical review. Applied Thermal Engineering. 2008;28:2109-25.

[35] Wen M-Y, Ho C-Y. Heat-transfer enhancement in fin-and-tube heat exchanger with improved fin design. Applied Thermal Engineering. 2009;29:1050-7.

[36] Xie H, Yu W, Li Y, Chen L. Discussion on the thermal conductivity enhancement of nanofluids. Nanoscale Research Letters. 2011;6:124.

[37] Eastman JA, Choi SUS, Li S, Thompson LJ, Lee S. Enhanced thermal conductivity through the development of nanofluids. Materials Research Society - Fall Meeting. Boston, MA1996.

[38] Xuan Y, Li Q. Heat transfer enhancement of nanofluids. International Journal of Heat and Fluid Flow. 2000;21:58-64.

[39] Xie H, Wang J, Xi T, Liu Y, Ai F, Wu Q. Thermal conductivity enhancement of suspensions containing nanosized alumina particles. Journal of Applied Physics. 2002;91:4568-72.

[40] Das SK, Putra N, Thiesen P, Roetzel W. Temperature Dependence of Thermal Conductivity Enhancement for Nanofluids. Journal of Heat Transfer. 2003.

[41] Kim P, Shi L, Majumdar A, McEuen PL. Thermal Transport Measurements of Individual Multiwalled Nanotubes. Physical Review Letters. 2001;87:215502.

[42] Choi SUS, Zhang ZG, Yu W, Lockwood FE, Grulke EA. Anomalous thermal conductivity enhancement in nanotube suspensions. Applied Physics Letters. 2001;79:2252-4.

[43] Eastman JA, Choi SUS, Li S, Yu W, Thompson LJ. Anomalously increased effective thermal conductivities of ethylene glycol-based nanofluids containing copper nanoparticles. Applied Physics Letters. 2001;78:718-20.

[44] Ding Y, Alias H, Wen D, Williams RA. Heat transfer of aqueous suspensions of carbon nanotubes (CNT nanofluids). International Journal of Heat and Mass Transfer. 2006;49:240-50.

[45] Chen L, Xie H, Li Y, Yu W. Nanofluids containing carbon nanotubes treated by mechanochemical reaction. Thermochimica Acta. 2008;477:21-4.

[46] Hwang Y, Lee JK, Lee CH, Jung YM, Cheong SI, Lee CG, et al. Stability and thermal conductivity characteristics of nanofluids. Thermochimica Acta. 2007;455:70-4.

[47] Xue Q, Xu W-M. A model of thermal conductivity of nanofluids with interfacial shells. Materials Chemistry and Physics. 2005;90:298-301.

[48] Han Z. Nanofluids with enhanced thermal transport properties [Doctor of Phylosophy]. Maryland, United States of America: University of Maryland; 2008.

[49] Wang X-Q, Mujumdar AS. A review on nanofluids - part I: theoretical and numerical investigations. Brazilian Journal of Chemical Engineering. 2008;25:613-30.

[50] Choi SU-S. Nanofluid technology : current status and future research1998.

[51] Lamas B, Abreu, B., Fonseca, A., Martins, N., and Oliveira, M.S.A. A Critical Analysis on the Thermal Conductivity Prediction Models for CNT based Nanofluids. Aveiro, Portugal: University of Aveiro; 2012.

[52] Martin CA, Sandler JKW, Shaffer MSP, Schwarz MK, Bauhofer W, Schulte K, et al. Formation of percolating networks in multi-wall carbon-nanotube,Äìepoxy composites. Composites Science and Technology. 2004;64:2309-16.

[53] Patel H, Anoop K, Sundararajan T, Das S. Model for thermal conductivity of CNT-nanofluids. Bulletin of Materials Science. 2008;31:387-90.

[54] Xie H, Chen L. Adjustable thermal conductivity in carbon nanotube nanofluids. Physics Letters A. 2009;373:1861-4.

[55] Munson-McGee SH. Estimation of the critical concentration in an anisotropic percolation network. Physical Review B. 1991;43:3331-6.

[56] Biercuk MJ, Llaguno MC, Radosavljevic M, Hyun JK, Johnson AT, Fischer JE. Carbon nanotube composites for thermal management. Applied Physics Letters. 2002;80:2767-9.

[57] Nan CW, Shi Z, Lin Y. A simple model for thermal conductivity of carbon nanotube-based composites. Chemical Physics Letters. 2003;375:666-9.

[58] Shenogin S, Bodapati A, Xue L, Ozisik R, Keblinski P. Effect of chemical functionalization on thermal transport of carbon nanotube composites. Applied Physics Letters. 2004;85:2229-31.

[59] Xue L, Keblinski P, Phillpot SR, Choi SUS, Eastman JA. Two regimes of thermal resistance at a liquid--solid interface. The Journal of Chemical Physics. 2003;118:337-9.

[60] Nan C-W, Liu G, Lin Y, Li M. Interface effect on thermal conductivity of carbon nanotube composites. Applied Physics Letters. 2004;85:3549-51.

[61] Xue QZ. Model for the effective thermal conductivity of carbon nanotube composites. Nanotechnology. 2006;17:1655.

[62] Xue QZ. Model for thermal conductivity of carbon nanotube-based composites. Physica B: Condensed Matter. 2005;368:302-7.

[63] Sastry NNV, et al. Predicting the effective thermal conductivity of carbon nanotube based nanofluids. Nanotechnology. 2008;19:055704.

[64] Koo J, et al. A nonlinear effective thermal conductivity model for carbon nanotube and nanofiber suspensions. Nanotechnology. 2008;19:375705.

[65] Kumar DH, Patel HE, Kumar VRR, Sundararajan T, Pradeep T, Das SK. Model for Heat Conduction in Nanofluids. Physical Review Letters. 2004;93:144301.

[66] Deng F, Zheng Q-S, Wang L-F, Nan C-W. Effects of anisotropy, aspect ratio, and non-straightness of carbon nanotubes on thermal conductivity of carbon nanotube composites. Applied Physics Letters. 2007;90:021914-3.

[67] Deng F, Zheng Q. Interaction models for effective thermal and electric conductivities of carbon nanotube composites. Acta Mechanica Solida Sinica. 2009;22:1-17.

[68] Pfautsch E. Forced Convection in Nanofluids Over a Flat Plate [Master of Science]. Missouri, United States of America: University of Missouri; 2008.

[69] He Y, Jin Y, Chen H, Ding Y, Cang D, Lu H. Heat transfer and flow behaviour of aqueous suspensions of TiO2 nanoparticles (nanofluids) flowing upward through a vertical pipe. International Journal of Heat and Mass Transfer. 2007;50:2272-81.

[70] Amrollahi A, Rashidi AM, Lotfi R, Emami Meibodi M, Kashefi K. Convection heat transfer of functionalized MWNT in aqueous fluids in laminar and turbulent flow at the entrance region. International Communications in Heat and Mass Transfer. 2010;37:717-23.

[71] Garg P, Alvarado JL, Marsh C, Carlson TA, Kessler DA, Annamalai K. An experimental study on the effect of ultrasonication on viscosity and heat transfer performance of multi-wall carbon nanotube-based aqueous nanofluids. International Journal of Heat and Mass Transfer. 2009;52:5090-101.

[72] Silva BAA. Caracterização de Nanofluidos do Ponto de Vista Termo-físico [Master Thesis]. Aveiro, Portugal: University of Aveiro; 2010.

[73] Abreu B, Lamas, B., Fonseca, A., Martins, N., and Oliveira, M.S.A. Convective Heat Transfer Characterization of Carbon Based Nanofluids under Laminar Flow Conditions. Aveiro, Portugal: University of Aveiro; 2012.

[74] Kolade B, Goodson KE, Eaton JK. Convective Performance of Nanofluids in a Laminar Thermally Developing Tube Flow. Journal of Heat Transfer. 2009;131:052402-8.

[75] Liu Z-H, Liao L. Forced convective flow and heat transfer characteristics of aqueous drag-reducing fluid with carbon nanotubes added. International Journal of Thermal Sciences. 2010;49:2331-8.

[76] Pak BC, Cho YI. Hydrodynamic and heat transfer study of dispersed fluids with submicron metallic oxide particles. Experimental Heat Transfer. 1998;11:151-70.

[77] Xuan Y, Roetzel W. Conceptions for heat transfer correlation of nanofluids. International Journal of Heat and Mass Transfer. 2000;43:3701-7.

[78] Li Q, Xuan Y. Convective heat transfer and flow characteristics of Cu-water nanofluid. Science in China Series E: Technological Sciences. 2002;45:408-16.

[79] Ding Y, Wen D. Particle migration in a flow of nanoparticle suspensions. Powder Technology. 2005;149:84-92.

[80] Mansour RB, Galanis N, Nguyen CT. Effect of uncertainties in physical properties on forced convection heat transfer with nanofluids. Applied Thermal Engineering. 2007;27:240-9.

[81] Kang C, Okada M, Hattori A, Oyama K. Natural convection of water,Äìfine particle suspension in a rectangular vessel heated and cooled from opposing vertical walls (classification of the natural convection in the case of suspension with a narrow-size distribution). International Journal of Heat and Mass Transfer. 2001;44:2973-82.

[82] Okada M, Suzuki T. Natural convection of water-fine particle suspension in a rectangular cell. International Journal of Heat and Mass Transfer. 1997;40:3201-8.

[83] Putra N, Roetzel W, Das S. Natural convection of nano-fluids. Heat and Mass Transfer. 2003;39:775-84.

[84] Khanafer K, Vafai K, Lightstone M. Buoyancy-driven heat transfer enhancement in a two-dimensional enclosure utilizing nanofluids. International Journal of Heat and Mass Transfer. 2003;46:3639-53.

[85] Kim J, Kang YT, Choi CK. Analysis of convective instability and heat transfer characteristics of nanofluids. Physics of Fluids. 2004;16:2395-401.

[86] Saeedi MKaA. Numerical investigation into the convective heat transfer of CuO nanofluids flowing through a straight tube with uniform heat flux. Indian Journal of Science and Technology. 2012;5.

[87] Maı®ga SEB, Nguyen CT, Galanis N, Roy G. Heat transfer behaviours of nanofluids in a uniformly heated tube. Superlattices and Microstructures. 2004;35:543-57.

[88] Yimin Xuan KY, Qiang Li. Investigation on flow and heat transfer of nanofluids by the thermal Lattice Boltzmann model. Progress in Computational Fluid Dynamics. 2005;5.

[89] Heris SZ, Esfahany MN, Etemad G. Numerical Investigation of Nanofluid Laminar Convective Heat Transfer through a Circular Tube. Numerical Heat Transfer, Part A: Applications. 2007;52:1043-58.

[90] Behzadmehr A, Saffar-Avval M, Galanis N. Prediction of turbulent forced convection of a nanofluid in a tube with uniform heat flux using a two phase approach. International Journal of Heat and Fluid Flow. 2007;28:211-9.

[91] Mohammed HA, Bhaskaran G, Shuaib NH, Abu-Mulaweh HI. Influence of nanofluids on parallel flow square microchannel heat exchanger performance. International Communications in Heat and Mass Transfer. 2011;38:1-9.

[92] Kalteh M, Abbassi A, Saffar-Avval M, Harting J. Eulerian,ÄìEulerian two-phase numerical simulation of nanofluid laminar forced convection in a microchannel. International Journal of Heat and Fluid Flow. 2011;32:107-16.

[93] Manca O, Mesolella P, Nardini S, Ricci D. Numerical study of a confined slot impinging jet with nanofluids. Nanoscale Research Letters. 2011;6:188.

[94] Heris SZ, Kazemi-Beydokhti A, Noie SH, Rezvan S. Numerical Study on Concective Heat Transfer of Al2O3/water, CuO/water ans Cu/water Nanofluids Through Square Cross-Section Duct in Laminar Flow. Engineering Applications of Computational Fluid Mechanics. 2012;6.

Chapter 2

Vibroelectronic Properties of Functionalized Single-Walled Carbon Nanotubes and Double-walled Boron Nitride Nanotubes

Metin Aydin and Daniel L. Akins

Additional information is available at the end of the chapter

1. Introduction

Carbon is the first element in group-IVof the periodic table and has a $1s^2 2s^2 2p^2$ electronic configuration, in which four valence electrons allow it to form a number of so-called hybridized atomic orbitals. Carbon atoms in elemental substances bond to each other covalently by the sharing of electron pairs, in which the covalent bonds have directional properties; this in turn provides carbon the capability to form various molecular and crystalline solid structures. The nature of the covalent bonds that are formed dictate the varied chemical and physical properties of carbon allotropes. Pure carbon-based materials not only exist as the commonly recognizeddiamond and graphite allotropes, but also more exotic entities such as fullerenes, carbon nanotubes (CNTs), and graphene; these latter allotropes having proven themselves important materials in nanotechnology.

The present chapter deals with single-walled carbon nanotubes (SWNTs), whose unique properties, as suggested above, derive from their distinctive structure. In SWNTs the carbon bonding that exists is akin to that that exists in graphite as opposed to that found in diamond. More specifically, diamond has a coordination number of four, with sp^3 hybridization, while, on the other hand, sp^2 hybridization exists in the planar layers of carbon atoms that give graphite its structure, and in the bonding that leads to the tubular structure of SWNTs. The sp^2 hybridization in graphite links carbon atoms in a two-dimensional (2D) layer of hexagons that lead to each layer in the graphite structure, in the ideal case, forming a planar structure. Each carbon atom contributes 3 electrons to 3 equivalent sigma bonds within the plane and has 1 electron left in the perpendicular p_z orbitals; such electrons are delocalized over the entire plane, resulting in a π-electron orbital system that allows the fourth valence electron to essentiallymove freely over the plane. Within the layers, the carbon-carbon bond distance is similar to the bond length in benzene (i.e., the carbon atoms are strongly bound to each other and the carbon-carbon distance is about 0.14 nm), leading to a very large inplane value for Young's modulus. However, the distance between layers (ca., 0.34 nm) is sufficiently large that the layers are bounded to each other mainly by weak, long-range Van der Waals type interaction. The weak interlayer coupling gives graphite the property of a seemingly very soft material, a property that makes graphite suitable for use in pencils and in lubricants.

As a result of its intrinsic structure, the electrical conductivity of graphite is directionally-dependent. Delocalized π-electrons parallel to the planes essentially experience metallic conduction, while electron mobility perpendicular to the layered planes would typically be much lower, but with possibly significant temperature dependency, thereby imbruing graphite as semiconductor character as well. The directionality of the conductivity translates to a band structure that has a filled valence bandand an empty conduction band separated by an energy gap. These bands, in one picture, would result from bonding and antibonding molecular π-orbitals that can be conceptualized in terms of energy lowering and energy raising combination of the perpendicular p_z atomic orbitals. The π-bonding orbitals would be fully occupied while the π-antibonding orbitals would be unoccupied, with the gap being the energy difference between the top and bottom of the respective orbitals. Because of the larger distance between its layers, graphite may form intercalation compounds with added species that act as electron donors, with graphite acting as an electron acceptor, incorporating the donated electrons into the vacant conduction band; or as electron acceptors, where graphite donates electrons from the full valence band. In diamond, it is to be noted that all valence electrons are localized around the carbon atoms, hence, such a structural characteristic has profound effectson its electrical properties, with diamond being an insulator with a band gap around 6 eV.

We now move to a more focused discussion of carbon nanotubes. Carbon nanotubes (CNTs) were discovery in 1991 [1], their unique physical, chemical, and electronic properties have led to a variety of technological application in functional nanodevices, especially as transistors and sensors [2, 3, 4], [5, 6, 7], in heat conduction systems [8, 9], in specialty electronics [10, 11], molecular memories [12], optics [13, 14, 15], electrically excited single-molecule light sources [16, 17, 18, 19], to functionalized DNA [20, 21], high-performance adsorbent electrode material for energy-storage device [22], and protein functionalization [23, 24].

As it is well known, carbon nanotubes can be obtained by rolling up a defined projected area from within the hexagonal lattice of a graphene sheet in a seamless fashion such that all carbon–carbon (C–C) valences are satisfied, and the direction in which the roll up is performed transforms into the circumference of the tube. The projected area is in fact a homomorphic representation of a particular carbon nanotube [48(f-g)]. The roll-up vector is also termed the chiral vector, and is defined as $n\vec{a}_1 + m\vec{a}_2$, where $\vec{a}_1$ and $\vec{a}_2$ are the unit vectors of the hexagonal lattice, and n and m are the so-called chiral indices. An infinite number of nanotube geometries are possible, with a specific nanotube characterized by chiral indices (n,m), which, in turn, define the chiral angle θ and tube diameter (d_t); the latter is also dependent on the C–C bond length of the hexagonal lattice. For n = m, the nanotube is said to have the "armchair" conformation; for $n \neq 0$ and m = 0, the conformation is called "zigzag"; while for $n \neq 0$ and $m \neq 0$ the conformation is termed "chiral."The diameter of the nanotube normally has values that range up to several nanometers from ~0.4 nm, while nanotubes are usually several microns in length. It is to be noted that single-walled and multi-walled carbon nanotubes generally have properties that are significantly different, while double-walled carbon nanotubes (DWNTs) can be viewed as representing the key structure that defines the transition between SWNTs and MWNTs.

Carbon nanotubes can be metallic or semiconducting depending on their structure. This is due to the symmetry and the unique electronic structure of graphene. If the chiral indices are equal, n = m, the nanotube is metallic; if n−m is a multiple of 3, then the nanotube is semiconducting, with a very small band gap; otherwise, the nanotube is a moderate semiconductor [25]. Interestingly, some nanotubes have conductivities higher than that of copper, while others behave more like silicon.

As it is well known, the optical properties of nanotubes are implicitly connected with the absorption, photoluminescence, and Raman spectroscopy of nanotubes. Such optical measurements permit a reliable characterization of the quality of nanotube, such as chirality, size, and structural defect. In the case of Raman measurements, even though a large number of phonon modes of carbon nanotubes would be expected, most of them are Raman inactive due to the selection rules that emanate from the high symmetry properties of the nanotubes. The Raman spectrum of a carbon nanotube exhibits a few characteristic modes that can be used to determine the size of nanotubes and to classify the type of the nanotubes, such as semiconducting and metallic. For example, in the low frequency region, one type of characteristic vibration is called the radial breathing mode (RBM); this movement of the carbon atoms is in the radial direction with the same phase, and corresponds to vibration of the entire tube, which is strongly diameter dependent [48(f-g)]. The RBM gives precise information about the nanotube diameter and is typically found between 100 cm^{-1} and 500 cm^{-1}. Additionally, in the high energy range from (1000 to 2000 cm^{-1}), there are two important characteristic Raman bands: the defect induced disordered band (D-band) that appears between 1300 and 1400 cm^{-1}, and tangential modes (G-band) that lie in the range from ~1560 to ~1600 cm^{-1}. The D-band is present in all graphite-like carbons and originates from structural defects. Therefore, the intensity ratio of the G/D modes is conventionally used to quantify the structural quality of carbon nanotubes. The G-band corresponds to planar vibrations of carbon atoms and is present in most graphite-like materials (at around 1580 cm^{-1}). This tangential mode (G-band) in SWCNT is split into several peaks. The splitting pattern and intensity depend on the tube structure and excitation energy; they can be used, though with much lower accuracy compared to RBM mode, to estimate the tube diameter and whether the tube is metallic or semiconducting. [48(f-g)].

Chemical functionalization by bond formation or by coating the nanotubes with organic/inorganic molecules or by encapsulating a varieties of semiconductor particles, including CdSe and CdTe, may lead to efficient energy transfer between the molecules and nanotube, as well as lead to significant enhancement in the optical properties of the composite [26, 27, 28, 29]. As an example of effect on an optical property, it has been reported that when a squarylium dye isencapsulated into a carbon nanotube, increased chemical and thermal stability of squarylium molecules occur, which, since encapsulation of a dye quenches strong dye luminescence, allows measurement and analysis of the dye's Raman spectra[30]. Also, L. Alvarez *et al.*[31] have reported that while infrared spectroscopy (IR) might provide evidence of a significant positive charge transfer for an inserted oligothiophene, Raman spectra evince different behaviors depending on the excitation energy and relationship to the oligomer's (specifically, quaterthiophene) optical absorption energy. For example, at high excitation wavelength (far from the oligomer's resonance), radial breathing modes exhibit a significant blue-shift as a result of the encapsulation effect, while at low excitation wavelength, close to resonance with the oligomer absorption, both the G-band and the low-frequency modes vanish, suggesting a significant charge transfer between the oligomer and the nanotube.

CNTs are also widely used in the clinical and research medical arenas. They find application as superior drug delivery media, for health monitoring devices; as biosensing platforms for the treatment of various diseases; in chemical sensor devices, etc. [32, 33, 34, 35]. Functionalized-SWNTs (i.e., f-SWNTs) have been known to increase solubility and permit efficient tumor targeting/drug delivery; prevents SWNTs from being cytotoxic; and possibly altering the functioning of immune cells. Moreover, carbon nanotubes have enhanced solubility when functionalized with lipids that make their movement through the human body easier and reduces the risk of blockage of vital body organ pathways. Also, CNTs exhibit strong optical absorbance in certain spectral windows, such as the NIR (near-infrared); when functionalized within tumor cell with specific binding entities, the nanotubes have allowed the selective destruction of disease (e.g., cancer) cells with NIR in drug delivery applications.

More recently, boron nitride nanotubes (BNNTs) can be counted among the modified CNT that have been synthesized [36, 37, 38]. The electronic properties of boron nitride nanotubes differ from carbon nanotubes: while carbon nanotubes can be either metallic or semiconducting, depending on their chirality and radius [39], all boron nitride nanotubes (BNNTs) are found to be semiconducting materials with a large band gap[40]. And since the band gap is large, the gap energy is only weakly dependent on the diameter, chirality, and the number of walls of a multi-walled tube structure. Moreover, because of their semiconducting character, BNNTs like CNTs themselves are also very interesting materials for application in nanoscale devices, and have been considered alternatives to CNTs [41, 42]. Like CNTs the modification of the electronic properties of BNNTs by doping and functionalization is an important avenue for making nanodevices. The doped BNNTs nanotubes may exhibit a dramatic change relative to the pristine nanotube. Furthermore, because of the strong interactions between electrons and holes in BNNTs [43, 44], the excitonic effects in BNNTs have proven more important than in CNTs. Bright and dark excitons in BNNTs qualitatively alter the optical response [45].

For a better understanding of the physical and optical properties of nanotubes, quantum mechanical calculations have been extremely helpful. In this chapter, we provide theoretical results on double-walled boron nitride nanotubes (DWBNNTs) and functionalized nanotubes using DFT; this report extends the quantum chemical computational approach that we have used earlier [48(f-g)]. The results of calculations not only indicate the shift in the spectral peak positions of the RBM and G-modes in Raman spectra of DWBNNTs relative to their corresponding isolated SWBNNTs, but also indicatesa charge transfer from the outer-shell to the inner-shell when DWBNNTs are excited, as discussed in Section 3. Furthermore, the plots of the frequencies of vibrational radial breathing modes (RBM) versus $1/d_t$ for (2n, 0)&(n,0)-DWBNNTs exhibit a strong diameter dependence.

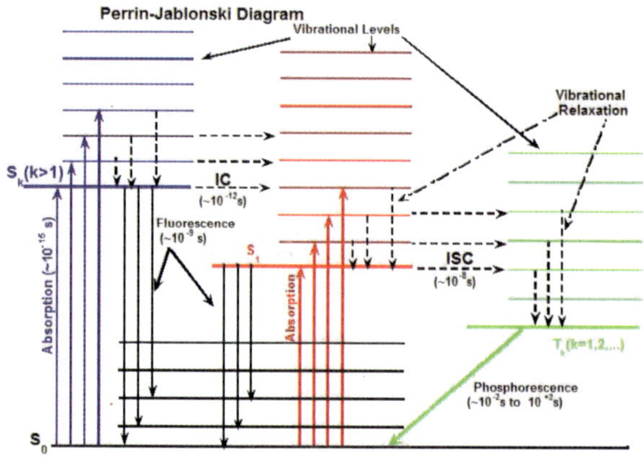

Figure 1. A general Perrin-Jablonski diagram for a fluorescent molecule, where S and T stand for singlet and triplet electronic states, respectively. IC and ISC represent "internal conversion" and "intersystem crossing", respectively.

For functionalized single-walled carbon nanotubes, we find that there should be a charge transfer process directed from the nanotube to an attached molecule, which is active in optical excitations. More generally, upon irradiation a system can undergo internal conversion (IC) and intersystem crossing (ISC) processes, in addition to photochemical and other photophysical processes. Transient intermediates are likely to form in the IC and ISC radiationless processes, herein referred to as "dark processes," which are not detected using conventional light absorption or emission spectroscopic methods. As seen from the combined Perrin-Jablonski diagram in Figure 1, for a typical molecule the emission of a photon from an electronically excited state to the ground state results in fluorescence in the region of 300 to 1500 nm. Photophysical processes for an isolated molecule occur as a result of transitions between the different internal energy states of that comprise the electronic states. A molecular system in the gas-phase or in the solution phase at room temperature is mostly expected to be in its ground state (S_0). The excitation of a molecular system from its ground state to an excited vibroelectronic state by absorption of a photon (occurring within ca. 10^{-15} second) is much faster than a emission of the photon from its excited electronic state (S_k, k>1) to its ground state (occurring in ca. 10^{-8} second). All of the excited molecular systems may not directly return back to their ground state by emission of a photon, $S_{k>0} \rightarrow S_0$ transition, but some of them may return back to their ground states (S_0) by internal conversion (CI), for instance, when the molecule is excited into a higher vibroelectronic state (S_k>1), it may undergo relaxation to the S_1 state (in 10^{-12} s) via vibrational coupling between these states before undergoing additional vibrational relaxation and returning to the lowest singlet electronic energy level (S_1), referred to as internal conversion. Subsequently, transition from S_1 to S_0 by emission of a photon (fluorescence) occurs. An alternate pathway for a molecule in the lowest energy S_1 state involves intersystem crossing (at rates that can compete with fluorescence) by the molecule into a triplet state T_1. From T_1, the molecule can undergo radiative de-excitation via a much slower process, which is known as phosphorescence ($T_1 \rightarrow S_0$ transition), such as illustrated by the Perrin-Jablonski diagram given in Figure 1.

It is to be noted that fluorescence resonance energy transfer (FRET) can be used to investigate intra- and/or intersystem energy transfer dynamics that might occur as one transitions fromsingle-walled nanotubes (SWNTs) to multi-walled nanotubes (MWNTs),or to the functionalized nanotubes (f-NTs). Such dark intermediates are expected to play crucial roles in IC and ISC processes and thus are fundamental to understanding mechanistic photochemistry of the functionalized-nanotubes and multi-walled nanotubes. We have used time-dependent DFT (i.e., TD-DFT) methods to determine the dark transient structures involved in radiationless processes for functionalized-SWCNTs and DWBNNTs. Also, we have calculated all possible singlet-triplet vertical electronic transitions and discussed these in terms of IC and ISC processes.

It is to be noted that CNTs have been shown to exhibit strong optical absorbances in certain spectral windows, such as the NIR (near-infrared). Moreover, when functionalized with tumor cell specific binding entities CNTs have facilitated the selective destruction of disease cells (e.g., cancer cell) in the NIR and play a significant role in drug delivery applications[46]. In the present chapter, we acknowledge the importance of calculating the IR spectra of both functionalized-SWCNTs and DWBNTS.

2. Results and discussion

Computational methods: The ground state geometries of single-walled carbon nanotubes (SWCNTs), double-walled carbon nanotubes (DWCNTs), single-walled boron nitride nanotubes (SWBNNTs), and functionalized-SWCNTs were optimized without symmetry restriction on the initial structures. Both structure optimization and vibrational analysis calculations were implemented using DFT with functionals, specifically, B3LYP, in which the exchange functional is of Becke's three parameter type, including gradient correction, and the correlation correction involves the gradient-corrected functional of Lee, Yang and Parr. The basis set of split valence type 6-31G, as contained in the Gaussian 03 software package[47], was used. The results of the calculations did not produce any imaginary frequencies. The vibrational mode descriptions were made on the basis of calculated nuclear displacements using visual inspection of the animated normal modes (using GaussView03) [47], to assess which bond and angle motions dominate the mode dynamics for the nanotube. The DFT method was chosen because it is computationally less demanding than other approaches as regards inclusion of electron correlation. Moreover, in addition to its excellent accuracy and favorable computation expense ratio, the B3LYP calculation of Raman frequencies has shown its efficacy in numerous earlier studies performed in this laboratory and by other researchers, often proving itself the most reliable and preferable method for many molecular species of intermediate size, including anions and cations [48]. In our calculations, hydrogen atoms have been placed at the end points of the unit cells. Furthermore, the time-dependent density functional theory at TD-B3LYP level were applied to calculate the vertical electronic transitions for the SWCNTs, SWBNNTs and functionalized (7,0)- and (10,0)-SWCNTs. For geometry optimization and calculations of electronic transitions, the 6-31G* basis set was used for sulfur atom (S) and the 6-31G basis set was used for the other atoms involved in the covalently functionalized nanotubes. It is worth nothing that the results of the calculated structural and spectroscopic properties of the double-walled boron nitride nanotubes (DWBNNTs) and the functionalized zigzag single-walled carbon nanotubes (f-(n, 0)-SWCNTs) used in this chapter have been submitted to elsewhere for publication.

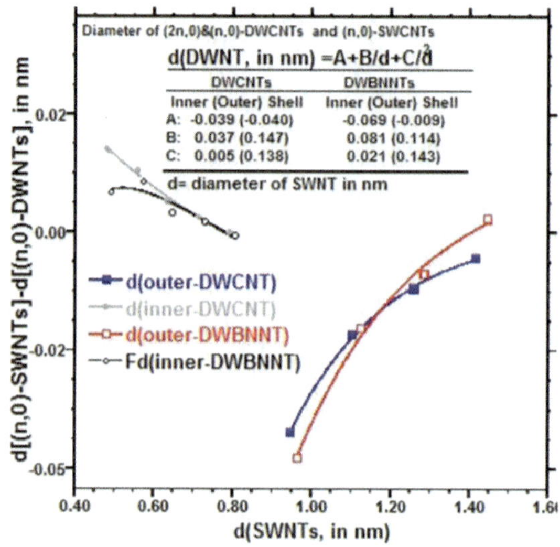

Figure 2. Calculated diameters of the double-walled carbon nanotubes, (2n,0)&(n,0)-DWCNTs, and double-walled boron nitride nanotubes, (2n,0)&(n,0)-DWBNNTs, for n = 6 to 10

3. Results and discussion

3.1. Structural results

Calculated diameters of the (0,n)&(0,2n)-DWCNTs (zigzag double-walled carbon nanotube) and (0,n)&(0,2n)-DWBNNTs (zigzag double-walled boron nitride nanotubes), for n = 6 to 10, were found to decrease for the inner-nanotube and increase for the outer-nanotube, referenced to the corresponding diameter of the zigzag single-wall nanotube ((0,n)-SWNT) which changes with n. A fit to the calculated individual tube diameters for each inner- and outer-shell of the DWCNTs and DWBNNTs using a functional form that depends inversely on single-walled nanotube's diameter: fit parameters are shown in Eq. 1a-2b

$$\begin{aligned} &D_t\left(\text{outer}-\text{shell}-\text{DWCNT}\right),\ \text{in nm}) = -0.040 + \frac{0.147}{d_t} + \frac{0.138}{d_t^2} \qquad a \\ &D_t\left(\text{inner}-\text{shell}-\text{DWCNT}\right),\ \text{in nm}) = -0.039 + \frac{0.037}{d_t} + \frac{0.005}{d_t^2} \qquad b \end{aligned} \tag{1}$$

$$\begin{aligned} &D_t\left(\text{outer}-\text{shell}-\text{DWBNNT}\right),\ \text{in nm}) = -0.009 + \frac{0.114}{d_t} + \frac{0.143}{d_t^2} \qquad a \\ &D_t\left(\text{inner}-\text{shell}-\text{DWBNNT}\right),\ \text{in nm}) = -0.069 + \frac{0.081}{d_t} + \frac{0.021}{d_t^2} \qquad b \end{aligned} \tag{2}$$

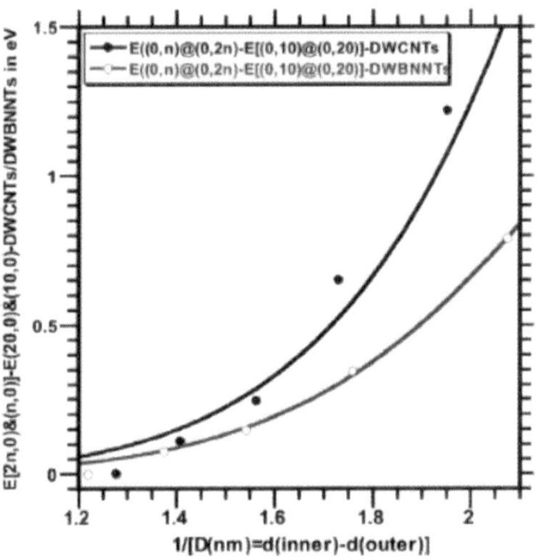

Figure 3. The diameter dependence of the curvature energies of the DWCNTs and DWBNNTs referenced to the global energies per hexagon of the (0,10)&(0,20)-DWCNTs/DWBNNTs is well fitted by a Lennard–Jones potential expression as given in Eqs. (3a) and (3b).

A comparison the diameters of the inner- and outer-shells of the DWNTs with their corresponding SWNTs diameters show that the inner-shells diameters decrease and the outer-shells diameters increased. These predictions explicitly indicate the existence of intertube interactions in DWCNT systems. As seen in Figure 3, the diameter dependence of the curvature energies of the DWCNTs and DWBNNTs referenced to the global energies per hexagon of the (0,10)&(0,20)-DWCNTs and -DWBNNTs is well fitted by a Lennard-Jones potential expression ($E_{LJ} = -\left(\frac{A}{r^6} - \frac{B}{r^{12}}\right)$ where parameters of the are A and B are van der Waals interaction parameters in Lennard-Jones potential) as given in equation 3a-b,

$$\Delta E(\text{DWBNNTs, in eV}) = -\left(\frac{0.503}{D_t(\text{nm})}\right)^6 \left\{1 - \left(\frac{0.266(\text{nm})}{D_t(\text{nm})}\right)^6\right\} \quad a$$

$$\Delta E(\text{DWCNTs, in eV}) = -\left(\frac{0.477}{D_t(\text{nm})}\right)^6 \left\{1 - \left(\frac{0.356(\text{nm})}{D_t(\text{nm})}\right)^6\right\} \quad b \qquad (3)$$

where $\Delta E\left(\frac{\text{DWCNTs}}{\text{DWBNNTs}}\right) = E[(2n,\ 0)\,\&\,(n,\ 0)] - E[(20,0)\,\&\,(10,0)]$ and $D_t = d_o(\text{outer shell}) - d_i(\text{inner shell}) D_t = d_t(\textit{outer shell}) - d_t(\textit{inner shell})$. The results of the calculations suggest that the DWNTs with large diameters can be much more easily formed than those with small diameters. When comparing the formation energy of the DWCNTs with the DWBNNTs, as shown in Figure 3, it can be seen that the formation of the DWBNNTs is favorable to that of DWCNTs due to the relatively strong interactions between the inner- and outer-shells in the case of the DWBNNTs. This finding also is supported by the calculated electron density, as discussed below, as well as the relative change in the tube diameters when going from the SWNT to the DWNT, as seen in equations 1-2. Furthermore, our ongoing calculations on the energetically stability of the DWBNNTs as function of the interwall distance (between inner- and outer-shells) indicates that the interwall distance around 0.34 nm is more stable, which are excellent agreement with the experimental observations by J. Cumings [59], which will be published elsewhere. However, at different experimental conditions, the DWBNNTs with small interwall distance such as (0,6)&(0,12)-DWBNNT might be formed at different experimentally conditions. The DWBNNTs with small interwall distance might be more interesting than other, in their optical applications.

Figures 4 A-B illustrate the calculated electron density of (12,0)&(6,0)-DWCNT and (0,n)&(0,2n)-DWBNNT, n= 6 and 8. For (12,0)&(6,0)-DWCNT, the geometry optimization, without any symmetry restriction, predicted ground state geometry has C_{2v} point group and the electronic state of ground state has singlet-A_1 symmetry. The plotted electron density showed that while first four highest occupied molecular orbitals (from HOMO to HOMO-4, of B_2, B_1 and $2E_2$ symmetries, respectively) involve both the inner- and outer-shell, the HOMO-5 with the $2E_1$ symmetry belongs to outer-shell only. The lowest unoccupied molecular orbital, LUMO (E_1) lies about 4.699 eV above the HOMO (B_2), and belongs to the inner-shell, while the next higher one (E_1) involves not only the inner- and outer-shell (lies 5.521 eV above the HOMO (A_1)), but also there is a significant sigma-bonding interaction between the inner and outer tubes in the excited state. For the (0,6)&(0,12)-DWCNT, the calculated electron density of (0,6)&(0,12)-DWCNT shows that the first four highest occupied molecular orbitals (from HOMO to HOMO-3, with the A_{1u}, A_{2g} and $2E_{1g}$ symmetries, respectively) belong to the outer-shell and the next higher occupied molecular orbitals, from HOMO-4 to HO-

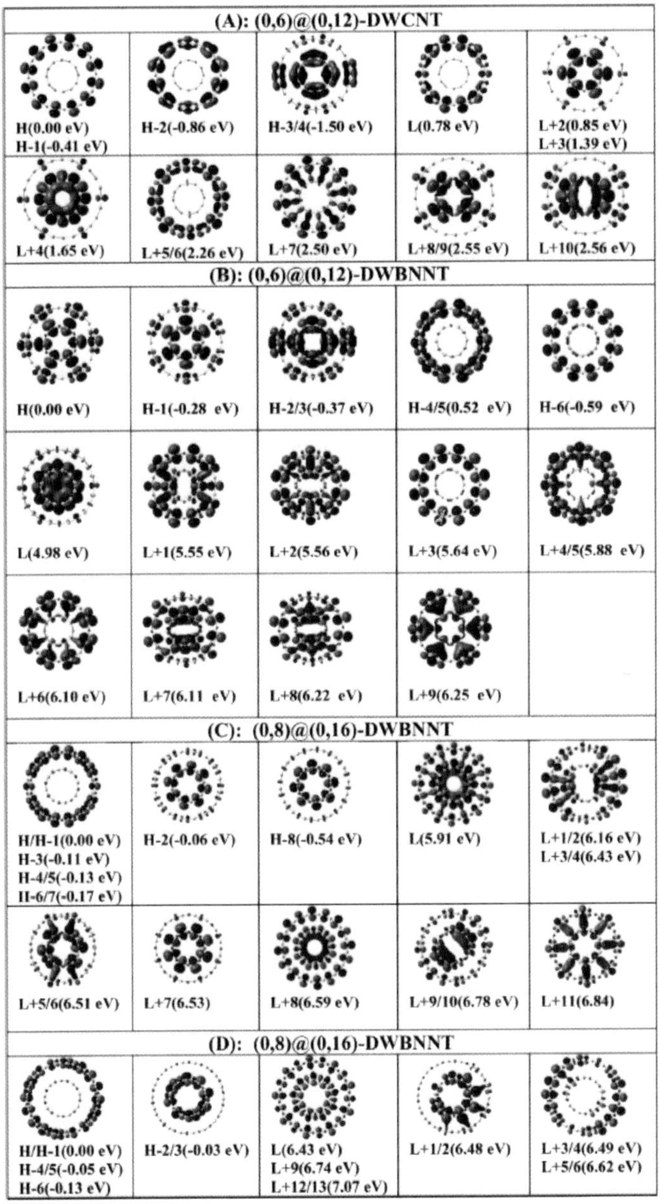

Figure 4. Calculated electron densities in the HOMO and LUMO states: A for the (0,6)&(0,12)-DWCNT, B for (0,6)&(0,12)-DWBNNT, C for (0,8)&(0,16)-DWBNNT, and D for (0,9)&(0,18)-DWBNNT.

MO-24, include both inner- and outer-shells of (0,6)&(0,12)-DWCNT. The lowest unoccupied molecular orbital, LUMO(E_{1u}) lies about 0.780 eV above the HOMO(A_{1u}) and belongs to the outer-shell, while the next one (with B_{2u} symmetry) belongs to the inner-shell, and lies 0.849 eV above the HOMO(A_{1u}). The calculated electron densities also indicate that an intratube (inner and outer tube) interaction may take place in the excited state, since the LUMO +7(A_{2u}), LUMO+8(E_{1u}), LUMO+10(E_{1g}) and LUMO+15(E_{1g}) lie about 2.494, 2.557, 2.563, 3.637 eV above the HOMO(A_{1u}), respectively.

The intratube σ-bonding interaction in the excited state of the (0,6)&(0,12)-DWBNNTs and DWCNT might lead to a probable intertube charge transfer, which can be observed by a significant change in the tangential modes (TMs) of resonance Raman spectra when the tube excited to its intratube charge transfer state. The TMs may not only provide information about the metallic or semiconducting character of nanotubes, but also about the inner-outer tube (intratube) charge transfer. Indeed, very recently, resonant Raman measurements [49], photoemission measurements, and theoretical calculations have provided evidence of charge transfer between the inner- and outer-shells of DWCNTs.

Given such a scenario, small sized-DWCNTs and DWBNNTs might be used as energy conversion systems due to charge transfer between intershells, which might be indicated by changes in of the Raman band intensities upon excitation in resonance with charge transfer between inner- and outer-shells.

3.2. Raman Spectra of Single-Walled and Double-Walled Boron Nitride Nanotube

We calculated Raman spectra for the zigzag single-walled boron nitride nanotube ((0,n)-SWBNNTs, n=6 to 19) and double-walled boron nitride nanotube, (n,0)&(2n,0)-DWBNNTs with n=6 to 9. While the Figs. 5A and 6 provide the calculated Raman spectra for the SWBNNTs and DWBNNTs, respectively, the Figure 7 provides the Raman spectra of the (0,8)&(0,16)-DWBNNT and isolated (0,8)- and (0,16)-SWBNNTs for the comparison. Furthermore, we provided the vibrational mode assignments and frequencies for the DWBNNT and isolated SWBNNTs in Tables 1. All assignments to motions of atoms or groups of atoms in Tables 1 have been accomplished through use of vibration visualization software (specifically, GaussView03). The results of the calculations are summarized below.

Zigzag-SWBNNTs: In the low frequency region (<500 cm^{-1}), the calculated Raman spectra the (0,n)-SWBNNTs (n=6 to 19) exhibited two Raman bands. One of them is known as the radial-breathing mode (RBM) and other is elliptical deformation mode (EDM). The RBM is an important mode for the characterization and identification of particular nanotubes, especially of their chiralities. The importance of the radial-breathing mode for the characterization of nanotubes derives from the inverse dependence of its frequency on the diameter of the nanotube. As seen in Figs. 5A-B, the radial breathing mode (RBM with A_{1g} symmetry, $\omega_{RBM}(A_{1g})$) and other Raman band (elliptical deformation mode (EDM) with E_{2g} symmetry, $\omega_{EDM}(E_{2g})$) have frequencies that inversely depend on a nanotube's diameter. A linear fit to the calculate RBM frequency dependence on nanotube diameter is provided; a linear equation, $\omega_{RBM}(A_{1g}) = 48.51 + \frac{183.54\ cm^{-1}.nm}{d_t(nm)}$, which is in excellent agreement with the results of the DFT within ± 1 cm^{-1}. However, the offset constant in the linear fitting equation (48.51 cm^{-1}) produce significant error for the (0,n)-SWBNNTs with large diameter because the RBM decreases with increasing tube diameter and RBM in the limit of infinite diameter yields to a simple translation of the BN sheet. The RBM frequency should therefore go to zero in this limit. Therefore, a curve fit may be obtained using a cubic equation such as $\omega_{RBM}(cm^{-1}) = \frac{307.36\ cm^{-1}.nm}{d_t(nm)} - \frac{97.87\ cm^{-1}.nm^2}{[d_t(nm)]^2} + \frac{24.12\ cm^{-1}.nm^3}{[d_t(nm)]^3}$, which reproduces the RBMs within a ± 3 cm^{-1} error range when, comparing with the calculated Raman spectra of the

SWBNNTs from (0,6) to (0,19) using the DFT technique and the RBM goes to zero in the limit of infinite diameter. An analytical expression for the other accompanying calculated low frequency bands (EDM of E_{2g} symmetry), which has lower frequency than the RBM, the best fit parameters carried out to third order in inverse diameter parameter is given by the equation: $\omega_{EDM}(E_{2g})\left(=113.64 + \frac{29.03\ cm^{-1}.nm}{d_t(nm)} - \frac{14.62\ cm^{-1}.nm^2}{[d_t(nm)]^2} + \frac{6.33\ cm^{-1}.nm^3}{[d_t(nm)]^3}\right)$, which reproduces exact calculated values of the EDMs. It is noting worth that, without offset constant, fitting equation (linear or high order) reproduces the calculated values of the EDMs within a large error range. The band is labeled as EDM for elliptical deformation, which derives from the predominate motions that define vibrational mode motions, as ascertained with the vibration visualization software mentioned earlier.

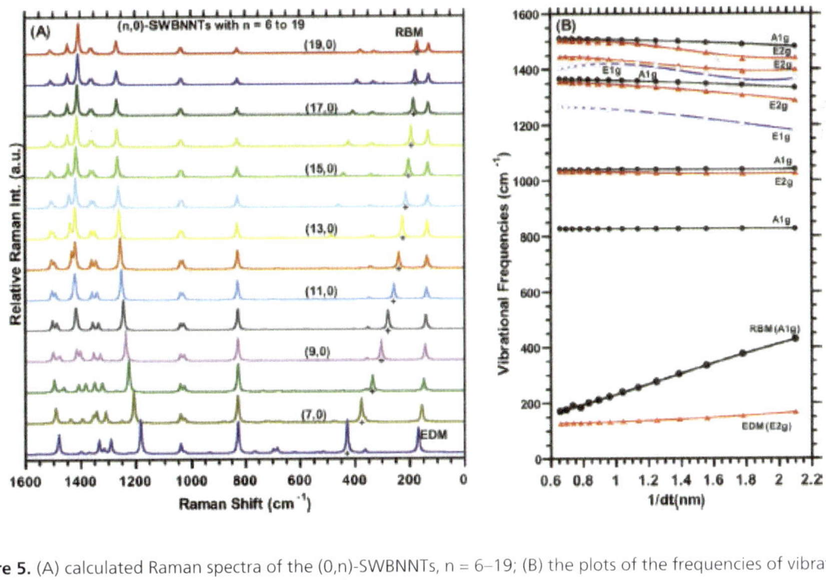

Figure 5. (A) calculated Raman spectra of the (0,n)-SWBNNTs, n = 6–19; (B) the plots of the frequencies of vibrational modes of symmetries A1g, E1g and E2g versus $1/d_t$.

The results of calculated Raman spectra of the (0,n)-SWBNNTs showed that: 1) the RBM of the frequency dramatically increases with decreasing the SWBNNTs diameter, which is not so surprising since the N–B–N bond strain and the sp^3 hybridization rapidly increases with decreasing SWBNNTs diameter; 2) as seen in Figure 5, for large sized SWBNNTs, the $\omega_{RBM}(A_{1g})$ and $\omega_{EDM}(E_{2g})$ mode frequencies converge. For instance, the calculated frequency separation between the RBM and EDM is found to be 3, 7, 21 and 43 cm^{-1}, when n has the values 26, 25, 22 and 19, respectively. Thus, one can anticipate the (0, 28)-SWBNNT would have unresolvable RBM and EDM bands for the experimental spectra. We can anticipate that the acquisition of Raman spectra for experimental samples consisting of large diameter SWBNNT with the purpose of characterizing the sample in terms of electronic properties and purity may be complicated by the existence of this EDM band, which, in general, can lead to apparent broadening of bands as well as the presence of additional bands that may lead to the erroneous conclusion that more than one type of SWBNNT is present in the sample. Of course, this issue is not expected to be of great significance since the synthesis routes that are presently in vogue do not lead to nanotubes with diameter as large as that corresponding to the (0,26) index. It is to be noted that the E_{2g} band has lower frequencies than the RBM, (see Figure 5A). This latter band is labeled as EDM for elliptical deformation, which derives from the predominate motions that define vibrational mode motions, as ascertained with the vibration visualization software mentioned earlier.

As regards other general conclusions that can be drawn from our calculations for the SWBNNTs, we have found that calculated Raman bands in the mid-frequency region exit nearly size-independent peak positions. As shown in Table 1 or Figs. 5A-B, in the high frequency region there are a few Raman bands of symmetries $E_{1g}/E_{2g}/A_{1g}$ that lie close to one another in frequency. For instance, the calculated Raman modes with symmetries of the A_{1g} (~1355 ± 10 cm^{-1}) and E_{2g} (~1330 ±25 cm^{-1}) approach one another in frequency with increasing diameter of the SWBNNT and then reach a constant values of 1365 and 1356 cm^{-1}, respectively, as seen in Table 1. A fitting equation indicated that these two Raman bands (with symmetries A_{1g} at ~ 1355 ±10 cm^{-1} and E_{2g} (~1330 ±25 cm^{-1}) first increase in frequency then approach a constant value of ~1366 and ~1360 cm^{-1}, respectively, with increasing diameter of the (0,n)-SWBNNT, n=25. Furthermore, the resonance Raman experiments [60,61] have been shown that there is only one strong band at 1355 ± 10 cm^{-1} in high energy region for the boron nitride nanotubes. Thus, the calculated these Raman bands at A_{1g} (~1355 ± 10 cm^{-1}) and E_{2g} (~1330 ±25 cm^{-1}) are not only in good agreement with experiments, but also the calculations suggest that only the Raman band(s) (of the symmetry of A_{1g} and/or E_{2g}) are theatrically enhanced by resonance excitation of the boron nitride nanotube.

Furthermore, the predicted shifts in the peak positions may result from the nanotube curvature effect as mentioned in Refs. 48(f-h), the curvature energy of the nanotube brings about dissimilar force constants along the nanotube axis and the circumference direction. Therefore, the nanotube geometry causes a force constant reduction along the tube axis compared to that in the circumferential direction. Consequently, the curvature effect might play crucial role in the shift of the peak positions of the G-band as well as the RBM band, as mentioned earlier. In addition, the calculated Raman band positions for bands at ~ 1240 ± 30 cm^{-1} are found to be slightly size dependent, exhibiting a slightly blue shift with increasing diameter of the SWBNNTs. This disorder induced mode is also important for the characterization and

	(0,6)	(0,7)	(0,8)	(0,9)	(0,10)	(0,11)	(0,12)	(0,13)	(0,14)	(0,15)	(0,16)	(0,17)	(0,18)	(0,19)	(0,6)& (0,12)	(0,7)& (0,14)	(0,8)& (0,16)	(0,9)& (0,18)
E_{2g}	167	155	147	143	139	137	135	134	133	131	130	130	129	128	246 156	206 147	170 139	152 130
Elliptical deformation (EDM) of both inner and outer tubes in the same phase																		
A_{1g}	428	376	335	303	277	256	239	224	212	201	192	184	177	171	497 256	416 226	354 200	310 179
Radial breathing of the outer tube only (RBM)																		
A_{1g}	826	827	826	827	827	827	827	827	827	827	827	827	827	828	820 831	823 832	823 833	823 832
Out-of-surface bending deformation of the NBN/BNB bonds on the tube																		
E_{2g}	1026	1024	1025	1026	1027	1029	1030	1031	1032	1033	1033	1034	1034	1035	1030	1027	1030	1034
BN stretching (in opposite phase) along tube axis																		
A_{1g}	1040	1039	1040	1040	1039	1040	1040	1040	1040	1040	1039	1039	1039	1039	1040	1034	1036 1036	1039 1044
BN stretching along tube axis only																		
E_{2g}	1181	1206	1223	1236	1244	1251	1255	1258	1261	1263	1265	1266	1267	1268	1253 1234	1206 1242	1243 1246	1238 1263
Asymmetric stretching vibrations and bending deformations of the BNB/NBN bonds																		
E_{1g}	1289	1308	1320	1329	1335	1340	1344	1347	1349	1351	1353	1354	1356	1356		1351	1341	1352
Bending deformation of the NBN/BNB bonds, including relatively weak BN bond stretching																		
A_{1g}	1333	1343	1349	1353	1355	1358	1360	1361	1362	1363	1363	1364	1364	1365	1371	1368	1358	1363
BN stretching and bending deformation of the NBN/BNB bonds along tube axis																		
E_{1g}	1370	1355	1382	1401	1413	1419	1419	1418	1416	1414	1412	1410	1408	1406	1366	1398	1373	1376
Asymmetric stretching vibrations and bending deformations of the BNB/NBN bonds.																		
E_{2g}	1399	1393	1407	1414	1417	1424	1432	1437	1441	1444	1443	1446	1447	1446			1420	1429
BN stretching and bending deformations of NBN/BNB bonds																		
E1g E2g	1442	1436	1461	1476	1486	1492	1495	1495	1499	1499	1501	1502	1503	1504	1413	1432	1434	1434
BN stretching, including bending deformations of NBN/BNB bonds																		
A_{1g}	1481	1490	1496	1500	1502	1504	1505	1507	1507	1508	1509	1509	1509	1510	1430 1511	1463 1517	1473 1517	1485 1519
BN stretching along tube axis, including bending deformations of NBN/BNB bonds																		

Table 1. DFT-calculated Raman vibrational frequencies (in cm^{-1}) and assignments for (0,n)-SWBNNT and (0,n)&(0,2n)-DWBNNTs at the B3LYP/6-31G level.

the defect on the nanotube as observed a broad feature around in the spectrum of the Al-modified MWBNNTs [63]. For example, in the resonance Raman enhanced spectrum, the relative intensity of the disorder mode increases relative to the intensity of the breathing and tangential modes since there is a defect on the nanotube surface as a result of chemical functionalization or caused by structural deformation. For the carbon nanotubes (CNTs), the experimental studies have showed that the increase in the intensity ratio (I_D/I_G) indicates an increase in the number of defects on the sidewall of the nanotube. This is expected result of the introduction of covalently bound moieties to the nanotube framework, in which significant amount of the sp^2 carbons is converted to sp^3 hybridization.

DWBNNTs: While Figure 6 provides the calculated nonresonance Raman spectra for the (0,n)&(0,2n)-DWBNNTs, with n ranging from 6 to 9; Figure 8 provides diagrams of the atomic motions associated with the vibrational frequencies for the (8,0)&(16,0)-DWBNNT used as a representative case. The calculations show that the frequencies of the radial breathing modes (RBMs) and tangential modes (TMs, known as G-mode) of (n,0)&(2n,0)-DWBNNT (with n=6 to 9) significantly differ from those calculated for the (0,n)-SWBNNTs (see Figure 7 and Table 1). The results of the calculations are summarized below. In the low frequency region, the calculated Raman spectra of these DWBNNTs exhibited two RBM modes resulting from the radial motion of the inner- and outer-shells, as shown in Figure 6, and both of these RBM modes are strongly diameter dependent. A large gap between RBMs in the Raman spectra of the DWBNNTs decreases with increasing diameter of the inner- and outer-shells (as seen in Figure 6). Comparing these calculated RBMs in the spectrum of the (0,8)&(0,16)-DWBNNT with their corresponding bands in the isolated (0,8)- and (0,16)-SWBNNTs spectra, as seen in Figure 7, we note that the RBMs at 335 cm^{-1} in the Raman spectrum of the (8,0)-SWBNNT and at 192 cm^{-1} in the (16,0)-SWBNNT spectrum are, respectively, upward shifted to 354 and 200 cm^{-1} in the spectrum of (0,8)&(0,16)-DWBNNT. Additionally, the RBMs for the (0,6)-SWBNNT(428 cm^{-1}) and for the (0,12)-SWBNNT (239 cm^{-1}) spectrum are, respectively, blue shifted to 497 and 256 cm^{-1} in the Raman spectrum of (0,6)&(0,12)-DWBNNT (see Table 1). The relative distances between RBMs in the spectra of (0,n)&(0,2n)-DWCNTs are greater than the separation between corresponding RBMs in Raman spectra of (0,n)- and (0,2n)-SWCNTs. For instance, the distance between the RBMs for (0,8)&(0,16)-DWBNNT is 154 cm^{-1}, this distance between the RBMs in the Raman spectra of the corresponding isolated (0,8)- and (0,16)-SWBNNTs is 143 cm^{-1}. A tentative fitting equation may be obtained as given in Equation 4a-b:

$$\omega_{\text{inner}}(\text{RBM, in cm}^{-1}) = \frac{181.27}{d_t(nm)} + \frac{37.00}{[d_t(\text{nm})]^2} - \frac{4.82}{[d_t(\text{nm})]^3} \qquad a$$

$$\omega_{\text{outer}}\left(\text{RBM, in cm}^{-1}\right) = \frac{237.34}{d_t(nm)} + \frac{65.65}{[d_t(\text{nm})]^2} - \frac{51.85}{[d_t(\text{nm})]^3} \qquad b \qquad (4)$$

where d_t stand for the shell diameter. The tentative fitting equations reproduced calculated RBMs within 0.5 cm^{-1} error range for both inner- and outer-tubes. Another Raman bands below RBM modes in the spectra of the SWBNNTs are blue-shifted relative to the corresponding peaks in the spectra of their corresponding DWBNNTs. For instance, these Raman features at 147 cm^{-1} in the spectra of (0,8)-SWBNNT and at 130 cm^{-1} in the spectrum of the (0,16)-SWBNNT are respectively blue-shifted to 170 and 139 cm^{-1} in the spectrum of the (0,8)&(0,16)-DWBNNT. Furthermore, in the mid-frequency region, the relatively weak intense peaks are centered 1036 (A_{1g}), 1030 (E_{2g}) and 823 (A_{1g}) cm^{-1} are predicted almost at the same positions in the spectra of both (0,8)- and (0,16)-SWBNNTs.

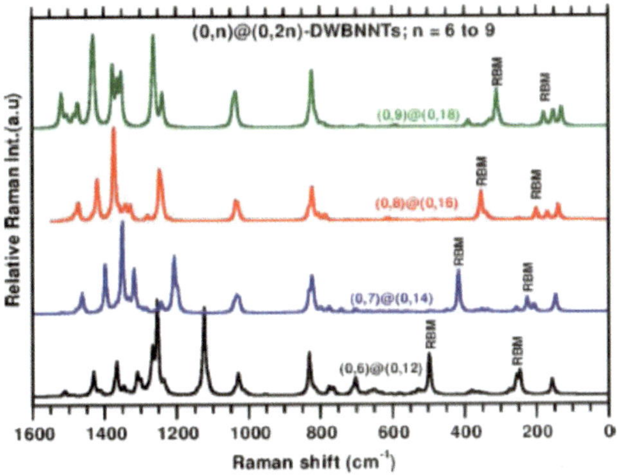

Figure 6. Calculated Raman spectra of the (0,n)&(0,2n)-DWBNNT, n = 0–9.

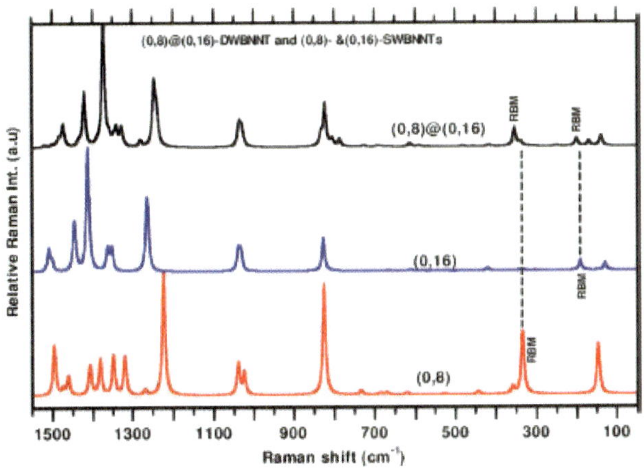

Figure 7. Calculated Raman spectra of the (0,8)&(0,16)-DWBNNT and (0,8)- and (0,16)-SWBNNTs.

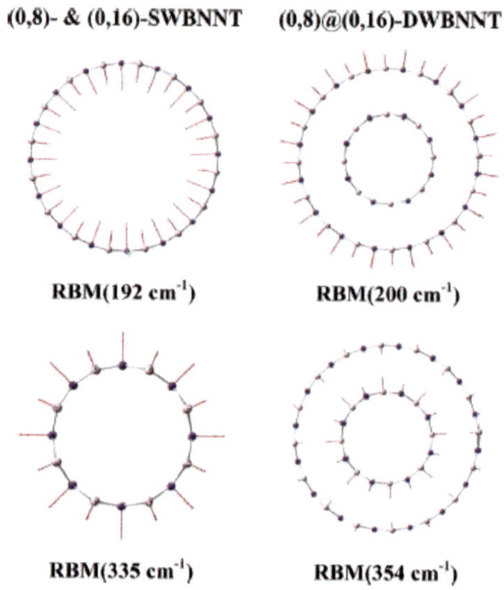

Figure 8. Calculated molecular motions for some vibrational bands of the (0,8)&(0,16)-DWBNNTs and (0,8)- and (0,16)-SWBNNTs.

In the high frequency region, comparing the Raman features in the spectra of the (0,8)&(0,16)-DWBNNTs with their band position in the corresponding (0,n)-SWBNNTs spectra, it can be seen that they are slightly shifted relative to SWBNNTs, as seen in Figure 7 for the (0,8)&(0,16)-DWBNNT. For instance, the Raman bands at 1434 (A_{1g}, relatively weak), 1420 (E_{1g}, medium intense), 1373 (E_{1g}, the most stronger one), 1358 (A_{1g}, relatively weak), and 1246 (E_{1g}, relatively strong) cm^{-1} in the spectrum of the (0,8)&(0,16)-DWBNNT correspond to the Raman features at 1461 (medium), 1407 (medium), 1382 (medium), 1349 (medium), and 1223 (the most stronger) cm^{-1} in the spectrum of the (0,8)-SWBNNT, and these are predicted at 1501 (relatively weak), 1443 (medium), 1412 (the most stronger), 1363 (relatively weak), and 1265 (relatively strong) cm^{-1} in the spectrum of the (0,16)-SWBNNT (see Figure 6), respectively.

Moreover, Y. Bando et. al. [62] have studied Raman spectra of the multi-walled boron (natural ^{11}B and isotope ^{10}B) nitride nanotubes (MWBNNT and MW^{10}BNNT). Their Raman spectra of the MWBNNT and MW^{10}BNNT showed only one strong Raman peak at 1366 and 1390 cm^{-1}, respectively, in the range of 1200 to 1500 cm^{-1}, which is assigned to a BN stretching deformation vibration mode. This measured Raman peak is in good agreement with our calculated Raman peak (E_{1g}) at 1373 cm^{-1} in the calculated nonresonance Raman spectrum of the (0,8)&(0,16)-DWBNNT, which is resulting from the BN stretching along tube axis, including bending deformation of the NBN/BNB bonds along tube axis. Additionally, Obraztsova and coworkers [63] have studied comparative Raman spectra of the multi-walled boron nitride nanotubes (MWBNNTs) samples before and after Al ion modifications have been investigated. Two features in the Raman spectra were observed: one at 1366 cm^{-1} that corresponds to in-plane vibrations between B and N atoms and broad feature around 1293 cm^{-1} in the spectrum of the Al-modified MWBNNTs. The broad peak around 1293 cm^{-1} is consistent with the calculated Raman feature around 1250 cm^{-1} in the spectra of the DW- and SW-BNNTs.

3.3. IR Spectra of Single-Walled and Double-Walled Boron Nitride Nanotube

Zigzag-SWBNNTs: Figure 9A provides calculated IR spectra for the (n,0)-SWNTs, where n ranges from 6 to 19. As evidenced in Figure 9, the calculated IR spectra exhibited seven peaks of symmetries E_1u and A_{1u} are slightly depend on the SWBNNTs diameter. In the range of 1000 to 1550 cm^{-1}, relatively very weak six IR features of symmetries E_{1u} are centered: ~ 1475 ± 25, ~1330 ± 30, ~1230 ± 30, ~1030 ± 5 cm^{-1}, and other two weak peaks with symmetry A_{1u} are centered ~1495 ± 15 and ~1350 ± 15 cm^{-1}. The strongest one with symmetry E_{1u} is centered 1395 ± 30 cm^{-1}. In the range of mid frequency, the calculated IR spectra of the (0,n)-SWBNNTs (n= 6 to 19) exhibited only one weak peak centered 805 ± 15 cm^{-1}. The analytical expressions for this calculated high frequency band as functions of third order in inverse of the (0,n)-SWBNNTs diameter are given by the equations: $\omega(cm^{-1}) = A + \frac{B}{d_t(nm)} + \frac{C}{[d_t(nm)]^2} + \frac{D}{[d_t(nm)]^3}$, where the parameters A(in cm^{-1}), B(in cm^{-1}.nm), C(in $cm^{-1}.nm^2$) and D(in $cm^{-1}.nm^3$) are respectively obtained such as: 1508.9, 10.3, -15.6, and 2.4 for the peak (A_{1u}) centered 1495 ± 15 cm^{-1}; 1515.7, -14.9, -2.4, and -3.6 for the peak (E_{1u}) centered 1475 ± 25 cm^{-1}; 1344.7, 130.0, -54.1, and -2.4 for the peak (E_{1u}) centered 1395 ± 30 cm^{-1}; 1357.2, 33.7, -41.0, and 10.2 for the peak (A_{1u}) centered 1350 ± 15 cm^{-1}; 1367.0, -4.9, -17.4, and 0.8 for the peak (E_{1u}) centered 1330 ± 30 cm^{-1}; 1264, 23.2, -37.3, and 3.8 for the peak (E_{1u}) centered 1230 ± 30 cm^{-1}; 1030.5, 28.6, -31.4, and 9.1 for the peak (E_{1u}) centered 1030 ± 5 cm^{-1}; and 824.8, 7.5, -22.3, and 4.1 for the peak (E_{1u}) centered 805 ± 15 cm^{-1}. The plots of the calculated IR features vs. inverse of the tube diameter are given in Figure 9B. In the low frequency region, the IR spectra exhibited many IR features; however, their intensities are extremely weak or vanish as seen in Figure 9A. Furthermore, we provided the vibrational mode assignments and frequencies for the IR spectra of the isolated zigzag-SWBNNTs in Tables 2.

DWBNNTs: While Figure 10 provides the calculated IR spectra for the (0,n)&(0,2n)-DWBNNTs, with n ranging from 6 to 9; Figure 11 provides calculated IR spectra of the (0,8)&(0,16)-DWBNNTs and isolated (0,8)- and (0,16)-SWBNNTs for comparison. The calculated spectra of the DWBNNTs 1517 (A_{1u}), 1474 (E_{1u}), 1434 (E_{1u}), 1373 (E_{1u}), 1347 (A_{1u}), 1238 (E_{1u}), 823 (E_{1u}) and 786 (E_{1u}) cm^{-1}, which are correspond the IR features at 1496, 1473, 1407, 1320, 1349, 1224, and 798 cm^{-1} in the spectrum of the (0,8)-SWBNNT; these are calculated at 1509, 1501, 1412, 1353, 1363, 1261, and 819 cm^{-1} in the spectrum of the (0,16)-SWBNNT, as seen in Table 2. Moreover, Y. Bando [62] have studied FTIR spectra of the multi-walled boron (natural ^{11}B and isotope ^{10}B) nitride nanotubes (MWBNNT and MW^{10}BNNT). Their FTIR spectra of the MWBNNT and MW^{10}BNNT revealed blue degraded strong IR peak at 1376 and 1392 cm^{-1}, respectively, which is assigned to a B-N stretching deformation vibration mode. This measured IR peak is in good agreement with our calculated IR peaks at 1367

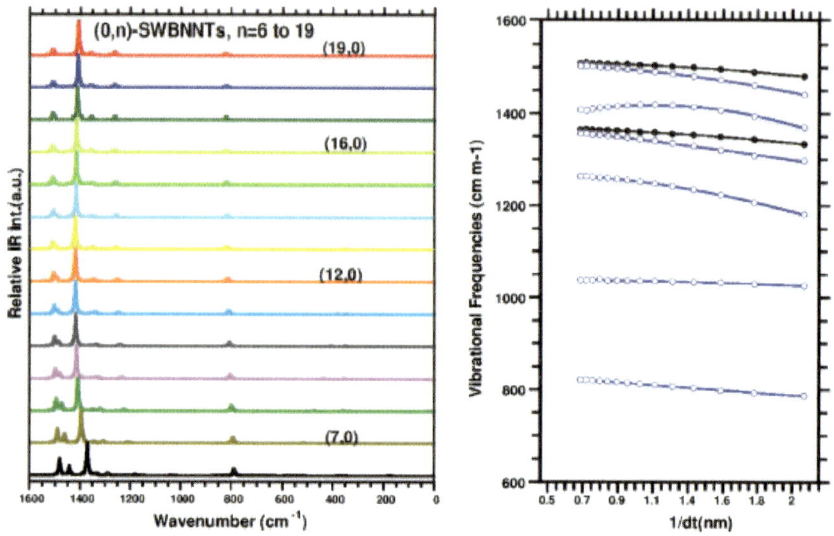

Figure 9. (A) calculated IR spectra of the (0,n)-SWBNNTs, n = 6–19 and (B) the plots of the frequencies of vibrational modes versus $1/d_t$.

cm^{-1} (E_{1u}, resulting from the bending deformation of the NBN/BNB bonds along tube axis) and 1424 cm^{-1} (E_{1u}, due to the BN stretching along tube axis, including bending deformations of NBN/BNB bonds). Author also observed a relatively weak and broad IR features at ~800 cm^{-1} and suggested that this IR peak is due to the existence of some B-O bonds in their BN nanotubes, see Figure 4 in Ref. [62]. However, our calculated IR spectra of the SWBNNTs and DWBNNTs exhibited IR feature with relatively weak around 800 cm^{-1} is as a result of the out-of surface bending deformation of NBN/BNB bonds on the boron nitride nanotube. Therefore, we suggest that this IR peak (~800 cm^{-1}) may originate from the boron nitride nanotube.

3.4. Electronic transition energies of DWBNNT and SWBNNTs

As mentioned in the introduction to this section, boron nitride nanotubes (BNNTs) can be viewed as modified CNT, but their electronic properties differ from carbon nanotubes. For instance, depending on their chirality and the radius, although carbon nanotubes can be either metallic or semiconducting, all boron nitride nanotubes (BNNTs) are semiconducting materials with a large band. And since the band gap is large, the gap energy is only weakly dependent on the diameter, chirality, and the number of the walls of the tube. Furthermore, owing to their semiconducting character, BNNTs, like CNTs, themselves are also very interesting materials for application in nanoscale devices, and have been considered alternatives to CNTs. The DWBNNTs as well as the doped BNNTs nanotubes may show a dramatic change relative to the isolated nanotube. On account of the strong interactions between elec-

	(0,6)	(0,7)	(0,8)	(0,9)	(0,10)	(0,11)	(0,12)	(0,13)	(0,14)	(0,15)	(0,16)	(0,17)	(0,18)	(0,19)	(0,6)& (0,12)	(0,7)& (0,14)	(0,8)& (0,16)	(0,9)& (0,18)
E_{1u}	788	793	798	804	807	811	813	815	817	818	819	820	821	821	764 813	775 820	786 823	795 824
Out-of-surface bending deformation of the NBN/BNB bonds on the tube																		
E_{1u}	1182	1207	1224	1235	1242	1248	1252	1255	1258	1260	1261	1262	1264	1264	1253	1197	1238	1257
Asymmetric stretching vibrations of the NBN/BNB bonds due to the motions of the N and B atoms along circumference direction.																		
E_{1u}	1298	1308	1320	1329	1335	1340	1344	1347	1349	1351	1353	1354	1356	1356	1366	1351	1373	1376
Bending deformation of the NBN/BNB bonds, including relatively weak BN bond stretching																		
A_{1u}	1333	1343	1349	1353	1355	1358	1359	1361	1362	1363	1363	1364	1364	1365	1332	1332	1347	1363
BN stretching and bending deformation of the NBN/BNB bonds along tube axis																		
E_{1u}	1370	1394	1407	1414	1418	1419	1419	1418	1416	1414	1412	1410	1408	1406	1439	1433	1434	1429
Asymmetric stretching vibrations and bending deformations of the BNB/NBN bonds.																		
E_{1u}	1442	1462	1473	1481	1486	1491	1493	1496	1498	1499	1501	1502	1503	1504	1488	1466	1474	1476
BN stretching, including bending deformations of NBN/BNB bonds																		
$A1_u$	1481	1490	1496	1500	1502	1504	1505	1507	1507	1508	1509	1509	1509	1510	1511	1517	1517	1519
BN stretching along tube axis, including bending deformations of NBN/BNB bonds																		

Table 2. DFT-calculated IR vibrational frequencies (in cm^{-1}) and assignments for (0,n)-SWBNNT and (0,n)&(0,2n)-DWBNNTs at the B3LYP/6-31G level.

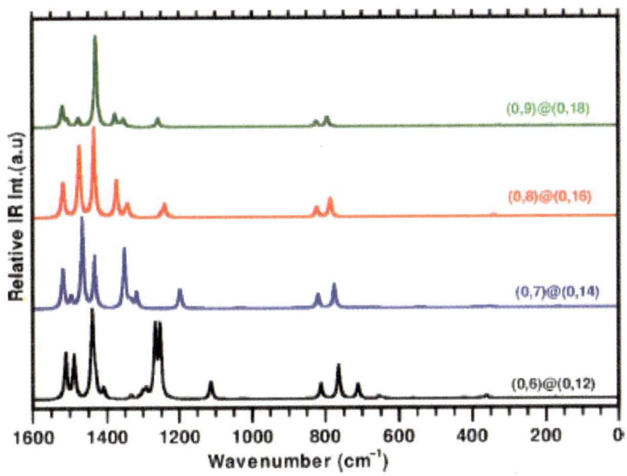

Figure 10. Calculated IR spectra of the (0,n)&(0,2n)-DWBNNT, n = 6–9.

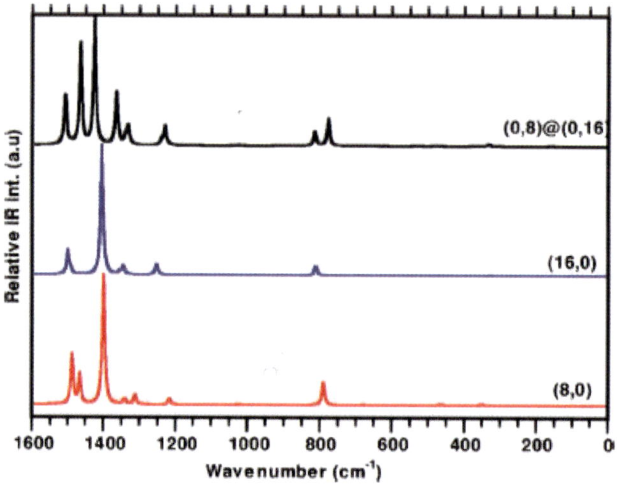

Figure 11. Calculated IR spectra of the (0,8)&(0,16)-DWBNNT and (0,8)- and (0,16)-SWBNNTs for comparison.

trons and holes in DWBNNTs, the excitonic effects in BNNTs is expected to be more important than in CNTs, since bright (dipole allowed) and dark (dipole forbidden) excitons in DWBNNTs can exhibit qualitatively different optical response. Therefore, the time-dependent DFT (i.e., TD-DFT) method has been applied to investigate the dark transient structures involved in radiationless processes for the DWBNNTs. In this section, we provide the calculated vertical electronic transitions of (0,6)&(0,12)-DWBNNT and (0,6)- and (0,12)-SWBNNTs using DFT and discuss these results in terms of IC and ISC processes.

The calculated vertical electronic transitions of (0,6)&(0,12)-DWBNNT and (0,6)- and (0,12)-SWBNNTs, as seen in Figure 12 and Table 3, indicated that the lowest electronic energy level (dipole forbidden) of the DWBNNTs are lower as much as about 0.4 eV relative to the (0,6)-SWBNNT and 1.5 eV relative to the (0,12)-SWBNNT. However, when we compare the lowest dipole allowed electronic transitions, the lowest dipole allowed electronic transitions of the DWBNNT are about 1.07 eV and 0.99 eV lower than that for (0,6)- and (0,12)-SWBNNTs, respectively.

The predicted dipole allowed electronc transitions, $S_0(A') \rightarrow S_7(A'')$ (4.90 eV) and $S_0(A') \rightarrow S_8(A')$ (4.91 eV), respectively,are due to the HOMO-4(A″)→LUMO(A′) and HOMO-5(A′)→LUMO(A′) transitions; $S_0(A') \rightarrow S_{10}(A'')$ (4.94 eV) is as a result of HOMO-6(A″)→LUMO(A′) transition; $S_0(A') \rightarrow S_{11}(A'')$ (5.12 eV) is mainly due to HOMO(A″)→LUMO+4(A″) and HOMO-3(A′)→LUMO+2(A′)transitions and $S_0(A') \rightarrow S_{12}(A'')$ (5.12 eV) is mainly because of HOMO-3(A′)→LUMO+1(A″) and HOMO(A″)→LUMO+5(A′) transitions. These calculated transitions, together with the plotted electron densities in the HOMOs and LUMOs, as seen in Figure 4B, indicated that first three of five dipole allowed electronic transitions of the

(0,6)&(0,12)-DWBNNT							**(0,12)-SWBNNT**			**(0,6)-SWBNNT**		
$S_0 \rightarrow S_n$				**T1→Tn (SCF Corrected)**			**$S_0 \rightarrow S_n$**			**$S_0 \rightarrow S_n$**		
Exc. St.#	Sym.	eV	f	SYM.	eV	f	Sym.	eV	f	Sym.	eV	f
1:	A"	4.35		A"	4.24		E_1	5.83		A"	4.72	
2:	A'	4.47		A'	4.59	0.0071	E_1	5.95		A'	4.86	
3:	A"	4.67		A"	4.59	0.0072	E_1	5.95		A"	4.86	
4:	A'	4.67		A"	4.93	0.0006	E_1	5.97	0.0871	A'	4.86	
5:	A'	4.89		A'	4.93	0.0007	E_1	5.97	0.0871	A"	5.71	
6:	A"	4.89		A'	4.96		E_1	6.01	0.0239	A'	5.71	
7:	A"	4.90	0.0334	A"	4.96		E_2	6.19		A'	5.83	
8:	A'	4.91	0.0331	A"	4.98		E_2	6.19		A'	5.89	0.0001
9:	A'	4.93		A'	5.01		E_1	6.30	0.8777	A"	5.89	0.0129
10:	A"	4.94	0.0003	A"	5.24		E_1	6.30	0.8777	A"	5.90	0.0316
11:	A'	5.12	0.0055	A'	5.24		A_1	6.36	0.0256	A'	5.90	0.0443
12:	A"	5.12	0.0058	A'	5.27	0.0088	E_1	6.39	0.0168	A"	5.94	0.0002
13:	A"	5.21		A"	5.27	0.0091	E_1	6.39	0.0168	A"	6.05	
14:	A'	5.21		A'	5.44	0.0251	A_2	6.43		A'	6.11	0.0062
15:	A"	5.25		A"	5.44	0.0234	E_2	6.51	0.5031	A"	6.11	0.0063
16:	A'	5.25		A'	5.52	0.0022	E_2	6.51	0.5031	A'	6.20	0.0001
17:	A"	5.26		A'	5.56		E_2	6.52		A"	6.20	0.0001
18:	A"	5.30		A"	5.65		E_2	6.52		A'	6.26	0.0092
19:	A'	5.37		A"	5.67		E_1	6.60	0.0012	A'	6.38	
20:	A"	5.37		A'	5.68		E_1	6.60	0.0012	A"	6.38	

Table 3. The calculated vertical electronic transitions, singlet-singlet ($S_0 \rightarrow S_n$ and triplet-triplet ($T_1 \rightarrow T_n$), of the (12,0)&(6,0)-DWBNNT and (12,0)- and (6,0)-SWBNNTs for comparison at the B3LYP/6-31G level of using DFT. Note that the SCF corrected triplet-triplet electronic transitions were calculated as the deference between the calculated global energies of the singlet and triplet sates added to triplet-triplet electronic transitions in order to comparing with the singlet-singlet transitions and where the letters S_0, T_1 and f are respectively the lowest energy level of the singlet, triplet states and oscillator strength.

(0,6)&(0,12)-DWBNNT, $S_0(A')\rightarrow S_7(A'')$/ $S_8(A')$/ $S_{10}(A')$, originating from the electron transfer from the outer-shell to the inner-shell. These results are clear evidence of the charge transfer from the other shell to the inner shell. The dipole allowed electronic transitions $S_0(A')\rightarrow S_{11}(A')$ shows the electron excited from both inner- and other-shells to mostly inner shells, also there is a significant sigma-bonding interactions between inner- and outher-shells. Finally, the $S_0(A')\rightarrow S_{11}(A')$ transition indicate that the transitions from both shells to the excited state mainly are due to sigma-bonding interactions. We also calculated the triplet-triplet transitions, which produce many dipole allowed transitions. The SCF corrected electronic transitions of the singlet-singlet and triplet-trpilet of the (0,6)&(0,12)-DWBNNT, together with the singlet-singlet transitions, are given in Figure 12. As seen in Figure 12 and Table 3, upon irradiation, there is the possibility of a system that can undergo internal conversion (IC) and intersystem crossing (ISC) processes via vibroelectronic coupling, besides the photochemical and other photophysical processes. The IC and ISC processes would able to be expected when taking account of the small distance between the electronic energy levels and range of the vibrational spectra of the DWBNNTs.

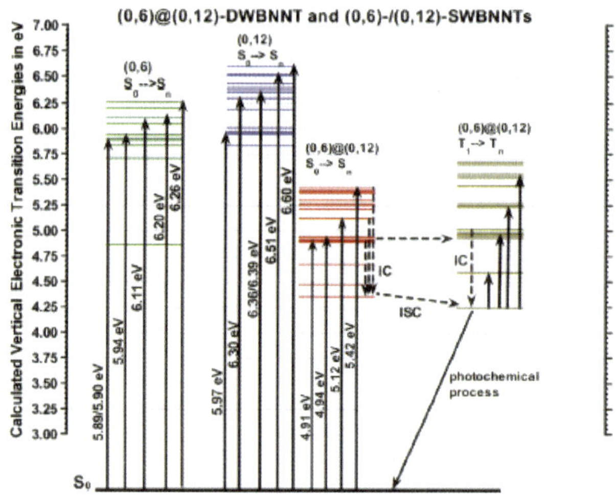

Figure 12. Calculated vertical electronic transitions, singlet–singlet (S0→Sn and triplet–triplet (T1→Tn) for (0,6)&(0,12)-DWBNNT and (0,6)- and (0,12)-SWBNNTs. The vertically solid arrow indicated dipole allowed transitions. The broken-arrows display possible internal conversion (IC) and intersystem crossing (ISC) processes.

C. H. Lee and coworkers [64] have measured absorption spectrum of the suspension of BNNTs in ethanol by using UV–visible absorption spectroscopy (HP 8453 Spectrophotometer). The authors observed three absorption bands at ~5.9 eV (very strong) and ~4.78 eV (weak), and ~3.7 eV (very weak) in the UV-visible spectrum and suggested that the band at about 4.75 eV originates from the intrinsic dark exciton absorption band; the relatively small band at ~ 3.7 eV was due to the defects of the boron nitride nanotubes (BNNTs), and the stronger band at 5.9 eV was as results of the optical band gap of BNNTs. For the (0,6)&(0,12)-DWBNNT, as seen in Table 3, our calculated electronic transitions produced a few dipole allowed electronic transitions below 5.37 eV such as: $S_0 \rightarrow S_7/S_8$ at 4.90 eV (with the f = 0.0334), $S_0 \rightarrow S_{10}$ at 4.94 eV (f = 0.0003), $S_0 \rightarrow S_{11}/S_{12}$ at 5.25 eV (f = 0.0055), which are in good agreement with this measured band at about 4.78 eV. Furthermore, for the (0,6)- and (0,12)-SWBNNTs, the calculations exhibited the lowest dipole allowed electronic transition around 5.9 eV, which is in accordance with the measured strong optical band at 5.9 eV. The lowest dipole forbidden transitions are predicted at 4.35, 4.72, and 5.83 eV for the (0,6)&(0,12)-DWBNNT, (0,6)- and (0,12)-SWBNNTs, respectively. Consequently, this experimentally measured UV-visible spectrum might be an evidence for the formation of the (0,6)&(0,12)-DWBNNT, the observed absorption band (at ~4.78 eV) may due to the $S_0 \rightarrow S_7/S_8$ (4.90 eV), not due to the intrinsic dark exciton as suggested by authors.

Furthermore, Figure 4A provides the calculated electron density of (0,6)&(0,12)-DWCNT (double-walled carbon nanotube), showing that the first four highest occupied molecular orbitals (from HOMO to HOMO-3 with the A_{1u}, A_{2g} and $2E_{1g}$ symmetries, respectively) belong to the outer-shell, and the next highest occupied molecular orbitals from HOMO-4 to HOMO-24 include both inner- and outer-shells of (0,6)&(0,12)-DWCNT. The lowest unoccupied molecular orbital LUMO (E_{1u}), lying about 0.780 eV above the HOMO (A_{1u}), belongs to the outer-shell, while the next one (B_{2u}) belongs to the inner-shell and lies 0.849 eV above the HOMO (A_{1u}). The calculated electron density also indicates that an intratube (inner and outer tube) interaction may possibly take place in the excited state: the LUMO+7 with A_{2u} symmetry and 2.494 eV above the HOMO (A_{1u}), LUMO+8 (E_{1u}; 2.557 eV), LUMO+10 (E_{1g}; 2.563 eV) and LUMO+15 (E_{1g}; 3.637 eV). The intratube CC σ-bonding interaction in the excited state may lead to an intertube charge transfer, which can be observed by a significant change in the tangential modes (TMs) of Raman spectra when the tube is excited to its intratube charge transfer state. The TM may provide information not only about the metallic or semiconducting character of nanotubes, but also on the inner-outer tube (intratube) charge transfer.

Similarly, the calculated vertical electronic transitions for the (0,n)&(0,2n)-DWBNNT and (0,n)- and (0,2n)-SWBNNTs, n= 8 and 9, at the same level of the theory. The calculated singlet-singlet ($S_0 \rightarrow S_n$ and triplet-triplet ($T_1 \rightarrow T_n$) electronic transitions are given in Table 4.

For the (0,8)&(0,16)-DWBNNT, the predicted dipole allowed electronc transitions, $S_0 \rightarrow S_2/ S_3$ (5.39 eV, mainly due to the H-1 →L and H→L) and $S_0 \rightarrow S_6$ (5. 47 eV, mainly due to the H-3 →L). These calculated transitions, in conjunction with the plotted electron densities in the HOMOs and LUMOs, as seen in Figure 4C, indicated that first three dipole allowed electronic transitions of the (0,8)&(0,16)-DWBNNT, $S_0 \rightarrow S_{2/3}/ S_6$ originating from the electron transfer from the outer-shell to the inner-shell. These results of the calculations provide not only clear evidence for the charge transfer from the other shell to the inner shell, but also there is a significant BB σ-bonding interaction between the inner- and outer-shells. As seen in Table 4, the lowest dipole allowed vertical electronic transition of the (0,8)&(0,16)-DWBNNT ($S_0 \rightarrow S_2$; 5.39 eV) lies 0.61 and 0.52 eV below the lowest allowed transitions of the $S_0 \rightarrow S_6$ and $S_0 \rightarrow S_4$ for the (0,8)- and (0,16)-SWBNNTs, respectively.

Furthermore, upon irradiation, a system can undergo internal conversion (IC) and intersystem crossing (ISC) processes, besides the photochemical and other photophysical processes. Transient intermediates are likely to form in the IC and ISC radiationless processes, which is also known as "dark processes". Our calculations also indicated that possibilities of the IC and ISC processes via vibroelectronic coupling, besides the photochemical and other photophysical processes. For instance, based on the calculated electronic transitions as seen in the Table 4, when the (0,8)&(0,16)-DWBNNTs are excited, all of the excited nanotubes may not directly return back to their ground state by emission of a photon, $S_{k>0} \rightarrow S_0$ transition, but some of them may return back to their ground states (S_0) by the IC (internal conversion), for instance, when the system is excited into a higher vibroelectronic state (S_6, 5.47 eV), it may undergo into the S_1 state (5.39 eV) via vibrational coupling between these two states before undergoing additional vibrational relaxation back to the lowest singlet electronic energy level (S_1), which is called internal conversion (IC), then, followed by transition from the second lowest singlet electronic energy level S_1(5.39 eV) to S_0 by emission of a photon is so-called fluorescence. An alternate pathway for a molecule in the S_1 state involves an intersystem crossing (ISC) by the nanotube into the lowest triplet electronic state T_1 (5.28 eV). From T_1, the nanotube can undergo radiative de-excitation via a much slower process, which is known as phosphorescence ($T_1 \rightarrow S_0$ transition) such as illustrated in Figure 12.

Likewise, for the (0,9)&(0,18)-DWBNNT, the calculations indicated that the lowest dipole allowed transition ($S_0 \rightarrow S_3$, 5.69 eV) lies 0. 18 and 0.21 eV below the lowest allowed transitions of the (0,9)- and (0,18)-SWBNNT. Additionally, as seen in Table 4, the calculations also indicated that the possibilities of the IC from the S_k (k=3,4,7-9, and 14) to S_1 as well as ISC proces from the singlet electronic state S_1(5.67 eV) to T_1 (5.71 eV) for the (0,9)&(0,18)-DWBNNT. The calculated dipole allowed vertical electronic transitions may be summarized as following: the transition $S_0 \rightarrow S_{3/4}$ (5.69 eV and f=0.1656) is predominantly as result of the electron excitation mostly from the outer shell to the inner shell (H→L+1, H-1→L, H-6→L+1/2), including excitations from inner shell to the outer (H-2→L+1/2 and H-3→L+1/2); $S_0 \rightarrow S_7$ (5.73 eV and f=0.0060 is mainly as result of the electronic excitation from the outer shell to the inner shell (H→L+1, H-1→L+2, H-6→L), including relatively weak contribution from inner shell to the outer (H-3→L+5 and H-2→L+6); and the transitions $S_0 \rightarrow S_{8/9}$ (5.74 eV and f=0.0007) and $S_0 \rightarrow S_{14}$ (5.78 eV and f=0.0077) are as result of the electronic excitation from the outer shell to the outer shell (H-5→L+3/4 and H-4→L+3/4), as shown in Figure 4 D.

The key conclusions on the calculated electronic spectra indicates that the first dipole allowed electronic transitions of the (0,n)&(0,2n)-DWBNNTs (n = 6,8, 9) lead to a charge transfer process from outer shell to the inner shell. Moreover, there is a significant intertube σ-bonding interactions between the inner- and outer-shells occurs with decreasing distance between the interwall of the DWBNNTs, in contrast, for the (0,9)&(0,18)-DWBNNT, there is a relatively weak contributions to the charge transfer process from the inner-shell to outer-shell.

	(0,8)&(0,16)-DWBNNT				(0,8)-SWBNNT		(0,16)-SWBNNT		(0,9)&(0,18)-DWBNNT				(0,9)-SWBNNT		(0,18)-SWBNNT	
	$S_0 \rightarrow S_n$		$T_1 \rightarrow T_n$		$S_0 \rightarrow S_n$		$S_0 \rightarrow S_n$		$S_0 \rightarrow S_n$		$T_1 \rightarrow T_n$		$S_0 \rightarrow S_n$		$S_0 \rightarrow S_n$	
n	eV	f	eV	f	eV	f	eV	f	eV	f	eV	f	eV	f	eV	f
1	5.39		5.28		5.61		5.79		5.67		5.71		5.77		5.79	
2	5.39	0.1039	5.32	0.0003	5.61		5.88		5.68		5.74	0.0002	5.86		5.87	
3	5.39	0.1039	5.32	0.0003	5.69		5.88		5.69	0.1656	5.77	0.0004	5.86		5.87	
4	5.46		5.37		5.80		5.91	0.1868	5.69	0.1652	5.78		5.87	0.0215	5.90	0.2880
5	5.46		5.37		5.87		5.91	0.1868	5.70		5.85	0.0021	5.87	0.0215	5.90	0.2880
6	5.47	0.0007	5.49		6.00	0.0255	5.94	0.0268	5.70		5.92	0.0002	5.95	0.0135	5.91	
7	5.47		5.50		6.00	0.0255	5.96		5.73	0.0060	5.94	0.0001	6.06		5.91	
8	5.47		5.50		6.02		5.96		5.74	0.0007	5.95	0.0058	6.06		5.93	0.0287
9	5.52		5.67		6.02		6.16	2.0492	5.74	0.0007	6.00	0.0018	6.06		6.06	
10	5.67		5.82	0.0594	6.04		6.16	2.0492	5.76		6.01	0.0038	6.06		6.06	
11	5.67		5.82	0.0594	6.07	0.0164	6.17		5.76		6.07	0.0021	6.09		6.11	2.5027
12	5.67		5.92	0.0167	6.11		6.17		5.78		6.15	0.0077	6.09		6.11	2.5026
13	5.69		5.92	0.0167	6.11		6.27	0.0432	5.78		6.16	0.0049	6.16		6.24	0.0165
14	5.69		5.94		6.14		6.27	0.0432	5.78	0.0077	6.23	0.0151	6.16		6.24	0.0165
15	5.74		6.00		6.14		6.31		5.83		6.24	0.0103	6.28		6.28	
16	5.74		6.00		6.16		6.31		5.83		6.26	0.0215	6.28		6.28	
17			6.03		6.17		6.34	0.0211			6.28	0.0065	6.30	0.0918	6.34	
18					6.18		6.35	0.0587					6.30	0.0918	6.34	
19					6.33	0.2616	6.35	0.0587							6.35	0.0334

Table 4. The calculated vertical electronic transitions, singlet-singlet ($S_0 \rightarrow S_n$ and triplet-triplet ($T_1 \rightarrow T_n$), of the (0,8)&(0,16)-DWBNNT and (12,0)- and (6,0)-SWBNNTs for comparison at the B3LYP/6-31G level of using DFT. Note that the SCF corrected triplet-triplet electronic transitions were calculated as the deference between the calculated global energies of the singlet and triplet sates added to triplet-triplet electronic transitions in order to comparing with the singlet-singlet transitions and where S_0 and T_1 is respectively the lowest energy level of the singlet and triplet states.

4. Covalently functionalized zigzag-SWCNTs

Carbon nanotubes have broad range of potential applications from medical to industry fields due to their unique structural, mechanical, and electronic properties, as mentioned in the introduction section. Different functionalization methods such as chopping, oxidation, wrapping and irradiation of the CNTs can lead to active bonding sites on the surface of the nanotubes.

In this section, we calculate, for covalently functionalized carbon nanotubes (f-CNTs), such parameters as the curvature energies referenced, IR and Raman spectra, and vertical electronic transitions. The latter one may be important to understand the optical mechanism for the charge transfer between functional group(s) and CNT as well as internal conversion and intersystem crossing, as well photochemical process that may occur.

The structure of the functionalized-single-walled carbon nanotubes, f-(n,0)-SWCNTs, constructed of functional group(s) covalently bound on the (n,0)-SWCNTs,of two unit cell length, has been investigated. The most stable of the geometry has been obtained by full optimization without any symmetry restriction. The optimized structure indicated that the cylindrical shape of the nanotube is altered to an elliptical form when two molecules attached to the surface of CNT; but the structure remains almost cylindrical with C_4 symmetry, when four functional groups are bound. When we used benzenesulfonic acid (ph-SO_3H; $C_6H_5SO_3H$) as a functional group that covalently bonds on the surface of the (n,0)-SWCNTs, n=6 to 12, the curvature energy per hexagon, ($\Delta E[f-(n,0)-SWCNTs]$),of the functionalized-(n,0)-SWCNT calculated relative to that of the corresponding isolated species is given by the following equation:

$$\Delta E[f-(n,0)-SWCNTs;in\ eV]=\frac{E[f-(n,0)-SWCNT]-E[f]}{2n}-\frac{E[(n,0)-SWCNTs]}{2n} \quad (9)$$

where E[f-(n,0)-SWCNTs], E[f] and E[(n,0)-SWCNTs] indicates the global energy of functionalized-(n,0)-SWCNT, isolated benzenesulfonic acid ($C_6H_5SO_3H$) and isolated (n,0)-SWCNT, respectively,The f and n stand for the functional group and chiral index of the zigzag-CNTs. The plot of the calculated relative curvature energy is given in Figure 13. As seen in the Figure 11, the relative curvature energy for the metallic and semiconducting CNTs are well separated. Based on the predicted value of the energies, the results suggested that the covalently functionalization of the SWCNT, with small diameters, are energetically more stable than that with large diameters for the metallic nanotubes. However, for semiconducting nanotubes, the functionalization of the tube is favorable, but the functionalization of the (11,0)-SWCNT is more favorable than (10,0)-SWCNTs. In order to make a correct overall assessment, we need to more data, at least for semiconducting zig-zag nanotubes.

4.1. Raman spectra of functionalized zigzag-SWCNTs

The calculated nonresonance Raman spectra for the covalently functionalized-(n,0)-SWCNTs with benzenesulfonic acid (-ph-SO3H) and the isolated (n,0)-SWCNTs (where n = 7 to 10), as well as the spectrum of the functionalized (7,0)-SWCNT with the carboxylic acid (-COOH), for comparison, are shown in Figure 14. Because of the similarity of the Raman spectra of the f-SWCNTs, here we only discuss the Raman spectra for the functionalization of the (7,0)-SWCNT with the benzenesulfonic acid and carboxylic acid, and the spectrum of the isolated (7,0)-SWCNT. The Raman spectra of both functionalized (7,0)-SWCNT exhibited many new features relating to the spectrum of the isolated (7,0)-SWCNT as well as shift in the peak positions. The predicted results are summarized below.

In the low energy region below 600 cm^{-1}, 1) one of the important Raman peak, which is the radial breathing mode (RBM), was predicted at 410 cm^{-1} in the isolated (7,0)-SWCNT shifted not only to 390 and 385 cm^{-1} in the spectra of the (7,0)-SWCNT functionalizes with benzenesulfonic acid and carboxylic acid, respectively, but also enhanced in both spectra; 2) the relatively peaks at 109 and 111 cm^{-1} result from the elliptical deformation of the carbon nanotube are respectively shifted to 75 and 121 cm^{-1} (in the spectrum of the (7,0)-SWCNT functionalizes with benzenesulfonic acid), and to 95 and 134 cm^{-1} in the Raman spectrum of the functionalization of the (7,0)-SWCNT with carboxylic acid, the intensity enhanced in both spectra of the functionalized tube; 3) a doubly degenerated peak predicted at 284 cm^{-1} (as a result of diagonal expansion of the tube) in the Raman spectrum of the isolated tube is split into well separated two peaks and appeared at about 250 and 306 cm^{-1} in the spectrum of each (7,0)-SWCNT functionalizes with benzenesulfonic acid and carboxylic acid; 4) a relatively very weak peak at 500 cm^{-1} in the spectrum of the isolated tube appeared at same position, but its intensity significantly enhanced, in the calculated both Raman spectra of the (7,0)-SWCNT functionalizes with benzenesulfonic acid and carboxylic acid, which is ; 5) many relatively weak Raman features (result from the out-of-plane structural deformation of the functional groups) appeared below 600 cm^{-1} as seen in the Figure 14 and 15. In the range from 600 to 1250 cm^{-1}, the Raman spectra of the f-(n,0)-SWCNT exhibited many relatively medium, weak and very weak new Raman peaks beside the peaks appeared at 760, 794 and 911 (very weak) cm^{-1} in the Raman spectra of the f-(n,0)-SWCNT. For instance, in the Raman spectrum of the (7,0)-SWCNT functionalizes with benzenesulfonic acid, the peaks with relatively intense at 1120 cm^{-1} (due to the structural deformation of the tube, including wagging of CH bonds of the benzene ring); at 1138 cm^{-1} (as a result of asymmetric CSO bond straching and OH bond wagging, including relatively weak bending deformation of the benzene ring); at 1142 cm^{-1} (structural deformation of the tube due to the CC streching, accompanied by wagging of Hs on the benzene ring), and the peak at 1185 cm^{-1} is owing to asymmetric CSO bond stretching and wagging of OH bond. The Raman spectrum of the (7,0)-SWCNT functionalizes with carboxylic acid exhibited relatively strong Raman features at 1122 cm^{-1} (caused by structural deformation of the nanotube, including OH bond wagging); 1146 cm^{-1} (by reason of asymmetric streching of CCO(H) bond, including tube deformation), and the calculated Raman peak at 1181 cm^{-1} is due to asymmetric streching of CCO bonds, including tube deformation. The Raman peaks at 760 cm^{-1} (due to expansion of the tube along the tube axis) and 795 cm^{-1} (as a result of out-of-surface bending deformation of the tube) in the Raman spectrum of the isolated (7,0)-SWCNT at the same positions of the f-SWCNT). A strong peak at around 1225 cm^{-1} in the spectra of the (7,0)-SWCNT and f-(7,0)-SWCNT is completely originates from the wagging of the CH bond at end of the tube. There are also many very weak Raman features appeared in this range from 600 to 1250 cm^{-1}. In the range from 1300 to 1800 cm^{-1}, two peaks at 1300 cm^{-1} (weak) and 1330 cm^{-1} (strong) in the spectrum of the (7,0)-SWCNT functionalizes with benzenesulfonic acid and with carboxylic acid, as a result of symmetric stretching of CCC bonds and bending deformations along tube axis, which correspond to a relatively weak and doubly degenerate Raman feature at 1305 cm^{-1}. A doubly degenerated peak (relatively very weak) at 411 cm^{-1} (result from asymmetric stretching of CCC bonds within the tube) in the Raman spectrum of the SWCNT is split into two weak peaks at about 1390 and 1405 cm^{-1} in the Raman spectrum of the functionalized (7,0)-SWCNT. The Raman peak with medium intense at 1486 cm^{-1}, resulting from CC bond stretching within the nanotube, corresponds to the peak at ~1504 cm^{-1} in the Raman spectrum of the functionalized (7,0)-SWCNT. The strongest and doubly degenerate Raman peak at 1574 cm^{-1} in the isolated (7,0)-SWCNT, resulting from asymmetric stretching of the CCC bonds along circumference direction of the tube, is blue shifted to nearly degenerated peak at 1590 and 1595 cm^{-1}, as a result of the CC bond stretching within the tube, in the Raman spectra of the f-(7,0)-SWCNT. In this range from 1300 to 1800 cm^{-1}, the Raman spectra of the (7,0)-SWCNT functionalizes with benzenesulfonic acid and with carboxylic acid showed many new Raman features. For example, the strongest peaks appeared at ~1380 and ~1390 cm^{-1} are as a result of asymmetric tube deformation due to the CC bonds stretching, which is not shown in the isolated (7,0)-SWCNT. The peaks at 1373 and 379 cm^{-1} in the spectrum (7,0)-SWCNT functionalizes with benzenesulfonic are mainly due to the asymmetric stretching of the OSO bond and wagging of the OH bond, including asymmetric stretching of the CCC bonds of the benzene ring. The peaks: at 1471 and 1482 cm^{-1}, which is the result of the CC bond stretching within the tube; at , 1548, and 1557 cm^{-1} is due to asymmetric CCC bond stretching within the tube, however, the peak at 1547 cm^{-1} is entirely due to symmetric stretching of the CC bonds of the benzenesulfonic acid . Furthermore, the predicted Raman peak at 1650 cm^{-1} is due to CC bond stretching of the benzene ring, including CH bond wagging on the benzene ring. A very weak peak at 1806 cm^{-1} is as a result of the CO stretching of the carboxylic acid only. As a result of the (7,0)-SWCNT functionalizes with benzenesulfonic acid and with carboxylic acid (f-(7,0)-SWCNT), the key conclusions on these calculated Raman spectra of the f-(7,0)-SWCNT are summarized below: 1) the RBM is red shifted as much as 25 cm^{-1}; 2) many new peaks appeared in the disorder (D) mode range from 1300 to 1450 cm^{-1}, which is due to the structural deformation of the tube and of the functional groups bound to the tube (7,0)-SWCNT); 3) the tangential (or G) mode is blue shifted as much as 20 cm^{-1}, as a result of the functional groups bound to the tube; 4) above the G-mode, appeared new Raman feature in the spectra of the f-(7,0)-SWCNT belong to the functional groups (benzenesulfonic acid and carboxylic acid); 5) the new Raman features are found to appear along the spectrum, which is owing to the combination of the structural deformation of the tube and the functional groups; 6) for the benzenesulfonic acid, while the CH bond stretching mode occurred range from 3200 to 3240 cm^{-1}, the OH bond stretching appear at 3703 cm^{-1}; for the carboxylic acid, the OH bond stretching is predicted at 3678 cm^{-1}; the CH bond stretching of the tube are predicted in the range from 3172 to 3200 cm^{-1}.; 7) the RBMs of frequency in the calculated Raman spectra of the functionalized (n,0)-SWCNT , (n=6 to 11) are slightly red-shifted relative to that for isolated SWCNTs as seen in Figure 15. The relative shift in frequency of the RBM decreases with increasing tube diameter.

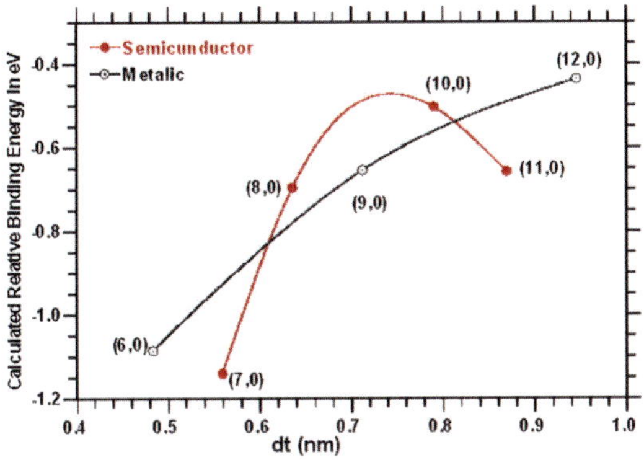

Figure 13. Calculated binding energies of the (n,0)-SWCNTs covalently functionalizes with the benzene sulfonic acid ((n,0)-SWCNTs-ph-SO_3H, n = 6–12). Energetically more stable covalently functionalized (n,0)-SWCNT (f-(n,0)-SWCNTs; n = 6–12) was predicted by using the equation: ΔE[f-(n, 0)-SWCNTs : in eV] = E[f-(n, 0)- SWCNT]/2n – [E(f) + E[(n, 0)-SWCNTs]/2n. Where ΔE[f-(n,0)-SWCNTs] is the energy difference between the total energy of the f-(n,0)-SWCNTs per the number of hexagons in the tube (E[f-(n,0)-SWCNT)/2n]) with reference to the total energy of their corresponding isolated (n,0)-SWCNTs per the number of hexagons in the tube (E[(n,0)-SWCNT)/2n]) and the total energy of the functional groups (E(f)/2n). The letters n and 2n stand for the chiral index of the zigzag-SWCNTs and the number of hexagon in the nanotube, respectively. See Section 3.1 for more detail.

It is worth nothing that the relative intensity of the peaks in the resonance Raman spectra significantly change. Because of the technical difficulty and calculation time, it is very difficult to calculate resonance Raman spectra. Furthermore, in the low frequency region below 600 cm^{-1}, there are many relatively very weak Raman peaks, which result from out-of-plane motion, or twisting of the phenyl group. These types of Raman bands of the functionalizated the CNTs may significantly enhanced in the resonance Raman spectrum (RRS) since there is a significant dipole-dipole interaction between the functional groups. This may play a crucial role and might be used as signature for the alignment of the CNTs in two dimensional networks, but also, the presence of additional bands may lead to the erroneous conclusion that more than one type of SWNT is present in the sample. For instance, the Raman band(s) resulting from out-of-plane motions are dramatically enhanced when dye molecule aggregate, and are referred to as J- or H- type aggregates.[48(a-d)]

New Raman peaks appeared around 1550 cm^{-1} due to the symmetric stretching of the CCC bonds and rocking of CH bonds in phenyl group of the benzenesulfonic acid. Several new Raman peaks result from only benzenesulfonic acid or combination of benzenesulfonic acid and nanotube dispersed throughout the spectrum. The Raman peak resulting from the stretching of CC sigma bonding between benzenesulfonic acid and SWCNTs is very weak and appear at about 1208 cm^{-1}. In the low frequency region, there are many relatively very weak Raman peaks below 600 cm^{-1}, which result from out-of-plane motion, or twisting of the phenyl group. These type of Raman bands of the functionalized-CNTs can play a crucial role and might be used as signature for the alignment of the CNTs in two dimensional networks. For instance, the Raman band(s) resulting from out-off plane motions are dramatically enhanced when dye molecule aggregate, and are referred to as J- or H- type aggregates.[48(a-d)] It is also worth that the calculations produced nonresonance Raman spectra which differ from the resonance Raman spectra in terms of intensity. Furthermore, the CH stretching of the end group of the CNT appear at around 3185 cm^{-1}, the CH stretching of the benzenesulfonic acid and OH stretching of the carboxyl group are, respectively, at about 3590 and 3680 cm^{-1}.

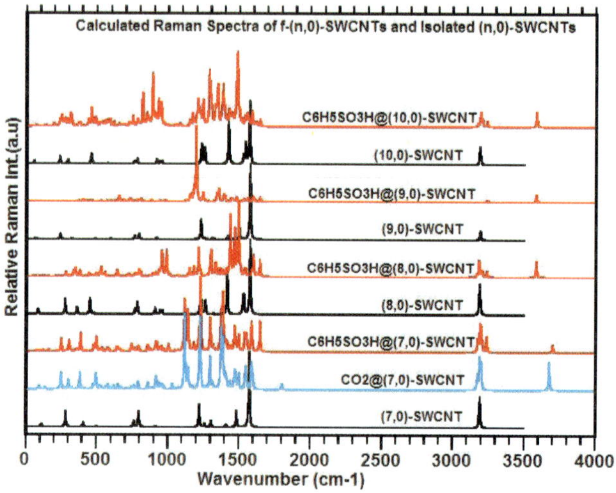

Figure 14. Calculated Raman spectra of the functionalized (n,0)-SWCNTs,benzenesulfonic acid, carboxylic acid, and isolated (n,0)-SWCNTs, n = 7 to 10.

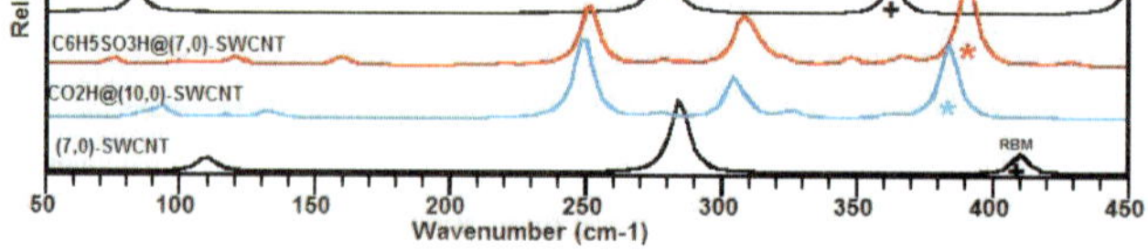

Figure 15. Calculated RBMs of frequencies in Raman spectra of the functionalized (n,0)-SWCNTs with benzenesulfonic acid and carboxylic acid, as well as isolated (n,0)-SWCNTs: n = 7 to 10.

The RBMs of frequency in the calculated Raman spectra of the functionalized SWCNT are slightly red-shifted relative to that for isolated SWCNTs as seen in Figure 15. The relative shift in frequency of the RBM decreases with increasing tube diameter.

4.2. IR spectra of functionalized SWCNTs

As provided in Figure 16, the predicted IR spectra of the (n,0)-SWCNT exhibits strong IR peaks centered at 890 and 845 cm^{-1}; however, the IR spectra of the functionalized (n,0)-SWCNTs display many new strong with relatively weak IR peaks dispersed through spectra, such as at 1650, 1275, 1150, 791, 570, 380, 143 cm^{-1}. Also, in range of 3000-4000 cm^{-1}, the CH and OH stretching modes of the benzenesulfonic acid and carboxylic acid are found to appear at around 1590 and 1670 cm^{-1}, respectively. The C=O bond resulting from C=O stretch of thecarboxyl groups, which is experimentally observed at 1782 cm^{-1} in the FTIR spectra of MWNT, after electron-beam irradiation by Eun-Ju Lee*et al.*[50], is predicted at 1800 cm^{-1} from the calculation.

The peaks found around 1650 cm^{-1} are mainly due to the C-C stretching and CCC bonding deformations; asymmetric and symmetric stretching of the O=S=O group in the benzenesulfonic acid group are found at 1275 and 1150 cm^{-1}, respectively; S-OH stretching appears at 780 cm^{-1}; bending deformation of the SO_3H,mimicking opening and closing of an umbrella,appears at 570 cm^{-1}; out-off plane motion of the phenyl group of the benzenesulfonic acidappears at 380 cm^{-1}; and twisting of the O=S=O bend appears at about 143 cm^{-1}.

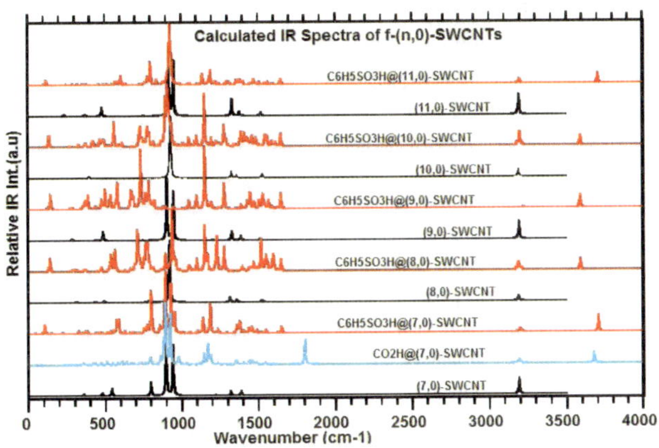

Figure 16. Calculated IR spectra of the (n,0)-SWCNTs functionalized with benzenesulfonic acid, carboxylic acid and isolated (n,0)-SWCNTs: n = 7 to 10.

4.3. Vertical electronic transitions of functionalized SWCNTs

We calculated the vertical electronic transitions for (n,0)-SWCNTs functionalized with benzenesulfonic acid. The functionalized-SWCNTs were constructed as two- and four- functional groups covalently attached to (7,0)/(9,0) and (12,0)/(8,0)-SWCNTs with length equivalent to two unit. Table 5 provides calculated electronic transitions of functionalized and isolated SWCNTs; selected calculated electron density for the HOMOs and LUMOs states involved in the electronic transitions are provided in Figure 17. The results of the calculations clearly indicate that both of the dipole allowed and forbidden electronic transitions are lowered as much as 0.8 eV relative to the transition energies of thecorresponding isolated SWCNT. Furthermore, the calculations also showed that below 2.5 eV there is no electron transfer from the nanotube to the functional group, or vice versa. However, the calculated electronic densities suggest that there would be intrasystem charge transfer between molecule and the nanotube. Because of the distance among the electronic energy levels is very small for some of the dipole allowed and forbidden electronic transitions, radiationless transitions are expected as a result of vibrational coupling or surface touching of the electronic potential energy surfaces. Coupling maybe very large and might lead to internal conversion (IC), again due to vibroelectronic coupling, which might be observable via fluorescence spectroscopic techniques, as discussed and illustrated in Figure 1 in the introduction section. We also would like to point out that while isolated SWCNTs exhibit one or a few dipole allowed electronic transitions below 2.5 eV, the functionalized SWCNTs produced many dipole allowed electronic transitions compared with the corresponding isolated SWCNTs, in addition to lowered electronic transitions.

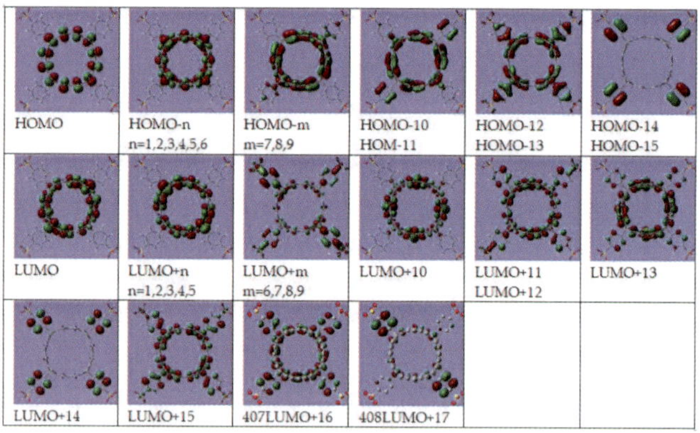

Figure 17. Calculated electron densities in the HOMO and LUMO states for the functionalized (12,0)-SWCNT with benzenesulfonic acid ($C_6H_5SO_3H$).

Calculated vertical electronic transitions, up to 2.53 eV, exhibited many dipole allowed and forbidden electronic transitions. The transitions up to 2.04 eV are due to transitions from the HOMOs of the SWCNT to the LUMO of the SWCNT. Above the 2.04 eV, calculation indicates the existence of charge transfer from the HOMOs of the SWCNT to the LUMOs of the benzenesulfonic acid ($-C_6H_4SO_3H$). For instance, the dipole allowed electronic transitions occur at 2.208, 2.232 and 2.523 eV, as a results of the transitions from the HOMOs of the (12,0)-

SWCNT to the molecule only: H→ L + 8, H→ L + 9 and H-1→ L + 9, respectively. As seen in Table 5, there are many dipole allowed electronic transitions from the HOMO of the SWCNT only to the LUMOs of both SWCNT and the benzenesulfonic acid. The results of the calculated vertical electronic transitions of functionalized nanotube ($C_6H_4SO_3H$@(12,0)-SWCNT) indicate that there is a charge transfer mechanism from the nanotube to the molecule above 2.0 eV. The small distance between the electronic transitions suggest that there would be internal conversion (IC) via vibrational coupling as much as 0.43 eVwhen taking account of the spectral range of the vibrational spectrum of the functionalized nanotube. These spectroscopic properties can be observable by quenching of the fluorescence and by Raman and IR spectroscopy. For the ($C_6H_4SO_3H$@(12,0)-SWCNT, the calculated electronic transitions up to 2.41 eV does not indicate any charge transfer process. However, when one examines the calculated energy levels of the HOMOs and the LUMOs it is possible for charge transfer processes to occur above 2.41 eV.

$S_0 \rightarrow S_k$	(12,0)-SWCNT		4F-(12,0)-SWCNT		(8,0)-SWCNT		4F-(8,0)-SWCNT	
k	T_e(eV)	f	T_e(eV)	f	T_e(eV)	f	T_e(eV)	f
1	0.54	0.0076	0.12	0.0001	0.77	0.0164	0.09	0.0001
2	0.82		0.15	0.0001	1.46	0.0006	0.11	
3	0.82		0.16	0.0001	1.46	0.0006	0.14	
4	1.27		0.42		2.44		0.31	0.0003
5	1.51		0.55	0.0111	2.51	0.3588	0.36	0.0059
6	1.51		0.55	0.0113	2.51	0.3588	0.55	0.0013
7	1.71	0.1082	0.65	0.0355	2.53		0.62	
8	1.87	0.6641	0.65	0.0355	2.77		0.79	
9	1.87	0.6641	0.74	0.0098	2.77		0.93	0.0462
10	2.31		1.10	0.0054	2.78		1.86	0.0049
11	2.31		1.10	0.0053	2.78		1.91	
12	2.56		1.37	0.0072	2.98	0.1295	1.96	0.0022
13	2.76		1.37	0.0071			2.00	
14	2.76		1.49				2.09	0.0002
15	2.81		1.50				2.12	0.0010
16	2.82		1.52	0.0059			2.18	
17	2.82		1.55				2.21	0.0031
18	2.94		1.63	0.0138			2.24	0.0043
19	2.94		1.63	0.0137			2.27	0.0236
20	2.99		1.70	0.0014			2.29	0.0157

21	2.99		1.70	0.0013			2.30	
22	3.03		1.78				2.36	0.0018
23	3.03		1.89				2.38	
24	3.18	0.0301	2.05	0.2005			2.41	0.0015

$S_0 \rightarrow S_k$	**(9,0)-SWCNT**		**2F-(9,0)-SWCNT**		**(7,0)-SWCNT**		**2F-(7,0)-SWCNT**	
k	T_e(eV)	f	T_e(eV)	f	T_e(eV)	f	T_e(eV)	f
1	0.56		0.27		0.91		0.39	
2	0.80		0.30	0.0144	1.25		0.79	0.0573
3	0.80		0.57		1.25		1.07	0.0067
4	0.93	0.0391	0.83	0.0087	1.45	0.0647	1.11	
5	2.32		1.34		2.52			
6	2.32		1.46	0.0307	2.53			
7	2.50		1.47	0.1432	2.93			
8	2.50		1.57		2.93			
9	2.64		1.64		3.02			
10	2.64		1.84		3.02			
11	2.72		1.88		3.08			
12	2.72		1.88	0.0268	3.08			
13			2.10					
14			2.11	0.0020				
15			2.19					
16			2.24	0.0272				

Table 5. Calculated vertical electronic transition energies (Te; in eV), $S_0 \rightarrow S_k$, of the mF-(n,0)-SWCNTs with that for the isolated (n,0)-SWCNTs for comparison with their oscillator strengths (f). Where m indicated the number of functional groups covalently bound to the (n,0)-SWCNTs and F symbolizes the benzenesulfonic acid used as functional group in this study.

5. Study of polyynes encapsulated into single-walled carbon nanotube

One-dimensional carbon atomic wires displaying sp hybridization have an attractive electronic and vibrational structure which severely affects their optical and transport properties. These kinds of structure have received researchers' interest because of their purely *sp*-hybridized carbon structure that is expected to display a completely different behavior than the more common sp^2 and sp^3 carbon structures. Polyyne molecules are linear carbon chains having alternating single and triple bonds, and ended by end atoms or groups. A. Milani et al. [51] have investigated the charge transfer in carbon atomic wires (polyynes) terminated by phenyl rings and its effects on the structure of the system using normal Raman and surface-enhanced Raman spectroscopy (SERS) techniques as well density functional theory (DFT) calculations forthe Raman modes. They reported that the occurrence of a charge transfer between polyynes and metal nanoparticles (both in liquids and supported on surfaces) is evidenced by Raman and SERS as a moderating of the vibrational stretching modes. They suggested that carbon wires alter their structure toward a more equalized geometry (i.e., all double bonds) as a consequence of the charge transfer. They also pointed out that these observations open potential perspectives for developing carbon-based atomic devices with tunable electronic properties. Therefore, it is necessary to carry out more experimental and theoretical investigation to get insight of them.

Even though the molecules like polyyne are very unstable at normal temperature and atmosphere conditions.[52, 53], it has beenreported that they are astoundingly stable inside single wall carbon nanotubes (SWCNT) even at high temperature (300 °C) [54, 55]. The Raman spectrum of the polyyne molecules exhibited two intense Raman shifts appear around 2000 -2200 cm^{-1}, which are labeled as α-bands and β-bands. The band positions of these two bands decrease in frequency with the increase in polyyne size. With the increasing chain lengths, while the frequency of the α-band almost linearly decreases, the position of β-bands is oscillating, and the difference between β-bands and α-bands in frequency shifts are dissimilar in polyyne molecules with different size.

Furthermore, L. M. Malard et al. [56] studied resonance Raman study of two polyyne molecules ($C_{10}H_2$ and $C_{12}H_2$) encapsulated inside the SWCNT using various different laser lines including the whole visible range. They indicated that the main Raman features associated with stretching modes of the linear chains in both samples ($C_{10}H_2$ @SWCNT and $C_{12}H_2$ @SWCNT) are strongly enhanced around 2.1 eV, while the optical absorption observed when these molecules are dispersed in isotropic medium [57] or in the gas phase[58] occurs above 4.5 eV. They concluded that dipole-forbidden (dark) transitions of the polyynes that become active as a result of a symmetry breaking when the molecules are encapsulated inside the SWCNT.

In this section, we will discuss the calculated results for the polyyne ($C_{10}H_2$) molecules encapsulated within (6,0)-SWCNT. Figure 18 and Table 6 provide the calculated electron density and energy levels of the molecular orbitals (MOs), HOMOs and LUMOs, of $C_{10}H_2$@(6,0)-SWCNT, respectively. The geometry optimization with/without symmetry restriction found the point group is respectively D_{6H} and D_{2H} symmetries. The structure with D_{2H} has the lowest energy as much as 0.19 eV than the structure with D_{6H} and both structure has the $^1A_{1G}$ electronic symmetry for the $C_{10}H_2$@(6,0)-SWCNT system.

For the isolated $C_{10}H_2$ (polyyne), predicted electronic symmetry is $^1\Sigma_{1G}$ and has the $D_{\infty H}$ point group. As seen in Figure 18, the plotted electron density showed that while three of first five highest occupied molecular orbitals (HOMO/HOMO-3/HOMO-4 with the A_g, B_{1u} and A_{1u} symmetries, respectively) only belong to the (6,0)-SWCNT, the HOMO-1 and HOMO-2 with the B_{2u} and B_{3u} symmetry belong not only to both of the $C_{10}H_2$@(6,0)-SWCNT and but also there is a significant bonding interaction between the polyyne molecule ($C_{10}H_2$) and

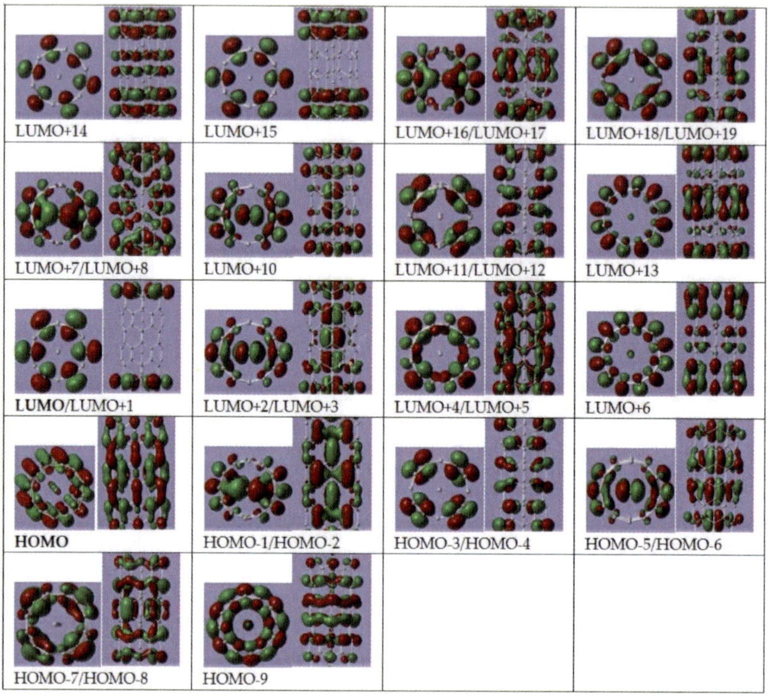

Figure 18. Calculated electron density of the molecular orbitals (MOs), HOMOs and LUMOs, of $C_{10}H_2$@(6,0)-SWCNT

(6,0)-SWCNT in the ground state. As seen in Table 6, the lowest unoccupied molecular orbital, LUMO (B_{3u})/LUMO + 1(B_{2g})/LUMO + 4 (A_g) LUMO + 5(B_{1g}) and LUMO + 6(B_{1u}) and lies about 0.43/0.43/0.89/0.89 and 1.66 eV above the HOMO (A_g) belong to the SWCNT only and the LUMO + 1(B_{3u})/LUMO + 2 (B_{2u}) belongs to the polyyne molecule and the SWCNT. However, the LUMO + 7(B_{3g})/LUMO + 8(B_{2g}) and LUMO + 9(B_{3g}) belong not only to both the $C_{10}H_2$@(6,0)-SWCNT (lies 1.99 /1.99 and 2.39 eV above the HOMO (A_g)), but there is a significant sigma bonding interaction in the excited states as seen in Figure 16.

The bonding interactions between $C_{10}H_2$ and (6,0)-SWCNT in the ground state leading to the increase the triple bond lengths and decrease the double C-C bond lengths within the polyyne molecule ($C_{10}H_2$) when encapsulated inside the (6,0)-SWCNT relative to its corresponding bond distance of the isolated single polyyne chain molecules ($C_{10}H_2$).

For instance, C-C bond distances in the encapsulated $C_{10}H_2$ molecules: 1.25974, 1.26769, 1.33655, 1.24771, 1.35216, 1.24771, 1.33655, 1.26769, 1.25974 Å and corresponding C-C bond distances in the isolated one: 1.22161, 1.35656, 1.23246, 1.34527, 1.23515, 1.34527, 1.23246, 1.35656, 1.22161 Å. These σ-bonding interactions between $C_{10}H_2$ and (6,0)-SWCNTs in the ground and excited states may be aspirant for the charge transfer between the molecule and

$C_{10}H_2$@(6,0)-SWCNTs			$C_{10}H_2$	
MOs	Sym.	ΔE(eV)	Sym.	ΔE(eV)
LUMO + 14	B_{3g}	3.47	SG_u	13.87
LUMO + 13	A_g	2.86	PI_g	13.24
LUMO + 12	B_{1u}	2.72	PI_g	13.24
LUMO + 11	A_u	2.72	SG_g	12.57
LUMO + 10	B_{2g}	2.39	SG_u	11.56
LUMO + 9	B_{3g}	2.39	PI_u	11.07
LUMO + 8	B_{2g}	1.99	PI_u	11.07
LUMO + 7	B_{3g}	1.99	SG_g	9.16
LUMO + 6	B_{1u}	1.66	SG_u	9.15
LUMO + 5	B_{1g}	0.89	PI_g	8.55
LUMO + 4	A_g	0.89	PI_g	8.55
LUMO + 3	B_{2u}	0.87	PI_u	6.11
LUMO + 2	B_{3u}	0.87	PI_u	6.11
LUMO + 1	B_{2g}	0.43	PI_g	3.90
LUMO	B_{3u}	0.43	PI_g	3.90
HOMO	A_g	-0.00	PI_u	0.00
HOMO-1	B_{2u}	-0.19	PI_u	0.00
HOMO-2	B_{3u}	-0.19	PI_g	-1.53
HOMO-3	B_{1u}	-0.23	PI_g	-1.53
HOMO-4	A_u	-0.23	PI_u	-2.86
HOMO-5	B_{3g}	-1.55	PI_u	-2.86
HOMO-6	B_{2g}	-1.55	PI_g	-3.87
HOMO-7	A_g	-1.91		

Table 6. Calculated energy levels ΔE(eV) of the molecular orbitals (MOs) for the $C_{10}H_2$@(6,0)-SWCNTs and $C_{10}H_2$ relative to their the highest molecular orbital (HOMO)

the SWCNT, which was observed between the polyyne and nanoparticles by the SERS as mentioned above.

The calculated vertical dipole allowed electronic transitions ($S_0 \rightarrow S_n$) of the $C_{10}H_2$@(6,0)-SWCNTs up to 0.52 eV are given in Table 7. Because of the technical difficulty, it was unable to calculate the higher electronic transitions that can provide more detailed information about internal conversion (IC) and inter system crossing (ISC). The lowest dipole allowed vertical electronic transitions $S_0(A_{1g}) \rightarrow S_7(B_{3u})$ as results of the HOMO-3→ LUMO + 1 and HOMO→ LUMO + 2 transitions and $S_0(A_{1g}) \rightarrow S_7(B_{2u})$ transition as a result of the HOMO-4->LUMO + 1 and HOMO→ LUMO + 3 transitions,the second lowest dipole allowed vertical electronic transitions $S_0(A_{1g})\rightarrow S_{11}(B_{3u})$ due to the HOMO-3→ LUMO + 1 and HOMO→ LUMO + 2 transitions, and $S_0(A_{1g})\rightarrow S_{12}(B_{2u})$ transition because of the HOMO-4→ LUMO + 1 and HOMO→ LUMO + 3 transitions clearly indicate that the existence of charge transfer from the SWCNT to the polyyne molecules when examine the electron density of the HOMO and LUMOs involved in these transitions. When we examine the calculated vertical electronic transitions together with the calculated energy levels of molecular orbitals (MOs) of the encapsulated polyyne molecule inside the SWCNT, the IC and ISC can be expected.

Based on these calculations, the molecule encapsulated inside the nanotubes (NTs) can be used as energy conversion systems as a consequence of charge transfer between them. This illustration also can reflect on the intensity of the Raman bands at the resonance excitation energy where the charge transfer takes place between the molecules or particle and the nanotubes.

		$S_0(A_{1G})\rightarrow S_n$						$S_0\rightarrow S_n$		
S_n	**Sym.**	**H→L**	**CI**	**T_e(eV)**	**f**	**S_n**	**Sym.**	**H→L**	**CI**	**T_e(eV)**
S_1	B_{2g}	H-"/>L+1	-0.84	0.06		S_{13}	B_{1g}	H-2-"/>L+3	-0.47	0.36
S_2	B_{3u}	H-"/>L	-0.83	0.06				H-1-"/>L+2	0.48	
S_3	B_{1g}	H-1-"/>L	0.61	0.08		S_{14}	A_g	H-2-"/>L+2	-0.49	0.37
S_4	A_g	H-2-"/>L	0.61	0.08				H-1-"/>L+3	0.49	
S_5	A_u	H-1-"/>L+1	0.62	0.09				H-"/>L+4	-0.15	
S_6	B_{1u}	H-2-"/>L+1	0.62	0.09		S_{15}	B_{1g}	H-2-"/>L+3	0.49	0.37
S_7	B_{3u}	H-3-"/>L+1	0.37	0.23	0.0001			H-1-"/>L+2	0.48	
		H-"/>L+2	0.56					H-"/>L+5	-0.15	
S_8	B_{2u}	H-4-"/>L+1	0.37	0.23	0.0001	S_{16}	A_g	H-2-"/>L+2	-0.13	0.41
		H-"/>L+3	0.57					H-1-"/>L+3	0.13	
S_9	B_{2g}	H-4-"/>L+3	-0.14	0.25				H-"/>L+4	0.60	
		H-3-"/>L	0.53			S_{17}	B_{1g}	H-2-"/>L+3	0.12	0.41
		H-3-"/>L+2	0.14					H-1-"/>L+2	0.12	
S_{10}	B_{3g}	H-4-"/>L	0.53	0.25				H-"/>L+5	0.60	
		H-4-"/>L+2	-0.14			S_{18}	B_{3g}	H-4-"/>L+2	-0.49	0.50
		H-3-"/>L+3	-0.14					H-3-"/>L+3	0.49	
S_{11}	B_{3u}	H-3-"/>L+1	0.40	0.30	0.0012	S_{19}	B_{2u}	H-2-"/>L+5	-0.51	0.51
		H-"/>L+2	-0.39					H-1-"/>L+4	0.52	
S_{12}	B_{2u}	H-4-"/>L+1	0.40	0.30	0.0012	S_{20}	A_u	H-4-"/>L+4	-0.44	0.52
		H-"/>L+3	-0.39					H-3-"/>L+5	0.44	

Table 7. The calculated vertical electronic transitions (Te; in eV) of $C_{10}H_2$@(6,0)-SWCNTs; $S_0(A1g)\rightarrow Sn$. where the f and CI stand for the oscillator strength and the configurationally interaction coefficients, respectively. The Letters H and L stands for HOMO and LUMO, respectively.

Author details

Metin Aydin[1] and Daniel L. Akins

1 Department of Chemistry, Faculty of Art and Sciences, Ondokuz Mayıs University, Samsun, Turkey

Center for Analysis of Structures and Interfaces (CASI), Department of Chemistry, The City College of The City University of New York, New York, USA

References

[1] Iijima, S. (1991). *Nature.*, 354, 56.

[2] Dhriti, Nepal., Jung-Inn, Sohn., Wilhelm, K. Aicher, Seonghoon, Lee., & Kurt, E. Geckeler. (2005). *Biomacromolecules*, 6(6), 2919.

[3] Karajanagi, S. S., Yang, H., Asuri, P., Sellitto, E., Dordick, J. S., & Kane, R. S. (2006). *Langmuir*, 22, 1392.

[4] Hod, Finkelstein., Peter, M. Asbeck, & Sadik, Esener. (2003). *3rd IEEE Conference on Nanotechnology (IEEE-NANO)*, 1, 441.

[5] Zhou, C., Kong, J., Yenilmez, E., & Dai, H. (2000). *Science*, 290, 1552.

[6] Misewich, J. A., Martel, R., Avouris, P., Tsang, J. C., Heinze, S., & Tersoff, J. (2003). *Science*, 300, 783.

[7] Nitzan, A., & Ratner, R. A. (2003). *Science*, 300, 1384.

[8] Wen, D. S., & Ding, Y. L. (2004). *International Journal of Heat and Mass Transfer*, 47, 5181.

[9] Masuda, H., Ebata, A., Teramae, K., & Hishiunma, N. (1993). Netsu Bussei (Japan) , 4, 227.

[10] Carissa, S. Jones, Xuejun, Lu, Mike, Renn, Mike, Stroder, & Wu-Sheng, Shih. (2009). *Microelectronic Engineering*, DOI: j.mee.2009.05.034.

[11] Jung, S. M., Jung, H. Y., & Suh, J. S. (2009). Sensors and ActuatorsB. , 139, 425.

[12] Rueckes, T., Kim, K., Joselevich, E., Tseng, G. Y., Cheung, C. L., & Lieber, C. M. (2000). *Science*, 289, 94.

[13] Michael, J. O'Connell, Sergei, M. Bachilo, Chad, B. Huffman, Valerie, C. Moore, Michael, S. Strano, Erik, H. Haroz, Rialon, Kristy. L., Boul, Peter. J., Noon, William. H., Carter, Kittrell, Ma, Jianpeng., Robert, H. Hauge, Weisman, R. Bruce, & Smalley, E. Richard. (2002). *Science*, 297, 593.

[14] Bachilo, S. M., Strano, M. S., Kittrell, C., Hauge, R. H., Smalley, R. E., & Weisman, R. B. (2002). *Science*, 298, 2361.

[15] Hartschuh, A., Pedrosa, H. N., Novotny, L., & Krauss, T. D. (2003). *Science*, 301, 1354.

[16] Maultzsch, J., Pomraenke, R., Reich, S., Chang, E., Prezzi, D., Ruini, A., Molinari, E., Strano, M. S., Thomsen, C., & Lienau, C. (2006). *Phys. Stat. Sol. B*, 243(13), 3204.

[17] Chang, E., Bussi, G., Ruini, A., & Molinari, E. (2004). *Phys. Rev. Lett.*, 92, 196401.

[18] Spataru, C. D., Ismail-Beigi, S., Benedict, L. X., & Louie, S. G. (2004). *Phys. Rev. Lett.*, 92, 077402.

[19] Perebeinos, V., Tersoff, J., & Avouris, P. (2004). Phys. Rev. Lett. , 92, 257402.

[20] Won-Il, Park, Kim, Hun-Sik, Kwon, Soon-Min, Hong, Young-Ho, & Jin, Hyoung-Joon. (2009). *Carbohydrate Polymers*, 77, 457.

[21] Meng, Lingjie, Fu, Chuanlong, & Lu, Qinghua. (2009). *Natural Science*, 19, 801.

[22] Hsieh, C. T., & Lin, Y. T. (2006). Microporous Mesoporous Mater. , 93, 232.

[23] Davis, J. J., Coleman, K. S., Azamian, B. R., Bagshaw, C. B., & Green, M. L. H. (2003). Chem Eur J. , 9(16), 3732.

[24] Poenitzsch, V. Z., Winters, D. C., Xie, H., Dieckmann, G. R., Dalton, A. B., & Musselman, I. H. (2007). J Am ChemSoc, , 129(47), 14724.

[25] Carbon nanotube science:. (2009). *Synthesis, Properties and Applications, by P.J.F. Harris (Cambridge University Press, Cambridge).*

[26] Zunfeng, Liu, et al. (2010). *Journal of Nanoscience and Nanotechnology*, 10, 5570.

[27] Istvan, Robel, Bruce, A, & Kamat, Prashant V. (2005). *Advanced Materials*, 17(20), 2458.

[28] Olek, M., Busgen, T., Hilgendorff, M., & Giersig, M. (2006). *Journal of Physical Chemistry B*, 110(26), 12901.

[29] Kazuhiro, Yanagi, Konstantin Iakoubovskii, Hiroyuki Matsui, Hiroyuki, Matsuzaki, Hiroshi, Okamoto, Yasumitsu, Miyata, Yutaka, Maniwa, Said, Kazaoui, Nobutsugu, Minami, & Hiromichi, Kataura. (2007). *J. Am. Chem. Soc.*, 129(16), 4992.

[30] Yuika, Saito, Kazuhiro, Yanagi, Norihiko, Hayazawa, Hidekazu, Ishitobi, Atsushi, Ono, Hiromichi, Kataura, & Satoshi, Kawata. (2006). *Jap. J. Appl. Phys.*, 45(12), 9286.

[31] Alvarez, L., Almadori, Y., Arenal, R., Babaa, R., Michel, T., Le Parc, R., Bantignies, J. -L, Jousselme, B. , Palacin, S. , Hermet, P., & Sauvajol, J. L. (2011). J. Phys. Chem. C, , 115(24), 11898.

[32] Srinivasan, C. (2008). *Current Science*, 94, 300.

[33] Hilder, T. A., & Hill, J. M. (2008). Current Applied Physics , 8, 258.

[34] Clendenin, J., Kim, J., & Tung, S. (2007). *Proc of 2007 2nd IEEE conference on Nanotechnology*, 1028.

[35] Bianco, A., Kostarelos, K., & Prato, M. (2005). *Current Opinion in Biotechnology*, 9, 674.

[36] Nasreen, G., Chopra, R. J., Luyken, K., Cherrey, Vincent. H., Crespi, Marvin. L., Cohen, Steven. G. Louie, & Zettl, A. (1995). Science. , 269, 966.

[37] Golberg, D., Bando, Y., Han, W., Kurashima, K., & Sato, T. (1999). *Chem. Phys. Lett.*, 308, 337.

[38] Tang, C. C., Bando, Y., Sato, T., & Kurashima, K. (2002). Chem. Commun. , 121, 290.

[39] Dresselhaus, M. S., Dresselhaus, G., & Eklund, P. C. (1996). Science of Fullerenes and Carbon Nanotubes. Academic Press, San Diegó, CA.

[40] Rubio, A., Corkill, J., & Cohen, M. L. (1994). Phys. Rev. B. , 49, 5081.

[41] Zhang, D., & Zhang, R. Q. (2003). Chem. Phys. Lett. , 371, 426.

[42] Mirzaei, M., & Hadipour, N. L. (2008). Physica E, , 40, 800.

[43] Park, C. H., Spataru, C. D., & Louie, S. G. (2006). *Physical Review Letters*, 96, 126105.

[44] Wirtz, L., Marini, A., & Rubio, A. (2006). *Physical Review Letters*, 96, 126104.

[45] Chunyi, Zhi, Yoshio, Bando, Chengchun, Tang, & Dmitri, Golberg. (2010). *Materials Science and Engineering R*, 70, 92.

[46] Kam, N. W. S., O'Connell, M., Wisdom, J. A., & Dai, H. (2005). *PNAS*, 102, 11600.

[47] Gaussian, Inc. *Carnegie Office Park-Bulding 6, Pittsburgh, PA106, USA.*

[48] (a) C. Guo, M. Aydin, H. R. Zhu, D. L. Akins, J. Phys. Chem. B 106 (2002) 5447; (b) H, Guo, X. Zhang, M. Aydin, W. Xu, H. R. Zhu, D. L Akins, Journal of Molecular Structure 689 (2004) 153; (c) M. Aydin, F. Jean-Mary, N. Stevens, D. L. Akins, J. Phys. Chem. B 108(2004) 9695; (d) W. Xu, M. Aydin, S. Zakia, D. L. Akins, J. Phys. Chem. B 108 (2004) 5588; (e) M. Aydin, J. R. Lombardi, J. Phys. Chem. A. 113 (2009) 2809; (f) M. Aydin, D.L. Akins, Vibrational Spectroscopy 53 (2010) 163; (g) M. Aydin, Vib. Spectrosc. 66 (2013) 30, (h) Metin Aydin and Daniel Akins (2011), Geometric and Spectroscopic Properties of Carbon Nanotubes and Boron Nitride Nanotubes, Electronic Properties of Carbon Nanotubes, Prof. Jose Mauricio Marulanda (Ed.), ISBN: 978-953-307-499-3, InTech, Available from: http://www.intechopen.com/books/electronic-properties-of-carbon-nanotubes/geometric-and-spectroscopic-properties-of-carbon-nanotubes-and-boron-nitride-nanotubes; (i) Metin AYDIN Photofragmentation Spectroscopy, Publisher: VDM Verlag, 2009; ISBN: 3639152891.

[49] Pfeiffer, R., Simon, F., Kuzmany, H., & Popov, V. N. (2005). *Phys. Rev.*, B72, 161404(R).

[50] Lee, Eun-Ju, Yoon, Jin-San, Kim, Mal-Nam, & Park, Eun-Soo. (1970). Preparation and Applicability of Vinyl Alcohol Group Containing Polymer/MWNT Nanocomposite

Using a Simple Saponification Method. *Carbon Nanotubes- Polymer Nanocomposites, Siva Yellampalli (Ed.)*, 978-9-53307-498-6, InTech, Available from:, http://www.intechopen.com/books/carbon-nanotubes-polymer-nanocomposites/preparation-and-applicability-of-vinyl-alcohol-group-containing-polymer-mwnt-nanocomposite-using-a-s.

[51] Milani, A., Lucotti, A., Russo, V., Tommasini, M., Cataldo, F., Li Bassi, A., & Casari, C. S. (2011). *J. Phys. Chem. C*, 115, 12836.

[52] Ravagnan, L., Siviero, F., Lenardi, C., Piseri, P., Barborini, E., Milani, P., Casari, C. S., Li Bassi, A., & Bottani, C. E. (2002). Phys. Rev. Lett. 89285506

[53] Casari, C. S., Li Bassi, A., Ravagnan, L., Siviero, F., Lenardi, C., Piseri, P., Bongiorno, G., Bottani, C. E., & Milani, P. (2004). *Phys. Rev. B*, 69075422.

[54] Daisuke, Nishide, Hirofumi, Dohi, Tomonari, Wakabayashi, Eiji, Nishibori, Shinobu, Aoyagi, Masashi, Ishida, Satoshi, Kikuchi, Ryo, Kitaura, Toshiki, Sugai, Makoto, Sakata, & Hisanori, Shinohara. (2006). *Chem. Phys. Letter*, 428, 356.

[55] Nishide, D., Dohi, H., Wakabayashi, T., Nishibori, E., Aoyagi, S., Ishida, M., Kikuchi, S., Kitaura, R., Sugai, T., Sakata, M., & Shinohara, H. (2006). Chem. Phys. Lett. , 428, 356.

[56] Malard, L. M., Nishide, D., Dias, L. G., Rodrigo, B. Capaz, Gomes, A. P., Jorio, A., Achete, C. A., Saito, R., Achiba, Y., Shinohara, H., & Pimenta, M. A. (2007). *Phys. Rev. B*, 76, 233412.

[57] Tabata, H, Fujii, M, Hayashi, S, Doi, T, & Wakabayashi, T. (2006). *Carbon*, 44, 3168, WakabayashiTTabataHDoiTNagayamaHOkudaKUmedaRHisakiISonodaMTobeY-MinematsuTHashimotoKHayashiS, 2007, *Chem. Phys. Lett.*, 433, 296.

[58] Pino, T, Ding, H, Güthe, F, & Maier, J. P. (2001). *J. Chem. Phys.*, 114, 2208.

[59] Cumings, A. Zettl, Chem. Phys. Lett. 316 (2000) 211.

[60] N. R. Arutyunyan, E. D. Obraztsova, M. Silly, P. Jaffrennou, B. A. Tretout, A. Loiseau, A. L. Chuvilin, Phys. Stat. Sol. (b) 243(13) (2006) 3316.

[61] S. Y. Bae, H. W. Seo, J. Park, Y. S. Choi, J. C. Park, S. Y. Lee, Chem. Phys. Lett. 374 (2003) 534.

[62] W.-Q. Han, H.-G. Yu, C. Zhi, J. Wang, Z. Liu,T. Sekiguchi, Y. Bando, Nano Lett. 8(2) (2008) 491.

[63] E. A. Obraztsova, D. V. Shtansky, A. N. Sheveyko, M. Yamaguchi, A. M. Kovalskii, D. Golberg, Scripta Mater. 67 (2012) 507.

[64] C. H. Lee, J. Wang, V. K. Kayatsha, J. Y. Huang, Y. K. Yap, Nanotechnology 19 (2008) 455605.

Chapter 3

Characterization of Laser-Induced Defects and Modification in Carbon Nanotubes by Raman Spectroscopy

Masaru Tachibana

Additional information is available at the end of the chapter

1. Introduction

Defects in single-wall carbon nanotubes (SWCNTs) have a great influence on their physical properties. In real SWCNTs, various types of defects such as vacancies, Stone–Wales defects, adatoms, or H–C complex are contained as shown in Figure 1. Such defects can be introduced at the stage of SWCNT growth or later on during device or composite production. They can be also created deliberately by chemical treatment or by irradiation with electron, ion, or laser light. Understanding the properties of such defects in SWCNTs is important for improving SWCNT growth methods, tailoring their physical properties, and controlling the irradiation-induced damages.

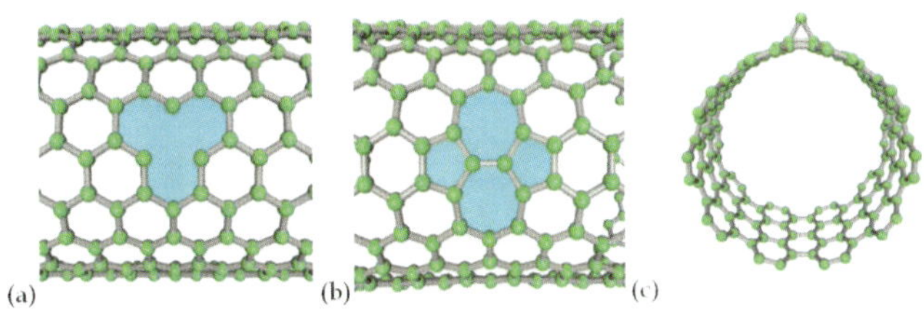

Figure 1. Schematic figures of typical defects such as (a) vacancy, (b) Stone-Wales defect, and (c) adatom in SWCNTs.

The irradiation with electron, ion, or laser light has been widely used for the studies not only on the properties of defects but also on the modification of CNTs. The nice reviews on electron and ion irradiation-induced effects in CNTs have been already published by Krasheninnikov *et al.* [1,2]. However, there are very few reviews on laser-induced defects. The laser irradiation can lead to the heating followed by burning. Such laser irradiation also gives rise to interesting effects such as the production of defects and the modification of CNTs [3-7]. This chapter presents our recent studies [8-11] on laser-induced effects in SWCNTs.

Resonant Raman spectroscopy is one of the most powerful tools for characterizing structural and electronic properties of SWCNTs [12]. In the Raman spectra, defect-induced phonon mode so-called D band is observed at around 1350 cm^{-1}. The intensity of the D band can be enhanced as the number of defects is increased in the SWCNT. Therefore, the D band has been used for the assessment of imperfection of SWCNTs and the understanding of the properties of their defects. Further finding Raman bands associated with defects can lead the Raman spectroscopy to a more effective tool for the characterization of defects.

This chapter presents our recent studies [8-11]on the characterization of laser-induced defects and modification in SWCNTs by Raman spectroscopy. This chapter consists of four parts as mentioned below: (1) Thermal relaxation of laser-induced defects in SWCNTs, (2) Phonon control in metallic SWCNTs by laser–induced defects, (3) Fine structure of D band related to laser-induced defects in SWCNTs, and (4) Formation of *trans*-polyacetylene from SWCNTs by laser irradiation.

2. Raman spectra of SWCNTs

The resonant Raman spectra of SWCNTs include two main features: a radial breathing mode (RBM) observed in the range of 50–350 cm^{-1} and a tangential mode (the so-called G band) observed in the range of 1450–1650 cm^{-1} [12].

The RBM is a signature for the presence of SWCNTs, and is observed as a peak or a multi-peak feature. In the RBM, as suggested by its name, all the C atoms are vibrating in the radial direction with the same phase, as if the tube are breathing. The atomic motion does not break the tube symmetry, that is, the RBM is a totally symmetric (A_1) mode. Since this particular vibrational mode only occurs in SWCNTs, it is used to distinguish SWCNTs from other sp^2 carbons such as graphite and graphene.

A very important characteristic is the RBM frequency (ω_{RBM}) dependence on the tube diameter (d_t):$\omega_{RBM} \propto 1/d_t$ [13]. Most of the RBM experiment results in the literature have been fitted with the relation:

$$\omega_{\mathrm{RBM}} = A / d_{\mathrm{t}} + B \tag{1}$$

with values for the parameters A and B varying widely from paper to paper. Thus, the RBM can give an easy and quick determination of the tube diameter. In addition, it is of important that the RBM peak intensity is a function of excitation energy. The RBM is intense when the incident light or the scattered light is in resonance with the SWCNT optical transition energies.

The G band (where the notation G comes from graphite) is related to the in plane C–C bond stretching mode in graphite and graphene. The G band is the Raman signature for all the sp^2 carbon materials, and is observed as a peak or multi-peak feature. The G band in SWCNTs is a more complex spectral feature. Due to the folding of graphene sheet into the SWCNT and the symmetry breaking effects associated with the nanotube curvature, G band splits into G^+ and G^-, which are related to atomic vibrations preferencially along (LO) and perpendicular (TO) to the tube (folding) axis, respectively, for semiconducting SWCNT. For metallic tubes, electron-phonon coupling softens the LO modes, so that G^+ and G^- are actually with TO and LO modes, respectively.

The G^- peak for metallic tubes is fitted by asymmetric and broad Breit-Wigner-Fano (BWF) line:

$$I(\omega)=I_0\frac{[1+(\omega-\omega_{BWF})/q\Gamma]^2}{1+[(\omega-\omega_{BWF})/\Gamma]^2} \quad (2)$$

where I_0, ω_{BWF}, Γ, and $1/q$ are intensity, renormalized frequency, broadening parameter, and the asymmetric parameter, respectively [14]. Note that ω_{BWF} and 2Γ are called the peak frequency and full width at half maximum (FWHM) of the BWF line, respectively. As clarified from Equation (2), $1/q$ determines the departure of the line shape from a symmetric Lorentzian line and, therefore, $|1/q|$ is a measure of the degree of the BWF coupling. Thus the asymmetric and broad BWF line is commonly used to distinguish metallic SWCNTs from semiconducting SWCNTs as fitted with symmetric Lorentzian lines.

Actually, due to the symmetry breaking effects associated with the nanotube curvature, G band in SWCNT generates up to six Raman-allowed G-band peaks corresponding to two totally symmetric A_1 modes, two E_1 modes and two E_2 symmetry modes. Three of each exhibit LO or TO-like vibration. Due to the delocalization effect and special resonance conditions, the A_1 modes usually dominante the G-band spectra.

In addtion, in the Raman spectra in SWCNTs, defect-induced phonon mode so-called D band is often observed at around 1350 cm^{-1} [12].The D band is a Raman signature of disorder in sp^2 carbons materials. The intensity of the D band can be enhanced as the number of defects is increased in the SWCNT. The D band has been used for the assessment of imperfection of SWCNTs and the understanding of the properties of their defects.

3. Thermal relaxation of laser-induced defects in SWCNTs

3.1. Laser irradiation for SWCNTs synthesized by electric arc-discharge method

SWCNTs synthesized by an electric arc-discharge method were used for laser irradition experiments. As-grown SWCNTs were purified by heating at 350°C for 90 min in air. A suspension of purified SWNTs in ethanol was prepared by ultrasonication. By drop-coating and air-drying the suspension, a SWCNT thin film was formed on a quartz substrate. The SWCNT film was irradiated with a 248 nm (~5.0 eV) pulsed KrF excimer laser in air. The irradiation fluence was approximately 3 J/(cm^2·pulse). The irradiation pulse number was selected to be only one because more than two pulses led to the breakdown of the SWCNTs.

3.2. Change in D band by laser irradiaton and thermal annealing

Figure 2 shows D and G bands in the Raman spectra for non-irradiated SWCNTs, irradiated SWNTs with a 248 nm pulsed excimer laser of 3 J/(cm^2·pulse) in air, and annealed SWCNTs at 673 K in a vacuum of 1 Pa for 240 min after the laser irradiation. The inset shows the close up of the D band. All of the spectra were normalized to the maximum intensity of the G band. Note that the Raman excitation was provided with a 532 nm (~2.33 eV) of a Nd:YVO4 laser where the laser power level in a focal spot of 1 μm in diameter on the sample was kept below 0.1 mW to prevent overheating the sample.

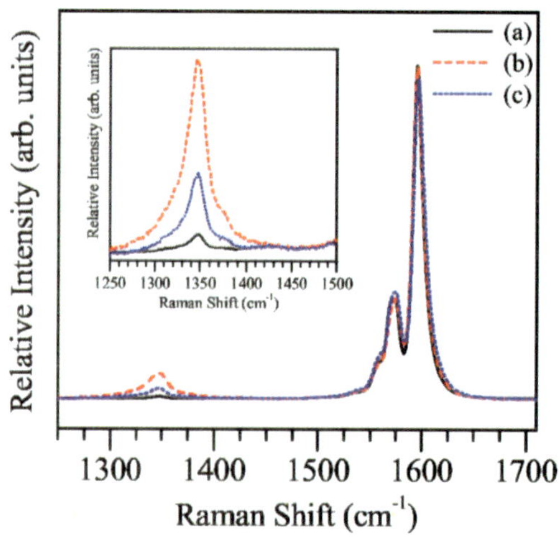

Figure 2. D and G bands in the Raman spectra of (a) pristine non-irradiated SWCNTs, (b) irradiated SWCNTs with a 248 nm pulsed excimer laser, and (c) annealed SWCNTs at 673 K for 240 min after the laser irradiation. The inset shows the close up of the D band. All of the spectra are normalized to the maximum intensity of the G band. [8]

It is found that the D band intensity significantly changes by laser irradiation and thermal annealing, while the spectral feature of the G band remains almost unchanged. For more clarifying the change in D band intensity, the relative intensity of the D band main peak at 1346 cm^{-1} to the G band main peak at 1593 cm^{-1} was defined as I_D/I_G. For non-irradiated SWCNTs as shown in Fig. 2(a), the I_D/I_G is 0.009. This value is very small and exhibits high quality SWCNTs. The I_D/I_G significantly increases due to laser irradiation as seen in Fig. 2(b). The I_D/I_G for the irradiated SWCNTs is 0.08, which is about ten times as much as that for the non-irradiated ones. Furthermore, when the irradiated SWNTs are thermally annealed at 673 K for 240 min, the I_D/I_G decreases and approaches that for non-irradiated ones as shown in Fig. 2(c). On the other hand, all peak frequencies, relative intensities, and FWHMs of RBM, D and G bands except for the intensity of the D band exhibit no significant change due to laser irradiation and thermal annealing. These spectral features mean that defects were successfully produced by the laser irradiation and relaxed by annealing with keeping the tubular structure of SWCNTs.

Let us consider the formation process of the laser-induced defects in SWCNTs. The knock-on energy of carbon atom into the direction perpendicular to the tube surface for an isolated SWCNT with a diameter over 1 nm is estimated to be 15–17 eV [15]. This energy is much higher than the irradiation energy of 248 nm (~5.0 eV) used in this experiment. This means that the formation of the laser-induced defects in SWCNTs would not be due to the physical knock-on phenomena. However, the increase of D band intensity related to the formation of defects clearly occurs for SWCNTs irradiated in air. The degree of the increase of D band intensity is much higher than those for SWCNTs irradiated in vacuum and Ar atmosphere. Therefore, the formation of the laser-induced defects in SWCNTs can be attributed to the chemical reaction with O_2 and H_2O in air by laser heating.

3.3. Analysis of recovery of D band by themal annealing

To examine the thermal relaxation of the laser-induced defects in SWCNTs, the time evolution of the relative intensity of D band in the irradiated samples at various annealing temperatures from 296 to 698 K was measured in the range of annealing times of 0 to 240 min. The typical annealing time evolution of the relative intensity of the D band at 573, 673, and 698 K are shown in Figure 3. Note that the *R* of the ordinate in the figure indicates the I_D/I_G normalized by that at t=0, i.e., before annealing for irradiated SWCNTs.

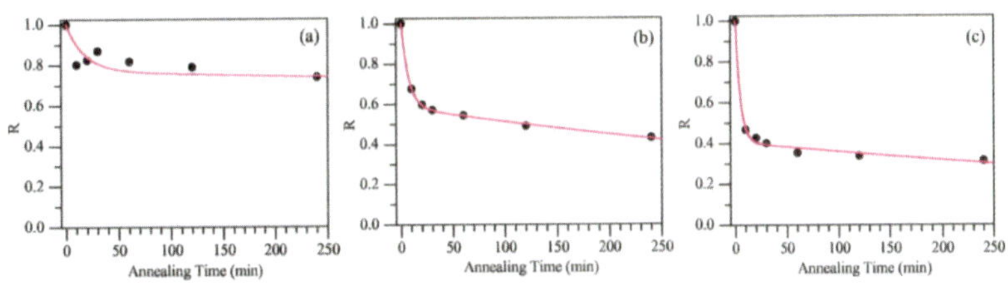

Figure 3. Time evolution of the *R* at annealing temperatures of (a) 573, (b) 673, and (c) 698 K for SWCNTs irradiated with a 248 nm pulsed excimer laser. Note that the *R* indicates the *ID*/ *IG* normalized by that at *t*=0, i.e., before annealing for the irradiated SWCNTs. Solid curves were obtained by the fitting analysis with Eq.(3). [8]

As shown in Fig. 3, the *R* decreases with increasing annealing time. For 573 K, the *R* gradually decreases and reaches 0.8 at 240 min. For 673 K, the *R* quickly decreases and reaches 0.6 at 20 min, and then slowly decreases and reaches 0.45 at 240 min. For 698 K, the *R* more quickly decreases and reaches 0.4 at 20 min, and then slowly decreases and reaches 0.3 at 240 min. On the other hand, no significant change in the *R* is observed for 296 K. These results indicate that the behavior of the R with annealing time strongly depends on annealing temperature. The feature of the dependence of the R on annealing time is clearly observed, especially at annealing temperatures of 673 and 698 K as shown in Figs. 3(b) and 3(c), respectively. From these behaviors, it is predicted that the thermal relaxation of laser-induced defects in SWCNTs includes two processes that have fast and slow rates.

As the previous analysis of the thermal relaxation kinetics of defects for graphite [16], the dependence of the R on annealing time for SWCNTs as shown in Fig. 3 was analyzed on the following three assumptions: (i) the I_D/I_G, or R is proportional to the density of defects in SWCNTs, (ii) some of the defects and others are annihilated by the fast and slow relaxation processes, respectively, and (iii) both processes occur independently. According to these assumptions, the annealing time evolution of the R can be expressed as

$$R(t) = A\exp(-k_I t) + (1 - A)\exp(-k_{II} t) \tag{3}$$

where A is the ratio of the defects relaxed by the fast relaxation process to the total defects in SWCNTs, k_I and k_{II} are the rate constants for the fast and slow relaxation processes, respectively, and t is annealing time. So the first and second terms on the right-hand side of Equation(3) correspond to the fast and slow relaxation processes, respectively. The annealing time evolution of the R as shown in Fig. 3 was fitted with Eq. (3). The fitted curves are indicated as solid lines in Fig. 3. As a result, the rate constants of k_I and k_{II} for the fast and slow relaxation processes, respectively, are determined as shown in Table I. For 296 K, k_I and k_{II} are estimated to be quite small ($<10^{-6}$) since no clear change in R was observed. The value of A runs from 0.23 through 0.59. As presented in Table I, both k_I and k_{II} show the strong temperature dependence.

Temperature (K)	k_I (s^{-1})	k_{II} (s^{-1})
296	$< 10^{-6}$	$< 10^{-6}$
573	9.5×10^{-4}	2.5×10^{-6}
623	1.5×10^{-3}	5.6×10^{-6}
648	2.9×10^{-3}	1.6×10^{-5}
673	2.4×10^{-3}	2.2×10^{-5}
698	3.5×10^{-3}	2.0×10^{-5}

Table 1. Thermal relaxation rate constants of k_I and k_{II} for the fast and slow processes, respectively, at annealing temperatures from 296 to 698 K for SWCNTs irradiated with a 248 nm pulsed excimer laser.[8]

3.4. Activation energies of thermal relaxation of laser-induced defects

Arrhenius plots of thermal relaxation rate constants of k_I t and k_{II} $t$in Table I are shown in Figure 4. From the slopes in Fig. 4, the activation energies for the fast and slow relaxation processes are determined to be 0.4 and 0.7 eV, respectively. According to the simulation on defects in

SWCNTs, the activation energy of the vacancy migration along the tube axis is calculated to be about 1 eV [17]. This value is close to that for the slow relaxation process as determined experimentally above. Such vacancy migration would also lead to the annihilation of the vacancies at the nanotube ends. Therefore, it is concluded that the slow relaxation process observed in this experiment corresponds to the vacancy migration along the tube axis.

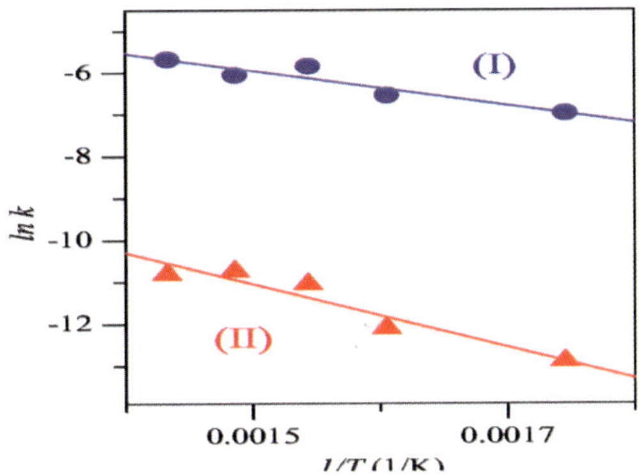

Figure 4. Arrhenius plots of the thermal relaxation rate constants of k_I(circles) and k_{II} (triangles) for the fast and slow processes, respectively, for SWCNTs irradiated with a 248 nm pulsed excimer laser, where T is the annealing temperature. [8]

On the other hand, two such relaxation processes have also been reported for graphite irradiated with He+ ions [16]. According to the report, the fast and slow relaxation processes correspond to the vacancy-interstitial recombination and vacancy migration in the graphene plane, respectively, which have the activation energies of 0.89 eV and 1.8 eV. This suggests that the slow relaxation process corresponds to the vacancy migration in both SWCNT and graphite. The activation energy of 0.7 eV of the vacancy migration for SWCNTs as deterimined experimentally above is much smaller than 1.8 eV for graphite. This smaller value for SWCNTs would be due to the curvature effect of nanotube that breaks the trigonal symmetry of a perfect graphene sheet. It is expected that, due to the curvature effect of nanotube, the activation energy of the vacancy-interstitial recombination for SWCNTs is also smaller than 0.89 eV for graphite. Thus, it is suggested that the fast relaxation process with the activation energy of 0.4 eV as determined in this experiment corresponds to the vacancy-interstitial recombination in SWCNTs.

In summary, laser-induced defects in SWCNTs can be introduced by the irradiation with a 248 nm pulsed excimer laser. The formation of defects might be related to thermal oxidation and burning by laser heating. Such laser-induced defects are thermally relaxed with two processes with fast and slow rates. The two relaxation processes show the strong temperature dependence. The activation energies of the fast and slow relaxation processes are determined to be 0.4 and 0.7 eV, respectively. These processes can correspond to vacancy-interstitial recombination and vacancy migration along the tube axis. Such relaxation processes with fast and slow rates for SWCNTs are similar to those for graphite irradiated with He+ ions. However, their activation energies for SWCNTs are smaller than those for graphite. The smaller activation energies for SWCNTs would be due to the effect of curvature of nanotube.

4. Phonon control in metallic SWCNTs by laser–induced defects

4.1. Change in G band for metallic nanotubes by laser irradiation and thermal annealing

SWCNT samples after laser irradiation and themal annealing were prepared by similar procedure as described in 3.1. Figure 5 show D and G bands in Raman spectra for (a) pristine SWCNTs, (b) laser-irradiated SWCNT swith a 248 nm (~5.0 eV) pulsed KrF excimer laser of approximately 5 J/(cm^2 ·pulse), and (c) annealed SWNTs at 400 °C in a vacuum of ~9 Pa for 60 min after the irradiation. Note that the spectra excitation was provided with a 632.8 nm (~1.96 eV) of a He–Ne laser where the laser power level in a focal spot of 1 μm in diameter on the sample was kept below 0.1 mW to prevent overheating the samples.

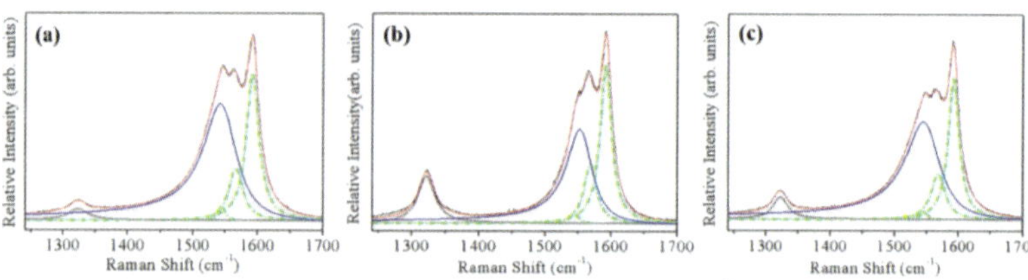

Figure 5. D and G bands in Raman spectra for (a) pristine SWNTs, (b) laser-irradiated SWNTs, and (c) annealed SWNTs after the irradiation, taken at E_{exc}=1.96 eV (λ_{exc}=632.8 nm). The *D* band was fittedwith one Lorentzian line. The *G* band was fitted with two Lorentzian (dotted) lines and one BWF (solid) line. [9]

D band was fitted with one Lorentzian line. On the other hand, G band was fitted with two Lorentzian lines and one asymmetric line that is Breit–Wigner–Fano (BWF) line [13] as given by Eq. (2). The values of the fitting parameters are listed in Table 2. In the G band for pristine SWCNTs in Fig. 5(a), one Lorentzian line at 1593 cm^{-1} corresponds to G$^+$ peaks for semiconducting SWCNTs and metallic SWCNTs. Note that G$^+$ peak associated with semiconducting SWCNTs is assumed to overlap with that with metallic ones. The other Lorentzian line at 1567 cm^{-1} corresponds to G$^-$ peak associated with semiconducting SWCNTs. The asymmetric and broad line at 1546 cm^{-1} corresponds to G$^-$ peak associated with metallic SWCNTs, which is largely downshifted, relative to G$^+$ one. These spectral components of the G band suggest that both semiconducting and metallic SWCNTs are resonant in the Raman spectrum of pristine SWCNTs taken with 1.96 eV in Fig. 5(a).

The D band for pristine SWCNTs was fitted with one Lorentzian curve at 1323 cm^{-1}. It is known that the intensity of the D band increases as the number of defects in SWCNTs is increased. The intensity ratio of D/G$^+$ peaks is often used as a measure of the defect density in

SWNTs. The relative intensity of the D band for pristine SWCNTs is 0.12. This value is quite small, comparable to those of high quality SWCNTs as reported so far. Thus, pristine SWCNTs used in this experiment have high quality or quite low defect density.

	pristine SWCNTs			laser-irradiated SWCNTs			annealed SWCNTs		
	ω (cm^{-1})	Γ (cm^{-1})	$I/I_{G^+(S+M)}$	ω (cm^{-1})	Γ (cm^{-1})	$I/I_{G^+(S+M)}$	ω (cm^{-1})	Γ (cm^{-1})	$I/I_{G^+(S+M)}$
G^+(S+M)	1593	24	1.0	1592	21	1.0	1593	21	1.0
G^-(S)	1567	26	0.36	1568	23	0.38	1568	27	0.33
G^-(M)	1546	63	0.78	1554	47	0.59	1548	65	0.69
D	1323	38	0.12	1322	31	0.42	1322	28	0.20

Table 2. Peak frequencies (ω), full widths at half maximum (Γ), and relative intensities ($I/I_{G^+(S+M)}$) of Lorentzian lines and Breit-Wigner-Fano lines used to fit *D* and *G* bands for pristine SWCNTs, laser-irradiated SWNTs, and annealed SWCNTs after the irradiation in Fig. 5. The relative intensities of the peaks are normalized by G^+ intense peak located at highest frequency. S and M indicate *G* peaks associated with semiconducting and metallic SWCNTs, respectively.

A significant change in the Raman spectrum was observed for laser-irradiated SWCNTs, as shown in Fig. 5(b). The intensity of the D band increased with the laser irradiation. Note that the frequency and linewidth remained almost unchanged even after the irradiation. These results mean that some specific defects were introduced in SWCNTs by the laser irradiation. The formation of defects might be related to thermal oxidation and burning by laser heating as discussed in 3.2. Moreover, it should be noticed that not only D band but also G band were affected by the laser irradiation. Especially, a significant change was observed for G^- peak associated with metallic tubes. The frequency of the G^- peak was upshifted by 8 cm^{-1}. Correspondingly, the linewidth of the G^- peak was reduced by 25%. In addition, the intensity ratio of the G^-/G^+ peak was reduced by 25%. On the other hand, the frequency, linewidth, relative intensity of G– peak associated with semiconducting SWCNTs remained almost unchanged even after the laser irradiation. Such behavior in the G^- peak associated with semiconducting SWCNTs is consistent with that in the G^- peak for the same SWCNTs taken with 2.33 eV in which only semiconducting SWNTs are resonant. Thus, the laser-induced defects significantly affect G^- peak associated not with semiconducting SWCNTs but metallic ones.

The G^- peak associated with metallic SWCNTs is due to the electron-phonon coupling as described in 2 [18,19]. The upshift of the frequency, the narrowing of the linewidth, and the reduction in the relative intensity for the G^- peak associated with metallic SWCNTs as seen in Fig. 5(b) imply the breaking of the electron-phonon coupling. Moreover, it should be noticed that D and G bands for the irradiated SWNTs recover the original ones after annealing in a vacuum of ~9 Pa at a sample temperature of 400 °C for 60 min, as shown in Fig. 5(c). As described in 3.4, the laser-induced defects such as vacancies can be annihilated by vacancy-interstitial recombination and vacancy migration to the nanotube end due to the thermal annealing [8]. Such annihilation of vacancies can be responsible for the recovery of the Raman spectral profile. Thus, the electron-phonon coupling can be reversibly controlled by the generation and annihilation of specific defects due to laser irradiation and thermal annealing.

4.2. Change in RBM for metallic nanotubes by laser irradiation and themal annealing

The change corresponding to that in D and G bands in Fig. 5 was also observed for radial breathing modes (RBMs) at the range of 150–200 cm^{-1} in Raman spectra for pristine SWNTs, laser-irradiated SWCNTs, and annealed SWCNTs after the irradiation, taken with E_{exc} =1.96 eV(λ_{exc}=632.8 nm), as shown in Figure 6. The RBMs are actually composed of a lot of peaks corresponding to various kinds of chiralities or diameters of SWNTs. For simplifying, the RBMs were fitted with five Lorentzian lines. The values of the fitting parameters are listed in Table 3.

The diameters d of SWCNTs resonantly contributing to the Raman spectrum for pristine SWCNTs in Fig. 6(a) are estimated to be d=1.44±0.2 nm from the corresponding RBM frequencies ω_{RBM} using the relation $\omega_{RBM}(cm^{-1}) = 234/d(nm) + 10$ [20], which has been found for typical SWCNT bundles. Moreover, according to the (revised) Kataura plot [21,22] with the excitation energy and SWCNT diameters estimated above, it is suggested that both semiconducting and metallic SWCNTs are resonant in the Raman spectrum in Fig. 6(a). Note that one RBM peak at the lowest frequency of 157 cm^{-1} is mainly associated with semiconducting ones whereas other four ones at higher frequencies of 168, 173, 186, and 196 cm^{-1} are associated with metallic ones. This result is consistent with the result of spectral components in G band as discussed in 4.1.

The RBM peaks associated with metallic SWCNTs changed after the irradiation, as shown in Fig. 6(b). Especially, the most intense RBM peak at 173 cm^{-1} drastically decreased with the irradiation. On the other hand, no significant change was observed for only RBM peak at 157 cm^{-1} associated with semiconducting SWCNTs. These results mean that the resonant off for the Raman excitation of 1.96 eV occurs for metallic SWCNTs. This also suggests that the change in the electronic structure for metallic SWCNTs occurs due to the laser-induced defects.

Vacancy defects can cause a bandgap opening in metallic SWCNTs due to the breaking of the symmetry [23,24]. Such metal-semiconductor transition has been also experimentally demonstrated by the measurements of electrical properties for metallic SWCNTs with the introduction of defects [25,26]. Therefore, the change in the electronic structure with the bandgap opening can be responsible for the resonant off for metallic SWCNTs. Such change in the electronic structure for metallic SWCNTs due to the laser-induced defects is also consistent with the change in the corresponding G^- band, i.e., the breaking of the electron-phonon coupling, as discussed in 4.1.

Moreover, as seen in Fig. 6(c), the thermal annealing also leads to the recovery of the RBMs to original ones, as D and G band in Fig. 5. This recovery can be also explained by that in the electronic structure due to the thermal annihilation of laser-induced defects such as vacancies as discussed in 3.4.

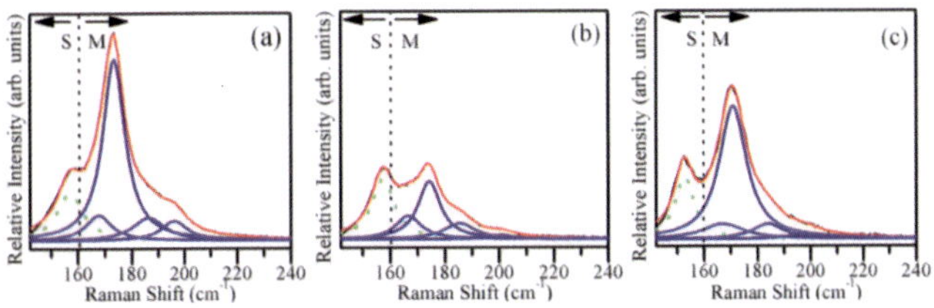

Figure 6. RBMs in Raman spectra for (a) pristine SWCNTs, (b) laser-irradiated SWCNTs, and (c) annealed SWCNTs after the irradiation, taken with E_{exc}=1.96 eV(λ_{exc}=632.8 nm). The RBMs were fitted with five Lorentzian lines. One (dotted) line at the lowest frequency is associated with semiconducting SWCNTs. Other (solid) lines at higher frequencies are associated with metallic SWCNTs. [9]

In summary, laser-induced defects influence not only D band but also G^- peak associated with metallic SWCNTs, which is attributed to the electron-phonon coupling with Kohn anomaly. The upshift and narrowing of the G^- peak occur due to the laser irradiation. The G^- peak can recover to the original one due to the thermal annealing. The electron-phonon coupling for metallic SWCNTs can be reversibly controlled by the generation and annihilation of specific defects due to the laser irradiation and thermal annealing.

	pristine SWNTs			laser-irradiated SWNTs			annealed SWNTs		
	ω (cm⁻¹)	*Γ (cm⁻¹)*	*I/I_{S1}*	*ω (cm⁻¹)*	*Γ (cm⁻¹)*	*I/I_{S1}*	*ω (cm⁻¹)*	*Γ (cm⁻¹)*	*I/I_{S1}*
S1	157	12	1.0	157	10	1.0	152	9	1.0
M1	168	12	0.52	167	11	0.43	167	19	0.53
M2	173	10	3.2	174	10	0.95	171	12	3.0
M3	186	15	0.59	185	14	0.36	184	14	0.40
M4	196	12	0.42	201	10	0.07	198	14	0.13

Table 3. Peak frequencies (ω), full widths at half maximum (Γ), and relative intensities (I/I_{S1}) of Lorentzian lines used to fit radial breathing modes (RBMs) for pristine SWCNTs, laser-irradiated SWCNTs, and annealed SWCNTs after the irradiation in Fig. 6. The relative intensities of the peaks are normalized by one located at lowest frequency associated with semiconducting SWCNTs. S and M indicate RBM peaks associated with semiconducting and metallic SWCNTs, respectively.

5. Fine structure of D band related to laser-induced defects in CoMoCAT SWCNTs

5.1. Heating and laser irradiation for CoMoCAT SWCNTs

As-received CoMoCAT SWCNTs (SWeNT® CG 100, SouthWest NanoTechnologies, Inc.) were used for heating and laser irradiation experiments. A suspension of SWCNTs in ethanol was prepared by ultrasonication. By drop-coating and air-drying the suspension, a SWCNT thin film was formed on a quartz substrate. For heating experiments, the film samples were annealed at 350 °C for 90 min in air. For laser irradiations, the samples were irradiated with a 532 nm (~2.33 eV) from a Nd:YVO_4 laser for 180 min. The irradiation power level in a focal spot of 1 μm in diameter on the sample was kept at ~20 mW. The heating and laser irradiation experiments were also carried out in a vacuum of~4.5 Pa and a dynamic vacuum of ~3.5×10^{-4} Pa, respectively.

5.2. Change in D band by heating

Figure 7 shows D and G bands in the Raman spectrum for a pristine CoMoCAT SWCNT sample. The corresponding radial breathing modes (RBMs) are also shown in the inset in the figure. Note that the spectrawere taken with E_{exc}=2.33 eV(λ_{exc}=532 nm) where the laser power level in a focal spot of 1 μm in diameter on the sample was kept below 0.1 mW to prevent overheating the sample.The spectral peaks are fitted with Lorentzian lines. From the RBM frequencies, the mean diameter of pristine SWCNTs is estimated to be ~0.8 nm, which corresponds to typical mean diameter of CoMoCAT ones [27]. Note that the diameter is estimated using ω [cm^{-1}]=234/d [nm]+10 where d is the SWCNT diameter and ω is the RBM frequency [12]. As shown in Fig. 7, the D band in pristine SWCNTs is fitted with two Lorentzian lines at 1313 and 1355 cm^{-1} which are denoted by D_1 and D_3, respectively. The relative intensities of the D components to the most intense G peak at 1594 cm^{-1} in each spectrum are compared in order to clarify the change in the D band.

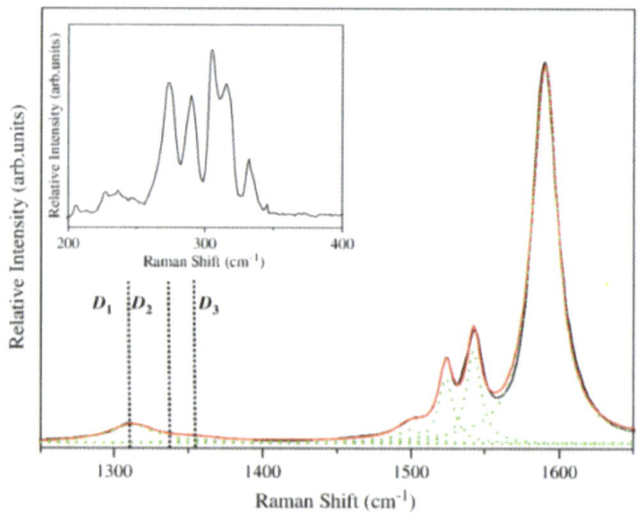

Figure 7. D and G bands in the Raman spectrum for a pristine CoMoCAT SWCNT sample. The corresponding radial breathing modes (RBMs) are also shown in the inset. [10].

Figure 8 shows *D* and *G* band in Raman spectra for CoMoCAT SWCNT samples heat-treated at 350 °C for 90 min in (a) air and (b) a vacuum of 4.5 Pa. The corresponding radial breathing modes (RBMs) are also shown in the inset. Note that these spectra were taken with E_{exc}=2.33 eV where the laser power level in a focal spot of 1 μm in diameter on the sample was kept below 0.1 mW to prevent overheating the sample. As shown in Fig. 8(a), the heat-treatment in air leads to the significant increase of the relative intensity of the D_3 component. The corresponding change is also seen in *G* band and RBMs. The broadening of *G* band occurs. This means that the imperfection of SWCNTs increases. Namely, a lot of defects are introduced into SWCNTs. In addition, the higher frequency peaks in the RBMs are disappeared. This means that the degradation of SWCNTs with smaller diameters occurs. On the other hand, no significant change in these Raman bands is observed in SWCNTs heat-treated in a vacuum of ~4.5 Pa as shown in Fig. 8(b). Therefore, it is suggested that the change in the Raman bands by the heat-treatment in air is due to the thermal oxidation.

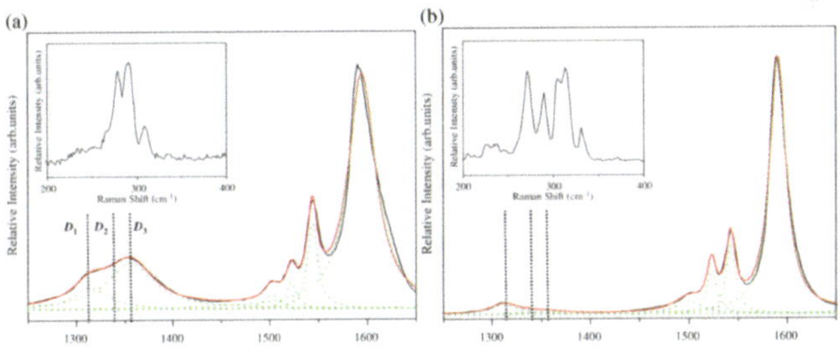

Figure 8. D and G bands in Raman spectra for CoMoCAT SWCNT samples heat-treated at 350 °C for 90 min in (a) air and (b) a vacuum of ~4.5 Pa. The corresponding radial breathing modes (RBMs) are also shown in the inset. [10]

Actually, pristine SWCNT sample contains impurities such as amorphous carbon, water, and C–H complex. The thermal oxidation gives rise to open-end structures of SWNTs or holes in the walls [28]. The extension of the oxidation process can generate C=O, C–O–C, and C–OH [29]. In addition, the corresponding *D* band frequency is theoretically predicted to be more than 1353 cm^{-1} [29]. This value is in good agreement with 1355 cm^{-1} of D_3. Thus, the D_3 component can be related to defects such as amorphous carbon and oxides as discussed above.

5.3. Change in D band by laser irradiation

Figure 9 shows *D* and *G* bands in Raman spectra for CoMoCAT SWCNT samples irradiated with E_{laser}=2.33 eV from a Nd:YVO_4 laser with 183 kW/cm^2-for 180 min in (a) air and (b) a dynamic vacuum of~3.5×10^{-4} Pa. The corresponding radial breathing modes (RBMs) are also shown in the inset. Note that these spectra were also taken with E_{laser}=2.33 eV where the laser power level in a focal spot of 1 μm in diameter on the sample was kept below 0.1 mW to prevent overheating the sample. As shown in Fig. 9(a), the laser irradiation in air leads to the increases of the relative intensities of not only D_3 but also D_1. The increase of the D_3 is due to the thermal oxidation, similar to that by the heat-treatment in air as discussed in 5.2, since the laser irradiation occurs the local heating. It should be noted that the D_1 intensity increases by 78%, relative to the pristine one. However, no significant change is seen in *G* and RBMs. This means that no significant degradation of SWCNTs occurs. These behaviors are quite different from those by the heat-treatment in air as discussed in 5.2. In the laser irradiation in air, H_2 and O_2 radical species can be produced [29,30]. These with water can give rise to C–H complex on the side walls, at defect sites, or ends of SWCNTs [30,31]. Therefore, the D_1 can be related to C–H complex produced by the laser irradiation in air.

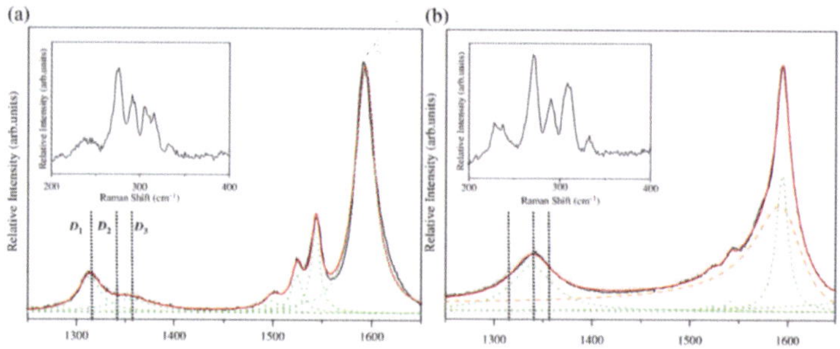

Figure 9. D and G bands in Raman spectra for CoMoCAT SWCNT samples irradiated with E_{laser}=2.33 eV from a Nd:YVO laser for 180 min in (a) air and (b) a dynamic vacuum of~3.5×10^{-4} Pa. The corresponding radial breathing modes (RBMs) are also shown in the inset.[10]

On the other hand, the laser irradiation in a dynamic vacuum of ~3.5×10^{-4} Pa leads to the appearance of a new *D* component at ~1340 cm^{-1}, denoted by D_2, as shown in Fig. 9(b). The peak profile is quite similar to that of step edges in graphite [32]. Therefore, the D_2would be related to the edges of SWCNTs cut by the laser irradiation in the dynamic vacuum. Such cutting might be due to absorbates-assisted burning by laser heating. In addition, it is of interest that the decrease of the D_1 intensity, relative to the pristine one, is observed. The C–H complex formed on SWCNTs is released by thermal annealing at more than 200 °C [30]. The laser irradiation in this experiment gives rise to the local heating. This will lead to the dehydration of SWCNTs so that the D_1 intensity can be decreased. The dehydration is also accompanied by the increases of the intensities of thelower frequency peak at 230 cm^{-1} in the RBMs and of the broad Breigt–Wigner–Fano line in the *G* band, associated with metallic SWCNTs, as seen in Fig. 9(b).

In summary, *D* band in CoMoCAT SWCNTs is composed of three components D_1, D_2, and D_3 at ~1313, 1340, and 1355 cm^{-1}, respectively. These components are attributed to different kinds of defects introduced by heating and laser irradiation. Such insight on the fine structure of the *D* band will play a role for more detailed understanding of *D* band and the identification of defects in SWCNTs.

6. Formation of *trans*-polyacetylene from CoMoCAT SWCNTs by laser irradiation

6.1. Laser irradiation for CoMoCAT SWCNTs

As-received CoMoCAT SWCNTs (SWeNT@CG 100, SouthWest NanoTechnologies, Inc.) were used in this experiment. A suspension of SWCNTs in ethanol was prepared by ultrasonication. The suspension was dropped on a clean quartz substrate and allowed to be air-dried at room temperature. The SWCNTs samples prepared in above procedure were used for laser irradiation experiments. The samples, which were exposed to air for less than 1 h before laser irradiation, are called "short air-exposure" ones. Some of samples were kept in air at room temperature for more than six months before laser irradiation. They are called ''long air-exposure'' ones.

Laser irradiation experiments were carried out using a micro-Raman systemequipped with mirrors, attenuators, a 100× microscope objective, a holographic notch filter, a single grating spectrometer (1800 1/mm grating), and a charge coupled device detector. In the laser irradiation experiments, all attenuators were removed. A 532 nm (~2.33 eV) from a cw $Nd:YVO_4$ laser was used to irradiate the samples. The laser beam was focused on the sample through the 100× microscope objective, with spot size of 1 μm. The laser power level on the sample was kept at 17.8 mW/μm^2. The irradiation time was 1 h.

6.2. Irradition effect for "short air-exposure" CoMoCAT SWCNTs

Figure 10 shows D and G bands in Raman spectra for a ''short air-exposure''CoMoCAT SWCNT sample (a) before and (b) after laser irradiation of a 532 nm with 17.8 mW/μm^2 for 1 h in air. The corresponding RBMs are also shown in the inset in the figure. Note that the spectra excitation was also provided with λ_{exc}=532 nm where the laser power level in a focal spot of 1 μm in diameter on the sample was kept below 0.2 mW/μm^2 to prevent overheating the sample.

The spectral peaks are fitted with Lorentzian lines. From the RBM frequencies in Fig. 10(a), the mean diameter of SWCNTs is estimated to be 0.8 nm, which corresponds to typical mean diameter of CoMoCAT ones [27]. Note that the diameter is estimated using ω [cm^{-1}] = 234/d [nm] + 10 where d is the SWCNT diameterand ω is the RBM frequency [20]. The D band at 1300–1400 cm^{-1} and G band at 1500–1700 cm^{-1} are associated with defects and tangential modes of SWCNTs, respectively. The D band can be fitted with two Lorentzian lines at 1313 and 1346 cm^{-1}. The G band can be fitted with three Lorentzian lines at 1523, 1543, and 1591 cm^{-1}.

After laser irradiation, the increase of relative intensity ofD band is observed as seen in Fig. 10(b). Especially, the lower-frequency component increases. The increase of D band intensity can be attributed to the oxidation and/or hydrogenationof SWCNTs [10,29,30] as discussed in 5. In addition, the RBMs exhibit the decrease of higher-frequency components by laser irradiation. This means that smaller diameter SWCNTs are degraded by thermal oxidation due to laser heating. On the other hand, no significant change in G band is observed as seen in Fig. 10(b).

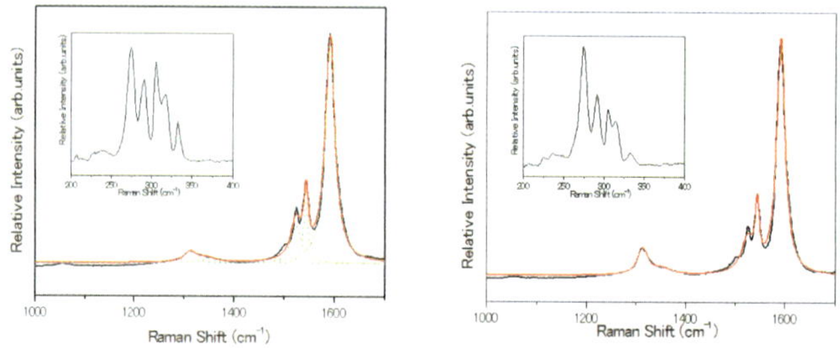

Figure 10. D and G bands in Raman spectra for a "short air-exposure"CoMoCAT SWCNT sample (a) before and (b) after laser irradiation of a 532 nm with 17.8 mW/μm^2 for 1 h in air. The corresponding RBMs are also shown in the inset in the figure. [11]

6.3. Irradiation effect for "long air-exposure" CoMoCAT SWCNTs

Figure 11 shows D and G bands in Raman spectra for a "long air-exposure"CoMoCAT SWCNT sample (a) before and (b) after laser irradiation of a 532 nm with 17.8 mW/μm^2 for 1 h in air. The corresponding RBMs are also shown in the inset in the figure.Note that the spectra excitation was also provided with λ_{exc}=532 nm where the laser power level in a focal spot of 1 μm in diameter on the sample was kept below 0.2 mW/μm^2 to prevent overheating the sample. As seenin Fig. 11(a), the "long air-exposure" sample also exhibits clearD and G bands, and RBMs, fitted by Lorentzian lines, as pristine ones. The peak profiles of G band and RBMs are similar to those of pristine one. On the other hand, the relative intensity of D band increases compared with that of pristine one in Fig. 10(a). The D band intensity is similar to that of the irradiated "short air-exposure" one in Fig. 10(b). This means that the oxidation and/or hydrogenation occur in "long air-exposure" ones even before laser irradiation. This is due to the exposure of air with water vapor for the long time more than six months.

It should be noted that a significant change in Raman spectra is observed for the irradiated "long air-exposure"sample as seen in Fig. 11(b). Namely, new intense peaks appearat 1138 cm^{-1} and 1514 cm^{-1} for the irradiated "long air-exposure" sample. These spectral features are quite similar to those of *trans*-polyacetylene, as have been reported so far [33–36]. According to the model calculation [37], the lower-frequency peak of 1138 cm^{-1} is assigned to a coupled C–C stretching and C–H bending vibration. The higher-frequency peak of 1514 cm^{-1} is assigned to a C–C stretching vibration. In addition, for short polyacetylene chains, the peak position around 1150 cm^{-1} strongly depends on the chain length, and shifts to lower frequency with increasing chainlength [37]. Thus, the appearance of intense peaks of 1138 and 1514 cm^{-1} means that *trans*-polyacetylene-like structural units are formed in irradiated "long air-exposure"samples.

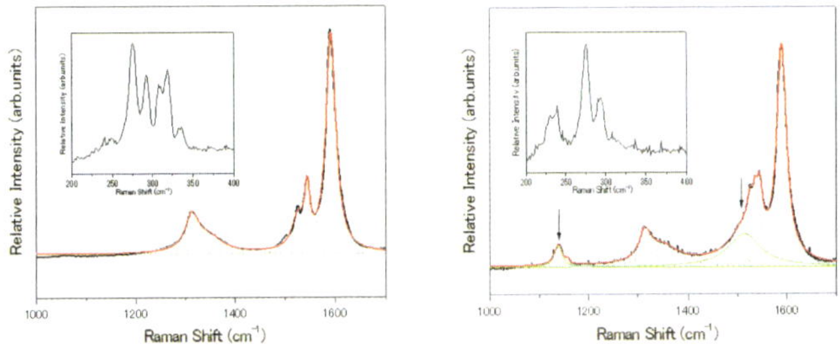

Figure 11. D and G bands in Raman spectra for a "long air-exposure"CoMoCAT SWCNTsample (a) before and (b) after laser irradiation of a 532 nm with 17.8 mW/μm^2 for 1 h in air. The corresponding RBMs are also shown in the inset in the figure. The arrows indicate peaks at 1138 and1514 cm^{-1}associated with *trans*-polyacetylene. [11]

6.4. Formation of polyacetylene from SWCNTs

In "long air-exposure" samples, a lot of water vapor can be absorbed. To remove the absorbed water, the "long air-exposure"samples were annealed at 150 °C for 1 h under a dynamic vacuum of 8.7×10^{-7} Pa. In order to examine the effect of absorbed water on the formation of *trans*-polyacetylene, laser irradiation experiments were also carried out for the annealed "long air-exposure" samples. As a result, for the annealed samples, no clear peak at around 1140 and 1540 cm^{-1} corresponding to *trans*-polyacetylene were appeared even after the laser irradiation. This means that the absorbed water plays a role for the formation of *trans*-polyacetylene.

Let us consider the formation process of *trans*-polyacetylene from "long air-exposure" samples by laser irradiation. It is well-known that CoMoCAT SWCNTs(SWeNT CG 100) include metal catalysts [27]. The catalysts are coated by a carbon layer, and deactivated. The water-can remove the carbon layer at high temperature, and revive the catalytic activity [38]. In "long air-exposure" samples used in this experiment, the water vapor can be sufficiently absorbed on catalysts. The water can remove the carbon layer under laser heating and revive catalytic activity. Consequently, as observed in graphite, graphene and CNT[39-43], the cutting of nanotubes can be realized by catalytic hydrogenation of carbon atoms, in which metal particle dissociate carbon atoms in CNTs, and then the dissociated carbon atoms react with H_2 to create hydrocarbon species such as CH_4. The cutting can be also accompanied with the formation of C–H species on the nanotubes as shown in Figure 12. Such cutting process would lead to the formation of polyacetylene-like structural units.

In summary, *trans*-polyacetylene is formed from CoMoCATSWCNT samples including metal catalysts and absorbed water by laser heating in air. The formation process might be related to the cutting of SWCNTs due to the catalytic hydrogenation of carbon atoms with laser heating, although the detailed mechanism is not yet understood. This shows a new use of the laser irradiation for the formation of functional materials from SWCNTs.

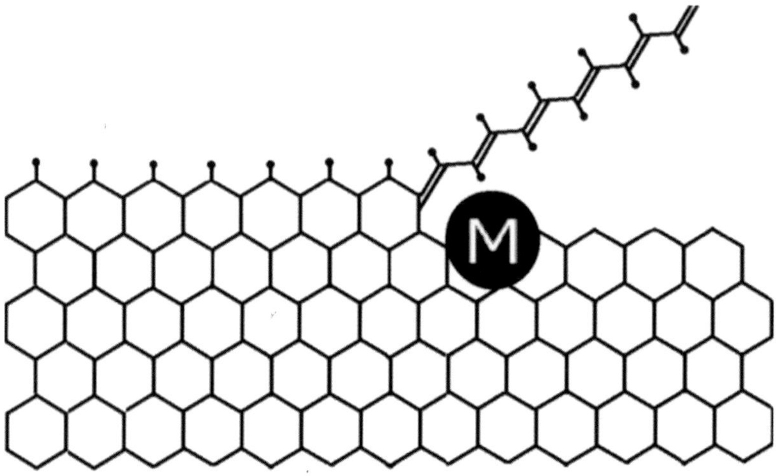

Figure 12. Schematic drawing of the formation of *trans*-polyacetylene by the cutting of SWCNT edge due to the catalytic hydrogenation with laser irradiation.

7. Summary

This chapter presented the characterization of laser-induced defects in SWCNTs by Raman spectroscopy. The laser irradiation with heating followed by burning can produce defects such as vacancies and interstitials in SWCNTs. These defects greatly influence electronic structures and phonon properties especially in metallic nanotubes. They can also be thermally relaxed by vacancy-interstitial recombination and vacancy migration along the tube axis with activation energy of 0.4 eV and 0.7 eV, respectively. This means that the electronic structures and phonon properties in metallic nanotubes can be reversibly controlled by the generation and annihilation of specific defects due to laser irradiation and thermal annealing. In addition, it was presented that the fine structure of Raman D band can be related to specific defects such as C-H complex on nanotubes and nanotube edges produced by laser irradiation. This can lead the Raman spectroscopy to a more effective tool for the characterization of defects in SWCNTs. Finally, it was presented that the laser irradiation can give rise to the formation of trans-polyacetylene from SWCNTs. The formation process might be related to the cutting of SWCNTs due to the catalytic hydrogenation of carbon atoms with laser heating, although the detailed mechanism is not yet understood. This shows a new use of the laser irradiation for the formation of functional materials from SWCNTs. Thus, the laser irradiation is useful for not only the understanding of the properties of defects in SWCNTs but also the modificaton of SWCNTs, as electron and ion irradiations.

Acknowledgements

The author acknowledges the contributions to the works presented here by all past and present collaborators, especially Dr. Takashi Uchida, Dr. Hironori Kawamoto, Mr Ken-ichi Kato, Mr. Dongchul Kang, Ms. Mari Hakamatsuka, Ms. Nagisa Hosoya, Mr. Noriaki Nemoto, and emeritus Prof. Kenichi Kojima. The works were supported in part by Strategic Research Projects in Yokohama City University and MEXT/JSPS KAKENHI.

Author details

Masaru Tachibana*

Address all correspondence to: tachiban@yokohama-cu.ac.jp

Department of Nanosystem Science, Yokohama City University, Japan

References

[1] Krasheninnikov, A. V., & Banhart, F. (2007). Engineering of nanostructured carbon materials with electron or ion beams. *Nat. Mater*, 6, 723-733.

[2] Krasheninnikov, A. V., & Nordlund, K. (2010). Ion and electron irradiation-induced effects in nanostructured materials. J. Appl. Phys , 107, 071301-1-70.

[3] Suzuki, S., & Kobayashi, Y. (2007). Healing of Low-Energy Irradiation-Induced Defects in Single-Walled Carbon Nanotubes at Room Temperature. *J. Phys. Chem. C*, 111(12), 4524-4528.

[4] Zandian, B., Kumar, R., Theiss, J., Bushmaker, A., & Cronin, S. B. (2009). Selective destruction of individual single walled carbon nanotubes by laser irradiation. *Carbon*, 47(5), 1292-1296.

[5] Jeschke, H. O., Diakhate, M. S., & Garcia, M. E. (2009). Molecular dynamics simulations of laser-induced damage of nanostructures and solids. *Applied Physics A: Materials Science & Processing*, 96(1), 33-42.

[6] Kumar, P., Panchakarla, L.S., & Rao, C.N.R. (2011). Laser-induced unzipping of carbon nanotubes to yield graphenenanoribbons. *Nanoscale*, 3, 2127-2129.

[7] Mases, M., Noël, M., Dossot, M., Mc Rae, E., & Alexander, V. (2011). Laser-induced damage and destruction of HiPCO nanotubes in different gas environments. *physica status solidi*, 248(11), 2540-2543.

[8] Uchida, T., Tachibana, M., & Kojima, K. (2007). Thermal relaxation kinetics of defects in single-wall carbon nanotubes. *J. Appl. Phys*, 101, 084313-1-4.

[9] Kang, D., Kato, K., Kojima, K., & Uchida, T. (2008). Tachibana M. Phonon control in metallic carbon nanotubes due to laser-induced defects. *Appl. Phys. Lett*, 93, 133102-1-3.

[10] Kang, D., Hakamatsuka, M., Kojima, K., & Tachibana, M. (2010). Influence of heating and laser irradiation on the Raman D band in single-wall carbon nanotubes. *Diamond & Related Materials*, 19, 578-580.

[11] Hakamatsuka, M., Yoshimura, H., & Tachibana, M. (2011). Formation of trans-polyacetylene from single-wall carbon nanotubes. *Carbon*, 46, 1869-1872.

[12] Jorio, A., Dresselhaus, M., Saito, R., & Dresselhaous, G. F. (2011). Raman Spectroscopy in Graphene Related Systems. *Weinheim: Wiley-VCH.*

[13] Bandow, S., Asaka, S., Saito, Y., Rao, A. M., Grigorian, L., Richte, E., & Eklund, P. C. (1998). Effect of the Growth Temperature on the Diameter Distribution and Chirality of Single-Wall Carbon Nanotubes. *Phys. Rev. Lett*, 80, 3779-3782.

[14] Brown, D. M., Jorio, A., Corio, P., Dresselhaus, M. S., Dresselhaus, G., Saito, R., & Kneipp, K. (2001). Origin of the Breit-Wigner-Fanolineshape of the tangential G-band feature of metallic carbon nanotubes. *Phys. Rev. B*, 63, 155414-155421.

[15] Crespi, V. H., Chopra, N. G., Cohen, M. L., Zettl, A., & Louie, S. G. (1996). Anisotropic electron-beam damage and the collapse of carbon nanotubes. *Phys.Rev. B*, 54, 5927-5931.

[16] Asari, E., Kitajima, M., Nakamura, K. G., & Kawabe, T. (1993). Thermal relaxation of ion-irradiation damage in graphite. *Phys. Rev. B*, 47, 11143-11148.

[17] Krasheninnikov, A. V., Lehtinen, P. O., Foster, A. S., & Nieminen, R. M. (2006). Bending the rules: Contrasting vacancy energetics and migration in graphite and carbon nanotubes. *Chem. Phys. Lett*, 418, 132-136.

[18] Lazzeri, M., Piscanec, S., Mauri, F., Ferrari, A. C., & Robertson, J. (2006). Phonon linewidths and electron-phonon coupling in graphite and nanotubes. *J. Phys. Rev. B*, 73, 155426-155431.

[19] Piscanec, S., Lazzeri, M., Robertson, J., Ferrari, A. C., & Mauri, F. (2007). Optical phonons in carbon nanotubes: Kohn anomalies, Peierls distortions, and dynamic effects. *Phys.Rev. B*, 75, 035427-035448.

[20] Milnera, M., Kurti, J., Hulman, M., & Kuzmany, H. (2000). Periodic Resonance Excitation and Intertube Interaction from Quasicontinuous Distributed Helicities in Single-Wall Carbon Nanotubes. *Phys. Rev. Lett*, 84, 1324-1327.

[21] Kataura, H., Kumazawa, Y., Maniwa, Y., Umezu, I., & Suzuki, S. (1999). Ohtsuka Y.,Achiba Y. Optical properties of single-wall carbon nanotubes. *Synth Met*, 103, 2555-2558.

[22] Jorio, A., Fantini, C., Pimenta, M. A., Capaz, R. B., Samsonidze, G. G., Dresselhaus, G., Dresselhaus, M. S., Jiang, J., Kobayashi, N., Gruneis, A., & Saito, R. (2005). Reso-

nance Raman spectroscopy (n,m)-dependent effects in small-diameter single-wall carbon nanotubes. *Phys. Rev. B*, 71, 075401-075411.

[23] Li, Y., Rotkin, S. V., & Ravaioli, U. (2004). Metal-semiconductor transition in armchair carbon nanotubes by symmetry breaking. *Appl. Phys. Lett*, 85, 4178-4180.

[24] , Y., Lehtinen, P. O., Foster, A. S., & Nieminen, R. M. (2004). Magnetic properties of vacancies in graphene and single-walled carbon nanotubes. *New J. Phys*, 6(68).

[25] Vijayaraghavan, A., Kanzaki, K., Suzuki, S., Kobayashi, Y., Inokawa, H., Ono, Y., Kar, S., & Ajayan, P. M. (2005). Metal–Semiconductor Transition in Single-Walled Carbon Nanotubes Induced by Low-Energy Electron Irradiation. *Nano Lett*, 5, 1575-1579.

[26] Park J.Y. (2007). Electrically tunable defects in metallic single-walled carbon nanotubes. *Appl. Phys. Lett*, 90(2), 023112-023114.

[27] Bachilo, S. M., Balzano, L., Herrera, J. E., Pompeo, F., Resasco, D. E., & Weisman, R. B. (2003). Narrow (n,m)-Distribution of Single-Walled Carbon Nanotubes Grown Using a Solid Supported Catalyst. *J. Am. Chem. Soc.*, 125(37), 11186-11187.

[28] Hou, P. X., Liu, C., & Cheng, H. M. (2008). Purification of carbon nanotubes. *Carbon*, 46(15), 2003-2025.

[29] Silva, A. M. D., Junqueira, G. M. A., Anconi, C. P. A., & Santos, H. F. D. (2009). New Insights on Chemical Oxidation of Single-Wall Carbon Nanotubes: A Theoretical Study. *Phys. Chem. C*, 113(23), 10079-10084.

[30] Zhang, G., Qi, P., Wang, X., Lu, Y., Mann, D., Li, X., & Dai, H. (2006). Hydrogenation and Hydrocarbonation and Etching of Single-Walled Carbon Nanotubes. *J. Am. Chem. Soc.*, 128(18), 6026-6027.

[31] Nikitin, A., Ogasawara, H., Mann, D., Denecke, R., Zhang, Z., Dai, H., Cho, K., & Nilsson, A. (2005). Hydrogenation of Single-Walled Carbon Nanotubes. *Phys. Rev. Lett.*, 95(22), 225507-225510.

[32] Cancado, L. G., Pimenta, M. A., Neves, B. R. A., Dantas, M. S. S., & Jorio, A. (2004). Influence of the Atomic Structure on the Raman Spectra of Graphite Edges. *Phys. Rev. Lett.*, 93(24), 247401-247404.

[33] Kuzmany, H., & Knoll, P. (1986). Resonance Raman cross-section and polarization of light scattered from trans-polyacetylene. *Raman Spectroscopy*, 17(1), 89-92.

[34] Takeuchi, H., Arakawa, T., Furukawa, Y., Harada, I., & Shirakawa, H. (1987). Density of vibrational states in trans-polyene: comparison with the infrared, Raman and neutron spectra of trans-polyacetylene. *J MolStruct*, 158, 179-193.

[35] Ferrari, A. C., & Robertson, J. (2001). Origin of the 1150-cm 1 Raman mode in nanocrystalline diamond. *Phys Rev B*, 63(12), 121405-121408.

[36] Kuzmany, H., Pfeiffer, R., Salk, N., & Günther, B. (2004). The mystery of the 1140 cm 1 Raman line in nanocrystalline diamond films. Carbon ., 42(5-6), 911-917.

[37] Owens F.J. (2005). Effect of nanosizing on some properties of one dimensional polyacetylene chains. *Physica E*, 25(4), 404-408.

[38] Yamada, T., Maigne, A., Yudasaka, M., Mizuno, K., Futaba, D. N., Yumura, M., Iijima, S., & Hata, K. (2008). Revealing the secret of water-assisted carbon nanotube synthesis by microscopic observation of the interaction of water on the catalysts. *Nano Lett*, 8(12), 4288-4292.

[39] Tomita, A., & Tamai, Y. (1974). An optical microscopic study on the catalytic hydrogenation of graphite. *J PhysChem*, 78(22), 2254-2258.

[40] Datta, S. S., Strachan, D. R., Khamis, S. M., & Johnson, A. T. C. (2008). Crystallographic etching of few-layer graphene. *Nano Lett.*, 8(7), 1912-1915.

[41] Campos, L. C., Manfrinato, V. R., Sanchez-Yamagishi, J. D., Kong, J., & Jarillo-Herrero, P. (2009). Anisotropic etching and nanoribbon formation in single-layer graphene. *Nano Lett.*, 9(7), 2600-2604.

[42] Ci, L., Song, L., Jariwala, D., Elías, A. L., Gao, W., Terrones, M., & Ajayan, P. M. (2009). Graphene shape control by multistage cutting and transfer. *Adv Mater.*, 21(41), 4487-4491.

[43] Elías, A. L., Botello-Méndez, A. R., Meneses-Rodríguez, D., González, V. J., Ramírez-González, D., Ci, L., Muñoz-Sandoval, E., Ajayan, P. M., Terrones, H., & Terrones, M. (2010). Longitudinal cutting of pure and doped carbon nanotubes to form graphitic nanoribbons using metal clusters as nanoscalpels. *Nano Lett.*, 10(2), 366-372.

Section 2

Structural Properties

Chapter 4

Synthesis, Atomic Structures and Properties of Boron Nitride Nanotubes

Takeo Oku

Additional information is available at the end of the chapter

1. Introduction

Since the development of boron nitride (BN) nanotubes (Chopra et al. 1995), various types of BN nanostructured materials have been reported because of the great potential for using materials with low dimensions in an isolated environment. Many studies have been reported on BN nanomaterials and single crystals such as nanotubes (Golberg et al. 2000, Mickelson et al. 2003), bundled tubes, nanocorns, nanohorns, nanocapsules, nanoparticles, BN clusters, and BN metallofullerenes, which are expected to be useful as electronic devices, field-effect transistors (Radosavljevi et al. 2003), high heat-resistant semiconductors, insulator lubricants, nanowires (Tang et al. 2002), magnetic nanoparticles, gas storage materials (Lim et al. 2007), and optoelectronic applications including ultraviolet light emitters. Theoretical calculations on BN nanomaterials such as nanotubes (Rubio et al. 1994), cluster-included nanotubes, BN clusters, BN metallofullerenes, cluster solids, nanohorns, and hydrogen storage have also been carried out for prediction of the properties. By controlling the size, layer numbers, helicity, compositions, and included clusters, these cluster-included BN nanocage structures with bandgap energy of ~6 eV (Watanabe et al. 2004) and nonmagnetism are expected to show various electronic, optical, and magnetic properties as shown in Fig. 1. The differences between BN and carbon nanomaterials (Oku et al. 2009) are summarized as shown in Table 1.

The present review shows BN nanotubes synthesized by arc melting and thermal annealing methods. They were characterized by high-resolution electron microscopy (HREM), and their properties were investigated and discussed. In order to confirm the atomic structures and to investigate stabilities and electronic states, total energy calculations were carried out by molecular mechanics and molecular orbital calculations. These studies will give us a guideline for the synthesis of the BN nanotubes, which are expected for the future nanoscale devices.

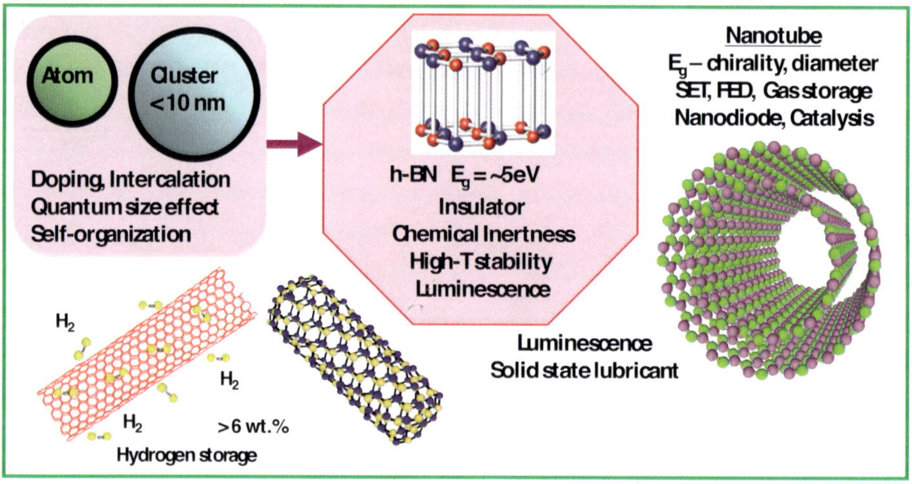

Figure 1. Structures and properties of BN nanotubes

	BN	C
Structure	4-, 6-, 8-membered rings	5-, 6-, 7-membered rings
Oxidation resistance	~900°C	~600 °C
Electronic property (E_g)	Insulator (~6 eV)	Metal-semiconductor (0~1.7 eV)
Band structure	Direct transition	Indirect transition

Table 1. Differences between BN and carbon (C) nanotubes

2. Synthesis of BN nanotubes

2.1. Arc-melting of boride powders

The purpose of the present work was to prepare the BN nanotubes by arc-melting YB_6 powder in nitrogen and argon gas atmosphere. Yttrium (Y) had been reported to show excellent catalytic properties for producing single-walled carbon nanotubes (Saito et al. 1995). In the present work, YB_6 was selected to take advantage of this excellent catalytic effect (Narita & Oku, 2003). It is not necessary to prepare the boride-rod if the YB_6 powder is used. To understand the formation mechanism of BN nanotubes, HREM and electron dispersive X-ray spectroscopy (EDX) were carried out.

The YB_6 powder (4.0 g, 99.6%, Kojundo Chemical Lab. Co., Ltd) was set on a copper mold in an electric-arc furnace, which was evacuated down to 1.0×10^{-3} Pa. After introducing a mixed gas of Ar (0.025 MPa) and N_2 (0.025 MPa), arc-melting was applied to the samples at an accelerating voltage of 200 V and an arc current of 125 A for 10 s (. Arc-melting was performed with a vacuum arcmelting furnace (NEV-AD03, Nissin Engineering Co., Ltd). Samples for HREM observation were prepared by dispersing the materials on holey carbon grids. HREM observation was performed with a 300 kV electron microscope (JEM-3000F). To confirm the formation of BN fullerene materials, EDX analysis was performed by the EDAX system.

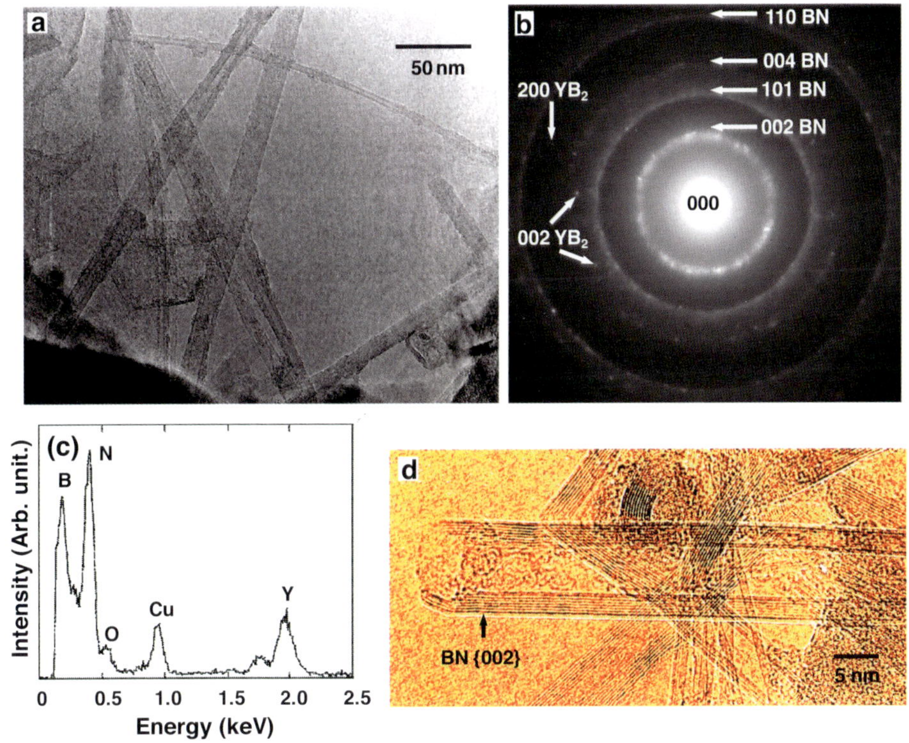

Figure 2. a) TEM image of BN nanotubes. (b) Electron diffraction pattern and (c) EDX spectrum of BN nanotubes with YB_x nanoparticles. (d) HREM images of BN nanotubes.

Low magnification image of BN nanotubes produced from YB_6 powder by arc-melting is shown in Fig. 2(a). In Fig. 2(a), length and width of the multi-wall BN nanotubes are in the range of 4–6 mm and 4–10 nm, respectively. An electron diffraction pattern of BN nanotubes with YBx nanoparticles indicate the existence of BN and YB_2, as shown in Fig. 2(b). In Fig. 2(a), {002} and {200} reflections of YB_2 are observed. Figure 2(c) is an EDX spectrum of BN nanotubes, and strong peak of boron, nitrogen, and Y are observed. Weak peak of copper is due to the HREM grid. The EDX results showed the composition ratio of the BN nanotubes was B/N = 1.1:1. A HREM image of a multi-walled BN nanotube is shown in Fig. 2(d).

A HREM image of BN nanotube in Fig. 3(a) shows that the BN nanotube has asymmetry layer-arrangements. The layer interval on one side of the tube is 0.34 nm. Other side is in the range of 0.34–0.70 nm, which is larger than the {002} of ordinary hexagonal BN (0.34 nm). {100} planes of YB_2 are observed after the formation of BN nanotubes at the end of it, as shown in Fig. 3(b). Amorphous B with opened-tip BN nanotube is also formed at the same time by arc-melting YB_6 powder, as shown in Fig. 3(c). A novel BN nanotube is shown in Fig. 3(d). A wavy BN layer is formed into BN nanotube by most internal BN layers.

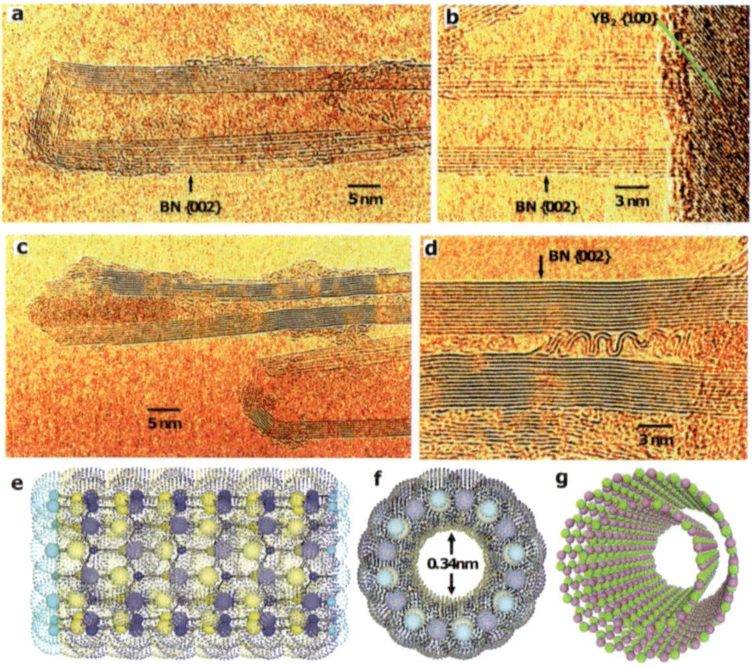

Figure 3. HREM images of (a) BN nanotube and (b) BN nanotube with YB_2 compounds. HREM images of (c) amorphous B with open-tip BN nanotube and (d) a wavy BN nanolayers in BN nanotube. Van der Waals force distribution in BN nanotube: (e) perpendicular to and (f) along the nanotube axis. Structure model of double-walled BN nanotube.

Figures 3(e) and 3(f) is a schematic illustration of Van der Waals force distribution of BN nanotube: (e) perpendicular to and (f) along the nanotube axis. The structure corresponds to the center of the BN nanotube. There is a space with a diameter of 0.34nm inside the BN nanotube, which would be expected to be a container for atomic storage. Figure 3(g) is an atomic structure model of double-walled BN nanotube.

	1/I	2/II	3	4	5	6	7	8	9	10	11	12	13/III	14/IV	15/V	16/VI	17/VII	18/VIII
1	H																	He
2	Li	Be											B ×	C	N	O	F	Ne
3	Na	Mg ×											Al ●	Si ×	P	S	Cl	Ar
4	K	Ca	Sc	Ti =●	V ●	Cr	Mn	Fe ●=	Co ●	Ni ●	Cu ●	Zn	Ga ○	Ge ●	As	Se	Br	Kr
5	Rb	Sr	Y =	Zr =	Nb =●	Mo	Tc	Ru	Rh	Pd ●	Ag	Cd	In	Sn	Sb	Te	I	Xe
6	Cs	Ba	La-Lu =	Hf =	Ta =	W =	Re	Os	Ir	Pt	Au ●	Hg	Tl	Pb	Bi	Po	At	Rn
7	Fr	Ra	Ac-Lr	Rf	Db	Sg	Bh	Hs	Mt									

Figure 4. Catalysis metals for BN fullerene nanomaterials confirmed by experiments on arc-method (=, BN nanotube; ●, BN nanocapsule; ○, BN nanocage; ×, non BN fullerene nanomaterials).

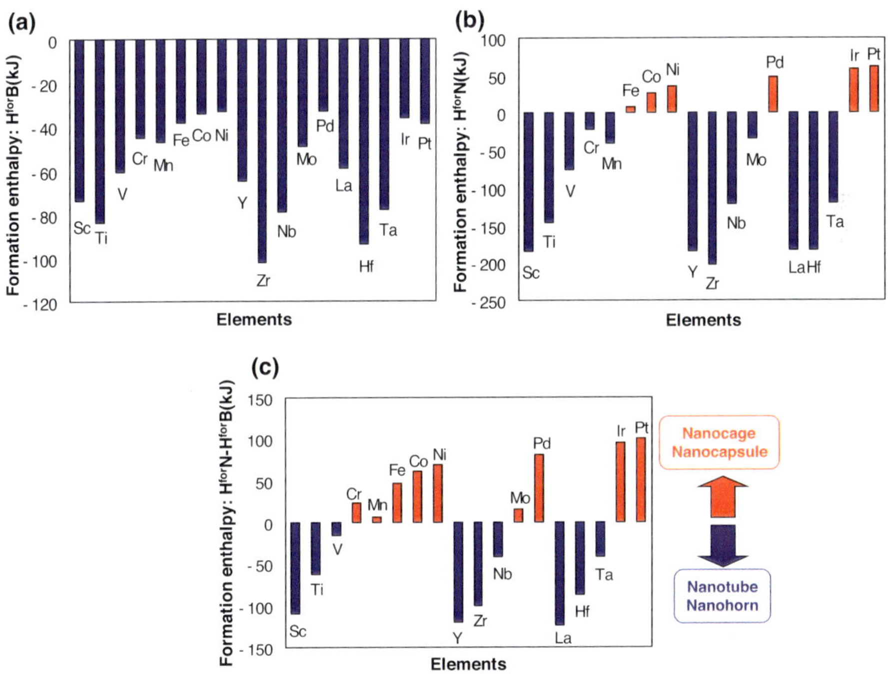

Figure 5. a) Formation enthalpy with boron ($H^{for}B$) and (b) nitrogen ($H^{for}N$); (c) Difference of formation enthalpy ($H^{for}N - H^{for}B$).

In the present work, yttrium worked as a good catalytic element to produce BN nanotubes. Catalytic metals for the formation of BN nanotubes, nanocapsules, and nanocages, which were confirmed by experiments on arc method, are summarized in Fig. 4 as periodic table. It has been reported that Zr, Hf, Ta, W, Nb, and La can be good catalytic metals for synthesis of BN nanotubes (Narita et al. 2003). On the other hand, other metals could not form BN nanotubes, although BN nanocapsules or nanocages were formed.

For some metals, formation enthalpies with boron ($H^{for}B$) and nitrogen ($H^{for}N$) are indicated in Fig. 5(a) snf 5(b), respectively. The data were from theoretical calculations (Oku et al. 2004). Difference of formation enthalpy ($H^{for}N$ - $H^{for}B$) is also shown in Fig. 5(c). The difference of formation enthalpy ($H^{for}N$ - $H^{for}B$) is very important for the formation of BN fullerene nanomaterials. Because, reactivity with nitrogen and boron is decided by this enthalpy. Basically, BN nanotubes are formed when rare earth metals are used as catalytic metals, such as Y, Zr, Nb, Hf, Ta, W and La. These elements have minus enthalpy, as shown in Fig. 3c. It means that catalytic elements for synthesis of BN nanotubes should be selected from those with minus formation enthalpy ($H^{for}N$ - $H^{for}B$). From the present guideline, Sc element could be a good catalytic element to form BN nanotubes.

In the present work, the Y worked as a good catalytic element to produce BN nanotubes. Schematic illustration of the formation mechanism of BN nanotubes is shown in Fig. 6. First, ion and radical gas that consist of Y, B, and N elements would be produced by arc melting. This ion gas would be cooled by collision with Ar and N_2 gas. In this process, Y and B ions form particles of Y + B compound, which are semi-liquid state. Since B atoms become supersaturated on cooling, Y + B particles separate out B atoms on the surface. As a result, Y + B particles are covered with amorphous B. Some amorphous B would be separated from the surface of Y + B particles. BN nanolayers are formed between separated amorphous B and surface of Y + B particle. N that is necessary to form BN nanotube is provided from environmental gas. Also, B of Y + B particle would be used to form BN nanotube, because the YB_{6-x} compound is thermodynamically more stable than YB_6. In the present work, BN nanotubes with YB_2 particles are formed. Closed or opened tips of BN nanotubes would be formed by cooling rate. If enough time is not given to the formation of BN nanotubes, amorphous B with opened-tip BN nanotubes would be formed, as shown in Fig. 3(c).

Some of the multi-walled BN nanotubes have asymmetry layer-arrangements as shown Fig. 3(a) and 3(b). This asymmetry layer-arrangement comes from the difference of layer-arrangement of B and N atoms. In the case of hexagonal BN, B atoms infallibly exist just above the N atoms with the layer interval of 0.34 nm. However, in case of BN nanotube, some N atoms are close to the N atoms of other layers, because each BN layer of multi-walled BN nanotube has different diameter or chirality. In such case, since lone-pair of N atoms reacts against each other, BN layers have large layer interval at this part. On the other hand, a part that B atoms exist just above the N atoms keeps the layer interval of 0.34 nm. As a result, some BN nanotubes form asymmetry layer-arrangements.

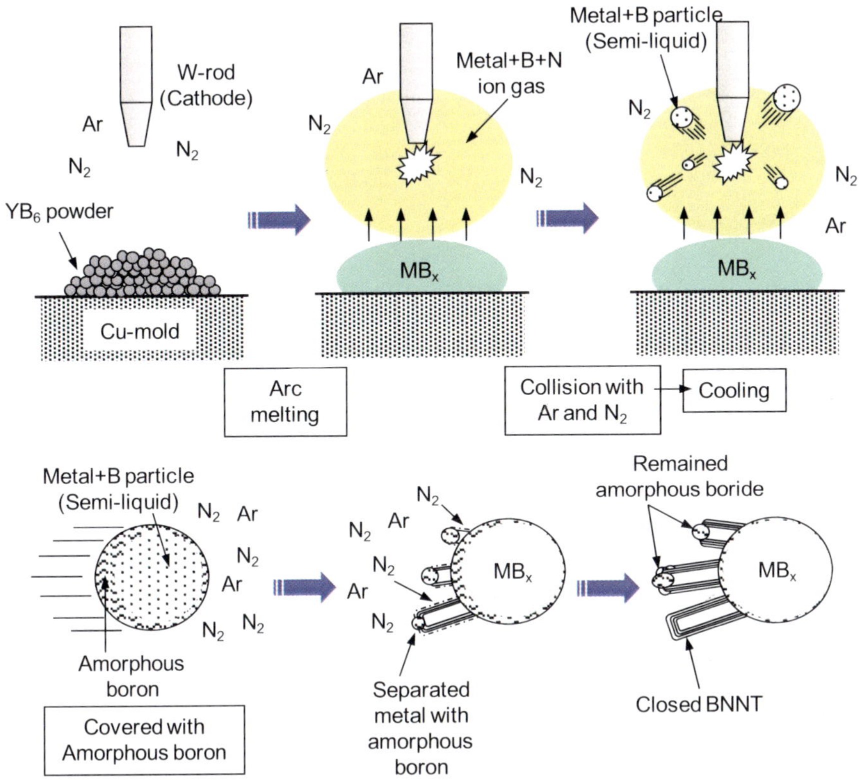

Figure 6. Schematic illustration of the formation mechanism of BN nanotubes.

2.2. Mass production of BN nanotubes

BN nanotubes have been synthesized by arc-discharge method, as decribed in the previous section. However, the arc-discharge method is not suitable for mass production because of limitation of the plasma area, and it is difficult to control nanotube size and the number of BN layers. The purpose is to synthesize BN nanotubes by ordinary thermal annealing, and to investigate the nanostructures. An Ellingham diagram of nitride metals for N_2 gas per mol was thermodynamically calculated by HSC Chemistry (Outokumpu Research Oy. Poli, Finland) software as shown in Fig. 7.

Fe_4N particles would be reduced to α-Fe completely by annealing with boron, because boron reacted with nitrogen more easily compared to Fe. Similarly, several nitrides would be reduced to pure metals by reaction with boron. In the present work, Fe was selected for the BN nanotube formation, and a mixture powder of Fe_4N/B was used for the synthesis (Koi et al. 2008).

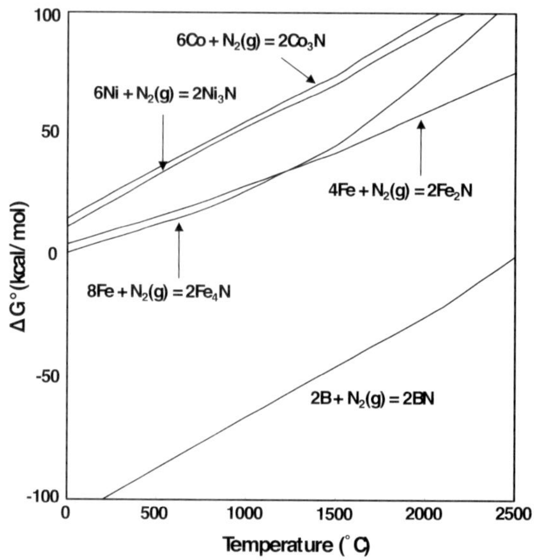

Figure 7. Ellingham diagram of Fe, Ni and Co nitrides for a N_2 molecule.

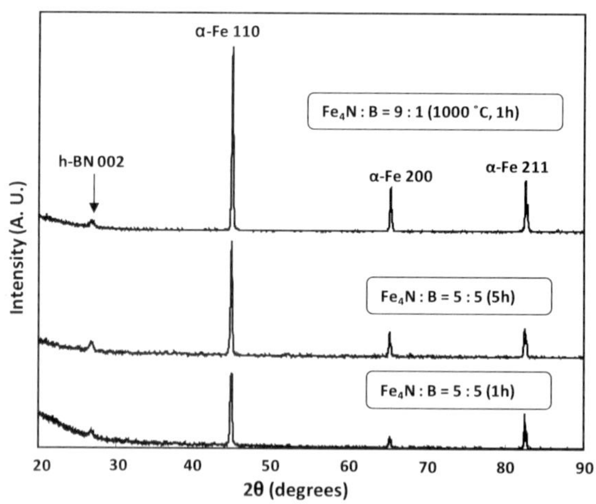

Figure 8. X-ray diffraction patterns of the annealed samples of Fe_4N:B, which are WR of Fe_4N:B = 5:5 annealed at 1000 °C for 1 h, (b) WR of Fe_4N:B = 5:5 annealed at 1000 °C for 5 h and (c) WR of Fe_4N:B = 9:1 annealed at 1000 °C for 1 h, respectively.

X-ray diffraction patterns of annealed samples of Fe_4N/B with various weight ratio (WR) of Fe_4N:B annealed at 1000 °C are shown in Fig. 8. Peaks of h-BN and α-Fe were confirmed for all samples, and no peak of Fe_4N and B was observed. Average diameters of Fe particles were measured to be 20~30 nm, which were calculated from halfwidths of α-Fe (110) by using the Scherrer's equation.

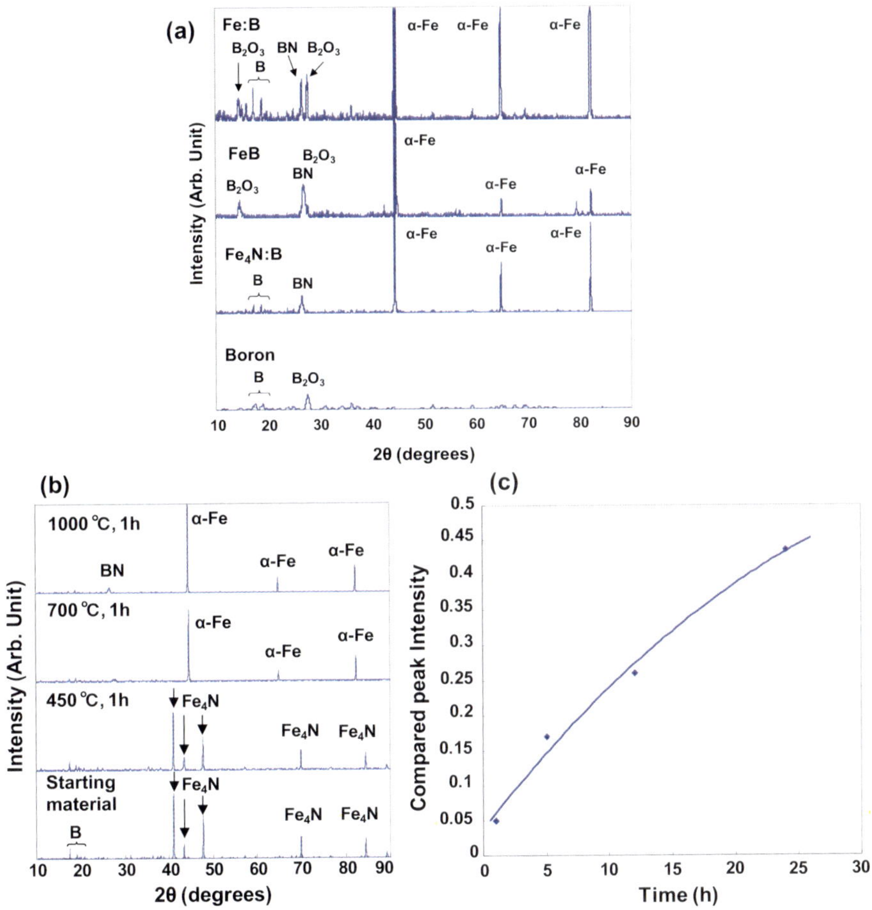

Figure 9. a) X-ray diffraction patterns of (a) various starting materials for BN nanotube formation after annealing at 1000 °C for 1 h. (b) X-ray diffraction patterns of samples at elevated temperatures. (c) Intensity change of BN as a function of annealing time. (=peak of BN/Fe)

To understand growth mechanism of BN nanomaterials, Fe_4N/B, Fe/B, FeB, B was used as starting materials, and the structures of BN nanomaterials were compared. Four-types of mixture powders (Fe/B, FeB, Fe_4N/B and B) were used as starting materials for BN synthesis. Particle sizes of Fe (purity of 99.5%, Mitsuwa's Pure Chemicals, Osaka, Japan), FeB (99%, Kojundo Chemical Laboratory (KCL) Co. Ltd., Saitama, Japan), Fe_4N (99.9%, KCL) and B (99%, KCL) were about 5, 850, 50, and 45 lm, respectively. After Fe/B and Fe_4N/B (Weight ratio [WR] = 1:1, respectively) were well mixed in a triturator, the samples were set on an alumina boat and annealed in the furnace. The furnace was programmed to heat at 6 °C /min from a room temperature to 450, 700, and 1000 °C and hold for 1–24 h, and then cooled at 3 °C /min to a room temperature. Nitrogen pressure was 0.10 MPa, and its gas flow was 100 sccm.

X-ray diffraction patterns of samples are shown in Fig. 9(a). Diffraction peaks of hexagonal BN and a-Fe were observed for each sample except for a sample synthesized from boron powder. Diffraction peaks of B_2O_3 were also observed for each sample except for a sample synthesized from Fe_4N/B powder. X-ray diffraction patterns of samples synthesized from Fe_4N/B were investigated at various temperatures and time. In Fig. 9(b), Fe_4N was reduced to Fe by boron at temperatures in the range of 450–700 °C, and BN was obtained at 1000 °C. Figure 9(c) shows intensity change of BN as a function of annealing time. A large amount of BN was obtained as time advances because Fe_4N would be sufficiently reduced to Fe.

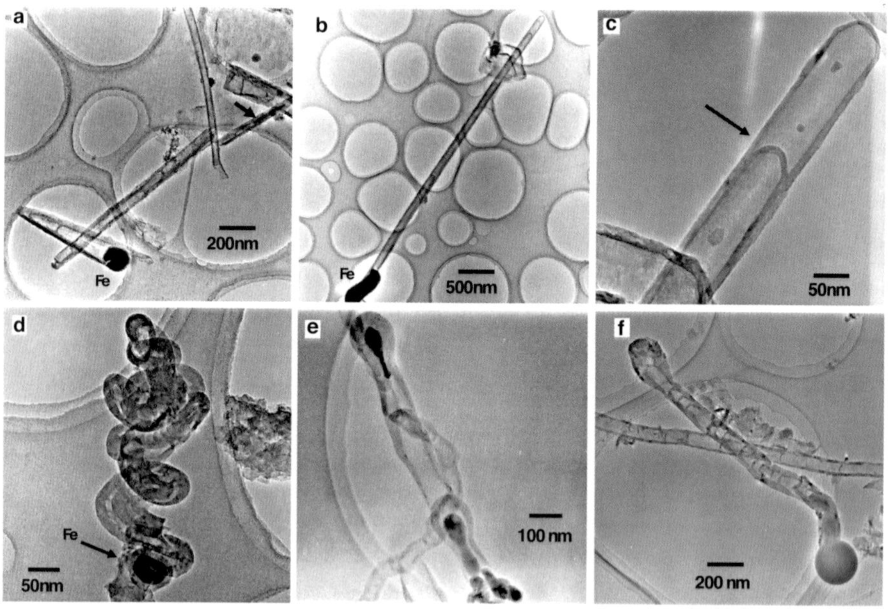

Figure 10. TEM images of BN nanotubes. (a) BN nanotubes and nanohorn. (b) BN nanotube with Fe nanoparticle. (c) Enlarged image of cap of (b). (d) BN nanocoil. (e) Bamboo-type BN nanotubes with Fe nanoparticles. (f) Bamboo-type nanotubes.

Phases of the samples were determined by X-ray diffraction, which showed peaks of hexagonal BN and α-Fe. Large amounts of BN nanotubes were produced, and Fig. 10(a) is a typical transmission electron microscope (TEM) image of the samples. BN nanohorn and nanotubes are observed, and lengths and widths of BN nanotubes were approximately 1–10 mm and 40–200 nm, respectively. A Fe nanoparticle is observed at the root area of a BN nanohorn. A nanotube shown by an arrow is a Fe-filled BN nanotube. Figure 10(b) is a TEM image of BN nanotube with a Fe nanoparticle, and the length is more than 2 μm. Figure 10(c) is a high magnification image of Fig. 10(b), and the BN nanotube has a bamboo-type structure, as indicated by an arrow. BN nanocoil was also produced, as shown Fig. 10(d), and a Fe nanoparticle is observed as indicated by an arrow. In the case of using magnetic materials as the catalysis metal for BN nanotubes, the magnetic nanoparticles move or rotate with the change of magnetic field, which arises from a coil heater, in the process of reaction. Therefore, it is considered that BN nanocoils were produced. High WR of Fe_4N would be suitable for synthesis of BN nanocoils because the frequency of moving is high with increasing of the amount of magnetic nanoparticles. Bamboo-type BN nanotubes were also observed, as shown in Fig. 10(e) and 10(f). Nanoparticles were observed at the root of the nanotubes, which would be closely related with BN nanotube growth.

Figure 11(a) is a TEM image of BN nanotubes with bamboo-structures. Lengths and widths of BN nanotubes are approximately 5–10 μm and 40–200 nm, respectively. In addition, iron nanoparticles were often observed at the tip of nanotubes, as shown in Fig. 11(b). Enlarged images of a tip and an interface between the Fe nanoparticle and the nanotube are shown in Fig. 11(c) and 11(d), respectively. In Fig. 11(c), amorphous structures (AM) and lattice fringes of Fe_2B {200} are observed near the growth point of BN layers. The amorphous structure would be boron-rich phase formed from reaction with Fe_4N. At the interface between the Fe particle and BN nanotube in Fig. 11(d), lattice fringes of Fe {110} are observed, and the BN {002} layers are inclined from the nanotube axis indicate by z-axis.

A small amount of nanocrystalline Fe_2B compounds were observed at the tip of the BN nanotube (Fig. 12). Chemical formulas that Fe_4N reacts with B, and generates Fe and BN in the experiments can be proposed as follows:

$$Fe_4N + 3B = BN + 2Fe_2B \tag{1}$$

Fe_2B and dissolution of boron were obtained, and BN was produced in the reaction expressed as eq. (1) because Fe_2B is thermodynamically more stable than Fe_4N. Although the

Fe_2B is stable to 1389 °C, the Gibbs-Thompson effect shown that the melting occurs at a significantly lower temperature compared to values in the standard phase diagram. Therefore, fluid-like Fe_2B can be attained more easily. In the next process, the reaction expressed as eq. (2) would take place.

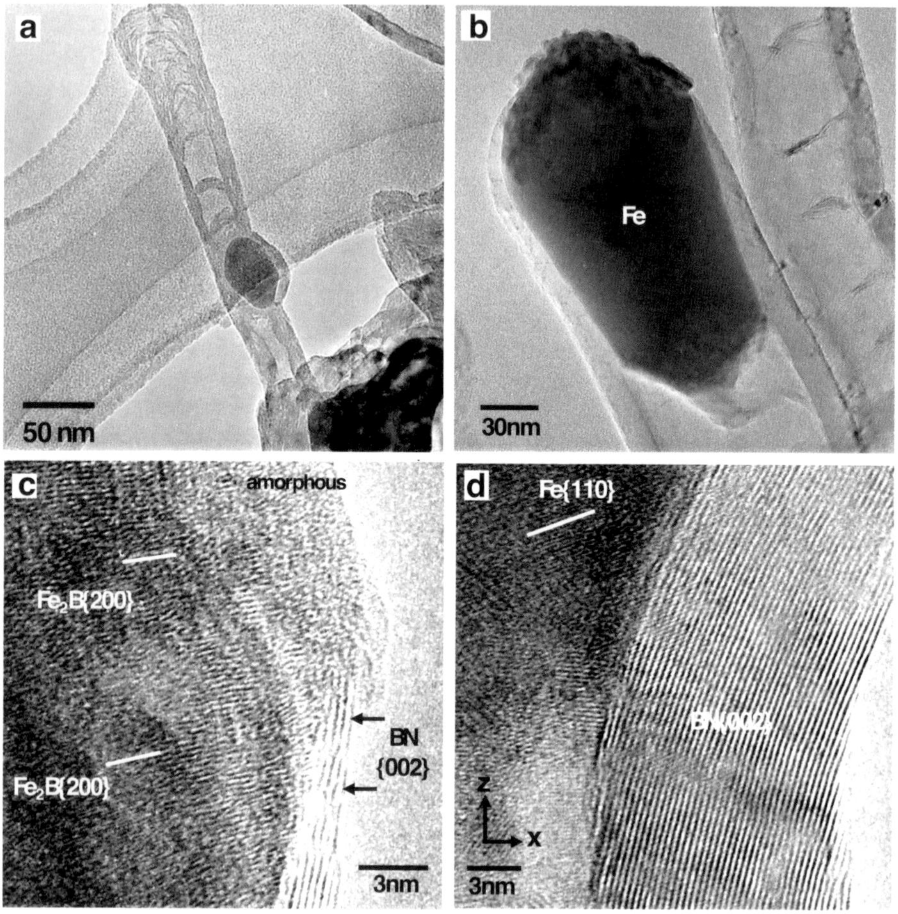

Figure 11. Low magnification images of (a) BN nanotubes with bamboo-structures and (b) iron nanoparticle at a tip of nanotube. Enlarged images of (c) a tip and (d) an interface between the Fe nanoparticle and the nanotube.

$$2Fe_2B + N_2(g) = 2BN + 4Fe \quad (2)$$

Boron in liquid-like Fe_2B started to segregate on the surface of the particle. The boron would react with N_2 gas, and BN was produced. α-Fe in liquid-like Fe_2B is epitaxially grown to the [110] direction, and Fe nanowires were produced in the reaction of eq. (2). In adeition, high WR would be mandatory for the formation of Fe-filled BN nanotubes. As the results of these reactions, the [110] of Fe is parallel to the BN nanotube axis.

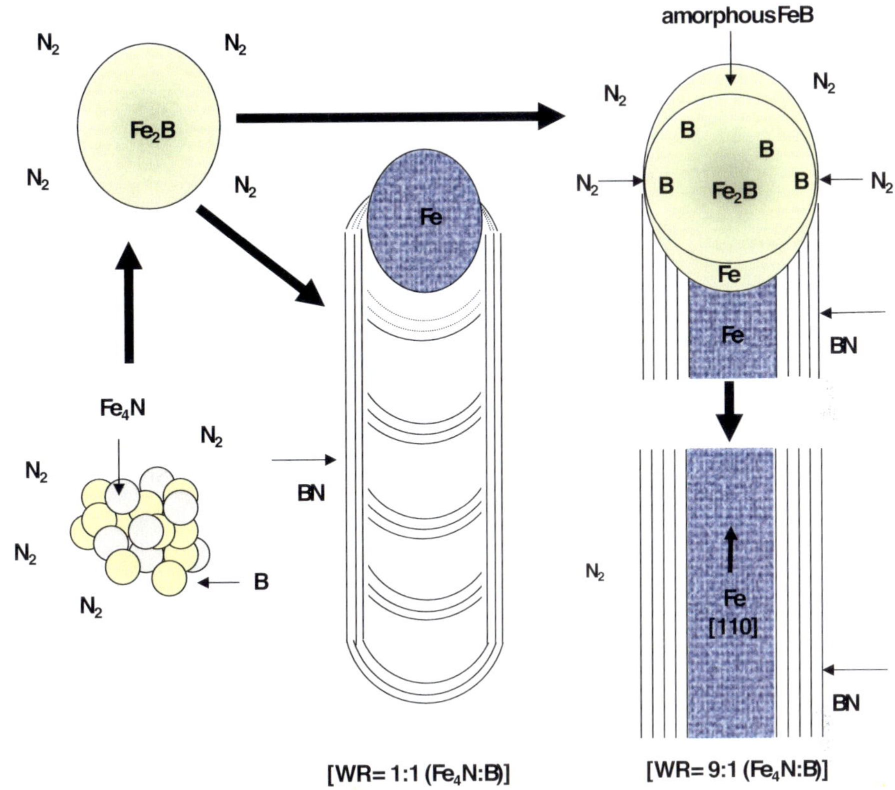

Figure 12. Schematic illustration of the formation mechanism of bamboo-type structure and Fe-filled BN nanotube.

Gibb's energy on each formula is calculated as -89:4 and -23:2 kcal for the formulas (1) and (2) at 1000 °C, respectively. These negative values would stand for correctness of the proposed formulas. It is considered that a formation of Fe-B compounds might plays an important role for growth of the BN nanotubes, and that amorphous boron might change to BN and Fe_2B on the surface of the Fe4N nanoparticles. When magnetic materials are used as catalysis metals for BN nanotube formation, the magnetic nanoparticles would move around by magnetic field of a coil heater during the reaction process. Then, segments of BN {002} layers were produced in the tubes, which results in formation of bamboo structures as shown in Fig. 12. The interval of the BN layer segments might be related to the amount of iron nanoparticles, and further studies are expected on the control of the bamboo structure.

2.3. Purification of BN nanotubes

Selective synthesis and purification methods for BN nanotubes are required to use them as devices, and an efficient method for purification of BN nanomaterials is required. The key steps in purification of BN nanomaterials in the present work would be HCl, HNO_3 and pyridine treatment (Koi et al. 2008).

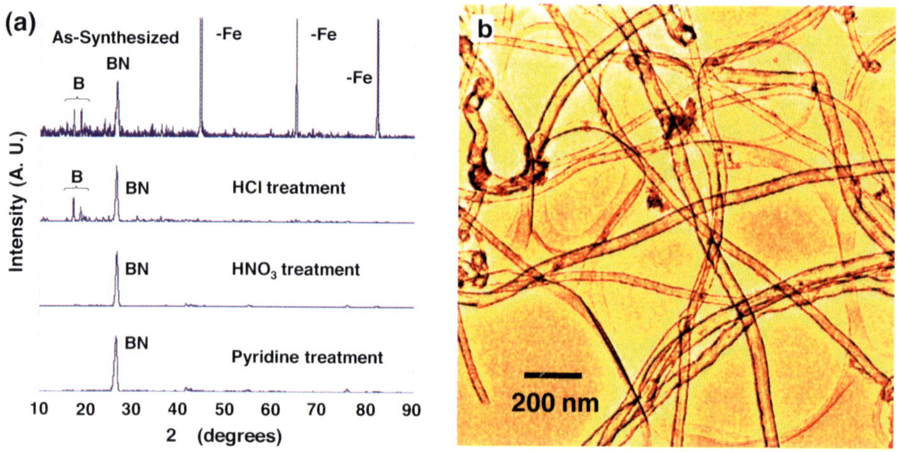

Figure 13. a) X-ray diffraction patterns of samples after synthesis, HCl treatment, HNO_3 treatment, and pyridine treatment. (b) TEM image of samples after pyridine treatment.

As-produced soot synthesized from Fe_4N/B via the above method was purified by the following steps. The as-produced soot were poured in 4 M HCl solution and stirred for 4 h at a room temperature. The green color of the solution provides an indication of the dissolution of Fe ions. After HCl treatment, the samples were poured in 1 M HNO_3 solution and stirred for 30 h at 50 °C. The yellow color of the solution provides an indication of the dissolution of boron. After both acid treatment, the solution was filtered and rinsed with deionized water until the pH of the filtrate became neutral and dried. Then, the samples were poured in pyridine to eliminate bulk BN, and high purity BN nanotubes with a cup-stacked structure were obtained by collecting supernatant.

X-ray diffraction patterns in a purification process are shown in Fig. 13(a). Diffraction peaks of hexagonal BN, boron and α-Fe are observed for the sample at annealed at 1000 °C for 1 h as shown in Fig. 13(a). It is found that Fe was removed after HCl treatment, and boron was removed after HNO_3 treatment. After pyridine treatment, a strong peak of BN was obtained as shown Fig. 13(a). Figures 13(b) show a TEM image of the sample, and there is no obvious change of the structure during the purification process, and BN nanotubes with small sizes were obtained after pyridine treatment. It is believed that bulk size of BN was eliminated and high purity BN nanotubes were obtained by pyridine treatment. Purification of BN nanotubes were carried out by HCl, HNO_3 and pyridine treatment to remove non-BN nanotubes such as metal catalysts, boron oxides and unreacted boron.

2.4. Nanotube growth from iron-evaporated boron

The purpose is to synthesize BN nanotubes by a normal thermal annealing method. To synthesize BN nanotubes, a Fe thin film was selected and used as a catalyst for nanotube growth in the present work. Boron (B) powders with a particle size of 45 μm (99%, Kojundo Chemical Laboratory) were used as starting materials. B powder was pressed at 100 kg mm^{-2} into pellets with the size of 4 mm height and 15 mm in diameter. Fe with a thickness of *ca.* 10 nm was evaporated on the compact at ~10^{-6} torr, and the Fe would have an island structure. The samples were set on an alumina boat and annealed in a nitrogen atmosphere. The furnace was programmed to heat at 6 °C/min from a room temperature to 1000 °C and hold for 1 h, and then cooled at 3 °C/min to a room temperature. N_2 gas pressure was 0.10 MPa, and its gas flow was 100 sccm.

SEM image of surface of the Fe-evaporated B compact after annealing is shown in Fig. 14(a). Agglomerated BN nanotubes with diameters in the range of 10–20 nm are observed, and they have a network-like structure. Fig. 14(b) is a TEM image of BN nanotubes which were removed from the pellet. Diameters and lengths of BN nanotubes are in the range of 10–20 nm and 100–500 nm, respectively, and the diameters agree well with those of SEM images in Fig. 14(a). One of the typical BN nanotubes is shown in Fig. 14(c), and a nanotube axis is indicated by z. Fig. 14(d) is a Fourier filtered HREM image of center of the same BN nanotube in Fig. 14(c), and hexagonal net planes of BN nanotube are observed clearly in the image of Fig. 14(d). A hexagonal BN ring is shown in Fig. 14(d), and the BN has a zigzag-type structure, as shown in Fig. 14(e).

Growth of carbon nanotubes was explained as a model of vapor-liquid-solid (VLS) mechanism [19]. In this model, hydrocarbon such as methane is resolved in catalyst metal nanoparticles. Supersaturated solid solution of carbon in catalyst metal was precipitated as carbon nanotubes. BN nanotube growth might be explained in a similar model. Schematic illustration of growth mechanism of BN nanotubes was proposed as shown in Fig. 15. Supersaturated solid solution of B in Fe nanoparticles was formed and reacted with N_2 gas. BN nanotubes grow from these sites, and the diameter of nanotubes depends on the particle size. Fe nanoparticles are easy to be separated from BN because Fe begins to react with BN from 1350 °C, and BN nanotubes would grow as shown in Fig. 15. Oriented BN nanotubes might be obtained when Fe nanoparticles are uniformly dispersed on surface of B.

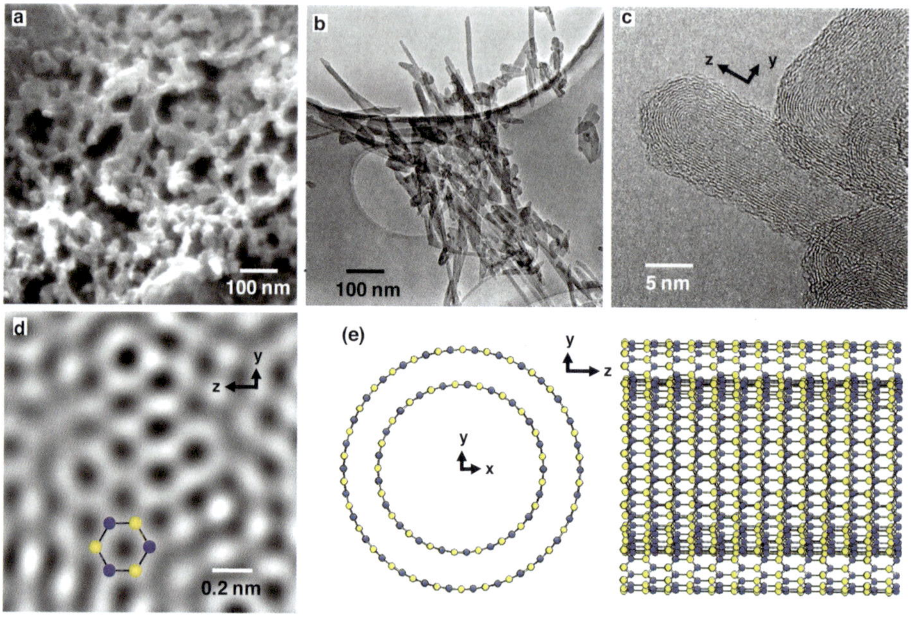

Figure 14. a) SEM and (b) TEM images of BN nanotubes grown from the Fe/B pellet. (c) HREM image of BN nanotube. (d) Enlarged image of the center of BN nanotube in (c). (e) Atomic structure model of zigzag-type BN nanotube.

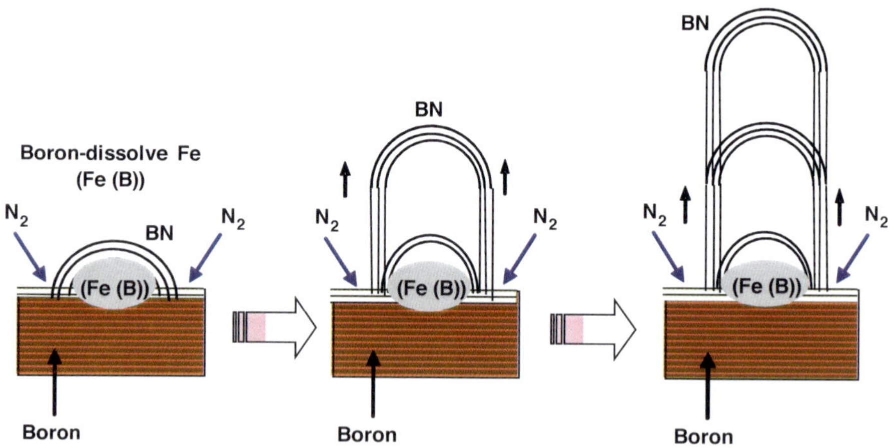

Figure 15. Schematic illustration of the growth mechanism of BN nanotubes.

3. Atomic structures of BN nanotubes

3.1. Chiralities of BN nanotubes

A low magnification TEM image of BN nanotubes produced from YB_6/Ni powder is shown in Fig. 16(a) (Oku & Narita 2004). The lengths and diameters of BN nanotubes are ~5 μm and 3–50 nm, respectively. Fig. 16(b) is an EELS spectrum of BN nanomaterials including BN nanotubes. Two distinct absorption features are observed at 188 and 401 eV, which correspond to boron K-edge and nitrogen K-edge onsets, respectively. The fine structure of boron in the EELS spectrum shows the hexagonal bonding between boron and nitrogen, which is indicated by presence of a sharp π^* peak and the shape of the σ^* peak. The EELS spectrum also shows the weak σ^* peaks of B and N, which indicate the spherical structure of BN nanomaterials.

A HREM image of a $B_{36}N_{36}$ cluster inside a BN nanotube is shown in Fig. 16(c). The BN nanotube has a multiwalled structure, and a diameter of the most inner tube is 1.75 nm. An atomic structure model of the center of Fig. 16(c) is shown in Fig. 16(d). Diameter and chirality of the BN nanotube are 1.747 nm and (22, 0), respectively. This kind of peapod-type self-organized structure would be useful for the nanoscale devices. Another HREM image of BN nanotubes with a bundled structure is shown in Fig. 16(e), and an atomic structure model observed from three different directions is shown in Fig. 16(f). There are some spaces among the BN nanotubes, and the space would be useful for gas storage such as hydrogen.

Figure 17(a) is a HREM image of a quadruple-walled BN nanotube. In the present work, all HREM images were taken close to the Scherzer defocus ($\Delta f_S = -41.2$ nm), which is an optimum defocus value of electron microscope, in order to investigate the atomic structures in detail. HREM observations and electron diffraction analysis on BN nanotubes have been reported, and direct observations of nanotube chirality were tried in the present work. An enlarged HREM image is shown in Fig. 17(b), which indicates lattice fringes in the BN nanotubes.

A filtered Fourier transform of Fig. 17(b) showed that this nanotube had a zigzag-type structure as shown in Fig. 17(c) (Oku 2011). A HREM image with clear contrast processed after Fourier noise filtering is shown in Fig. 8d. The intervals of the bright and dark dots are 0.14 nm, which corresponds to the structure of h-BN rings, as shown in Fig. 17(e). Layer intervals of each tube are 0.35 nm, as shown in Fig. 17(f). Diameters of each nanotube are 2.8, 3.5, 4.2, and 4.9 nm from the inside to outside.

Another HREM image of BN nanotube produced from YB_6 powder is shown in Fig. 18(a). Width of the multiwalled BN nanotube is 8.5 nm. The BN nanotube consists of nine layers and has asymmetry layer arrangements. Layer distances are in the range of 0.34–0.51 nm, which is larger than that of {002} of ordinary h-BN (0.34 nm). Diameters of the first and second internal nanotubes are 1.7 nm and 2.6 nm, respectively. Hexagonal net planes of BN nanotube are observed in an enlarged image of Fig. 18(b). Figure 18(c) is a filtered Fourier transform of Fig. 18(b), which indicates 002 and 100 reflections of BN structure. Inverse Fourier transform of Fig. 18(c) is shown in Fig. 18(d), which indicates the lattice fringes of hexagonal networks clearly. A h-BN ring is shown in Fig. 18(d), and the BN has an armchair-type structure.

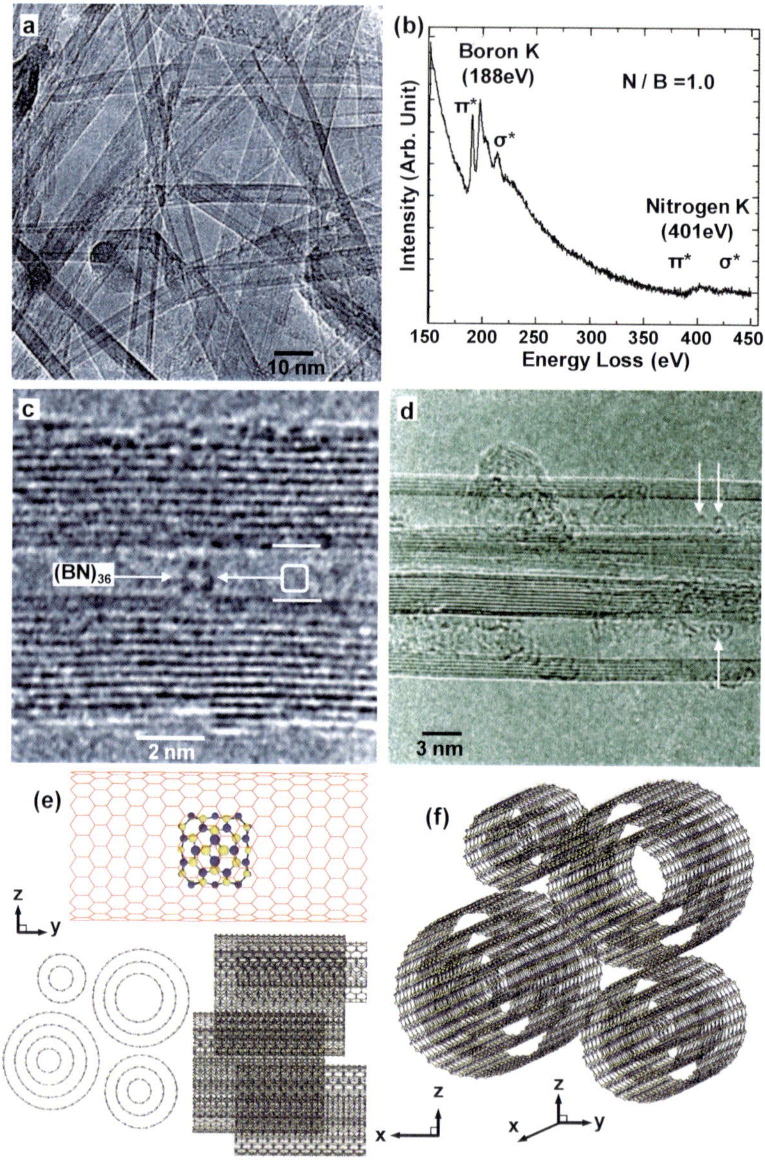

Figure 16. a) TEM image and (b) EELS spectrum of BN nanotubes. (c) HREM image of $B_{36}N_{36}$ cluster in BN nanotube. (d) HREM image of bundled BN nanotubes. BN clusters are indicated by arrows. (e) Structure model of the center of (c). (f) Atomic structure model from three different directions for bundled BN nanotubes

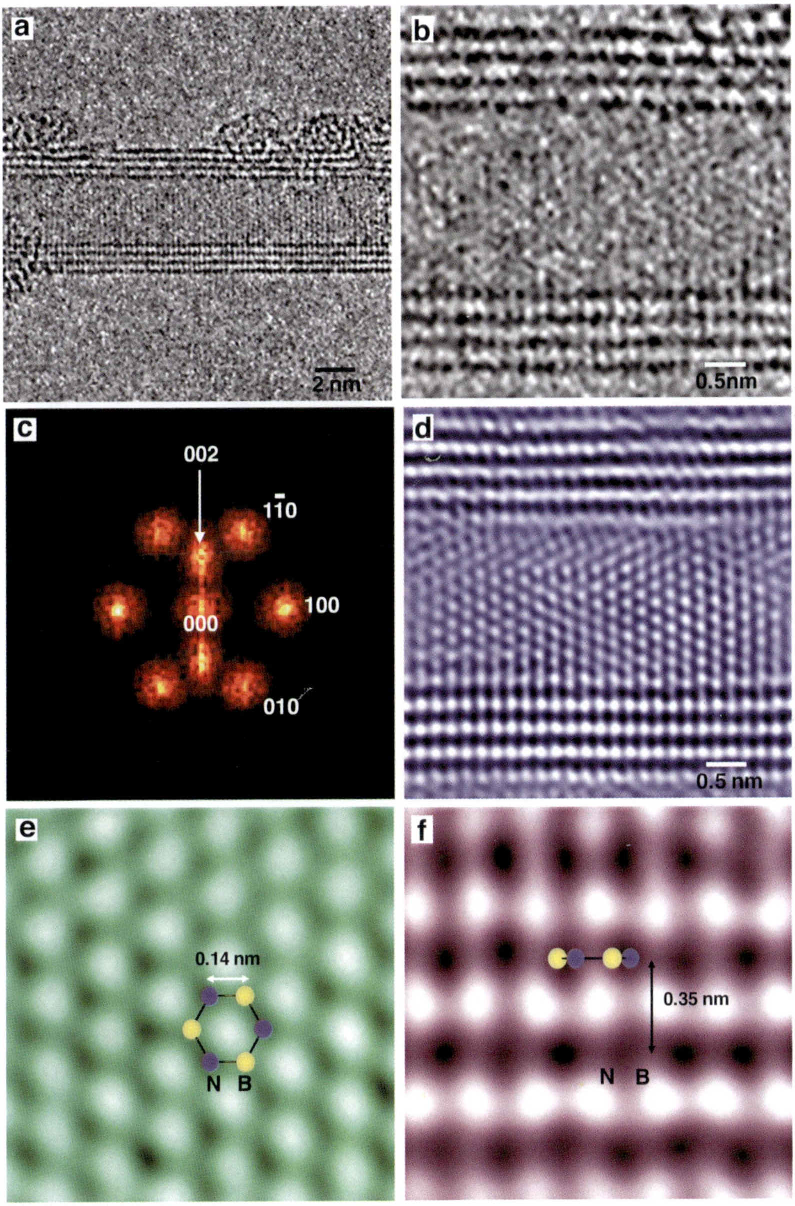

Figure 17. a) HREM image of zigzag-type BN nanotube. (b) Enlarged HREM image of (a). (c) Filtered Fourier transform of (b). (d) Inverse Fourier transform of (c). Enlarged images of center (e) and edge (f) of the BN nanotube in (d).

Atomic structure models were proposed from observed diameters of BN nanotubes, which were based on layer intervals of 0.34–0.35 nm. The chirality of (n, m) is derived from the equation

$$d_t = \frac{\sqrt{3}a_{B\text{-}N}\sqrt{n^2 + nm + m^2}}{\pi} \tag{3}$$

The d_t means a diameter of BN nanotube with nm scale, and the a B-N corresponds to the nearest distance of boron and nitrogen atoms. For the BN nanotubes, the value of a B-N is 0.144 nm. When a BN nanotube has a zigzag structure, the value of m is zero.

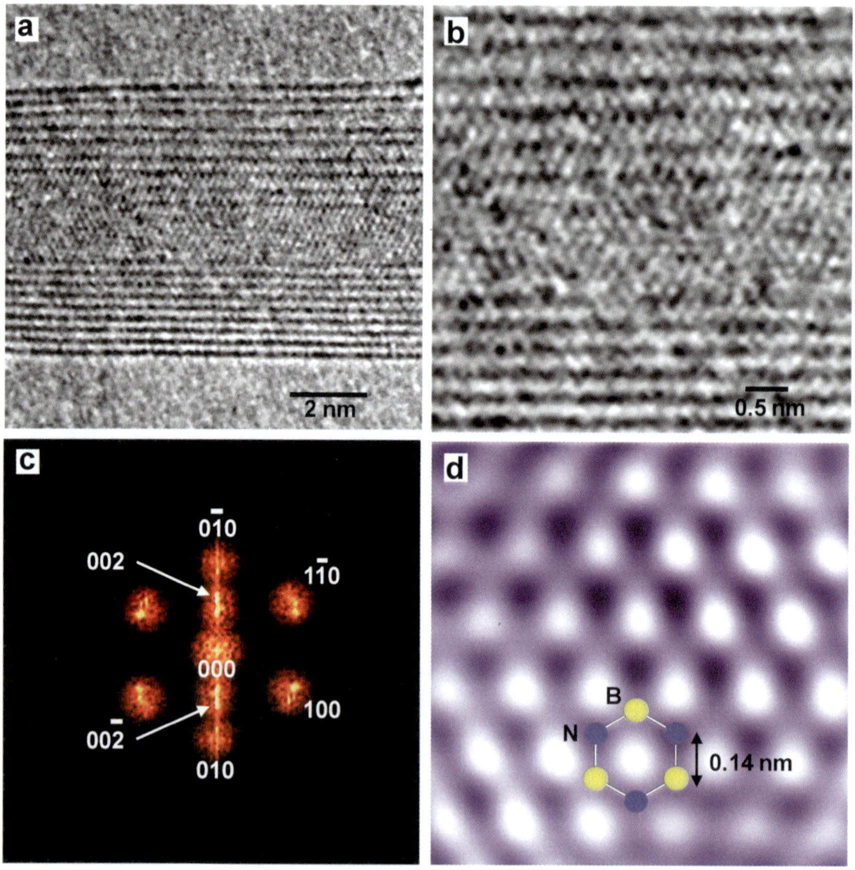

Figure 18. a) HREM image of armchair-type BN nanotube. (b) Enlarged HREM image of (a). (c) Filtered Fourier transform of (b). (d) Inverse Fourier transform of (c).

Figure 19(a) shows a proposed structure model of the quadruple-walled BN nanotube. Chiralities of each zigzag BN nanotube are (35, 0), (44, 0), (53, 0), and (62, 0) from the inside to outside. These chiralities were derived from (3). The arrangement of boron and nitrogen atoms was reversed at each layer, as boron atoms exist just above the nitrogen atoms while maintaining the layer intervals of 0.35 nm. Calculated images of the proposed model as a function of defocus values are shown in Fig. 19(b). Contrast of hexagonal rings was clearly imaged at the defocus values in the range of −40 to −50 nm, and these simulated images agree well with the observed HREM image of Fig. 17(d).

A proposed structure model of double-walled BN nanotube corresponding to Fig. 18 is shown in Fig. 19(c). Chiralities of the BN nanotube are (13, 13) and (19, 19) for the first and second layers, respectively. Layer intervals of lattice fringes of {002} planes are accorded with observed ones in Fig. 18(a). Based on the projected structure model, image calculations were carried out for various defocus values, as shown in Fig. 19(d) and a HREM image calculated at −40 nm agrees well with the experimental data of Fig. 18(d).

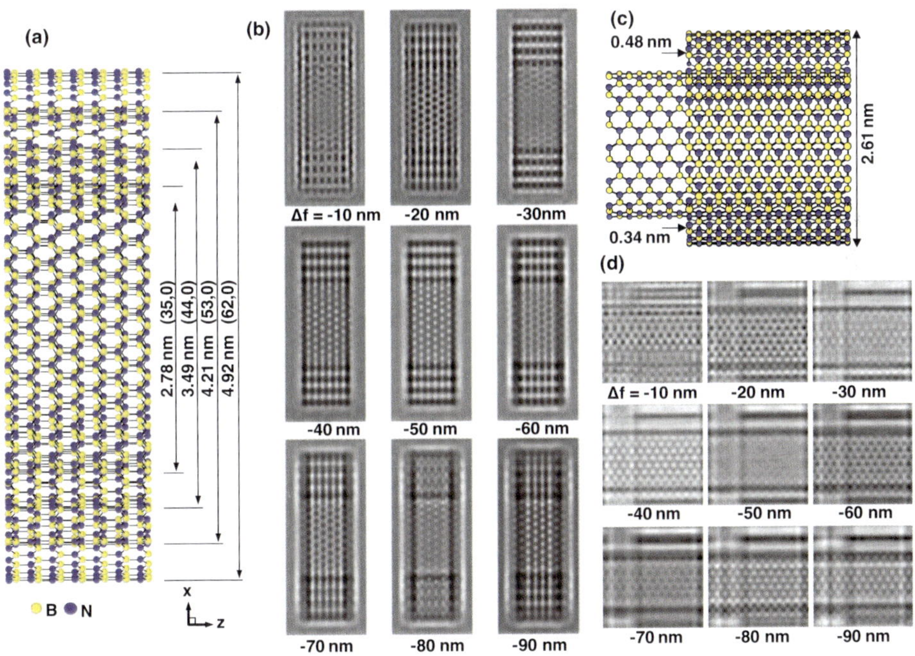

Figure 19. a) Proposed structure model of quadruple-walled BN nanotube. Chiralities of zigzag BN nanotubes are (35, 0), (44, 0), (53, 0), and (62, 0) from inside to outside. (b) Calculated images of the proposed model (a) as a function of defocus values. (c) Proposed structure model of doublewalled BN nanotube. Chiral vectors of nanotube are (13, 13) and (19, 19) for the first and second layers, respectively. (d) Calculated images of the proposed model (c).

3.2. BN nanotubes with cup-stacked structures

Figure 20(a) shows TEM image of BN nanotubes with a cup-stacked structure after purification process (Oku et al. 2007). Diameters and lengths of the BN nanotubes are in the range of 40-100 nm and 5-10 μm, respectively. Fe nanoparticles and bulk BN was eliminated during the process. An enlarged image of one of the BN nanotubes is shown in Fig. 20(b), which shows a cup-stacked structure as indicated by lines of BN {002}. Figure 20(c) is an electron diffraction pattern of Fig. 20(b). 002 reflections of BN are splitting in Fig. 20(c), which indicates that the BN nanotube has a cup-stacked structure and the cone angle between the BN layers at both nanotube walls is ~20°. Most of BN nanotubes (~90%) have this cup-stacked structure with cone angle of ~20°, and normal structures with a cone angle of 0° were sometimes observed (~10%). An optical absorption spectrum of BN nanotubes is shown in Fig. 20(d). In Fig. 20(d), a strong peak is observed at 4.8 eV, which would correspond to the energy gap of BN nanotubes. A broad, weak peak is also observed around 3.4 eV, which is considered to be impurity level (oxygen or hydrogen) of the BN layers. Comparable data (4.5-5.8 eV) were reported for other optical measurements (Lauret et al. 2005).

A HREM image of edge of the nanotube side wall in Fig. 20(b) is shown in Fig. 21(a), and a cup-stacked structure was observed. Edge structures are observed as indicated by arrows, and the BN {002} planes are inclined compared to nanotube axis (z-axis). Figure 21(b) is a processed HREM image after Fourier filtering of nanotube center of Fig. 21(b), and hexagonal arrangements of white dots are observed, which would correspond to BN six-membered rings. From these observations, a structure model for BN cup-structure was proposed, which consists only of h-BN rings, as shown in Fig. 21(c) and (d).

Based on the structure model of a four-layered cup-stacked $B_{2240}N_{2240}$ nanotube, an image calculation was carried out as shown in Fig. 21(e). Enlarged calculated HREM images of the edge and the center of the BN nanotube in Fig. 21(e) are shown in Fig. 21(f), 21(g), respectively. These calculated images agree with the experimental data of Fig. 21(a), 21(b), respectively.

As shown in Fig. 20(c), BN layers are often inclined compared to nanotube axis, which are called cup-stacked nanotubes. A HREM image and Fourier filtered image of nanotube wall of bamboo-type BN nanotube with cup-stacked structures (WR = 1:1) is shown in Fig. 22(a) and 22(b), respectively. The nanotube axis is indicated by z-axis. BN {002} layers are inclined compared to the nanotube axis, and the cone angle between the BN layers at both nanotube walls is ~36° (Nishiwaki et al. 2005). An enlarged image of nanotube center is shown in Fig. 22(c), and a HREM image with clear contrast was processed after Fourier noise filtering as shown in Fig. 22(d), which shows hexagonal arrangements of white dots.

A structure model for $B_{494}N_{494}$ cup-layer was proposed, which consists only of hexagonal BN rings. A structure model and calculated HREM images of four-fold walled $B_{1976}N_{1976}$ nanotube with a cup-stacked structure are shown in Fig. 22(e) and 22(f), respectively. The calculated images (Fig. 22(f)) at defocus values of 40 and 50 nm have similar contrast of the HREM images in Fig. 22(b) and 22(d).

In order to investigate the stability of the cup-stacked structure, four types of nanotubes are considered, as shown in Fig. 23. Atomic structure models of double-walled BN nanotubes with zigzag-type and armchair-type structures, respectively, are shown in Fig. 23(a) and 23(b). Atomic structure models of four-layered, cup-stacked BN nanotubes with different cone angles are shown in Fig. 23(c) and 23(d). The values of these structures were summarized as in Tables 2 and 3. Total energies of these four-type structures indicates that BN multilayered nanotubes with and without a cup-stacked structure would be stabilized by stacking h-BN networks.

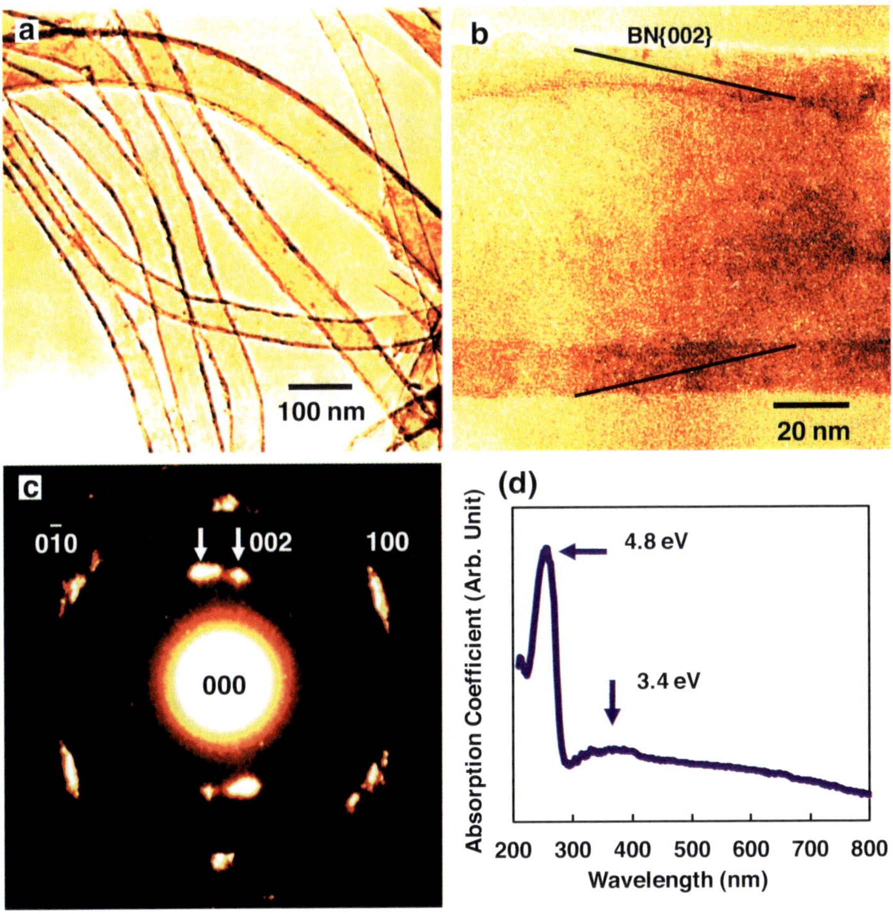

Figure 20. a) TEM image of BN nanotubes after purification. (b) Enlarged image of BN nanotube with cup-stacked structure. (c) Electron diffraction pattern of (b). (d) Optical absorption spectrum of BN nanotubes.

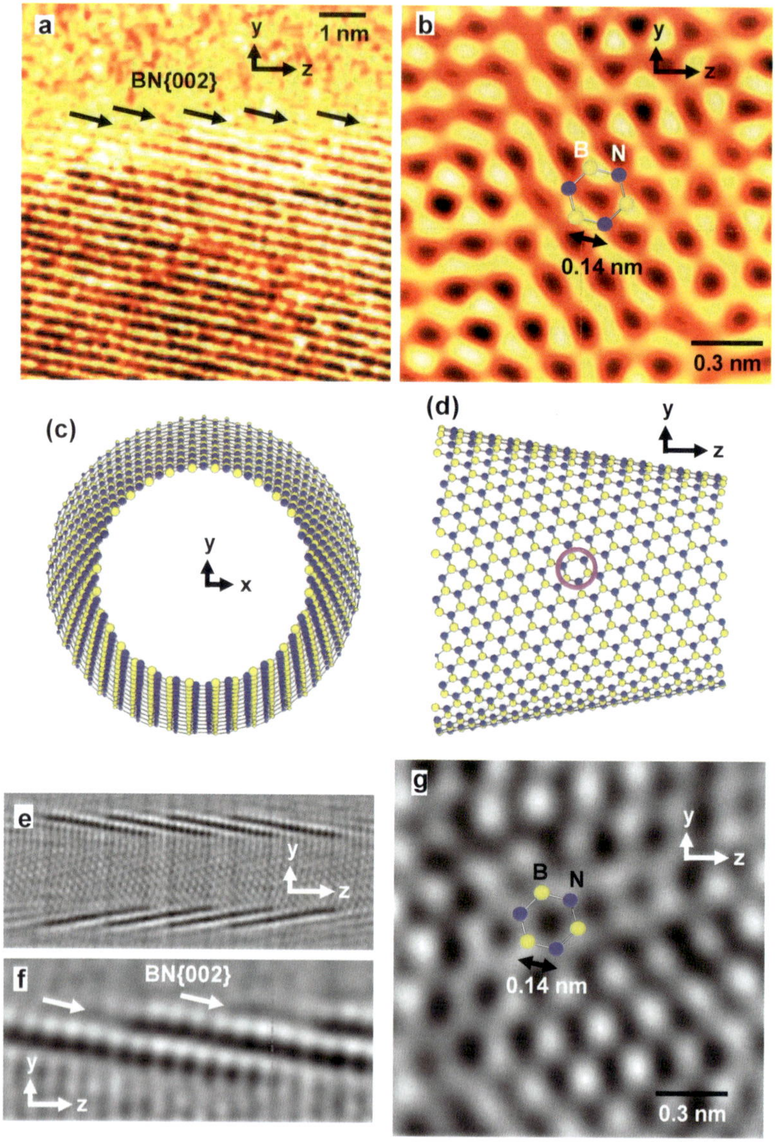

Figure 21. a) HREM image of edge of the BN nanotube wall in Fig. 20(b). (b) Processed HREM image after Fourier filtering of the nanotube center of Fig. 20(b). Proposed model of the BN cup structure projected along (c) the z-axis (nanotube axis) and (d) the x-axis. (e) Calculated HREM image of four-layered, cup-stacked BN nanotube at defocus values of −40 nm. Enlarged image of (f) edge and (g) center of BN nanotube in (e).

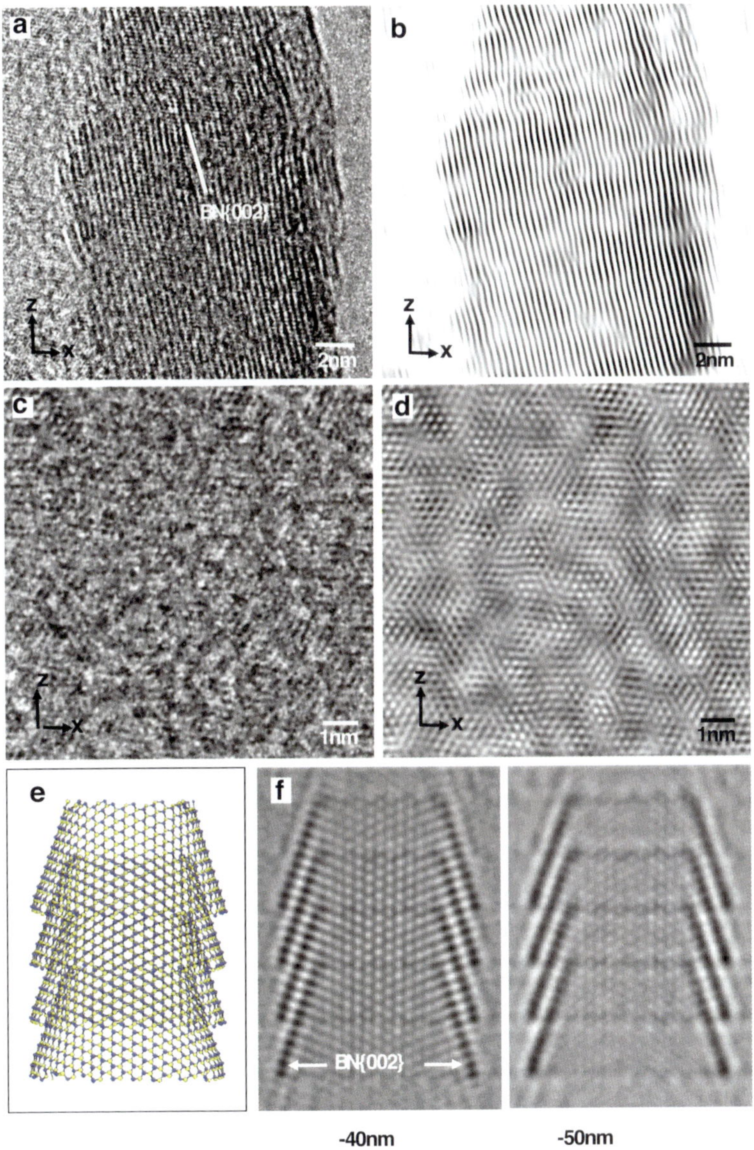

Figure 22. a) HREM image of nanotube wall of bamboo-type BN nanotube with cup-stacked structures. (b) Processed image after Fourier filtering of (a). (c) HREM image of nanotube center. (d) Processed image after Fourier filtering of (c). (e) Processed image after Fourier filtering of (c). (e) Structure model of four-fold walled $B1_{976}N_{1976}$ nanotube with a cup-stacked structure. (f) Calculated HREM images as a function of defocus values.

	$B_{273}N_{273}$	$B_{390}N_{390}$	$B_{273}N_{273}$ @$B_{390}N_{390}$	$B_{264}N_{264}$	$B_{384}N_{384}$	$B_{264}N_{264}$ @$B_{384}N_{384}$
Structure type	Zigzag	Zigzag	Zigzag	Armchair	Armchair	Armchair
Outer diameter (nm)		2.3	2.3		2.2	2.2
Inner diameter (nm)	1.6		1.6	1.5		1.5
Number of layers	1	1	2	1	1	2
Total energy (kcal/mol)	459.2	701.5	556.0	466.6	693.2	779.3
Total energy (kcal/ mol·atom)	0.841	0.899	0.419	0.883	0.902	0.601

Table 2. Calculated values for various BN nanotubes

	$B_{560}N_{560}$	$B_{1120}N_{1120}$	$B_{2240}N_{2240}$	$B_{494}N_{494}$	$B_{988}N_{988}$	$B_{1976}N_{1976}$
Corn angle (°)	20	20	20	36	36	36
Outer diameter (nm)	3.4	3.4	3.4	4.2	4.2	4.2
Inner diameter (nm)	2.4	2.4	2.4	2.4	2.4	2.4
Number of layers	1	2	4	1	2	4
Total energy (kcal/mol)	31.456	-287.924	-936.415	895.1	1269	2062
Total energy (kcal/ mol·atom)	0.028	-0.129	-0.209	0.906	0.642	0.522

Table 3. Calculated values for various BN nanotubes with a cup-stacked structure.

Distance between BN layers of nanotubes with a cup-stacked structure in a HREM image was found to be ~0.35 nm, and the basic structure model was constructed based on this observation. Geometry optimizations at molecular mechanics level result in the interlayer distances of ~0.38 nm. Comparing the empirical total energies of all the considered structures, a cup-stacked structure ($B_{2240}N_{2240}$) with cone angle of 20° was found to be the lowest in energy, which indicates the high stability of this structure.

The BN nanotubes with cup-stacked structures in the present work would also be one of the candidates for atomic and gas storage, as well as carbon nanotubes. Cone angles of BN cup-stacks were measured to be ~36°, which agreed well with that of the model in Fig. 22(e) (38°). Cone angles of carbon nanotubes with a cup-stacked structure were reported to be in the range of 45–80° (Endo et al. 2003). The cause of the different cone angles of the present cup-stacked BN nanotubes would be due to the different stacking of BN layers along c-axis (B-N-B-N...) from carbon layers. The cone angles might also depend on the shape of catalysis particles, as shown in Fig. 11(b).

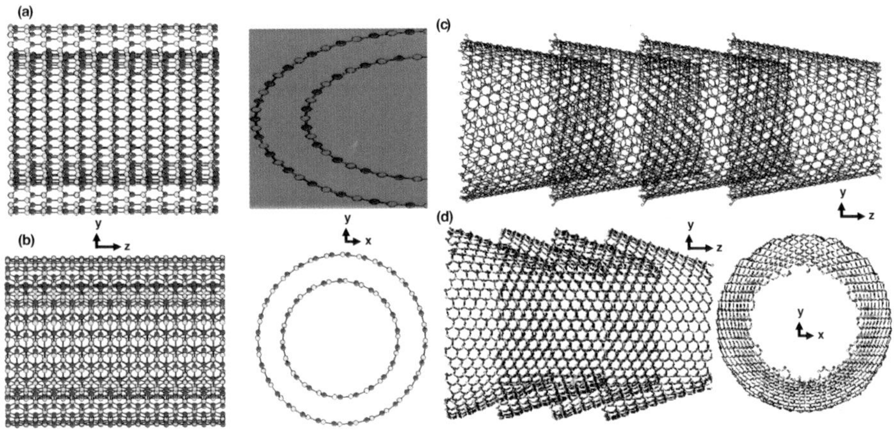

Figure 23. Atomic structure models of double-walled BN nanotubes with (a) zigzag-type and (b) armchair-type structures. Atomic structure models of four-layered, cup-stacked BN nanotubes with cone angles of (c) 20° and (d) 36°

3.3. STM observation of BN nanotube

Although the network structure of carbon nanotubes has already been observed by scanning tunneling microscopy (STM) (Wilder et al. 1998), only few works on the STM observation of the hexagonal plane of BN nanotubes have been reported because of the insulating behavior. The STM image of BN nanotubes on highly oriented, pyrolytic graphite (HOPG) is shown in Fig. 24(a) (Oku et al. 2008). Three BN nanotubes are observed in the image, and the smallest one is selected for enlarged observation and electronic measurements. The nanotube axis is indicated as the z -axis. An enlarged image of the surface of the BN nanotube is shown in Fig. 24(b). The surface of the BN nanotubes is indicated by arrows. A lattice image of the BN nanotubes is observed, and an enlarged STM image of the BN nanotubes is shown in Fig. 24(c). Hexagonal arrangements of dark dots are observed, which correspond to the size of the sixmembered rings of BN. Current-voltage (I-V) measurements were also carried out for the BN nanotubes, as shown in Fig. 24(d). The I–V curve indicates an onset voltage at 5.0 V, which agreed with optical measurement of Fig. 20(d), and is almost comparable to the energy gap of BN nanomaterials. Comparable data were also reported for other STM measurements (Ishigami et al. 2005, Wang et al. 2005).

4. Metal nanowires encapsulated in BN nanotubes

Several studies have been reported on metal-filled BN nanomaterials. Nanowires constructed from magnetic materials, especially Fe, Co and some Fe-based alloys are of interest, because they are likely to be used in nanoelectronics devices, magnetic recording media and biological sensors. However, the oxidation- and corrosionresistances of surface are weak point of the metallic nanowires. BN nanocables are of potential use for nanoscale electronic devices and nanostructured ceramic materials because of providing good stability at high temperatures with high electronic insulation in air. Therefore, metal-filled BN nanomaterials would have significant advantages for technological application. Although it is reported that Fe-filled BN nanotube could be achieved (Golberg et al. 2003), they still have some problems such as little production and low yield because it is difficult to exist in directly fabricating BN nanocable with metal cores to the poor wetting property of BN to metal.

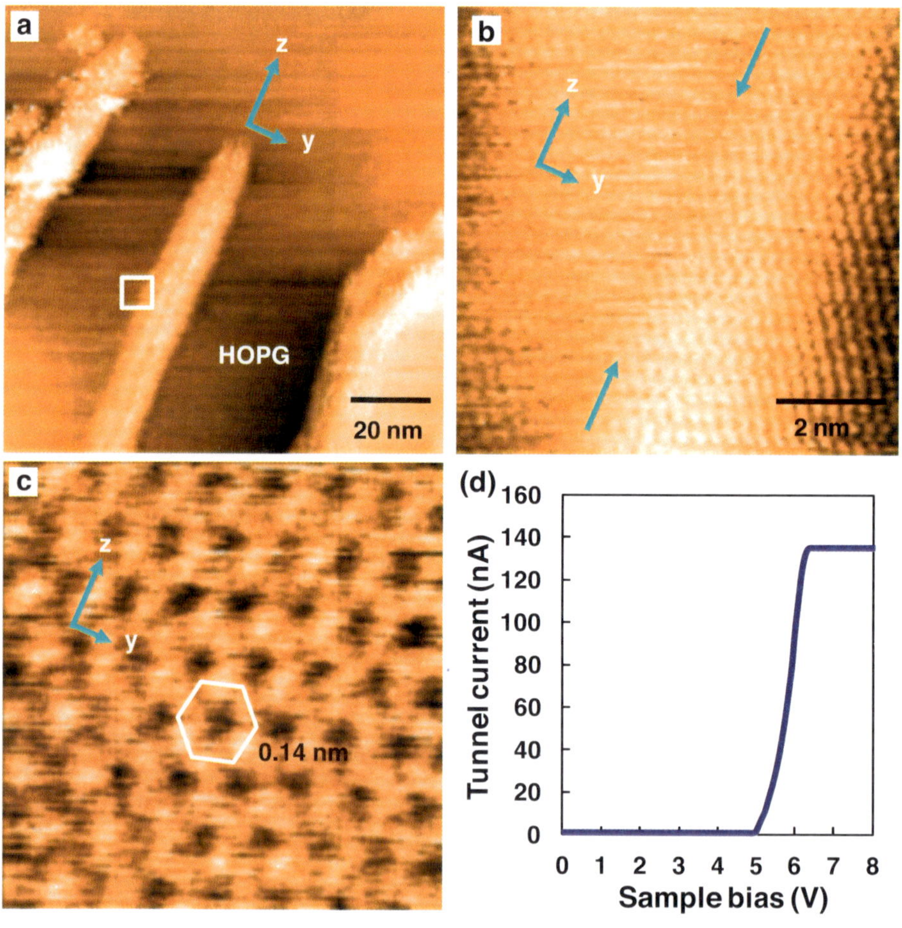

Figure 24. a) STM image of BN nanotubes on HOPG. (b) Enlarged image of the surface of the BN nanotube indicated by a square in (a). (c) Enlarged STM image of the BN nanotube. (d) I - V characteristic of the single BN nanotube.

The purpose of the present work is to synthesis metal-filled BN nanotube and various BN nanomaterials and to investigate the morphology of Fe-filled BN nanotube by HREM, high-angle annular dark-field scanning transmission electron microscopy (HAADF-STEM), electron diffraction and energy dispersive X-ray spectroscopy (EDX). It is possible to use HAADF-STEM to detect single heavy atoms on alight support. Scattering is caused by the nucleus and follows roughly a Z^2 dependence. Fe-filled BN nanotubes could be observed by performing centrifugation. It is considered that centrifugation is effective in collecting Fe-filled BN nanotube because density of Fe is higher than that of BN nanomaterials. Formation mechanism of Fe-filled BN nanotube was proposed based on these results.

Fe_4N (99%, Kojundo Chemical Laboratory (KCL) Co. Ltd., Saitama, Japan) and boron (B) powders (99%, KCL) were used as raw materials. Their particle sizes were about 50 and 45 mm, respectively. After the Fe4N and B (weight ratio WR = 1:1) were mixed by a triturator, the samples were set on an alumina boat and annealed in the furnace. The furnace was programmed to heat at 6 °C/min from ambient to 1000 °C and hold for 1–5 h and then cooled at 3 °C /min to ambient temperature. Nitrogen pressure was 0.10 MPa, and its gas flow was 100 sccm. As-produced soot synthesized via the above method was centrifuged at 8000 rpm for 2 min, and supernatant liquid is removed. The remaining sediments were collected and observed.

Figure 25(a) and 25(b) are TEM and HAADF-STEM images of Fe-filled BN nanotubes (WR = 9:1), which were remaining sediment after centrifugation. The contrast in the TEM image is weak and direct observation of Fe-filled BN nanotubes is difficult. The same area imaged by HAADFSTEM shows excellent contrast and the morphology of Fe-filled BN nanotubes can be observed in detail. A great number of Fe-filled BN nanotubes were observed by HAADF-STEM. High WR of Fe_4N would be necessary for synthesis of Fe nanowires. TEM image of one of Fe-filled BN nanotubes is shown in Fig. 25(c). Figure 25(d) is an EDX spectrum of the Fe-filled BN nanotube. In Fig. 25(d), two peaks of boron, nitrogen are observed. This shows the atomic ratio of B:N = 46.5:53.5, which indicates formation of BN. A strong peak of Fe (0.70 keV) is also observed, while a Cu peak arises from the HREM grid. Figure 25(e) is an enlarged image of Fig. 3(c). Fig. 3(f) is an electron diffraction pattern of the Fe-filled BN nanotube. Strong peaks of BN nanotubes correspond to the planes of (002) of BN. Strong peaks are also indexed as metallic Fe with a bcc structure, and the incident beam is parallel to the [111] zone axis of α-Fe.

Figure 26(a) is an enlarged HREM image of Fig. 25(e), and Fig. 25(b) is filtered Fourier transform of Fig. 26(a) (Oku et al. 2007). Figure 26(c) is inverse Fourier transform of Fig. 26(b), and Fig. 26(d) is an enlarged image of Fig. 26(c). Figure 26(d) shows a lattice image of the bcc Fe-filled BN nanotube. The nanotube axis is parallel to the [110] direction of Fe, which indicates the bcc Fe is epitaxially grown to the [110] zone axis. The tubular layers around the nanowire have an average interlayer spacing of 0.34 nm, which corresponds to the (002) spacing of BN. Figure 26(e) is inverse Fourier transform of Fig. 26(b) using 000, Fe ($01\bar{1}$ and $0\bar{1}1$) reflections, and Fig. 26(f) is an enlarged image of Fig. 26(e). Several edge-on dislocations are observed as indicated by arrows, which would be due to lattice distortion produced during Fe-filled nanotube growth. This lattice distortion is also observed as expansion in the electron diffraction pattern of Fig. 3(f), as indicated by arrows. These unique structures would be suitable materials for nanoelectronics devices, magnetic recording media and biological sensors with excellent protection against oxidation and wear.

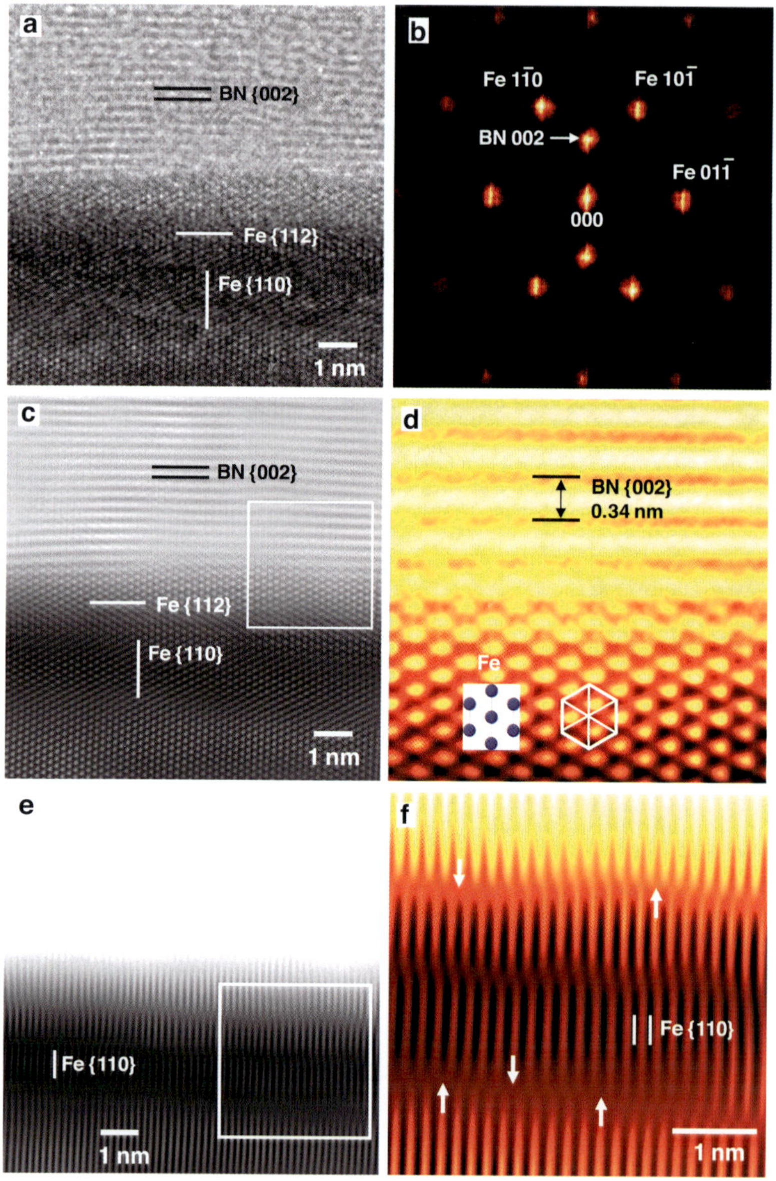

Figure 25. a) TEM and (b) HAADF images of Fe-filled BN nanotubes. (c) TEM image of Fe-filled BN nanotube. (d) EDX spectrum of Fe-filled BN nanotube. (e) Enlarged image of (c). (f) Electron-diffraction pattern obtained from (e).

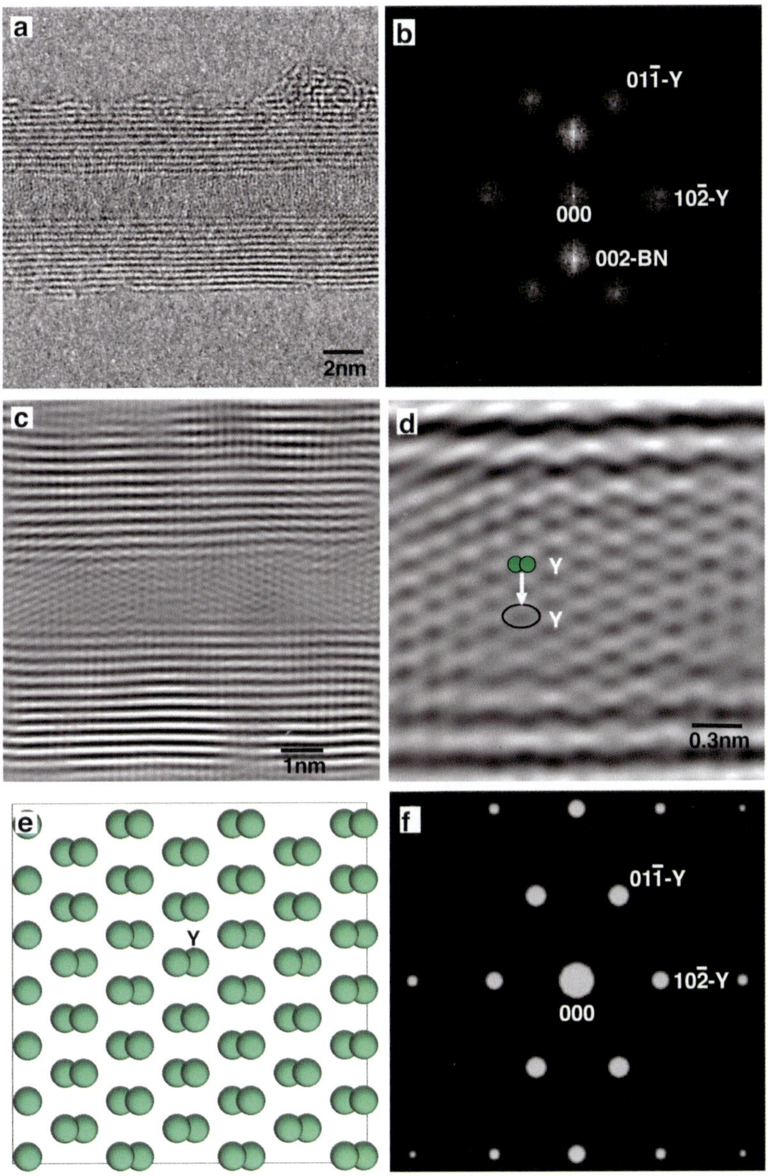

Figure 26. a) HREM image of Fe-filled BN nanotube. (b) Filtered Fourier transform of (a). (c) Inverse Fourier transform of (b). (d) Enlarge image of square in (c). (e) Inverse Fourier transform of (b) using 000, Fe 01$\bar{1}$ and Fe 0$\bar{1}$1 reflections. (f) Enlarged image of square in (e).

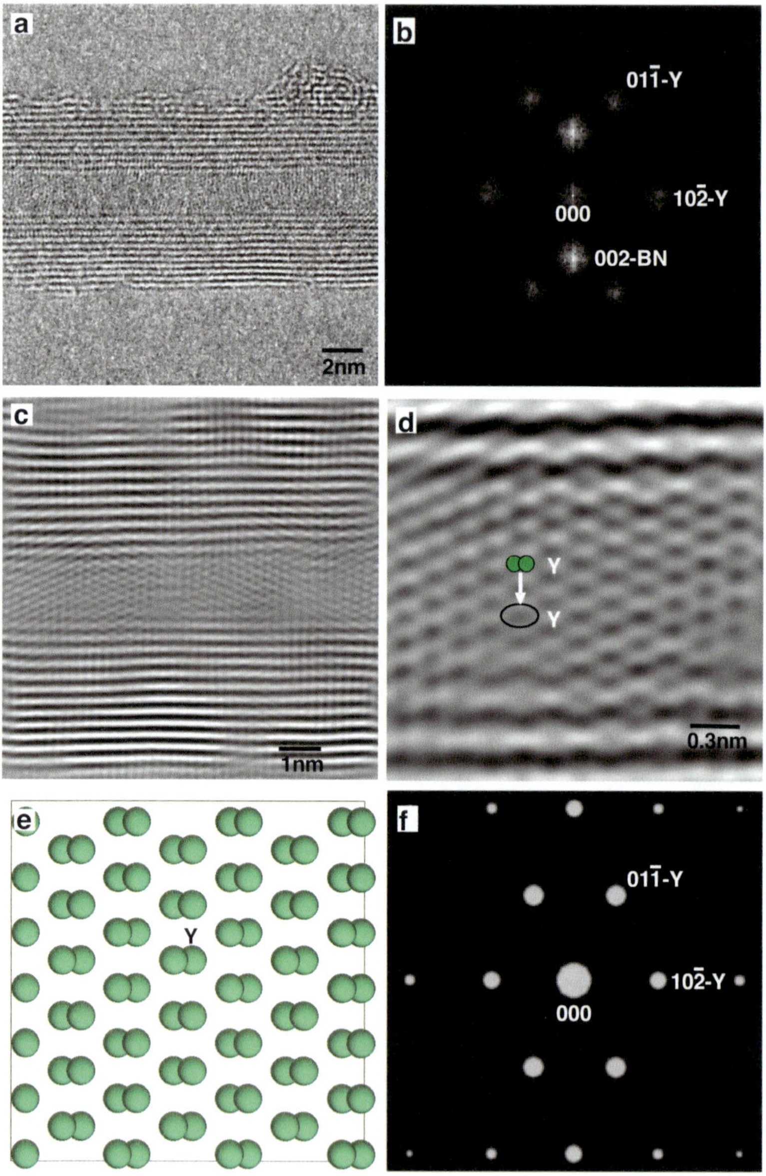

Figure 27. a) HREM image of BN nanotube synthesized from YB6 powder. (b) Filtered Fourier transform of (a). (c) Inverse Fourier transform of (b). (d) Enlarged image of (c). (e) Atomic structure model of yttrium along [101]. (f) Calculated diffraction pattern of (e).

A HREM image of a BN nanotube synthesized from YB_6 powder is also shown in Fig. 27(a), which was taken nearly at Scherzer defocus. Number of BN {002} layers is 12, and lattice fringes are observed in the BN nanotube. A filtered Fourier transform of Fig. 27(a) is shown in Fig. 27(b). Spots of BN 002 are observed as bright spots. In addition, reflections corresponding to the yttrium structure are observed and indexed with the incident electron beam along the [101] direction. Figure 27(c) is an inverse Fourier transform of Fig. 27(b), and BN{002} layers are clearly observed in the image. An enlarged image of Fig. 27(c) is shown in Fig. 27(d), which indicates lattice fringes at the center of the BN nanotube [91]. Lattice parameters of yttrium with a hexagonal structure, as determined by X-ray diffraction analysis, were a = 0.36474 nm and c = 0.57306 nm, which agrees well with the present lattice fringes (Oku et al. 2004). Dark contrast corresponds to yttrium atom pairs, as indicated in Fig. 27(d).

Based on the observations, an atomic structure model of yttrium along [101] was constructed as shown in Fig. 27(e), which indicates the yttrium atom pairs. Figure 27(f) is a calculated diffraction pattern of Fig. 27(e), and tense well with the observed Fourier transform of Fig. 27(b). Since YB_6 powders formed BN nanotubes in the present work, boron atoms were consumed preferentially. As a result, yttrium element would remain in the BN nanotube as a nanowire. These BN nanotubes with metal nanowires would be interesting nanomaterials for nanocables.

5. Conclusion

BN nanotubes with zigzag-, armchair-type and cup-stacked structures were synthesized and investigated by HREM, image simulation and total energy calculation. Hexagonal networks of BN nanotubes were directly observed by HREM in atomic scale, and chiralities of the BN nanotubes were directly determined from HREM images. Atomic structure models for quadruple- and double-walled nanotubes were proposed, and simulated images based on these models agreed well with experimental HREM images. Molecular mechanics calculations showed good stability of a zigzag-type structure compared to the armchair-type structure, which agreed well with the experimental data of disordered armchair-type BN nanotubes. BN nanotubes encapsulating a $B_{36}N_{36}$ cluster, and yttrium and Fe nanowires were also produced and confirmed by HREM and diffraction calculation.

Acknowledgments

The authors would like to acknowledge I. Narita, N. Koi, A. Nishiwaki, K. Suganuma, M. Inoue, K. Hiraga, M. Nishijima, R. V. Belosludov, and Y. Kawazoe for experimental help and useful advices.

Author details

Takeo Oku

The University of Shiga Prefecture, Japan

References

[1] Chopra, N., G., , Luyken, R. J., Cherrey, K., Crespi, V. H., Cohen, M. L., Louie, S. G., & Zettl, A. (1995). Boron nitride nanotubes. Science, , 269, 966-967.

[2] Endo, M., Kim, Y. A., Hayashi, T., Yanagisawa, T., Muramatsu, H., Ezaka, M., Terrones, H., Terrones, M., & Dresselhaus, M. S. (2003). Microstructural changes induced in "stacked cup" carbon nanofibers by heat treatment. Carbon, , 41, 1941-1947.

[3] Golberg, D., Bando, Y., Kurashima, K., & Sato, T. (2000). Ropes of BN multiwalled nanotubes. Solid State Communications, , 116, 1-6.

[4] Golberg, D., Xu, F. F., & Bando, Y. (2003). Filling boron nitride nanotubes with metals. Applied Physics A, , 76, 479-485.

[5] Ishigami, M., Sau, J. D., Aloni, S., Cohen, M. L., & Zettl, A. (2005). Observation of the giant stark effect in boron-nitride nanotubes. *Physical Review Letters*, 94, 056804-056801.

[6] Koi, N., Oku, T., Inoue, M., & Suganuma, K. (2008). Structures and purification of boron nitride nanotubes synthesized from boron-based powders with iron particles. *Journal of Materials Science*, 43, 2955-2961.

[7] Lauret, J. S., Arenal, R., Ducastelle, F., Loiseau, A., Cau, M., Attal-Tretout, B., Rosencher, E., & Goux-Capes, L. (2005). Optical transitions in single-wall boron nitride nanotubes. *Physical Review Letters*, 94, 037405-037401.

[8] Lim, S. H., Luo, J., Ji, W., & Lin, J. (2007). Synthesis of boron nitride nanotubes and its hydrogen uptake. Catalysis Today, , 120, 346-350.

[9] Mickelson, W., Aloni, S., Han, W. Q., Cumings, J., & Zettl, A. (2003). Packing C_{60} in boron nitride nanotubes. Science, , 300, 467-469.

[10] Narita, I., & Oku, T. (2003). Synthesis of boron nitride nanotubes by using NbB_2, YB_6 and YB_6/Ni powders. *Diamond and Related Materials*, 12, 1912-1917.

[11] Nishiwaki, A., Oku, T., Tokoro, H., & Fujii, S. (2005). Atomic structures and stability of boron nitride nanotubes with a cup-stacked structure. *Diamond and Related Materials*, 14, 1163-1168.

[12] Oku, T., & Narita, I. (2004). Atomic structures and stabilities of zigzag and armchair-type boron nitride nanotubes studied by high-resolution electron microscopy and molecular mechanics calculation. *Diamond and Related Materials*, 13, 1254-1260.

[13] Oku, T., Narita, I., Nishiwaki, A., & Koi, N. (2004). Atomic structures, electronic states and hydrogen storage of boron nitride nanocage clusters, nanotubes and nano-horns, Defects and Diffusion Forum, , 226-228, 113-141.

[14] Oku, T., Koi, N., Narita, I., Suganuma, K., & Nishijima, M. (2007). Formation and atomic structures of boron nitride nanotubes with cupstacked and Fe nanowire en-capsulated structures. Materials Transactions, , 48, 722-729.

[15] Oku, T., Koi, N., & Suganuma, K. (2008). Electronic and optical properties of boron nitride nanotubes. Journal of Physics and Chemistry of Solids, , 69, 1228-1231.

[16] Oku, T., Narita, I., Koi, N., Nishiwaki, A., Suganuma, K., Inoue, M., Hiraga, K., Mat-suda, T., Hirabayashi, M., Tokoro, H., Fujii, S., Gonda, M., Nishijima, M., Hirai, T., Belosludov, R. V., & Kawazoe, Y. (2009). Boron nitride nanocage clusters, nanotubes, nanohorns, nanoparticles, and nanocapsules, In: B-C-N nanotubes and related nano-structures. Y. K. Yap (Ed.), Springer., 149-194.

[17] Oku, T. (2011). High-resolution electron microscopy of nanostructured materials. Nanoscience & Nanotechnology-Asia, , 1, 59-75.

[18] Saito, Y., Okuda, M., Tomita, M., & Hayashi, T. (1995). Extrusion of single-wall car-bon nanotubes via formation of small particles condensed near an arc evaporation source. Chemcal Physics Letters, , 236, 419-426.

[19] Radosavljevi, M., Appenzeller, J., Derycke, V. R., Ph, Avouris. M., Loiseau, A., Co-chon, J., , L., & Pigache, D. (2003). Electrical properties and transport in boron nitride nanotubes. Applied Physics Letters, , 82, 4131-4133.

[20] Rubio, A., Corkill, J. L., & Cohen, M. L. (1994). Theory of graphitic boron nitride nanotubes. Physical Review B, , 49, 5081-5084.

[21] Tang, C. C., Bando, Y., & Sato, T. (2002). Synthesis and morphology of boron nitride nanotubes and nanohorns. Applied Physics A, , 75, 681-685.

[22] Wang, J., Kayastha, V. K., Yap, Y. K., Fan, Z., Lu, J. G., Pan, Z., Ivanov, I. N., Pure-tzky, A. A., & Geohegan, D. B. (2005). Low temperature growth of boron nitride nanotubes on substrates. *Nano letters*, 5, 2528-2532.

[23] Watanabe, K., Taniguchi, T., & Kanda, H. (2004). Direct-bandgap properties and evi-dence for ultraviolet lasing of hexagonal boron nitride single crystal. *Nature materials*, 3, 404-409.

[24] Wilder, J. W. G., Venema, L. C., Rinzler, A. G., Smalley, R. E., & Dekker, C. (1998). Electronic structure of atomically resolved carbon nanotubes. Nature, , 391, 59-62.

Chapter 5

Recent Progress of Plasma CVD for Structure Controlled Growth of Single-Walled Carbon Nanotubes

Toshiaki Kato and Rikizo Hatakeyama

Additional information is available at the end of the chapter

1. Introduction

One-dimensional single-walled carbon nanotubes (SWNTs) are potential materials for future nanoelectronics. Since the electronic and optical properties of SWNTs strongly depend on their structure, such as diameter and chirality, the selective synthesis of SWNTs with desired structures is a major challenge in nanotube science and applications. SWNT growth was first achieved by arc discharge in 1993. Several growth techniques have been developed since then, including laser ablation and chemical vapor deposition (CVD). Since it is possible to grow SWNTs at a specific position on a substrate by patterning a catalyst, CVD has attracted much attention in nanoelectronics applications. In general, CVD can be divided into two types: thermal CVD [1-4] and plasma CVD [5-7]. Due to the strong electric fields in plasma sheaths, nanotubes grown by plasma CVD tend to have an individually- and vertically-freestanding shape [5,8-10]. Thermal CVD decomposes carbon source gases using thermal energy. In contrast, in plasma CVD, the source gas decomposition is effectively carried out by electron impact with no additional thermal energy; hence, the growth temperature is significantly lower compared to that of thermal CVD. Despite these benefits of plasma CVD, it is difficult to control the structure of SWNTs by plasma CVD because there are many unknown factors in plasma, such as ion density, ion energy, radical species, radical densities, and sheath electric field, which restrict the potential application of plasma CVD in nanotube science. Based on our studies, SWNT growth by plasma CVD has been significantly improved in recent years. In this chapter, we give a brief overview of recent progress in SWNT growth by plasma CVD.

2. Freestanding single-walled carbon nanotube growth

The potential of plasma CVD for nanotube growth was first demonstrated by Ren et al. in 1998 [5]. Vertically- and individually-aligned multi-walled carbon nanotubes (MWNTs) are grown by plasma CVD. Since carbon nanotubes (CNTs) grown by thermal CVD are known to form a spaghetti-like entangled shape, the well-aligned growth of CNTs by plasma CVD makes it an attractive CNT-growth method that may solve the integration issue in CNT-based nanoelectronics. However, plasma CVD is limited to the production of MWNTs; SWNTs, which have superior electrical and optical characteristics compared with MWNTs, have not been successfully produced by plasma CVD. The growth of SWNTs by plasma CVD was first reported by our group in 2003 [11, 12]. SWNTs are grown by plasma CVD using a zeolite as a catalyst support. Zeolites are nanoporous materials known to maintain small catalyst particle sizes on their rough surfaces, even under high-temperature conditions. Thus, certain plasma effects might enhance catalyst particle aggregation during plasma CVD, which could be the main reason why SWNTs could not be grown by plasma CVD. It is thought that catalyst particle aggregation is enhanced due to high-energy ions attacking the catalyst. In general, ions in plasma are accelerated through the potential drop between space potentials in the plasma and substrate biases. The minimum value of this potential drop is determined by the electron temperature in the plasma. Thus, low electron-temperature plasma can significantly decrease the energy of ions arriving at the substrate. Since the diffusion region in plasma is known to have very low electron temperatures, we used the diffusion plasma to decrease the energy of ions attacking the catalyst to below a few eV. SWNT growth under the diffusion plasma region occurs on a flat substrate without using catalyst support materials [13, 14]. Thus, the critical element promoting catalyst aggregation is high-energy ion bombardment. Interestingly, SWNTs grown by diffusion-plasma CVD have the well-aligned freestanding form, i.e., all SWNTs are individually- and vertically-standing on the flat substrate. Figures 1a–d show a typical scanning electron microscope (SEM) image (Figure 1a), low-magnification (Figure 1b) and high-magnification (Figure 1c) transmission electron microscope (TEM) images, and Raman scattering spectra (Figure 1d) of freestanding SWNTs. Relatively high-quality SWNTs were grown with the individually freestanding form, and this alignment can be obtained by the plasma-sheath electric field. Based on numerical calculation, the rotation energy of the dipole moment in SWNTs is much higher than the thermal energy, which disturbs the tube alignment [14]. This indicates that individual SWNTs can be aligned along the electric field. Owing to their unique as-grown state, it is possible to directly detect photoluminescence (PL) spectra from the as-grown freestanding SWNTs on the substrate (Figure 1e) [15]. This is a remarkable advantage for optoelectrical applications and fundamental studies toward chirality control, which will be discussed later.

3. Growth kinetics of SWNTs in plasma CVD

In this section, we report a crucial finding of remarkable etching reaction of SWNTs during the plasma CVD, and key parameters for such etching reaction are also revealed with a numerical analysis of the experimentally established SWNT-growth equation. A reactive ion etching model is also developed to explain the etching reaction of SWNTs in plasmas[16].

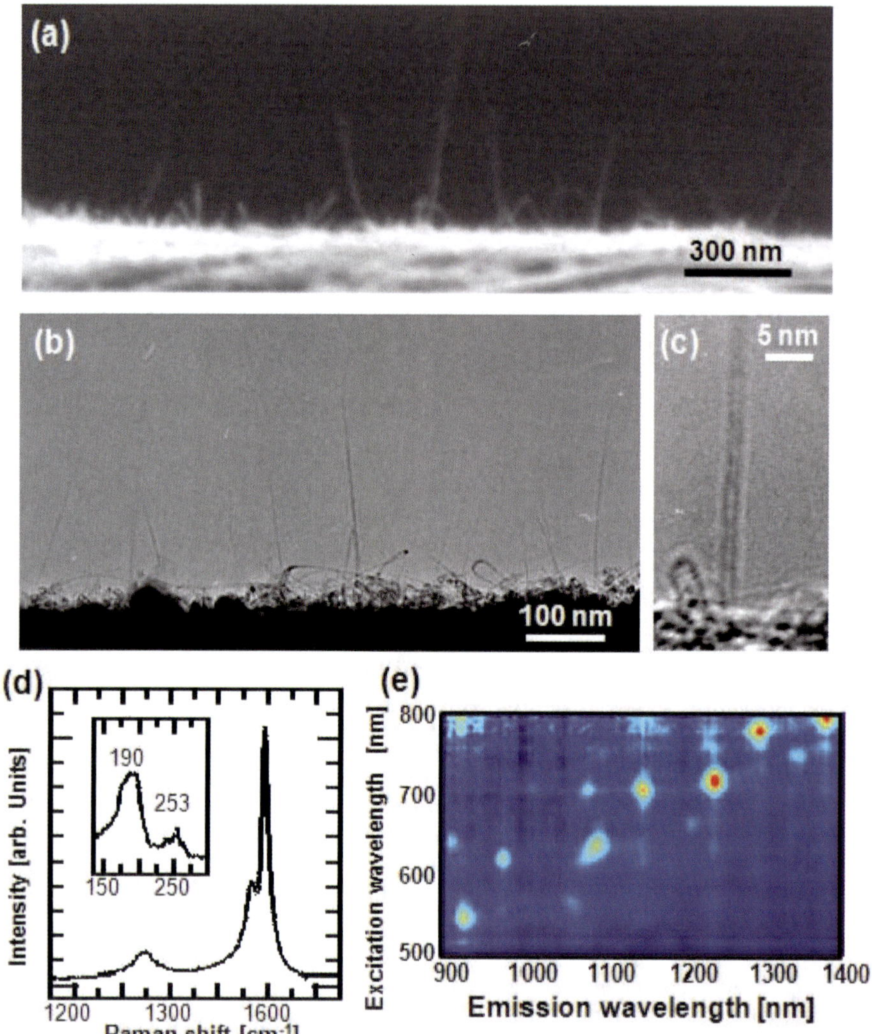

Figure 1. (a) SEM and (b) (c) TEM images of freestanding individual SWNTs. (d) Raman scattering spectrum of free-standing individual SWNTs. Inset of (d) is emphasis of the RBM region. (e) PLE map obtained from as-grown freestanding SWNTs without any dispersion process.

Figures 2a and b presents typical Raman scattering spectra of SWNTs as a function of growth time (t_g). Raman scattering spectroscopy has been known as one of the powerful tools to characterize the SWNTs structure such as diameter, chirality, quality, and so on. In addition to these structural information, the absolute value of G-band intensity (I_G) at 1593

cm^{-1} originating from a graphite nature in the SWNTs is sometimes utilized to discuss the amount of SWNTs. As information of the amount of SWNTs, therefore, we utilize the absolute value of I_G measured under the almost same experimental conditions; laser power: ~ 0.2 mW/μm², laser wavelength: 488 nm, laser spot size: 4 μm², accumulation time: 60 sec. When radio-frequency power (P_{RF}) is 40 W, I_G gradually increases with an increase in t_g (Figure 2a). On the other hand, I_G suddenly decreases when the growth time is longer than 50 sec under the 100 W P_{RF} condition (Figure 2b). These results indicate that the growth kinetics of SWNTs is strongly influenced by the plasma conditions, and several specific factors in plasmas cause the strong etching of SWNTs as shown in Figure 2b.

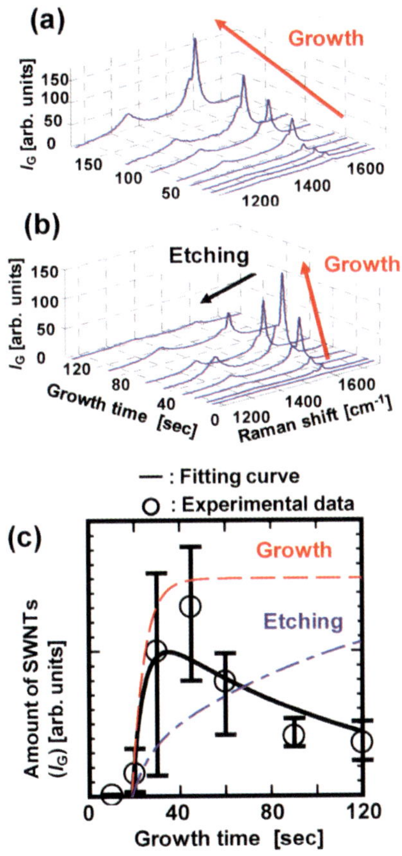

Figure 2. Raman spectra of SWNTs as a function of t_g. (a) P_{RF} = 40 W and (b) P_{RF} = 100 W, respectively. (c) The comparison between the experimental data and fitting curve of Eq. 1.

In the case of the thermal CVD, it has been reported that the growth kinetics of SWNTs can be expressed with a following equation (normal equation) [17]. $I_G = I_0[1-\exp\{-(t_g-\Delta t)/\tau_{gro}\}]$, where I_0, Δt, and τ_{gro} denote saturated I_G, incubation time, and relaxation time of the growth, respectively. Our experimental results of damage free growth (Figure 2a) well match with this equation, which denotes our estimation of SWNTs amount with I_G is reliable. However, there is obviously no formula which can describe the growth kinetics including the etching effect as described in Figure 2b. On purpose to express this phenomenon, therefore, we assume that the growth mode can be described with a following balance equation: $I_G = G(t_g) - E(t_g)$, where $G(t_g)$ and $E(t_g)$ are growth and etching functions, respectively. When the etching effect is weak and negligible, the growth kinetic has to be described with the above-mentioned normal equation, which means that $G(t_g)$ is the same as the normal equation. One of the most important points in this study is how to describe $E(t_g)$. Since carbon atoms are etched out only when atoms or molecules attach themselves to the carbon atoms in SWNTs, the probability of the etching reaction can be simplified in terms of an adsorption reaction. The Langmuir's adsorption isotherm is known as one of the most basic ones, and to be expressed by the following form: $d\theta / dt = \alpha P(1-\theta)$, where θ, t, α, and P indicate the percentage of covered area, reaction time, adsorption efficiency, and pressure of adsorbate, respectively. Actually, the Langmuir's equation has been utilized in the wide range of fundamental scientific studies to understand chemical adsorption reactions. In our study, θ corresponds to the etched area against the area of the graphite sheet of SWNTs, *i.e.* $\theta = E(t_g)/G(t_g)$. Since the solution of the Langmuir's equation is $\theta(t_g) = 1-\exp(-t_g/\tau_{etc})$, where $\tau_{etc} = 1/\alpha P$ is the relaxation time of the etching reaction, $E(t_g)$ results in the following equation: $E(t_g) = G(t_g)\{1-\exp(-t_g/\tau_{etc})\}$. According to the above mentioned equations, an advanced growth equation can be established as

$$I_G = I_0\left[1-\exp\left\{\frac{-(t_g-\Delta t)}{\tau_{gro}}\right\}\right]\left\{\exp\left(\frac{-t_g}{\tau_{etc}}\right)\right\} \tag{1}$$

Figure 2c shows a comparison between the experimental result of Figure 2b and fitting curve with Eq. (1). The fitting curve gives good agreement with the experimental result, indicating that the advanced equation established enables us to discuss a more detailed correlation between plasma parameters such as ion energy U_i and species density and growth parameters such as and etching efficiency ($k = 1/(\tau_{etc}-\tau_{gro})$).

Based on the advanced growth Eq. (1), we attempt to understand effects of U_i coming to the substrate during the SWNT growth. Figure 3a gives a counter plot of I_G as functions of t_g and U_i. Since U_i is determined by a potential drop between the plasma and substrate, the substrate bias voltage is changed to adjust U_i. When U_i is fairly low (~ 1 eV), I_G gradually increases and saturates with an increase in t_g. The similar tendency can also be found under the condition of U_i = 30 eV. When U_i = 10 eV and over 50 eV, on the other hand, a sudden decrease of I_G can be found after the specific t_g (t_g >150 ~ 200 sec). This decrease of I_G indicates that the etching reaction arises at these specific energy window of ions as similar to the result in Figure 2b. The I_G/I_D plot also supports the evidence of etching reaction in such a specific energy range, where I_D is D-band intensity around 1350 cm^{-1} in Raman spectroscopy. Only under the U_i condition of 10 eV and > 50 eV, the SWNTs quality remarkably decreases independently of t_g (Figure 3b). These lead us to conclude that several significant damages are likely caused via energetic ions ranging around 10 eV and over 50 eV.

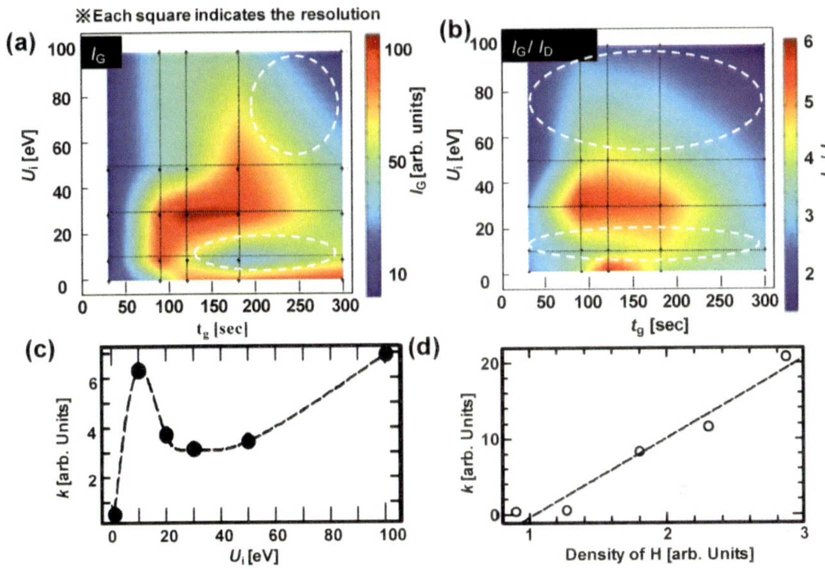

Figure 3. (a), (b) Contour plot of I_G (a) and I_G/I_D (b) as functions of U_i and t_g. Each dot indicates the condition, where experiments have been done. For visuall help, all of data between each dot are computationally compensated. (c) Etching efficiency k as a function of U_i. (d) Etching efficiency k as a function of relative density of H.

A further quantitative and practical analysis is also performed upon the fitting of experimental results with the advanced growth equation (Eq. 1). Figure 3c shows a plot of estimated k under the different U_i condition. As similar to the result in Figures 3a and b, the clear increment of k is recognized under the specific energy condition of U_i = 10 eV and > 50 eV.

In addition to the ions, there are many kinds of factors causing significant impacts on the structure of SWNTs in hydrocarbon plasmas. Especially, the density of radicals is much higher than that of ions in reactive plasmas. Thus we attempt to reveal a correlation of k with several radical densities. The relative densities of radicals are measured by an Actinomery method with optical emission spectroscopy (OES). When k is plotted as a function of H density, it is surprising that a clear liner correlation is found as displayed in Figure 3d. Although the dependence of H on etching reactions has already been mentioned by several groups without any direct measurement of H density, our time-evolution results combined with the systematical H-density measurement afford more direct evidence for the primal factor determining k during the growth of SWNTs.

Based on the above mentioned experimental results, a consistent reactive ion-etching model can be established as follows. In our study, the etching efficiency k is expressed by the following equation

$$k = 1/(\tau_{etc} - \tau_{gro}) \approx 1/\tau_{etc} = \alpha P \tag{2}$$

The linear dependence of k on H density shown in Figure 3d indicates that the density of H corresponds to in Eq. 2. Since the density of H is kept constant during the experiment of ion energy effects, on the other hand, the supplied amount of etching elements (P) must be constant. Hence, the variation of k depending on U_i (Figure 3c) originates from the difference of α. Under the specific U_i condition, the carbon-carbon bond is considered to be broken, and the efficiency of adsorption between the carbon and etching element (H) resultantly comes to increase. After all, the direct factor causing the etching of SWNTs is the H, and the specific energy of ions enhances those etching reactions by changing the adsorption efficiency between the carbon and hydrogen atoms.

4. Structure-controlled growth of SWNTs

The structure of SWNTs, including diameter and chirality, strongly influences their electrical and optical properties; therefore, it is important to precisely control the structure of SWNTs. Here, we discuss recent progress in the structure-controlled growth of SWNTs by plasma CVD.

4.1. Diameter control

The band gap is known to be inversely proportional to tube diameter; thus, controlling the tube diameter is very important for electrical and optical applications. Here, we present our experimental results for diameter tuning of SWNTs based on gas-phase control in plasma CVD [18].

Figures 4a–d show photoluminescence-excitation (PLE) maps of as-grown SWNTs produced at different gas pressures. Note that all PLE measurements were carried out immediately after the growth process to prevent the freestanding SWNTs from forming bundles, which cause significant PL changes [15]. Peaks in the PLE map at high growth pressures (Figure 4a) appeared in the range of long excitation and emission wavelengths. The peak positions shifted to the region of short excitation and emission wavelengths with decreasing growth pressure (Figures 4b–d). Since each peak corresponds to a different chirality in the sample, and smaller-diameter SWNTs appeared in the shorter wavelength region, the peak-position shifts in the PLE map indicate that the diameter distribution of SWNTs is strongly influenced by growth pressure. Thus, lower pressure results in smaller SWNT diameters. This diameter dependence on the growth pressure is also reflected in Raman scattering spectra of SWNTs grown at different growth pressures. Figure 4e shows that peak positions of the radial breathing mode (RBM) clearly shifted from higher to lower wavenumbers with increasing growth pressure. The RBM peak position and the SWNT diameter are known to have a close correlation, $\omega = 248/d$ [19], where ω and d are the RBM peak position (cm^{-1}) and diameter (nm), respectively. This result is fairly consistent with the PLE result shown in Figures 4a–d. The typical pressure range where SWNTs can be grown is from 30 Pa to 650 Pa and depends on the P_{RF} used for the plasma generation. Although the absolute intensity of the G-band in Raman scattering spectra decreased in the low- or high-pressure range, the G-band to D-band ratio was almost constant. This indicates that the quality of SWNTs should be independent of the pressure range, whereas the density of SWNTs depends on the pressure. When we increased the input P_{RF}, it was possible to grow SWNTs, even below 30 Pa, indicating that a lack of hydrocarbon supply is significant under low-pressure conditions. Hence, an additional input P_{RF} is required to increase the density of active species used for the growth of SWNTs.

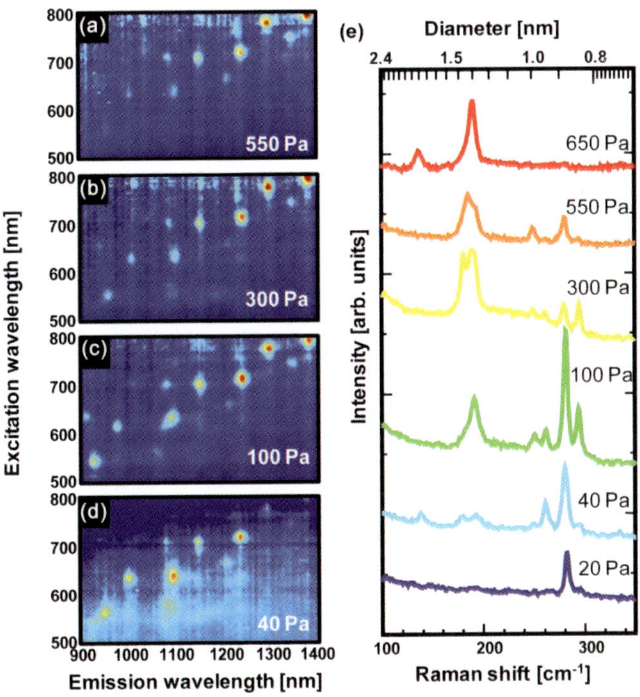

Figure 4. (a-d) PLE maps of as-grown freestanding SWNTs grown under different pressures, (a) 550 Pa, (b) 300 Pa, (c) 100 Pa, and (d) 40 Pa. (e) Growth pressure dependency of RBM in Raman scattering spectra of as-grown freestanding SWNTs.

Since the pressure during the heating and growth were the same in our growth process, the process pressure affected both the heating and the growth process. Based on this systematic investigation, we believe that the catalyst particle size increased due to aggregation after high-pressure annealing, which resulted in the growth of large-diameter SWNTs. The density of reactive hydrocarbon radicals and ions should increase under higher growth pressure conditions. Under high carbon supply conditions, a small catalyst can be deactivated by an oversupply of hydrocarbons, causing the population of small-diameter SWNTs to decrease. Therefore, the heating pressure is an important parameter that controls the catalyst particle size distribution, which directly influences the diameter of SWNTs. The pressure during plasma CVD is also important for a narrow SWNT diameter distribution [18].

4.2. Selective growth of semiconducting SWNTs

Field effect transistors (FETs) are one of the most promising applications of SWNTs. Although the high mobility and flexibility of SWNTs films can provide lots of opportunities to be utilized in various kinds of industrial applications, the low on/off current ratio in SWNT FETs caused by the mixture of metallic and semiconducting SWNTs restricts the practical use of SWNTs in FET applications. Recent progress in chemical separation enables us to fabricate good devices with on/off ratio: ~10^4 and effective gate mobility: ~ 52 cm^2/Vs. However, impurities and defects are sometimes introduced in chemically treated nanotubes during the separation process, which significantly decreases the device performance. Since as-grown SWNTs maintain the original high-quality with low impurity concentration, the selective growth of semiconducting SWNTs is desirable. Dai *et al.* reported the preferential growth of semiconducting SWNTs by plasma CVD. Although several similar reports with plasma CVD and thermal CVD have also been reported, the elucidation of this selective growth is still an open question and further investigations are needed. Here we discuss our recent findings that show a clear correlation between the performance of semiconducting devices fabricated by plasma CVD and their mean diameter [20], which might lead to a possible explanation for the preferential growth of semiconducting SWNTs by plasma CVD.

Figure 5a shows typical Raman scattering spectra of SWNTs grown under different growth temperatures. The high graphite (G)-peak to defect (D)-peak ratio indicates that the quality of SWNTs is comparable to other conventional CVD grown SWNTs. The RBM in a lower wave number region in Raman spectra exhibits the clear down shift with an increase in the growth temperature. The mean diameter of SWNTs is found to increase with growth temperature. This seems to be due to the catalyst particle size effect. Higher growth temperatures cause particle aggregation and result in the increase of the particle size, which can produce larger diameter SWNTs. A clear dependence is obtained from the plot of on current (I_{on}) vs. on/off ratio (I_{on}/I_{off}) as a function of SWNTs growth temperature. The on/off ratio of each device clearly decreases with a decrease in the growth temperature (Figure 5b). The concentration of the working devices, which have on/off ratios greater than 5, is counted and plotted as a function of the growth temperature (Figure 5c). Noticeably, the working device concentration is only 2.5 % in the case of 600 ºC (smaller diameter SWNTs), whereas more than 90 % of the devices work in the case of 800 ºC (larger diameter SWNTs). The density of SWNTs grown under the different growth temperatures is almost the same.

In order to explain the dependence of the working device concentration on the SWNT diameter, devices were irradiated by an Ar plasma, and a defect formation rate is estimated from the current change before and after the plasma treatment. In the case of small diameter SWNTs devices, the on/off ratio does not change, and on and off currents significantly decrease after the Ar plasma irradiation, whereas the on/off ratio increases with an increase in the Ar plasma irradiation time and the off current depression is significant compared to that of the on current in the case of large diameter SWNTs devices. Based on these results, the following model can be developed to explain the dependence of the working device concentration on the diameter. Due to the curvature effect, small diameter SWNTs are more unstable than large diameter ones. Hence, both metallic and semiconducting SWNTs are easily deformed by the Ar plasma irradiation without any difference in the tube metallicity. On the other hand, in the weak curvature range, the dependence of the defect formation rate on a unique metallicity appears, which might correlate with the reactivity, binding energy between carbon and carbon, and healing process. This model is consistent with the selective etching of metallic SWNTs by gas phase reaction, which was previously reported. Further detailed studies relating to the selective damage of metallic SWNTs might provide the possible answer for the preferential growth of semiconducting SWNTs by plasma CVD.

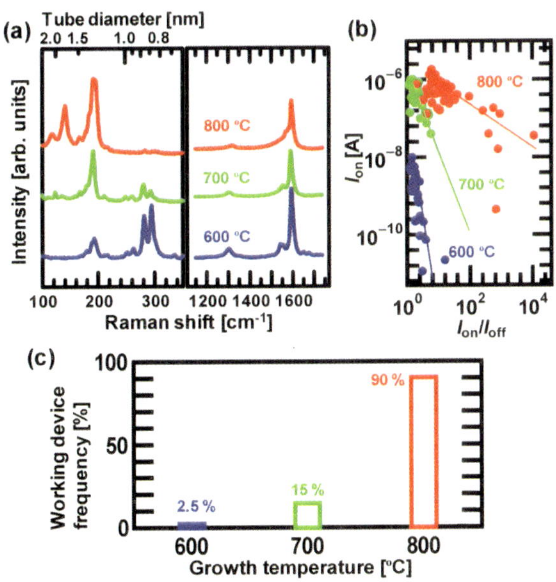

Figure 5. (a) Raman scattering spectra of SWNTs grown at different growth temperatures (laser excitation energy: 1.96 eV). (b) I_{on}-I_{on}/I_{off} plot of TFT devices with SWNTs grown at different growth temperatures. (c) Histogram of working device concentration of TFT devices with SWNTs grown at different growth temperatures.

4.3. Narrow-chirality distributed growth of SWNTs

The chirality of SWNTs directly determines their electronic and optical properties; thus, selective synthesis of SWNTs with desired chiralities is a major challenge in nanotube science and applications. In this session, we demonstrate the recent progresses of narrow chirality distributed growth of SWNTs by plasma CVD based on different two approaches, which focus on catalytic reaction and gas phase reaction.

4.3.1. Catalytic reaction control

Narrow chiratliy distributed growth of SWNTs is one of the most critical issue in the scientific field of SWNTs production stage. Some progress has been made with silica-supported CoMo [21] and zeolite-supported FeCo [22] catalysts. FeRu [23] and FeNi [24] catalysts have also been developed to achieve narrow chirality distributions. Interestingly, all syntheses resulting in narrow chirality distributions have involved magnetic catalysts. The main obstacle to research on intrinsic magnetic properties of SWNTs is residual ferromagnetic catalyst particles; thus, SWNT growth with nonmagnetic catalysts is beneficial. Despite recent improvements in SWNT growth with nonmagnetic catalysts [25-28], diameter and chirality (n,m) distribution control with nonmagnetic catalysts is still required for fundamental studies and a variety of applications [29].

Based on this background, we attempto to grow SWNTs with narrow chirality distributions using nonmagnetic catalyst [30, 31]. PLE mapping was used to assign (n,m) of SWNTs grown from the Au catalyst at different H_2 concentrations (Figures 6a–c). The total pressure was kept at 50 Pa by adjusting the pumping rate of the rotary pump throughout this experiment. Lower H_2 concentrations (0 and 3 sccm) led to larger diameters and wider (n,m) distributions with (6,5), (7,5), (7,6), (8,4), (8,6), and (8,7) (Figures 6a and b). On the other hand, the 7-sccm H_2 concentration yielded the narrowest (n,m) distribution with a dominant peak corresponding to the (6,5) tube (Figure 6c). The UV-Vis-NIR optical absorbance spectra of Au-plasma CVD SWNTs grown at the 7-sccm H_2 flow rate showed one dominant peak in the first van Hove E_{11} range (900–1400 nm) corresponding to SWNTs with (6,5) chirality (Figure 6d). Since clear metallic SWNT peaks were not observed in the UV-Vis-NIR spectra (Figure 6d), the concentration of metallic SWNTs was lower than that of the generally grown SWNTs. This is the first result showing narrow chirality distributions for SWNTs grown from a nonmagnetic catalyst [31].

To elucidate the effects of Au and plasma CVD on the narrow chirality distribution, other combinations of catalysts and CVD methods were systematically investigated. Based on the PLE analysis, SWNTs grown by the Fe catalyst with plasma CVD (Fe-plasma CVD) did not show a clear correlation between the H_2 flow rate and the chirality distribution, which was broader than that of SWNTs grown by Au-plasma CVD. This indicates that H_2-assisted Au catalyzation is a critical factor for achieving narrow chirality distributions, which is in good agreement with theoretical predictions. The first-principle calculation by Yazyev *et al.* reveals that coinage metals, such as Cu, Ag, and Au, produce narrow chirality distributions [32]. Ding *et al.* have reported that the SWNT diameter is larger on the surfaces of Fe, Co, and Ni particles than on Cu, Pd, and Au particles because of the different bond energies on the catalyst surfaces [33]. Based on these theoretical models, we can explain the effect of H_2-assisted Au catalyzation on the narrow chirality distirbution as follows. Since the binding energy of hydrocarbons on the Au surface is much weaker than on the Fe surface, it is difficult to achieve cap formation for large-diameter Au catalysts [33]. Additonal H_2 also enhances the etching of the carbon precursor from the catalyst surface, which strongly suppresses the growth of large-diameter SWNTs; hence, the chirality distribution of SWNTs grown from the Au catalyst should be narrower than those grown from the Fe catalyst. The stability of the cap structure is a possible reason why the (6,5) tube was dominant in the small-diameter Au-plasma CVD SWNTs. The number of cap structures, which satisfies the isolated pentagon rule, is highly limited for small-diameter SWNTs, and (6,5) is known to have a stable cap structure in this diameter range [22]. A comparison between Au-plasma CVD and Au-thermal CVD was also carried out. Although the chiraity distribution became relatively narrow for SWNTs grown by Au-thermal CVD under appropriate H_2 concentrations, it was much broader than that of SWNTs grown by Au-plasma CVD. Comparison of the Au-plasma CVD and Au-thermal CVD processes showed that there were two significant differences in the SWNT growth conditions: growth temperature and incubation time. The lower limit of growth temperature for Au-plasma CVD was 700 °C, which was lower than that of Au-thermal CVD by 50 °C. The initial SWNT growth occurred 1 min after the growth substrate was exposed to the plasma for Au-plasma CVD, whereas 15 min were required for the growth of SWNTs with Au-thermal CVD. These results suggest that the low-temperature and short-time growth with Au-plasma CVD prevents aggregation of catalyst particles during SWNT growth, which suppresses the growth of large-diameter SWNTs and results in a narrow chiratliy distribution.

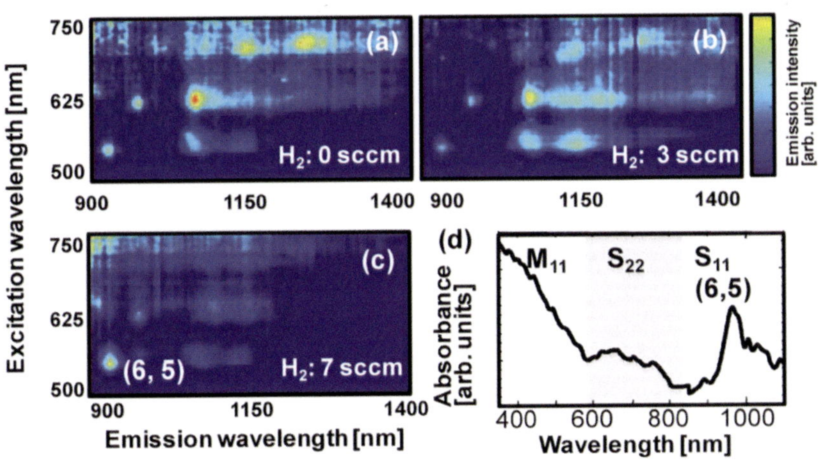

Figure 6. (a) PLE maps of SWNTs from Au catalyst by plasma CVD at (a) 0-sccm, (b) 3-sccm, and (c) 7-sccm H_2 flow rates, respectively. (d) UV-vis-NIR spectrum of SDS-dispersed SWNTs from Au catalyst at 7-sccm H_2 flow rate.

4.3.2. Gas phase reaction control

Recent progress in the in-situ TEM observation during the SWNT growth revealed that metal-catalyzed SWNT growth is initiated by the formation of a carbon cap structure on the surface of a catalytic nanoparticle with a certain incubation time (t_i) [34]. Although the detailed mechanism in this incubation period is still argued, it is expected that there might be correlations between the t_i and SWNT structures such as the d_t and chirality. When we assume that the t_i of the small d_t (or specific chirality) SWNTs is shorter than that of the larger (or other chiralities) one, it should be possible to selectively grow the narrow-d_t (or -chirality) distributed SWNTs by strictly controlling the t_g at their initial growth stage. Chemical vapor deposition (CVD), thermal CVD and plasma CVD, is one of the most promising SWNT production methods, which has advantages such as a large scale production and a direct site-assigned growth. In the case of thermal CVD, the SWNT growth gradually starts and stops after the initiation of feeding and pumping the hydrocarbon gas, respectively (Figure 7a). The growth time in thermal CVD includes some uncertainness, which makes it difficult to be used for a precise growth time control. In the case of plasma CVD, on the other hand, reactive ions and radicals are main species for the nanotubes growth, and the SWNT growth is carried out only when a plasma is generated. This suggests that the growth time can be controlled by timing an electric power supply used for the plasma generation (Figure 7b), and the precise t_g control on the order of micro second is possible in plasma CVD. Based on this strategy, we attempt to grow the narrow-d_t and -chirality distributed short SWNTs by precisely adjusting the t_g with time programmed-plasma CVD (TP-PCVD).

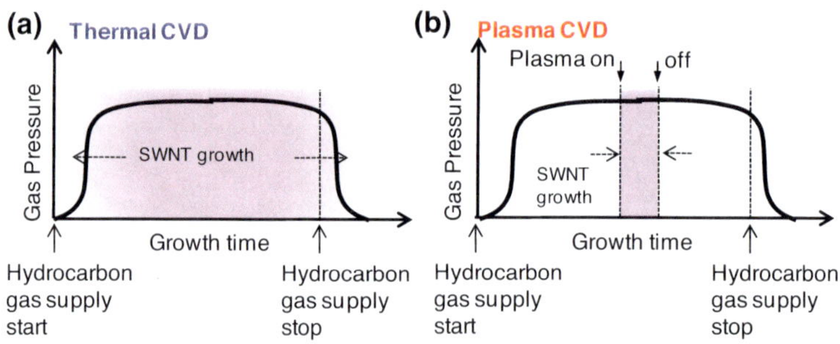

Figure 7. (a) (b) Time evolution of thermal (a) and plasma (b) CVD.

In order to analyze the chirality distribution of the short SWNTs in detail, we carry out the PLE map analysis. It is to be noted that since our plasma CVD grown SWNTs take on the freestanding form due to the strong electric field in the plasma sheath area during their growth, it is possible to observe PL signals from as-grown SWNTs on a substrate [15]. All the PLE measurements are carried out immediately after the growth to prevent SWNTs from forming thin bundles, which leads to causing the PL intensity change by exciton energy transfer between each tube [15]. From the density estimation of SWNTs with the direct

TEM observations, it is confirmed that abundant SWNTs exist in the area where the PLE measurement is carried out. Thus, the PLE map gives us macroscopic information in each sample. Figures 8a-c show the PLE maps of the as-grown SWNTs as a function of t_g. It is found that the d_t distribution just after the incubation (2 sec) is relatively narrow and the main diameter is about 0.8 nm with (6, 5), (7, 5), (7, 6), (8, 4), and (9, 2) dominant chiralities (Figure 8a). Then, the relatively large d_t (0.95 nm) SWNTs ((8, 6) and (9, 4)) initiate their growth at 5 sec (Figure 8b). The larger d_t SWNTs of (8, 7) are finally grown at 10 sec, where the chirality and d_t distributions are broad (Figure 8c). The small end of the d_t does not change, whereas the large end increases with an increase in the t_g. In more detail, when we plot the t_i as a function of the d_t, a clear dependence is obtained (Figure 9). It is to be noted that the t_i of each chirality is defined as the t_g when a clear PL signal is firstly observed (Figures 8a-c). The t_i increases with an increase in the tube diameter, which supports the validity of our basic concept.

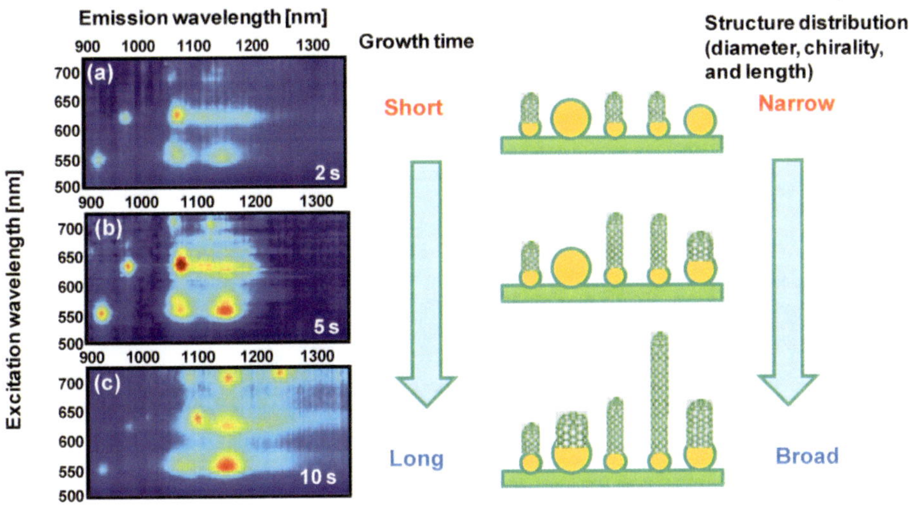

Figure 8. Chirality and diameter distribution of SWNTs as a function of growth time. (a)(b)(c) PLE-map of SWNTs produced by different growth time, which is 2 sec (a), 5 sec (b), and 10 sec (c).

To further narrow the initial d_t and chirality distributions, we adjust the other growth conditions. Figure 10 shows the PLE map dependence on the growth temperature of SWNTs grown for the very short growth time (2 sec). Interestingly, when we decrease the growth temperature down to 600 ºC the chirality distribution is very narrowed, and (7,6) and (8,4) SWNTs are predominantly grown (Figure 10c). At the lower growth temperature (≤ 580 ºC), SWNTs could not be grown for the very short growth time (2 sec). However, similar narrow-chirality distributed SWNTs are grown by extending the growth time until 15 sec. This indicates that the t_i for each d_t and chirality is sensitive to the growth temperature. It should be also emphasized that the chirality distribution just after the t_i is always narrow independent of other growth parameters. The narrow-chirality distributed SWNTs growth by TP-PCVD is reproducibly obtained. This is the first result of the direct growth of the short-length SWNTs with a narrow-chirality distribution [35].

The most important factor for the narrow-chirality distributed growth of short SWNTs by TP-PCVD is to control the d_t dependence on t_i for each chirality SWNT. If each t_i largely varies depending on the d_t or chirality of SWNT, it is possible to more selectively grow the specific chirality SWNTs by adjusting the growth time. As shown in Figure 10, the growth temperature is one of the key factors to cause the t_i variation for each d_t SWNT. The cap formation process might be one of the critical reasons for this temperature dependence. The central part of the graphene sheet formed on the top of catalyst surface has to lift off the catalytic particle for the growth of SWNTs. This can only happen if the kinetic energy per area at the interface between the graphene sheet and the catalyst (E_{kin}) is high enough to overcome the work of adhesion per area of graphite toward the catalytic particle (W_{ad}) ($E_{kin} > W_{ad}$)[36]. Small tubes have lower W_{ad}[37]. Since the E_{kin} should be proportional to the growth temperature, the producible type of SWNTs under the low growth temperature condition can be highly selected compared with that in the higher growth temperature.

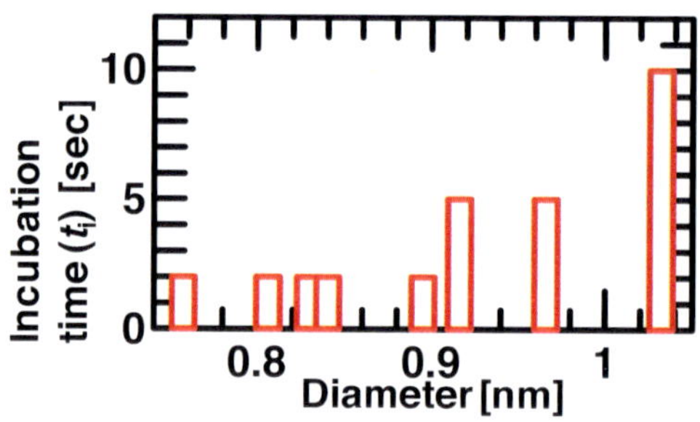

Figure 9. Plot of incubation time as a function of tube diameter.

As we have discussed above, the growth temperature is one of the key parameters to realize the narrow-chirality distribution of SWNTs by TP-PCVD. In order to realize the further precise control of the d_t and chirality distribution, it is inevitable issue to reveal the critical factor, which causes the d_t dependence on the t_i. The catalyst particle-size variation during the heating process might be one possibility. If the catalyst aggregation is enhanced during the growth process, the chirality distribution also changes. To investigate this issue, we carry out the PLE mapping measurement of SWNTs grown for the different preheating time. After heating the substrate up to 620 ºC, the temperature is kept for certain time (0 sec, 30 sec, and 60 sec), and then similar TP-PCVD is carried out. If the catalyst particle size distribution varies during the heating process and this is the critical factor of the d_t dependence on t_i, clear differences should appear in the PLE map of SWNTs grown under the different preheating conditions. In any preheating time, however, the d_t and chirality distribution does not show obvious changes. This indicates that the catalyst size distribution is almost the same during the short time growth (Figures 8a-c), and its effect is negligible for the d_t dependence on t_i (Figure 9).

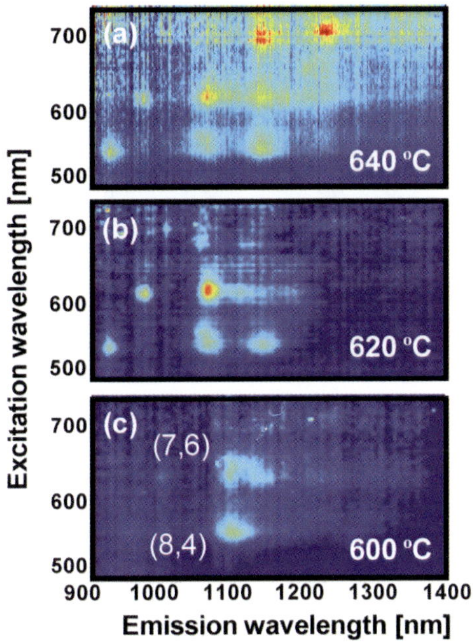

Figure 10. Growth temperature dependence of the PLE-maps of SWNTs grown for very short growth time (2sec). The growth temperature of (a), (b), and (c) is 640 °C, 620 °C, and 600 °C, respectively.

The other possibility to cause the d_t dependence on the t_i is a supersaturation time difference. Since the SWNTs growth is carried out following a supersaturation of carbons in a catalyst, it is expected that the small catalysts are rapidly supersaturated with carbon atoms prior to the case of the large catalysts. To confirm this effect, we carry out the PLE map measurement of SWNTs grown under the low hydrocarbon supply condition. If the supersaturation time difference is the critical factor to cause the d_t dependence on the t_i, the selectivity of d_t or chirality should be improved by decreasing the hydrocarbon supply. The amount of the hydrocarbon supply is controlled by adjusting P_{RF} used for the plasma generation. The t_i clearly increases up to 20 sec by decreasing the amount of hydrocarbon supply (P_{RF} = 25 W), which should be caused by the longer supersaturation period of carbons in the catalyst at their initial growth stage. However, various kinds of chirality species of SWNTs equally start their growth and the selective growth of narrow-d_t or -chirality distributed SWNTs are not observed under this low hydrocarbon supply condition. Although the t_i for whole SWNTs is sensitive to the hydrocarbon supply, the selectivity of d_t or chirality is found to be conducted by the other factors. A simple estimation based on the carbon atom numbers also supports the validity of this explanation. The soluble carbon atoms in a typical 1 nm Fe catalyst (Fe_{50}) is about 26 ~ 27 [38]. On the other hand, the number of carbon atoms constructing a 1 nm d_t and 100 nm length SWNT is about 2.35×10^4. Judging from this carbon atom numbers, it is difficult to selectively achieve the supersaturation only for the d_t = 0.8 nm with 100 nm length SWNT prior to the d_t = 0.95 nm SWNT (Figure 9).

5. Conclusions

Recent progress in SWNT growth was presented, with a special emphasis on plasma CVD. Due to the strong plasma-sheath electric field, it is possible to grow freestanding individual SWNTs by plasma CVD. Based on the time-evolution study and the detailed plasma parameter measurements, the growth kinetics of SWNTs in plasma CVD were well established. The concentration of semiconducting SWNTs in FET devices can be increased by tuning the mean diameter of SWNTs, and this effect is attributable to selective damage of metallic SWNTs during plasma CVD. Moreover, narrow chirality-distributed growth of SWNTs were also achieved by different two approaches. Au catalyzed plasma CVD with appropriate amount of hydrogen addition can realize preferential gorwth of (6,5) SWNTs. The narrow chirality distributed growth of SWNTs were also demonstrated with precise incubation time control by time-programmed plasma CVD.

Author details

Toshiaki Kato and Rikizo Hatakeyama

Department of Electronic Engineering, Tohoku University, Sendai, Japan

References

[1] Kong, J., Cassell, A. M., & Dai, H. (1998). Recent progress of plasma CVD for structure controlled growth of single-walled carbon nanotubes . *Chemical Physics Letters,* 567 EOF-574 EOF.

[2] Maruyama, S., Kojima, R., Miyauchi, Y., Chiashi, S., Kohno, M., & Low, . (2002). Recent progress of plasma CVD for structure controlled growth of single-walled carbon nanotubes . *Chemical Physics Letters,* 229 EOF.

[3] Dai, H., Rinzler, A. G., Nikolaev, P., Thess, A., Colbert, D. T., Smalley, R. E., & Single, . (1996). Recent progress of plasma CVD for structure controlled growth of single-walled carbon nanotubes . *Chemical Physics Letters,* 471 EOF-475 EOF.

[4] Tibbetts, G. G. (1992). Recent progress of plasma CVD for structure controlled growth of single-walled carbon nanotubes . *Carbon*, 30(3), 399-406.

[5] Ren, Z. F., Huang, Z. P., Xu, J. W., Wang, J. H., Bush, P., Siegal, M. P., & Provencio, P. N. (1998). Recent progress of plasma CVD for structure controlled growth of single-walled carbon nanotubes . *Science*, 282(6), 1105-1107.

[6] Chhowalla, M., Teo, K. B. K., Ducati, C., Rupesinghe, N. L., Amaratunga, G. A. J., Ferrari, A. C., Roy, D., Robertson, J., & Milne, W. I. (2001). Recent progress of plasma CVD for structure controlled growth of single-walled carbon nanotubes . *Journal of Applied Physics*, 90(10), 5308-5317.

[7] Delzeit, L., Nguyen, C. V., Stevens, R. M., Han, J., & Meyyappan, M. Growth of Carbon Nanotubes by Thermal and Plasma Chemical Vapor Deposition Processes and Applications in Microscopy. Nanotechnology (2002)., 13(3), 280-284.

[8] Hirata T., Satake N., Jeong G.-H., Kato T., Hatakeyama R., Motomiya K., and Tohji K., Magnetron-Type Radio-Frequency Plasma Control yielding Vertically Well-Aligned Carbon Nanotube Growth. (2003). *Applied Physics Letters*, 83(6), 1119-1121.

[9] Bower, C., Zhu, W., Jin, S., Zhou, O., & Plasma, . (2000). Recent progress of plasma CVD for structure controlled growth of single-walled carbon nanotubes . *Applied Physics Letters*, 77(6), 830-832.

[10] Merkulov, V. I., Melechko, A. V., Guillorn, M. A., Lowndes, D. H., & Simpson, M. L. (2001). Recent progress of plasma CVD for structure controlled growth of single-walled carbon nanotubes . *Applied Physics Letters*, 79(18), 2970-2972.

[11] Kato, T., Jeong, G., , H., Hirata, T., Hatakeyama, R., Tohji, K., Motomiya, K., & Single, . (2003). Recent progress of plasma CVD for structure controlled growth of single-walled carbon nanotubes . *Chemical Physics Letters*, 422 EOF.

[12] Kato, T., Jeong, G., , H., Hirata, T., Hatakeyama, R., & Tohji, K. Recent progress of plasma CVD for structure controlled growth of single-walled carbon nanotubes . Japanese Journal of Applied Physics(2004). A) LL1280., 1278.

[13] Kato, T., Hatakeyama, R., & Tohji, K. Diffusion Plasma Chemical Vapor Deposition Yielding Freestanding Individual Single-Walled Carbon Nanotubes on a Silicon-Based Flat Substrate. Nanotechnology (2006)., 17(9), 2223-2226.

[14] Kato, T., & Hatakeyama, R. Recent progress of plasma CVD for structure controlled growth of single-walled carbon nanotubes . Chemical Vapour Deposition (2006). , 12(6), 345-352.

[15] Kato, T., & Hatakeyama, R. (2008). Recent progress of plasma CVD for structure controlled growth of single-walled carbon nanotubes . *Journal of the American Chemical Society*, 130(25), 8101-8107.

[16] Kato, T., & Hatakeyama, R. (2008). Recent progress of plasma CVD for structure controlled growth of single-walled carbon nanotubes . *Applied Physics Letters*, 031502 EOF.

[17] Chiashi, S., Ph, D., thesis, Univ., & of, . Tokyo (2005).

[18] Kato, T., Kuroda, S., & Hatakeyama, R. (2011). Recent progress of plasma CVD for structure controlled growth of single-walled carbon nanotubes . *Journal of Nanomaterials*.

[19] Jorio A., Saito R., Hafner J. H., Lieber C. M., Hunter M., McClure T., Dresselhaus G., and Dresselhaus M. S. Structural (n,m) Determination of Isolated Single-Wall Carbon Nanotubes by Resonant Raman Scattering. Physical Review Letters 2001; 86(6) 1118–1121.

[20] Kato, T., & Hatakeyama, R. Recent progress of plasma CVD for structure controlled growth of single-walled carbon nanotubes . Journal of Nanotechnology 2010; (2010).

[21] Kitiyanan, B., Alvarez, W. E., Harwell, J. H., & Resasco, D. E. (2000). Recent progress of plasma CVD for structure controlled growth of single-walled carbon nanotubes . *Chemical Physics Letters*, 497 EOF-503 EOF.

[22] Miyauchi, Y., Chiashi, S., Murakami, Y., Hayashida, Y., & Maruyama, S. (2004). Recent progress of plasma CVD for structure controlled growth of single-walled carbon nanotubes . *Chemical Physics Letters*, 198 EOF.

[23] Li, X., Tu, X., Zaric, S., Welsher, K., Seo, W. S., Zhao, W., & Dai, H. (2007). Recent progress of plasma CVD for structure controlled growth of single-walled carbon nanotubes . *Journal of the American Chemical Society*, 129(51), 15770-15771.

[24] Chiang, W. H., & Sankaran, M. R. Linking Catalyst Composition to Chirality Distributions of As-Grown Single-Walled Carbon Nanotubes by Tuning NixFe1−x Nanoparticles. Nature Materials (2009). , 8(11), 882-886.

[25] Zhou, W., Han, Z., Wang, J., Zhang, Y., Jin, Z., Sun, X., Zhang, Y., Yan, C., & Li, Y. (2006). Recent progress of plasma CVD for structure controlled growth of single-walled carbon nanotubes . *Nano Letters*, 6(12), 2987-2990.

[26] Takagi, D., Homma, Y., Hibino, H., Suzuki, S., Kobayashi, Y., & Single, . (2006). Recent progress of plasma CVD for structure controlled growth of single-walled carbon nanotubes . *Nano Letters*, 6(12), 2642-2645.

[27] Takagi, D., Kobayashi, Y., Hibino, H., Suzuki, S., & Homma, Y. (2008). Recent progress of plasma CVD for structure controlled growth of single-walled carbon nanotubes . *Nano Letters*, 8(3), 832-835.

[28] Bhaviripudi S., Mile E., Steiner III S. A., Zare A. T., Dresselhaus M. S., Belcher A. M., and Kong J. CVD Synthesis of Single-Walled Carbon Nanotubes from Gold Nanoparticle Catalysts. Journal of the America Chemical Society 2007; 129(6) 1516–1517.

[29] , Y., Lehtinen, P. O., Foster, A. S., & Nieminen, R. M. (2004). Recent progress of plasma CVD for structure controlled growth of single-walled carbon nanotubes . *New Journal of Physics*, 68 EOF.

[30] Ghorannevis, Z., Kato, T., Kaneko, T., & Hatakeyama, R. Growth of Single-Walled Carbon Nanotubes from Nonmagnetic Catalysts by Plasma CVD. Japanese Journal of Applied Physics (2010). BA01-1-4.

[31] Ghorannevis, Z., Kato, T., Kaneko, T., Hatakeyama, R., & Narrow, . (2010). Recent progress of plasma CVD for structure controlled growth of single-walled carbon nanotubes . *Journal of the American Chemical Society*, 132(28), 9570-9572.

[32] Yazyev, O. V., & Pasquarello, A. Recent progress of plasma CVD for structure controlled growth of single-walled carbon nanotubes . Physical Review Letters (2008).

[33] Ding, F., Larsson, P., Larson, J. A., Ahuja, R., Duan, H., Rosen, A., & Bolton, K. (2008). Recent progress of plasma CVD for structure controlled growth of single-walled carbon nanotubes . *Nano Letters*, 8(2), 463-468.

[34] Hofmann, S., Sharma, R., Ducati, C., Du, G., Mattevi, C., Cepek, C., Cantoro, M., Pisana, S., Parvez, A., Cervantes-Sodi, F., Ferrari, A. C., Dunin-Borkowski, R., Lizzit, S., Petaccia, L., Goldoni, A., & Robertson, J. (2007). Recent progress of plasma CVD for structure controlled growth of single-walled carbon nanotubes . *Nano Letters*, 7(3), 602-608.

[35] Kato, T., & Hatakeyama, R. (2010). Recent progress of plasma CVD for structure controlled growth of single-walled carbon nanotubes . *ACS Nano*, 4(12), 7395-7400.

[36] Ding, F., Bolton, K., & Rosen, A. Recent progress of plasma CVD for structure controlled growth of single-walled carbon nanotubes . Journal of Physical Chemistry B (2004). , 108-17369.

[37] Kanzow, H., Lenski, C., Ding, A., & Single, . (2001). Recent progress of plasma CVD for structure controlled growth of single-walled carbon nanotubes . *Physical Review B*.

[38] Ding, F., Larsson, P., Larsson, J. A., Ahuja, R., Duan, H., Rosen, A., & Bolton, K. Recent progress of plasma CVD for structure controlled growth of single-walled carbon nanotubes . Nano Lett. (2008). , 8-463.

Chapter 6

Characterization of Carbon Nanotubes

Rolant Eba Medjo

Additional information is available at the end of the chapter

1. Introduction

Carbon is one of the most abundant and most fascinating elements on earth. It appears in several different forms or allotropes with widely different properties. Among these allotropes are diamond, graphite and amorphous carbon. It forms also a great variety of novel structures that are being discovered day after day these last years. The discovery of novel carbon allotropes or *carbon nanostructures* (CNSs) has attracted intensive attention due to their fundamental and technological interests [1, 2]. They exhibit unique structural and physical properties. Carbon nanostructures are promising to revolutionize several fields of fundamental science and contribute as major component of nanotechnology. Previous studies have shown that these nanostructures can be used in composite materials or in individual functional elements of nanodevices such as: hydrogen storage, nanomanipulation, medical usages and nonporous membranes [3]. Such devices imply precise demands as high aspect ratio, vertical alignment on flat plan and electronic conductivity among the properties of CNSs. Special attention is now devoted to electrons or field emission. But there still remains a wide range of unexplored potential applications in various nanotechnological areas such as aerospace, energy, automobile, medicine, or chemical industry, in which CNSs can be used as gas adsorbents, templates, actuators, composite reinforcements, catalyst supports, probes, sensors, nanopipes and nanoreactors. Besides the attractive aspects of carbon nanostructures, their synthesis is complex compared to other materials used in different domains of technology and different existing methods of synthesizing. These methods lead to CNSs with sometimes an important quantity of impurities incorporated, encapsulated or adsorbed, whose amount and types depend on techniques and parameters of synthesis.

Among carbon nanostructures, CNTs are the most important because they possess the most determinant properties for revolutionary applications. Proprieties exhibited by CNTs are in general given qualitatively by electron microscopies. However, SEM examination provides an overview of nanostructures, and a more accurate examination by TEM generally reveals many defects [4]. All these microscopies very often, mask some observations of CNTs arrangements. The controlling or characterization is also done qualitatively by very few other techniques, often used to more deeply investigate the morphology and the structure, such grazing-incidence small-angle X-ray scattering (GISAXS) [5]. This technique is semi-quantitative. It is part of those which allow an in situ and real time study of islands growing on a substrate and give access to their third dimension. This aspect is the most important when the use of imagery techniques becomes difficult and impossible. It gives structural and morphological information and correlations on CNTs and their mutual orientation. The obtained results are in agreement with the qualitative ones from SEM and TEM. GISAXS is an extension of the well knows SAXS (Small Angle X-ray Scattering). XANES spectroscopy has been a powerful tool that not only provides information on the local environment around carbon, such as diamond, carbon nitride and graphitic carbon [6-8], but also investigates the absorption and adsorption of hydrocarbon molecules [9], radicals and atoms with specific selectivity for the orientation of these compounds. This property of XANES spectroscopy is due to the angular dependence of the absorption transition [6]. XANES is a local probe, sensitive to chemical impurities, defects, chemical adsorption and curvature induced orbital rehybridization.

In this chapter are presented characterization techniques of carbon nanotubes notably electron microscopies and XANES spectroscopy. Later, electron microscopy images analysis is done. XANES spectra are quantitatively analyzed. It ends with a conclusion.

2. Characterization Techniques

Electron microscopy is an imaging technique that uses an electron beam in order to probe a material. Since the wavelength of an electron is much smaller than the wavelength of visible light, diffraction effects occur at much smaller physical dimensions. The microscope was the first tool by means of which a real study could be made of objects too small to be seen with the naked eye. From its crude beginning some 300 years ago, it has been developed into an instrument that is a credit to the inventive skill and analytical ability of those that have worked on it. The modern microscope is an instrument that approaches the "theoretical limit" of its performance, the information that can be extracted from High Resolution Transmission Electron Microscopy (HRTEM) is not straight-forward since the preparation of TEM samples may mask some observation of CNTs arrangements. A great variety of experiments has been carefully done to give the properties of the atoms, molecules, electrons, protons... regarded as the constituents of matter. The limitation on the performance of a microscope is set by its resolving power. Further characterization tools are desirable especially those which are not destructive for the samples to complete electron microscopy studies. Among these tools are cited:

- the scanning tunnelling microscopy (STM),
- the atomic force microscopy (AFM),
- the X-ray photoemission spectroscopy (XPS),
- the grazing incidence small angles X-ray scattering
- the X-ray absorption near-edge structure spectroscopy.
- the Raman spectroscopy.

It has been shown that nuclear magnetic resonance (NMR) and XANES have proved to be the most powerful ones in detecting and resolving the various bonding environments. The main drawback of NMR is that it requires detaching and powderizing the coating until reaching a relatively large amount of sample. XANES, instead, can be performed on a single coating without any sample preparation.

2.1. Electron Microscopy

Electron microscopy is an essential tool for characterizing any nanomaterial because it allows direct observation of size, shape, structure. The local structure of the CNSs can be investigated at the nanometer and subnanometer level by TEM, SEM, AFM and STM. TEM and SEM are useful tools to check the exfoliation of bundles and the purity of the material. However TEM and SEM produce damages on the sample due to the use of the electron beam.

Electron Microscopy was developed due to the limitations of Light Microscopy which is limited by the physics of light to 500 or 1,000 times magnification and a resolution of 0.2 micrometer. In the early of the 1930 decade, this theoretical limit had been reached and there was a scientific desire to see the fine details of the interior structures of organic cells. Electron Microscopy is a scientific technique used to examine objects on a very fine scale. This thin specimen is irradiated with an electron beam of uniform current density.

This required 10,000 times magnification which was not possible to be obtained using Light Microscopes.

Electron Microscopes operate exactly as their optical counterpart except that they use a focused beam of electrons instead of light to "image" the specimen and gain information about its structure and composition.

There are two types of electron microscopy, namely the transmission electron microscopy and the scanning electron microscopy.

The information that TEM or SEM examination can yield are the following:

- The topographical information: the surface features of an object or "how it looks", its texture.
- The morphological information: the shape and size of the particles making up the object.
- The composition information: the elements and compounds that the object is composed of and the relative amounts of them

The relationship is direct between these information and the material properties.

The crystallographic information itemizes how the atoms are arranged in the object and provides direct relation between these arrangements and materials properties is given only by TEM.

There are two main constrains for the sample observation in Electron Microscopy. The first is that the sample should be stable in the vacuum. The second, the sample must be under the electron beam. Other specific limitations have to be fulfilled in each Electron Microscopy.

The critical point of the study of CNSs is their direct observation limitation by the use of HRTEM. The information that can be extracted is not straight-forward since preparation of TEM specimens may mask the observation nanostructure arrangements. For example, during ion beam milling for cross-section imaging preparation, high ion bombardment may destroy the structure due to surface amorphization. These drawbacks may be overcome by the reduction of the ion bombardment in the final steps of the thinning process or by studying plan-view specimens when deposition on soluble substrates is possible.

Another problem in HRTEM analysis comes from projection of artefacts, which may preclude the resolution of structural features. This can be partially solved by considering selective area electron diffraction (SAED), since any overlapping does not affect the characteristic lattice spacing in the diffraction pattern and information on the degree of ordering can be derived from the brightness and width of the diffraction pattern.

Finally, an additional drawback of TEM measurements for studying films is the destructive character entailed. Further identification of CNTs with additional characterization tools is required, especially if they are not destructive for the sample. In the next section, we present XANES spectroscopy.

2.2. X-ray absorption near edge structure spectroscopy

X-ray absorption near-edge structure also called NEXAFS (near-edge X-ray absorption fine structure) exists in the energy level of 50 eV above the absorption edge. It includes the unoccupied part of band structure just above the Fermi level. Thus, certain aspects of electronic structure of detected element can be revealed.

X-ray absorption spectroscopy in general, has gained much of its reputation of being a powerful analytical and research tool, mostly due to the use of synchrotron radiation sources. In this kind of spectroscopy, interactions between photons and matter are studied by measuring the photoabsorption cross-section. The absorption of X-ray creates the photon-induced excitations of an electron from a core state to an empty state above the Fermi level. This absorption is measured as a function of photon energy, close to a core level binding energy. The intensity of X-ray beam passing through a sample (Figure 1b) of thickness d is given by the Beer Lambert law:

$$I = I_0 e^{-\mu(E)d} \tag{1}$$

where is the beam intensity hitting the sample and I the intensity transmitted through the material.

XAS provides valuable information about the electronic structure by probing unoccupied states above the Fermi level (Figure 1a). XAS technique has two very important aspects which are the site and the symmetry selectivity. The site symmetry is due to the specific binding energy of the core electron and the localized character of the excitation. The technique is also a local probe since the excitation is localized and the dipole selection rule is applicable and gives the symmetry dependence of each feature of the spectra.

This X-ray absorption cross section μ(E) is most generally given by the Fermi's golden rule

$$\mu(E) \propto \sum_f |\langle f | H | i \rangle|^2 \delta(E_i - E_f + h\nu) \quad (2)$$

where, and denote the initial and final states and their energies. In XAS matrix elements, the final state is localized.

- H is the whole system Hamiltonian, in the dipole approximation,
- The dipole selection rules can be considered.

Thus, equation (2) becomes:

$$\mu(E) \propto \sum_f |\langle f | \vec{\varepsilon}.\vec{r} | i \rangle|^2 \delta(E_i - E_f + h\nu) \quad (3)$$

Only the dipole contribution to the total cross-section has been considered in Equation (3). By exciting an atom using an X-ray source, the electrons configuration of the atom is changed; one electron, usually a core-shell electron, or more electrons populate unoccupied bound or continuum states (figure 1). The success of this spectroscopy lies in the fact that the photoelectron acts as very sensitive probe that can "feel" the charge distribution and the arrangement of the neighbouring atoms around the absorbing atom, or, in the other words, it can feel the chemical environment of the neighbouring atoms. When the photon energy is not high enough, the photoeffect in one of the core shells can occur. It results in the step like shape of the absorption spectrum: the increased photoabsorption cross-section due to the knocking-out of an electron is called absorption edge.

One way of understanding this excitation process in a bound atom is describing it by means of multiple scattering (MS): the photoelectron's wave is scattered on atoms surrounding the absorbing atom. The cross-section of a bound atom therefore depends on the positions and types of the neighbours and is different from that of an isolated atom. The photoelectrons can populate either unoccupied bound states or low-lying continuum states. The part of the spectrum (Figure 2) concerned is the XANES.

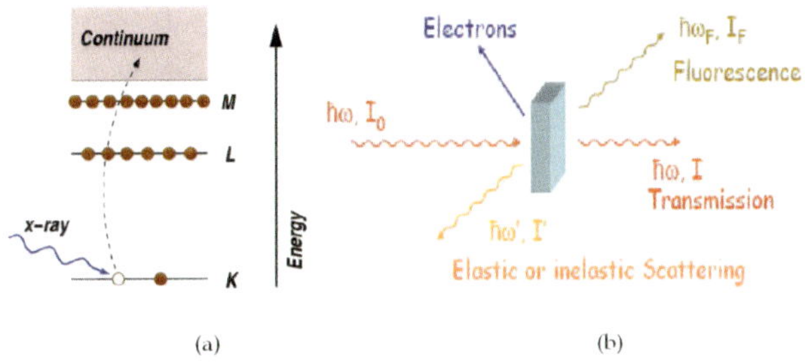

Figure 1. a) The absorption of X-ray by an atom promoting a core-level electron (K) into the continuum, (b) X-ray radiation and matter interactions [10].

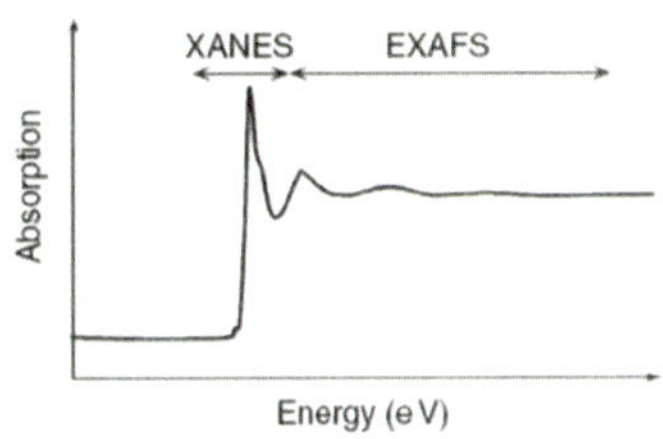

Figure 2. Schematic illustration of an X-ray absorption spectrum, showing the structured absorption that is seen both within 50 eV of the edge (the XANES) and for several hundred to >1,000 eV above the edge (the EXAFS) [11].

On its high energy end, XANES extends up to Extended X-ray Absorption Fine Structure (EXAFS) [12]. The limiting energy that divides XANES from EXAFS is by no means exactly defined since the transition from the one regime to the other is smooth. As a rule, near-edge structure ends approximately where the electron wavelength equals the distance from the absorbing atom to its nearest neighbours [12, 13], which usually means about 40–50 eV above the edge. In the XANES regime, the electron's kinetic energy is small and the scattering on the neighbouring atoms tends to be strong, while the effect of the scatterers becomes smaller at higher energies; in EXAFS region, the photoelectrons are only weakly scattered. Albeit EXAFS represents the most understood part of XAS, XANES is the only part used in this study to characterize carbon nanostructures and its theoretical study is the aim of the next paragraph.

When attention is paid to scattering form of potentials of several objects (systems of particles), each of them makes a non-zero contribution only within a spherically non-overlapping scattering region of finite radius (Figure 3a). The total potential V is given by:

$$V = \sum_i \upsilon^i \tag{4}$$

Outside the spherical regions – in the interstitial region – the potential is set to zero. This approximation presented in Figure 3b is known as the muffin-tin approximation. The scattering parameters of each of the scatterers, namely the scattering amplitudes and phase shifts, are determined separately for each scatterer and are therefore pure atomic quantities. The propagation of a photoelectron in such muffin-tin potential V is described by the Hamiltonian:

$$H = H_0 + V \tag{5}$$

is the kinetic-energy operator. The stationary solution with the energy E is given by:

$$(E - H_0)|\psi\rangle = V|\psi\rangle \tag{6}$$

Solving the "inhomogeneous" Equation (6) by means of the Green's functions and to be more specific, if $\langle r \mid \varphi\rangle$ is the solution of the "homogeneous" part of Equation (6)

$$(E - H_0)\langle r|\varphi\rangle = 0 \tag{7}$$

The free-electron Green's function G_0 is defined by relation

$$(E - H_0)G_0(r,r';E) \equiv (\Delta + k^2)G_0(r,r';E) = \delta(r - r') \tag{8}$$

then the general solution of Equation (6) is given as a sum of the solution of the *homogeneous* equation and the particular solution [14, 15] is given below:

$$\langle r|\psi\rangle = \langle r|\varphi\rangle + \int d^3r' G_0(r,r';E)\langle r'|V\psi\rangle \tag{9}$$

For the photoelectron only weakly scattered by the potential V (XANES), the solution $\langle r \mid \psi\rangle$ is close to the free-electron solution$\langle r \mid \varphi\rangle$. Furthermore, when V is identically zero everywhere, the exact equality$\langle r \mid \psi\rangle=\langle r \mid \varphi\rangle$ holds, as expected. It is clear from Equations (7) and (9) that there are solutions of the total Hamiltonian, that $\mid \psi\rangle$ have the same energy as $\mid \varphi\rangle$ only when elastic scattering is considering.

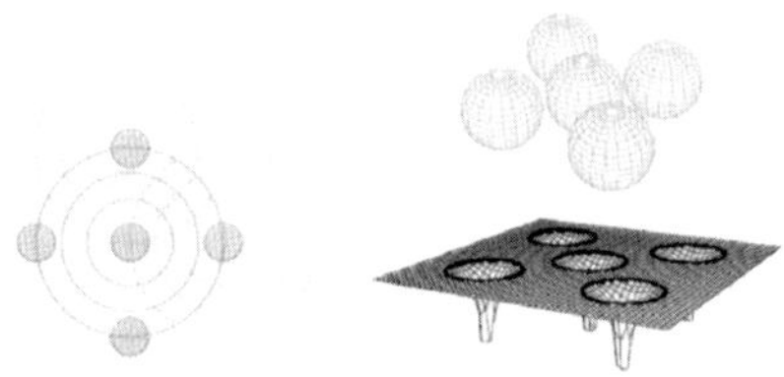

Figure 3. a) The wave function of the photoelectron is scattered on the neighbouring atoms (b) the muffin-tin potential consisting of non-overlapping spherical regions [14].

The formal solution of the operator on equation (8) is given by the Lippman-Schwinger equation [15, 16],

$$|\psi\rangle = |\varphi\rangle + \frac{1}{E - H_0 \pm i\eta} V |\psi\rangle \tag{10}$$

The operator $\frac{1}{E - H_0 \pm i\eta}$ can be considered as $\frac{1}{E - H_0}$ modified by an imaginary term in the denominator $i\eta$with ηinfinitesimally small and positive. The comparison of equations (9) and (10) reveals the following equality (11)

$$G_0^{\pm}(r, r'; E) = \langle r | \frac{1}{E - H_0 \pm i\eta} | r' \rangle \tag{11}$$

The operator $\frac{1}{E - H_0 \pm i\eta}$ can be considered as $\frac{1}{E - H_0}$ modified by an imaginary term in the denominator $i\eta$with ηinfinitesimally small and positive. The comparison of equations (11) and (12) reveals the following equality (11)

and then

$$G_0^{\pm} = \frac{1}{E - H_0 \pm i\eta} \tag{12}$$

And obviously,

$$\langle r | G_0^{\pm} | r' \rangle = G_0^{\pm}(r, r'; E) = \frac{-1}{4\pi} \frac{\exp(\pm ik(r - r')}{|r - r'|} \tag{13}$$

where$k=\sqrt{E}$, $G_0^+$and G_0^- describe how outgoing and incoming spherical waves propagate in free space, respectively. The Lippman-Schwinger equation can be solved formally if the transition operator T is introduced [15, 16],

$$V|\psi\rangle = T|\varphi\rangle \tag{14}$$

The propagator of the whole system can be defined as:

$$G(E) = \frac{1}{E - H_0 - V + i\eta} \tag{15}$$

One can deduce that:

$$G = G_0 + G_0 T G_0 = G_0 + G_0 V G_0 \tag{16}$$

and

$$T = V + VG_0V = V + VGV \tag{17}$$

The total potential is $V=\sum_i \upsilon^i$ and the potentials $\upsilon^i$are non-overlapping, the transition operator of individual atoms is:

$$t^i = \upsilon^i + \upsilon^i G_0 \upsilon^i \tag{18}$$

The propagator matrix G written in equation (16) can be extended by:

$$G = G_0 + G_0 T G_0 + G_0 T G_0 T G_0 + \ldots \tag{19}$$

where

$$T = \sum_i t^i \tag{20}$$

This last equation is a geometric series and therefore:

$$G = (1 - G_0 T)^{-1} G_0 \tag{21}$$

In the multiple scattering theory, the hard work is in computing the Green's function G. The function G describes all the possible ways for a photoelectron to interact with the surrounding atoms, G_0 is the function that describes how an electron propagates between two points in space. T is the parameter that denotes how a photoelectron scatters from a neighbouring atom.

The multiple scattering formalism used with the paradigm shift theorem, leads to another formulation of the Fermi's Golden law given by Equation (21):

$$\mu(E) \propto -\frac{1}{\pi} \operatorname{Im}\langle i|\vec{\varepsilon} \bullet \vec{r}' G(r',r;E)\vec{\varepsilon} \bullet \vec{r}|i\rangle \tag{22}$$

This last expression of the Fermi's Golden rule is very efficient, because there is no more sum over final states. This law is, in some cases, expressed with the wave functions that can be determined from electronic-structure calculations. In band structure calculations in crystals for example, the eigenstates and eigenvalues of the total Hamiltonian H can be obtained using either Bloch and Wannier functions or Linear Combinations of Atomic Orbitals (LCAO) methods [17]. In that case, the state vectors $|i\rangle$ and $|f\rangle$ represent the stationary solutions of electrons moving in a periodic potential.

Conversely, the situation can be faced using the MS theory. According to the paradigm shift theorem, there is a relation between the Green's function $G(r', r, E)$ and the final state vector:

$$n(E) = -\frac{1}{\pi} \operatorname{Im} G(r',r,E) = \sum_f |f\rangle \delta(E-E_f)\langle f| \tag{23}$$

As with the total Green's function $G(E)$ which consists of the free propagator $G_0(E)$ and the term describing the propagation of the scattered wave, the total density of states can also be divided into two parts as shown in equation (23). The "atomic" part, $n_0(E)$ and the change due to the scattering, $n_{scat}(E)$:

$$n(E) = n_0(E) + n_{scat}(E) \tag{24}$$

It is clear that the density of final states is changed because of the presence of the scatterers. Thus, the behaviour of the physical system can be reinterpreted. The solutions can be understood either as stationary solutions of the total Hamiltonian or as solutions that are changed by the multiple scattering from surrounding atoms.

3. Electron microscopy observations

3.1. SEM Observations

The SEM morphology images are performed on an XL30S-FEG PHILIPS working at 3 kV. SEM images clearly illustrated in different figures, show that the CNTs synthesized in this study display widely different morphologies according to the values and the concentrations of variable parameters. Within medium plasma power (around 150 W); carbon nanotubes are grown with graphitic planes in a parallel direction to the fiber axis. These samples however display different mutual orientations.

Highly oriented films are obtained under optimized conditions as represented in figure 4a, consisting in the evaporation of TM which is Co or mixture of Co and Fe. The pressure of the chamber is 15 mbars. CNTs obtained are very appropriate for applications like electric field emission devices, as they require samples aligned perpendicularly on a flat substrate. But those obtained in figure 4b, are poorly oriented. These figures clearly show a general view of poorly oriented CNTs, with important physical defects. They can have another area of applications. Carbon nanostructures defects are useful in many fields of applications.

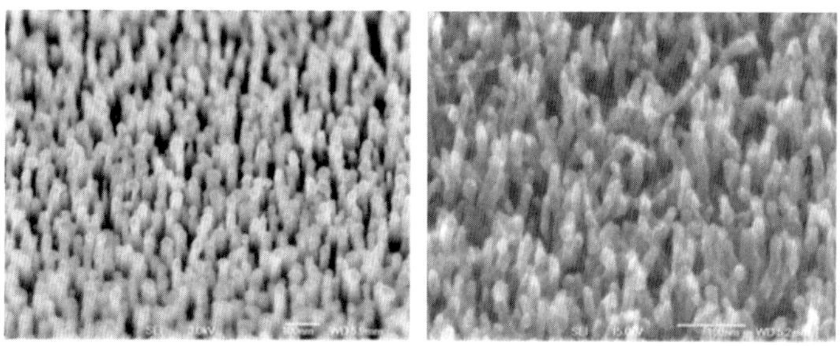

(a) (b)

Figure 4. SEM images: (a) highly oriented carbon nanotubes obtained under optimized conditions (b) poorly oriented carbon nanotubes

Figure 4(b) shows poorly oriented nanotubes films obtained with more defects. Anyway the presence of hot filaments heated around 2200 K must be stressed. They provide hydrogen radicals that are very reactive towards all kinds of amorphous carbon. Thus, carbon surrounds not only the particle but is also spread onto the surface of the sample. Figures 5 and 6 show the impact of the concentration of ammonia in synthesis chamber. In the particular case of CNTs, Table 1 shows important differences with respect to the ammonia concentration. First of all, the density decreases when the ammonia concentration increases. Secondly, the length distribution changes. The different values are: 300 to 450 nm for 0% of ammonia concentration, for 1% it is 250 to 500 nm and 800 to 3000 nm for 3%. The morphological defects are also related to ammonia concentration. They are important at 0 and 3%.

Figure 5. SEM images with low and high magnification at 0% NH_3

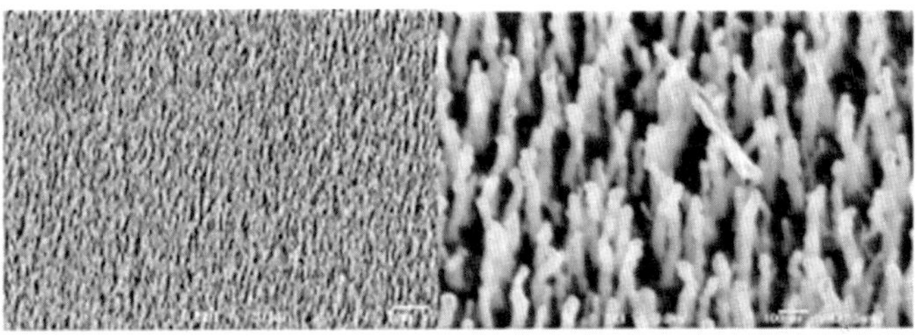

Figure 6. SEM images with low and high magnification at 1% NH_3

Sample	% NH_3	Density (μm^{-2})	Mean diameter (nm)	Length distribution
1	0	290	20 ± 5	300 - 450
2	1	50	20 ± 5	250 - 500
3	3	3	20 ± 5	800 - 3000

Table 1. Main experimental morphological and correlation data determined from SEM and TEM observation for CNTs.

In spite of the fact that SEM analysis gives valuable information in morphological and structural characterization of CNTs, it is however not sufficient to establish the ultimate nature of carbon nanotubes. It is easy to confuse only on the basis of SEM observations carbon nanotubes from nanofibers. Thus, one proceeds to TEM analysis of the samples to have deeper information on obtained CNTs.

3.2. TEM observations

The samples are examined by TEM in order to complete the SEM study by controlling the density and the morphology of CNTs deposited. TEM observations are performed on a TOPCON 002B microscope operating at 200 keV. Each sample is scratched with a diamond tip and the material is directly pulled onto the membrane for observations. The carbon membrane is drilled with holes in order to get more accurate observations. TEM micrographs clearly illustrate that nanotubes obtained display widely different morphologies according to some variable parameters. And it is possible to control the morphology. Within the medium value of the plasma power, as shown with SEM study, carbon nanotubes are yet grown. These samples however display different mutual orientations. The highly oriented films are obtained under optimized conditions and poorly and medium oriented films are also obtained, showing more defects.

One of the very important observations is that each carbon nanotube grown has TM nanoparticles at its top or its base [18, 19]. The TEM observations show at the same time that the surface of the nanotubes exhibits an amorphous structure due to wall surface defects (Figure 5a and Figure 5b). The outer diameters of the CNTs are directly determined from TEM images with high accuracy. Thus the mean outer diameter is 20 nanometers the smallest being 10 nm corresponding surely to SWCNTs or to a few number of walls. The upper values, may account for outer diameters of MWCNTs.

The analysis of TEM images lead to extensive values of length, varying from 100 to 400 nanomaters, as illustrated in Table 1. The density of nanotubes spreads also in a very large range. It ranges between 350 to 1000 μm. Figure 7 shows CNTs completely detached from the substrate with catalyst paticles on one end (figure 8), and the graphitic end of attachment to the substrate.

Unfortunately, TEM images of CNTs do not give the opportunity to determine the exact number of walls of each sample of CNTs (Figure 7a). This deficiency is the result maybe of defects. In this way, mutually aligned tubes of different densities are obtained, depending on the ammonia concentration in the reactive gas mixture. TEM images of CNTs show that the mean diameter is 25 nm and few micrometers in length. From the TEM image one could see that CNTs are very thin like a broadcast needles on the floor. The absence of TEM analysis can lead to wrong conclusion in the case of carbon nanomaterials. It is the most important and most reliable technique for correctly identifying the nature and the form of carbon nanomaterials. However, it remains some missing information. TEM images of CNTs and CNFs clearly distinct, but it is quite difficult to know the exact number of walls.

HRTEM is now capable of imaging individual atoms in nanostructures with subangström resolution as shown by Cowley and Liu [20]. The continue development of new tools is critical to the pace of further progress in nanoscience and technology, they provide the "eyes" to see and the "fingers" to handle nanostructures. In the nearer term, the greater need is to provide laboratory researchers with the instruments and tools to discover and investigate new chemical, physical, and biological phenomena and applications.

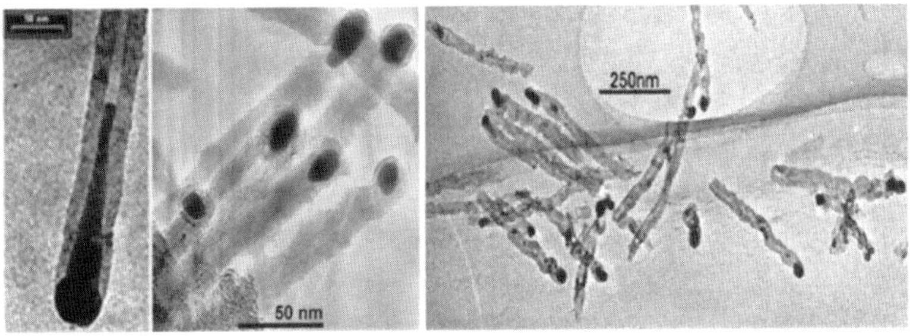

(a) (b) (c)

Figure 7. CNTs TEM images with TM particles inserted at one end: (a) CNT highly oriented, (b) CNTs medium oriented and (c) CNTs sample synthesized with 1% of NH3. The two ends of each CNT are well observed detached from the substrate. One end has graphitic structure (half fullerene) and the other has a TM particle inserted in the tube.

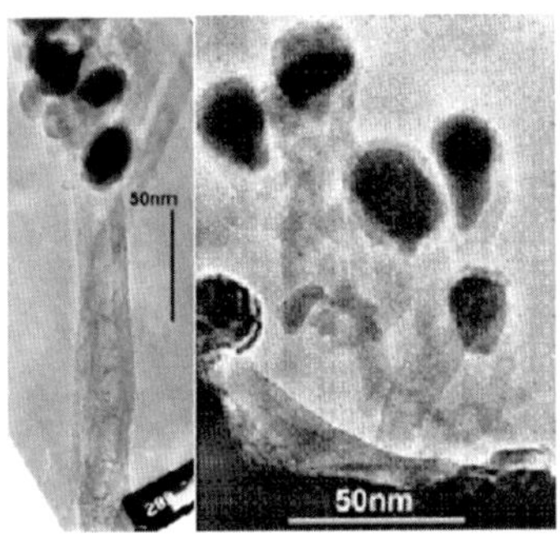

(a) (b)

Figure 8. CNTs with metal particles on top: a) highly oriented and b) poorly oriented.

In the longer term, those tools will evolve into inexpensive, easy-to-use sensors and/or diagnostic devices with numerous applications.

3.3. SEM and TEM Characterization and discussions

A perfect CNT is an abstraction, because the hexagonal structure sp^2 forms layers with different types of alterations as shown in literature. These alterations can come from the growth mode, from the deposit on a substrate, or they can be the result of a heat or chemical treatment. An important consequence of these defects on surface morphology that needs to be pointed out is the roughness of CNTs surface. The defects are in general very important because they can modify the electronic properties of the nanostructure, and so, can be seen as a feature that can influence nanostructures applications [21]. Generally, a defective site has a high chemical reactivity, in other words, this site is chemical reactions favoured [18]. Thus CNTs and CNFs with poor orientation, and which have many defects are not appropriate for applications in field emission devices, but can have other important applications like functionalization. In most cases, TEM investigations and seldom in SEM's, it is shown that the catalyst metal particles are attached to the nanostructures top end (nanofibers or nanotubes), or found inside the nanostructures (nanofibers, nanotubes and nanoparticles) or at last in the sidewalls for all varieties obtained (figures 7 and 8).

According to aiming applications, the presence of ultrafine cobalt or cobalt/iron nanoparticles of various diameters at the top of nanostructures can have negative effect. In particular in the case of emission of electrons or electric field, the particles need to be carried off. Thus, the purification of nanostructures needs to be one of crucial parts of synthesis process.

Single walled carbon nanotubes come often as tightly bundles of single walled nanotubes entangled as curly locks is often seen in literature. The packing of the nanotubes inside a bundle is not observed in specific case helping therefore to differentiate SWCNTs from MWCNTs. CNTs are simply observed in SEM images as tubes forest with poorly, medium or highly orientation depending on synthesis conditions, in particular the concentration of ammonia. It is possible that CNTs grown by HF PE CCVD be only MWCNTs. The second hypothesis is that defects or the presence of TM particles or other contaminants in nanostructures sidewalls, preclude SWCNTs present to gather in bundles. The third assumption can be that SWCNTs are present in samples, but these CNTs with lowest outer diameter which is 10 nm cannot form bundle because of orientation problems. The CNTs with the lowest outer diameter are depicted on two samples namely Nanot 29 and FLN 2. Nanot 29 is poorly oriented while FLN 2 has the high orientation as seen in figure 7. But in both cases, nanotubes are not gathering in bundles. Partially, the conclusion is that the degree of orientation is not precluding nanotubes to form bundles. But the absence of bundles in the case of eventual SWCNTs may be induced by defects or the attachment of catalytic nanoparticles. This situation may be solved by additional chemical or heat treatment stage for purification of obtained samples before any electron microscope imaging.

TEM gives direct insight into the structure of carbon nanomaterials and can help to identify the material or the phase correctly. Without observations by TEM, one may lead to wrong or incorrect conclusions. It is the most important and most reliable technique for correctly identifying the nature and the form of carbon nanomaterials, in spite of the fact that the information that can be extracted is not straight-forward, since the preparation of TEM samples may mask observation of the nanostructures of lower size, or the possibility of the projection of artefacts. TEM has become useful for *in situ* microscopy, for observing dynamic processes at the nanoscale nanomeasurements which directly correlate physical properties with structures, holographic imaging of electric and magnetic fields, quantitative chemical mapping at subnanometer resolution and for ultra-high resolution imaging. The main characteristics of CNTs obtained are given in Table 2.

Understand of the growth mode of CNSs in general is among the imperatives in their characterization and can lead to the growth modelling. In the case of CNTs or CNFs, two dominant growth modes have been observed in SEM and TEM images. The tip growth mode and the base growth mode, which are in agreement with literature. In tip growth mode, the transition metal nanoparticle catalyzing the growth of the carbon nanostructure remains at the top of the nanostructures. The adhesive forces between the substrate and the catalyst nanoparticles seem typically too small and the particles are lifted up as the CNTs or CNFs grow. The process takes place till the temperature is upper, before the cooling.

Sample	Carbon nanostructure	Outer diameter (nm)	Inner diameter (nm)	Lenth (nm)	Density (μm)
I Nanot29	CNTs (poorly oriented)	10	/	/	/
II Nanot31	CNTs	30	9	375	400
III Nanot42	CNTs (highly oriented)	25	5	400	349
IV FLN1	CNTs (medium oriented)	/	/	100	/
V FLN2	CNTs (highly oriented)	10	4	187	1000

Table 2. Main characteristic parameters of the carbon nanotubes synthesized.

4. XANES Characterization of CNTs

4.1. Assignments of peaks of CNTs

Owing to the alignment of nanotubes with a specific orientation of the σ bonds, it is expected that the absorption on the C K-edge, using total electron yield (TEY) and partial electron yield (PEY), knowing that the last is less surface-sensitive, will present angular selectivity when considering the specific $\pi \rightarrow \pi^*$ transition. The potential for the use of XANES as an unambiguous method of determining carbon nanostructures quality, orientation and contamination is due to the fact that one measures directly the unfilled electronic states and thus the chemical bonding state of the target atoms.

The assignment of features of XANES spectra of our samples of CNTs leads to the following tables (table 3 and table 4).

Annealed and potassium-contaminated CNTs at grazing incidence Nanot29_OE29				
Peak name	**Peaks**	**Binding energy (eV)**	**Kinetic energy (eV)**	**Final-state band and Brillouin-zone**
A	285.50	285.08	1199.32	π_0 near Q
	286.53	287.14	1197.26	free-electron-like interlayer states + adsorption
A''	288.40	288.52	1195.88	free-electron-like interlayer states + adsorption
A'''	288.70	289.68	1194.72	free-electron-like interlayer states + adsorption
A''''	290.70	290.60	1193.80	free-electron-like interlayer states + adsorption
C-H Exc.	291.76	291.44	1192.96	Exciton
B	292.65	292.20	1192.20	σ_1, σ_2: $\Gamma \rightarrow Q$
C	295.50	294.55	1189.85	π_0 or π_1 near Γ
D	297.80	298.05	1186.35	
P''	297.51	296.35	1188.05	Potassium L_2 level
E	303.50	301.75	1182.65	σ_7 near Q
F	307.50	306.50	1177.90	σ_9 near Q
G	308.50	309.72	1174.68	
H	316.50	314.10	1170.30	π_4 near Q
I	329.00	329.00		
J	333.00	333.00		
P	296.60	296.95	1187.45	Potassium L_2, L_3 level
P'	299.83	299.37	1185.03	Potassium L_2-L_1 levels

$R = (A+C+H+P'')/(B+D+E+F+G) = 0.863$

$R=(\pi/\sigma)_{tot}$

Table 3. Main features' parameters for annealed and potassium-contaminated CNTs XANES spectrum at grazing incidence.

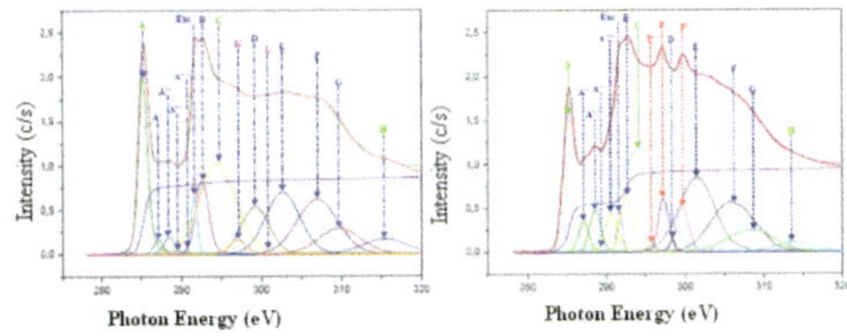

(a) MWCNTs (b) SWCNTs

Figure 9. CNTs experimental and calculated superimposed XANES spectra at grazing incidence from unannealed samples: (a) without potassium contamination, (b) with potassium contamination.

Annealed and potassium-contaminated CNTs at normal incidence Nanot29_OE36				
Peak Name	**Peaks**	**Binding energy (eV)**	**Kinetic energy (eV)**	**Final-state band and Brillouin-zone**
A	285.50	285.20	1199.20	π_0 near Q
A'	286.53	287.31	1197.09	free-electron-like interlayer states + adsorption
A''	288.40	288.67	1195.73	free-electron-like interlayer states + adsorption
A'''	288.70	289.68	1194.72	free-electron-like interlayer states + adsorption
A''''	290.70	290.35	1194.05	free-electron-like interlayer states + adsorption
C-H Exc.	291.76	291.64	1192.76	Exciton
B	292.65	292.40	1192.00	σ_1, σ_2: $\Gamma \rightarrow Q$
C	295.50	294.67	1189.73	π_0 or π_1 near Γ
D	297.80	298.30	1186.10	
P''	297.51	296.20	1188.20	Potassium L_2 level
E	303.50	301.85	1182.55	σ_7 near Q
F	307.50	306.54	1177.86	σ_9 near Q
G	308.50	310.00	1174.40	
H	316.50	314.30	1170.10	π_4 near Q
I	329.00	329.00		
J	333.00	333.00		
P	296.60	297.20	1187.20	Potassium L_2, L_3 level
P'	299.62	299.37	1184.78	Potassium L_2-L_1 levels
R = (A+C+H+P'')/(B+D+E+F+G) = 0.962				

Table 4. Main features' parameters for annealed and potassium-contaminated CNTs XANES spectrum at normal incidence.

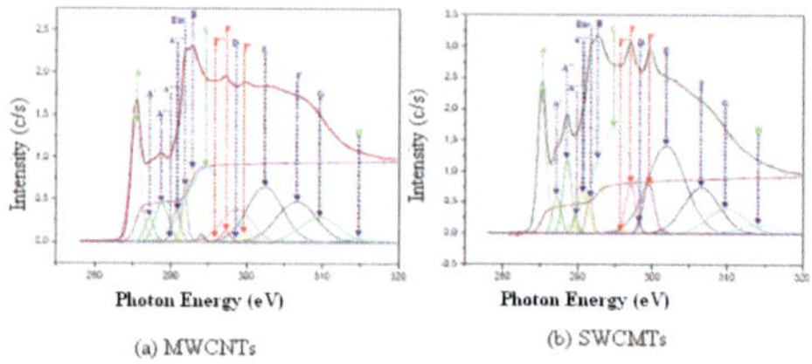

Figure 10. Contaminated CNTs experimental and calculated superimposed XANES spectra at normal incidence from annealed (550°C) samples: (a) sample is sensitive to thermal annealing; (b) sample is less sensitive to thermal treatment.

It is elementally selective by the tunability of the synchrotron X-ray source and sensitive to the bond order according to optical dipole selection rules. In the light of the XANES and HOPG spectra used as a starting point model, a prior annealing of the samples prevents the increase of intensity in the free-electron-like interlayer states region of the spectra, clarifying that these features are not intrinsic.

4.2. Orientation tendency parameter of CNTs R(α)

The parameter R(α) deduced from the fitting of the carbon K-edge absorption spectra by the ratio of the intensity of π^*-type features (A+C+K+H) over σ^*-type features (B+D+E+F+G+L), is defined to determine more quantitatively the respective contributions of the σ^* and π^* transitions at incidence angle α, and it is given by the relation below:

$$R(\alpha) = \frac{A+C+H+K}{B+D+E+F+G+L} \tag{25}$$

or

$$R(\alpha) = \frac{\sum \pi}{\sum \sigma} \tag{26}$$

This ratio or parameter calculated in Table 3 and Table 4, is indicative of the orientation tendency of the CNTs orbitals, and by the way, the graphite layer. In the related case, R(α) equals 0.868 in grazing incidence and 0.962 in normal incidence. It is expected to be maximum and minimum whens the XANES spectrum is recorded at normal and grazing incidence, respectively. In good agreement, R(α) in grazing incidence and R(α) in normal incidence are quoted to 0.868 and 0.962, respectively.

A prior annealing of the samples has prevented the increase of intensity in the free-electron-like interlayer states band region of the spectra, clarifying that these features are not intrinsic as shown elsewhere [8, 9]. But there is not total extinction observation of the σ^* band at 285.5

eV at normal incidence, the residual intensity of that peak at α=0 is probably due to either the uncompleted polarization of the synchrotron X-ray light beam or sample misalignment. The R(α) parameter in Table 9 equals 0.078. Thus, unambiguously the π orbitals lie perpendicular to the graphene sheet.

As a general trend, the *R(α)* parameter values at normal incidence are greater than those at grazing incidence for CNTs contrary to those of the HOPG [24, 25]. The obtained CNTs are classified in three groups according to its values according to table 5:

- Those with poor orientation,
- The other with medium range orientation,
- The last group which is formed of high CNTs.

For poorly oriented CNTs, the discrepancy between the two values (NI and GI) is too small.

It increases towards high oriented CNTs according to Table 5 bellow.

CNTs	CNTs1	CNTs2	CNTs3
GI	0.778	0.962	0.975
NI	0.732	0.882	0.665
ΔR	0.046	0.079	0.310
Observation	Poor Oriented	Medium Oriented	High Oriented

Table 5. The values of *R(α)* of three samples of CNTs. The discrepancy of these values at normal incidence (NI) and grazing incidence (GI) allows the classification of CNTs in: poor, medium and high oriented.

5. CNSs contamination

If the defects like: topological defects (the occurrence of pentagons and heptagons), the sp^3 hybridized carbons atoms and incomplete bonding that have slight changes that can be neglected, are not taken into account, it is observed that some features present in Figure 9 (a) are not found in Figure 9 (b), especially peaks K and L, and are replaced by P, P′ and P″. These new peaks are not intrinsic to CNTs. They are the result of contamination which can be considered as accidental adsorption of atoms, molecules or radical compounds, in agreement with SEM and TEM analysis, where it is not found bundles of CNTs as it may be, according to literature. Among the reasons of the presence of non-intrinsic features in XANES spectra is the presence of TM particles as proved by SEM and TEM. Actually, it is known that the features attributed to the so-called free-electron-like interlayer states in the graphite and other carbon nanostructures are also due to contamination [8, 9]. The peaks P, P′ and P″ are assigned to adsorbed potassium atoms according to features parameters [9, 23].

6. Conclusion

We investigate qualitatively and quantitatively the properties of carbon nanotubes synthesized by HF PE CCVD on SiO_2/Si(100) substrate using electron microscopies and X-ray absorption spectroscopy near-edge. According to SEM and TEM images and XANES spectra, CNTs are highly oriented under optimized conditions, notably when the ammonia concentration is 1% of gases mixture, but when this concentration is different (0 or 3%), samples are full of defects. Highly oriented CNTs are obtained under optimized conditions consisting of the evaporation of Co or mixture of Co and Fe or other TM element. The pressure of the chamber is 15 mbars. The CNTs obtained are very appropriate for applications like electric field emission devices when ammonia concentration is 1% of gases mixture, because such devices need CNSs samples oriented perpendicular to the plan of the substrate. The analysis of TEM images reveals that the length of each CNT varies from 100 to 400 nm, and their density is included between 350 and 1000 μm. The mean outer diameter depends on the size of catalyst particle. Its value is around 20 nm while the smallest value is 10 nm corresponding surely to SWCNTs.

Perfect CNTs are an abstraction or a creation of the mind, because of the hexagonal structure of sp^2 carbon atoms, which form the graphite layers, with always alterations. These defects are important because they often modify the electronic properties of the nanostructures, and can influence their applications. SWCNTs come often as tightly bundles entangled as curly locks is seen in literature, but these packing of the nanotubes inside bundles is not observed in specific case helping to differentiate SWCNTs from MWCNTs. CNTs are simply observed in SEM images as tubes with poorly, medium or highly oriented, depending on synthesis conditions. The degree of orientation is not precluding nanotubes to form bundles. But the absence of bundles in the case of eventual SWCNTs may be induced by defects or the attachment of catalytic nanoparticles. In particular, in the case of emission of electrons or electric field, these nanoparticles need to be removed. Thus, the purification of nanostructures needs to be one of crucial parts of synthesis process. This situation may be solved by additional chemical or heat treatment of obtained sample before electron microscope imaging and spectroscopic.

The contamination is due to atoms, radicals or molecules adsorbed of many species present or formed during synthesis phase, among them are oxygen (O); water and TM (catalyst). But the spectral features P, P′ and P′′ observed in many spectra, are assigned to 7potassium (K) contamination. This potassium present in samples might come from the beam lines contamination. These results elucidate that the CVD is one of the best techniques for synthesis of the CNSs for multiple purposes.

As a summary, the morphology and the structure of CNTs obtained by HF PE CCVD on SiO_2/Si(100) depend widely on the transition metal used as catalyst, and the experimental parameters during the growth process. It appears that synthesized CNTs can be used for many purposes according to growth conditions which determine their properties. Those with electronic defects are appropriated for functionalization because of their high chemical reactivity. Those with small outer diameter are also characterized by high chemical reactivity and can also be used in functionalization. On the other hand, CNTs with highly oriented configuration are very ideal in field or electrons emission devices manufacture.

Acknowledgements

The author would like grateful acknowledge Pr Motapon Ousmanou, Pr Mane Mane J., Dr Ben-Boli G. H. and Dr Tiodjio Sendja B. for many helpful discussions.

Author details

Rolant Eba Medjo*

Address all correspondence to: emeroch@yahoo.fr

Department of Physics, Faculty of Science, University of Douala, Republic of Cameroon

References

[1] Dresselhaus, M. S., Dresselhaus, G., & Eklund, P. C. (1996). *Science of Fullerene and Carbon Nanotubes*, Academic Press.

[2] Ebbesen, T. W. (1997). *Carbon Nanotubes: Preparation and Properties*, Chemical Rupper Corp, Boca Raton, FL.

[3] Zuttel, A., & Sudana, P. (2002). *International Journal of Hydrogen Energy*, 27, 203.

[4] Taschner, C., Pacal, F., Leonhardt, A., Spatenka, P., Bartsch, K., Graff, A., & Kaltofen, R. (2003). *Surface and Coatings Technology*, 81, 174.

[5] Thiodjio Sendja, B., Eba Medjo, R., Mane Mane, J., Ben Bolie, G., Diop, D., & Owono Ateba, P. (2010). A GISAXS Study of Angular dependence of Carbon Nanotubes grown on a plain substrate by dc HF CCVD process. *Phys. Scr.*, 82, 025601.

[6] Rosenberg, R. A., Love, P. J., & Rehn, V. (1986). Polarization-dependent C(K) near-edge x-ray-absorption fine structure of graphite. *Phys. Rev. B*, 33, 4034.

[7] Shimoyama, Iwao, Wu, Guohua, Tetsuhiro, Sekiguchi, & Yuji, Baba. (2001). Study of electronic of graphite-like carbon nitride. *Journal of Electron Spectroscopy and Related Phenomena*, 114-116, 841-848.

[8] Zhong, J., Liu, C., Wu, Z. Y., Mamatimin, Kurash I., Cheng, H. M., Gao, B., & Liu, L. (2005). XANES Study of Carbon Based Nanotubes. *High Energy Physics and Nuclear Physics*, 29, 97.

[9] Eba Medjo, R., Thiodjio Sendja, B., Mane Mane, J., & Owono Ateba, P. (2009). A study of carbon nanotube contamination by XANES spectroscopy. *Phys. Scr.*, 80, 045601.

[10] NIST. *Physical Reference Data,* http://physics.nist.gov/PhysRefData/XrayMassCoef/cover.html.

[11] Joly, Y. http://www.ned.cnrs.fr.

[12] Koningsberger, D. C., & Prins, R. (1988). X-Ray Absorption- Principles, Applications, Techniques of EXAFS, SEXAFS and XANES. *Chemical Analysis*, 92, Wiley.

[13] Koningsberger, D. C., & Prins, R. (1992). X-Ray Absorption- Principles, Applications, Techniques of EXAFS, SEXAFS and XANES. *Wiley-Interscience*, New-York.

[14] Mihelic, A. (2002). http://www.p-ng.si/~arcon/xas/xanes/xanes-theory.pdf.

[15] Sakurai, J. J. (1994). *Modern Quantum Mechanics*, Rev. Ed., Addison-Wesley.

[16] Merzbacher, E. (1970). *Quantum Mechanics*, John Willey.

[17] Ashcroft, N. W., & Mermin, N. D. (1976). *Solid state physics*, Saunders College.

[18] Eba Medjo, R., Thiodjio Sendja, B., Mane Mane, J., & Owono Ateba, P. (2009). XAS study of orientated carbon nanotube films. *Phys. Scr.*, 80, 055602.

[19] Mane Mane, J. (2007). *Habilitation à Diriger des Recherches*, Université Louis Pasteur de Strasbourg.

[20] Cowley, J. M., & Liu, J. (1993). Contrast and resolution in REM, SEM and SAM. *Surface Science*, 298, 456.

[21] Bonard, J-M. (2005). Carbon nanostructures by Hot Filament Chemical Vapor deposition: Growth, properties, applications. *Thin Solid Films*, 501, 8-14.

[22] Stöhr, J. (1992). *NEXAFS Spectroscopy*, Springer, Berlin.

[23] Sasaki, S. (1984). KEK, National Laboratory for High Energy Physics. *Report*, 83, 22.

[24] Eba Medjo, R. (2011). *Structural and morphological characterization of carbon nanostructures synthesized by chemical vapour deposition using spectroscopic techniques and electron microscopies*, Thesis, University of Yaounde I.

[25] Eba Medjo, R. (2011). Carbon Nanotubes Synthesis. *In: Marulanda, J. M. (ed) Carbon Nanotubes Applications on Electron Devices*, Rijeka: Intech, 3-36, Available from, http://www.intechopen.com/books/carbon-nanotubes-applications-on-electron-devices/carbon-nanotubes-synthesis, accessed 01 august 2011.

Chapter 7

Carbon Nanotubes Under Simple Tension and Torsion – Molecular/Structural Mechanics and the Finite Element Method

Najib A. Kasti

Additional information is available at the end of the chapter

1. Introduction

The intended applications of carbon nanotubes have steadily increased since their discovery by Ijima [4]. They range from the nanoscale, as in the tip of an atomic electron microscope, to the macroscale, as in the preliminary design of the space elevator cable.

In modeling CNTs, molecular as well as quantum mechanics have been the primary tools for analysis. Also, closed form expressions were developed to study the response of CNTs in different environments [3,8,12].

On the other hand, some attempted to use structural mechanics, and built corresponding finite element models, to study the behavior of CNTs, as evident in several publications [5-8]. Some of these publications simplified the property relations between molecular mechanics and structural mechanics. They assumed the structural bending stiffness EI/a to be a constant and set it equal to the molecular bond bending stiffness C.

However, in [1], with a simple proof, we showed that the main assumption used by various authors to equate the element bending stiffness to the bond bending stiffness ($C= EI/a$) does not hold. In addition, in our previous publications (Kasti [1,2]), we related some of the mechanical properties used in molecular and structural mechanics.

In [1], we derived an expression for the axial deformation of zigzag CNTs that accounts for the axial and bending structural stiffnesses under simple tension. While molecular mechanics uses the bond angle between two bonds to describe bond bending deformations, structural mechanics uses the bending within one 3D frame element for this definition. Comparing the deformation equation in structural mechanics to the equivalent equation derived for molecular mechanics, leads us to a "consistent" frame bending stiffness for an infinitely long zigzag CNT. For large diameter tubes, the frame bending stiffness tends to half the bond bending stiffness. This later case is representative of a graphene sheet. For small diameters, EI/a changes with the bond bending stiffness C, the torsional angle φ and the lattice translational index n. The expression for the axial deformation was then expanded to include the torsional stiffness of 3D frame elements. This provided us with an equation for the "elastic modulus" of an infinitely long zigzag CNT. It included the axial, bending and torsional deformations. Then, in [2], we extended the work to the case of simple torsion of zigzag carbon nanotubes. The expression relating the molecular bond bending stiffness C and the structural bending stiffness EI/a was derived. It was found to be different from the case of simple tension. The structural bond bending stiffness was both load and chirality dependent. However, for the particular configuration of a graphene sheet, the relation of simple tension was recovered, namely $EI/a=C/2$. We concluded the work by presenting the expression for the deformation of the tube when axial, bending and torsional structural stiffnesses are accounted for. We noticed in this case of simple torsion that the axial stiffness couples with the bending and torsion stiffnesses, unlike simple tension.

In a recent paper (Chen et al., [9]), the radial elastic modulus of the original Molecular Structure Mechanics model (MSM) was compared to the one from the Molecular Dynamics (MD) simulation. In that paper, it was pointed to the fact that a modification to the original MSM model was suggested in our previous paper (Kasti, [1]).

In this chapter, we extend our previous work to armchair carbon nanotubes under simple tension. In addition, we summarize the equivalent results for zigzag CNTs under simple tension and torsion.

We start with a brief review of molecular and structural mechanics and we refer to the work of Chang and Gao [3]. Then, the relation between the structural bending stiffness EI/a and the molecular bond bending stiffness C is derived for the case of simple tension. This shows that EI/a depends on the bond bending stiffness C and the torsional angle φ. In the limit of an infinite tube radius, which represents a graphene sheet, we recover the previous relation, i.e., EI/a tends to $C/2$. Finally, an expression for the Young's modulus is presented that accounts for the axial, bending and torsional deformations. We conclude the chapter with numerical simulations that validate the results.

2. Bond Energies and the Finite Element Method

This section deals with the molecular and structural mechanics formulations of bond energies. Also, a short review of the finite element method is presented as it applies to the modeling of carbon nanotubes subjected to mechanical loading.

2.1. Characterization of the Atomic Structure and Molecular/Structural Mechanics of Carbon Nanotubes (CNTs)

The geometry of a CNT could be described with the pair (n,m), the lattice translational indices, and the bond length a. In general, the diameter d of a CNT is defined using the expression

$$d = \frac{\sqrt{3}a}{\pi}\sqrt{n^2 + m^2 + nm} \tag{1}$$

For zigzag CNTs (Fig. 1a), the value of m is zero. In this case, the diameter is simply given by the formula $d = \frac{\sqrt{3}an}{\pi}$. However, for armchair CNTs (Fig. 1b), the value of m is equal to n. In this case, the diameter is given by the formula $d = \frac{3an}{\pi}$.

The bond energies between carbon atoms include the stretching U_a, bending U_b and torsional U_t energies. For small distortions from equilibrium, these energies could take the forms:

$$\begin{aligned} U_a &= \frac{1}{2}K(r - r_o)^2 \\ U_b &= \frac{1}{2}C(\Theta - \Theta_o)^2 \\ U_t &= \frac{1}{2}C_t(\Phi - \Phi_o)^2 \end{aligned} \tag{2}$$

where K, C, and C_t are the bond stretching, bending and torsional stiffnesses.

r, Θ, and Φ are the stretched position, bending and torsional angles, respectively.

The subscript "$_o$" refers to the initial equilibrium configuration.

As far as structural mechanics, the linear elastic deformation is assumed to be the combination of axial, bending and torsional deformations. Their corresponding strain energies are expressed as:

$$\begin{aligned} U_{axial} &= \frac{1}{2}\int EA\left(\frac{du}{dx}\right)^2 dx \\ U_{bending} &= \frac{1}{2}\int EI\left(\frac{d\theta}{dx}\right)^2 dx \\ U_{torsion} &= \frac{1}{2}\int GJ\left(\frac{d\phi}{dx}\right)^2 dx \end{aligned} \tag{3}$$

where EA, EI, and GJ are the axial, bending and torsional stiffnesses; and u, θ and φ are the axial, bending and torsional deformations, respectively.

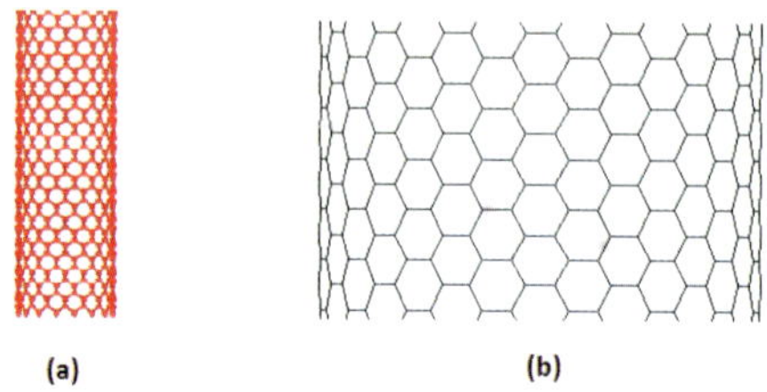

Figure 1. Carbon Nanotubes: (a) Zigzag, (b)Armchair.

2.2. Review of the Finite Element Method for 3D Space Frames

For linear elastic behavior of 3D space frames, the axial, bending and torsional strain energies can be expressed as in equation (3).

The bonding between two carbon atoms is modeled by placing a 3D space frame element between them, Fig. 2. When this procedure is repeated throughout the tube, a finite element mesh is obtained with the carbon atoms becoming the nodes in the mesh.

Each node is assumed to have six degrees of freedom, three translational and three rotational.

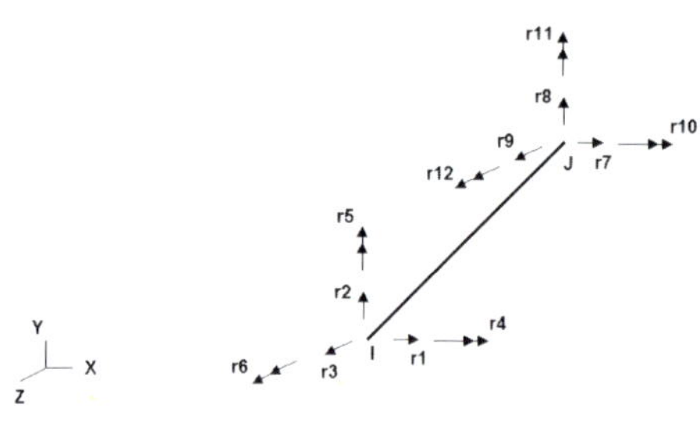

Figure 2. Space frame element.

The stiffness matrix K relating the degrees of freedom to their corresponding forces and moments at both ends of a 3D frame element is a 12x12 matrix.

$$K = \begin{bmatrix} K_{ii} & K_{ij} \\ K_{ij}^{T} & K_{jj} \end{bmatrix} \tag{4}$$

3. Work of Chang and Gao [3]

Chang and Gao derived closed form expressions for carbon nanotubes subjected to simple tensile loading using molecular mechanics.

Representing units of zigzag and armchair carbon nanotubes are shown in Fig. 3 with α and β being the internal angles. Equivalent equations to the ones of Chang and Gao will be derived in the next section for the armchair CNT.

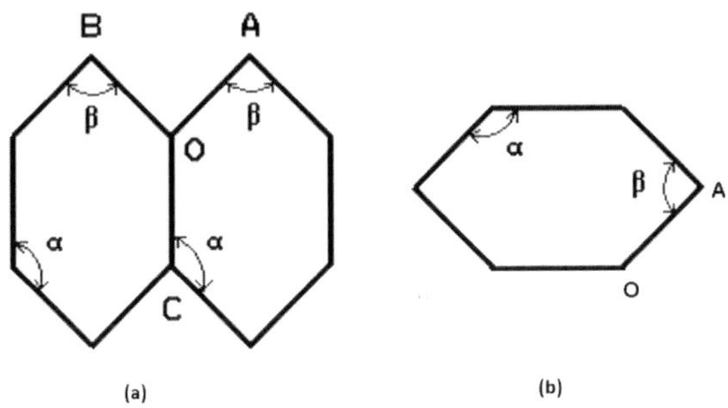

Figure 3. a) Two units of a zigzag CNT. (b) One unit of an armchair CNT.

4. Axial and Bending Stiffnesses of Armchair CNTs: Molecular versus Structural Mechanics

We start by expressing the results of Chang and Gao [3] in a more suitable form using the principle of minimum total potential energy. We will split the approach into bond stretching and bond bending deformations.

4.1. Bond Stretching – Molecular/Structural Mechanics

Due to bond stretching, it is easy to verify that $K = EA/a$.

We start with the molecular energy expression

$$\Pi = \sum_1 \frac{1}{2} K(u - u_o)^2 - \sum_2 F\Delta \tag{5}$$

where F is the axial load applied to a single carbon atom, Δ is the deflection of the end of the tube, Σ_1 is a summation over the number of bonds and Σ_2 is the summation over the number of atoms with applied loads.

Expression (5) takes the following form in structural mechanics:

$$\Pi = \sum_1 \frac{1}{2}(EA / a)(u - u_o)^2 - \sum_2 F\Delta \tag{6}$$

Since both K and EA/a are conjugate to the axial deformation between carbon atoms in the energy equation, they represent the same axial stiffness. Thus,

$$K = EA / a \tag{7}$$

To determine the tube deformation, we let n_u be the number of vertical units of Fig. 3 in a carbon nanotube and n_v equal to $(2n_u-1)$. The deflection at the end of the tube due to axial bond deformations can be expressed as:

$$\Delta = \frac{2n_v F \sin^2(\alpha / 2)}{(EA / a)} \tag{8}$$

4.2. Bond Bending – Molecular Mechanics

Due to bond bending, the total potential energy is written as:

$$\prod = \sum_1 \frac{1}{2} C(\alpha - \alpha_0)^2 + \sum_2 \frac{1}{2} C(\beta - \beta_0)^2 - \sum_3 F\Delta \tag{9}$$

Let n_x be the number of units along the circumference and n_v the number of vertical units. Then, for an infinite cylinder with no end effects,i.e., all units have the same deformation, Π will be equal to:

$$\prod(\alpha, \beta, \Delta) = \frac{1}{2}[2n_x(2n_v - 1)]C(\alpha - \alpha_0)^2 + \frac{1}{2}[4n_x(2n_v - 1)]C(\beta - \beta_0)^2 - (2n_x F)\Delta$$

Minimizing the total potential energy with respect to α gives

$$d\Pi / d\alpha = n_x[(2n_v - 1)C(\alpha - \alpha_0) + 2(2n_v - 1)C(\beta - \beta_0)d\beta / d\alpha - Fd\Delta / d\alpha] = 0$$

Since $\cos\beta = -\cos(\pi/2n)\cos(\alpha/2)$ and $\cos\varphi = \tan(\alpha/2)/\tan(\beta)$, we get $d\beta/d\alpha = \cos\varphi/2$, where φ is the torsion angle between the planes of adjacent units of an armchair nanotube (Fig. 3).

The vertical deformation of the tube can be expressed as:

$$\Delta = H - H_0 = (2n_v)a(\sin(\alpha/2) - \sin(\alpha_0/2)) \tag{10}$$

where H_0, H are the initial and current heights of the tube, respectively. Differentiating Eq. (10) with respect to α

$$d\Delta \Big/ d\alpha = \frac{1}{2}(2n_v)a\cos(\alpha \big/ 2)$$

Solving for $\Delta\alpha$ by minimizing the total potential energy, we get

$$\Delta\alpha = \alpha - \alpha_0 = Fa\cos(\alpha/2)/(2C + C\cos^2\phi) \tag{11}$$

Substituting $\Delta\alpha$ in Eq. (10) above, we get the following expression for the vertical deflection of an infinitely long CNT (due to bond bending only)

$$\Delta = \frac{(2n_v)Fa^2\cos^2(\frac{\alpha}{2})}{4C + 2C\cos^2\phi} \tag{12}$$

In the next Section, we will derive an equivalent expression in terms of the material properties of structural mechanics. This will allow us to deduce a relation between *EI/a* and *C*.

4.3. Bond Bending - Structural Mechanics: Armchair Carbon Nanotube under Simple Tension

Let *O* and *A* be two atoms on the CNT with coordinates $(R\cos(-\theta/2), R\sin(-\theta/2), 0)$ and $(R\cos(-\pi/n+\theta/2), R\sin(-\pi/n+\theta/2), a\sin(\alpha/2))$, respectively. The angle θ is given by $2R^2(1-\cos\theta)=a^2$.

Also, let *u* be the unit vector from *O* to *A* expressed as:

$$\boldsymbol{u} = u_1\boldsymbol{i} + u_2\boldsymbol{j} + u_3\boldsymbol{k} \tag{13}$$

where *i*, *j* and *k* are the unit vectors in the Cartesian coordinate system.

For an infinite cylinder and due to symmetry, the radial and tangential rotations at *O* and *A* are zero.

In addition, due to multiple symmetries (for *n*=4,8,12,16,..), we assume the displacement and rotation fields at *O* and *A* take the following forms:

$$\begin{aligned} \boldsymbol{\delta r_o} &= \delta \boldsymbol{i} \\ \boldsymbol{\delta r_A} &= \delta \sin(-\pi / n)\boldsymbol{i} + \delta \cos(-\pi / n)\boldsymbol{j} + \Delta \boldsymbol{k} \end{aligned} \quad (14)$$

and

$$\begin{aligned} \boldsymbol{\theta_o} &= \theta \boldsymbol{k} \\ \boldsymbol{\theta_A} &= -\theta \boldsymbol{k} \end{aligned} \quad (15)$$

where (δr_o, δr_A) and (θ_o, θ_A) are the displacements and rotations vectors at *O* and *A* respectively. In this work, small displacements and rotations are assumed. Without loss of generality, the tensile load at each carbon atom is assigned a value of one.

Following a similar procedure to Kasti [1], one can show that θ is equal to zero.

And, the only force in the inclined member OA in Fig. 3 is the vertical force F and the moment is (1/2)F.a.sin(α/2).

Due to multiple symmetries (for *n*=4,8,12,16,..), and after some simplifications, the axial deflection at the end of the whole CNT is given by:

$$\Delta = \frac{(2n_v)Fa^2\cos^2(\frac{\alpha}{2})}{12(EI / a)} \quad (16)$$

and the corresponding elastic modulus Y_s

$$Y_s = \frac{8EI\sin(\frac{\alpha}{2})}{a^3\cos^2(\frac{\alpha}{2})} \quad (17)$$

Zigzag CNTs under simple tension

The corresponding axial deformation at the end of a zigzag CNT under simple tension is given by [1]:

$$\Delta = \frac{(n_v + 1)Fa^2\sin^2\alpha}{(EI / a)[24 - 36\gamma / (1 + \gamma)]} \quad (18)$$

$$\text{with} \quad \gamma = 2\left(1-\cos\left(\pi / n\right)\right) / \left(3-2\left(R / a\right)^2 \sin^2\left(\pi / n\right)\right) \tag{19}$$

Zigzag CNTs under simple torsion

The tangential deformation at the end of a zigzag CNT unit under simple torsion is given by following expression for Δ_B [2]:

$$\Delta_B = -\frac{Fa^2(1+2\cos^2\left(\frac{\pi}{2n}\right))}{48.(\frac{EI}{a}).\cos^2(\frac{\pi}{2n})}. \tag{20}$$

5. Discussion and Validation of the Results obtained in Section 4 – Armchair CNTs

Comparing the molecular mechanics expression Eq. (12) with the structural mechanics Eq. (16), we obtain

$$\frac{EI}{a} = \frac{C}{6}[2+\cos^2\phi] \tag{21}$$

Thus, in general, the bending stiffness to be used in the structural mechanics varies with the bond bending stiffness C and torsional angle φ.

For long CNT tubes with large diameters, $cos^2(\varphi) \rightarrow 1$ and $EI/a \rightarrow C/2$, which is the same result derived by Kasti [1] for the particular case of a graphene sheet.

To validate the closed form solution Eq. (16), we compared the axial deformation of member OA (Fig. 3) and the change in radius to the results from a finite element model in ABAQUS [14]. The results are shown in Tables 1 and 2 below for C=1.42 nN.nm.rad^{-2}. The accuracy obtained is excellent.

Lattice Translational Index, n	Molecular/Structural Mechanics	ABAQUS
4	5.9844	5.9849
8	5.9477	5.9470
12	5.9358	5.9358
16	5.9310	5.9306

Table 1. Vertical deformation of member OA (x10^{-4}).

Lattice Translational Index, n	Molecular/Structural Mechanics	ABAQUS
4	-1.3538	-1.3539
8	-2.6402	-2.6401
12	-3.9383	-3.9383
16	-5.2403	-5.2402

Table 2. Change in radius ($x10^{-3}$).

Zigzag CNTs under simple tension

For zigzag CNT under simple tension, the equivalent stiffness is given by [1]:

$$\frac{EI}{a} = \frac{4C + 8C\cos^2\phi}{24 - 36\gamma/(1+\gamma)} \tag{22}$$

Thus, in general, the bending stiffness to be used in the structural mechanics varies with the bond bending stiffness C, torsional angle φ and γ.

For long CNT tubes with large diameters, $\cos\varphi \rightarrow 1$, $\gamma \rightarrow 0$ and $EI/a \rightarrow C/2$, which is the same result for the particular case of a graphene sheet.

Zigzag CNTs under simple torsion

For a zigzag CNTs under simple torsion, the corresponding stiffness is given by [2]:

$$\frac{EI}{a} = \frac{C}{6}\left[1 + 2\cos^2\left(\frac{\pi}{2n}\right)\right]. \tag{23}$$

Thus, in general, the bending stiffness to be used in structural mechanics varies with the bond bending stiffness C and lateral translational index n.

For long CNT tubes with large diameters, $n \rightarrow \infty$, $\cos(\frac{\pi}{2n}) \rightarrow 1$ and $EI/a \rightarrow C/2$.

6. Deformation of Armchair CNTs due to Axial, Bending and Torsional Structural Stiffnesses

In Sections 4 and 5, a closed form expression was developed for the deformation of infinitely long armchair CNT under simple tension. It included the axial and bending stiffnesses of 3D frame elements. In this Section, we study the effect of the torsional stiffness of 3D space frames.

6.1. Bond Bending and Torsion - Structural Mechanics: Armchair Nanotube under Simple Tension

Similar work to Kasti [1] will show that the torsional stiffness does not enter the expression for the deformation of an infinitely long armchair carbon nanotube under simple tension.

When the axial, bending and torsional deformations are combined, we obtain the following formula:

$$\Delta = \frac{2n_v F \sin^2(\alpha/2)}{(EA/a)} + \frac{(2n_v)Fa^2\cos^2(\frac{\alpha}{2})}{12(EI/a)} \quad (24)$$

To validate the closed form expression of Eq. (24), we compared the vertical deformation of member OA (Fig. 3) and the change in radius to the results from a finite element model in ABAQUS. The results are shown in Tables 3 and 4 below for K=652nN.nm^{-1} and C=1.42 nN.nm.rad^{-2}. The accuracy obtained is excellent.

Lattice Translational Index, n	Molecular/Structural Mechanics	ABAQUS
4	1.7686	1.7686
8	1.7500	1.7500
12	1.7461	1.7461
16	1.7446	1.7446

Table 3. Vertical deformation of member OA (x10^{-3}).

Lattice Translational Index, n	Molecular/Structural Mechanics	ABAQUS
4	-0.53131	-0.53136
8	-0.96001	-0.96001
12	-1.4088	-1.4088
16	-1.8634	-1.8634

Table 4. Change in radius (x10^{-3}).

The contribution of each of the bond stiffnesses (axial, bending and torsion) to the total vertical deformation of an armchair carbon nanotube is shown in the following example.

Two long carbon nanotubes (40 armchair carbon units) with lattice translational indices "n" equal to 4 and 16, respectively, are modeled using MSC/Nastran [15]. The tubes are supported at the bottom and subjected to tensile loading at the top.

The resulting vertical deformations are compared to the closed form solution of Eq. (24), as shown in Fig. 4. In spite of the difference in boundary conditions between the closed form solution and the finite element modeling, the errors in the results are less than 3%.

Zigzag CNTs under simple tension

Going through the same manipulations as for an armchair CNT, the vertical deformation at the end of a zigzag CNT under simple tension that accounts for bending and torsional deformations can be expressed as [1]:

$$\Delta = \frac{(n_v + 1)Fa^2 \sin^2 \alpha}{(EI / a)[24 - 36\gamma / (1+\gamma)]} \tag{25}$$

where F is the load applied at a carbon atom, n_u is the number of vertical units of Fig. 3 and $n_v = (2n_u - 1)$. However, in this case, γ takes on the following expression:

$$\gamma = 2\left(1 - \cos(\pi / n)\right) / \left(3 - 2(R/a)^2 \sin^2(\pi / n)(1 - \lambda)\right) \quad \text{and} \quad \lambda = GJ / EI \tag{26}$$

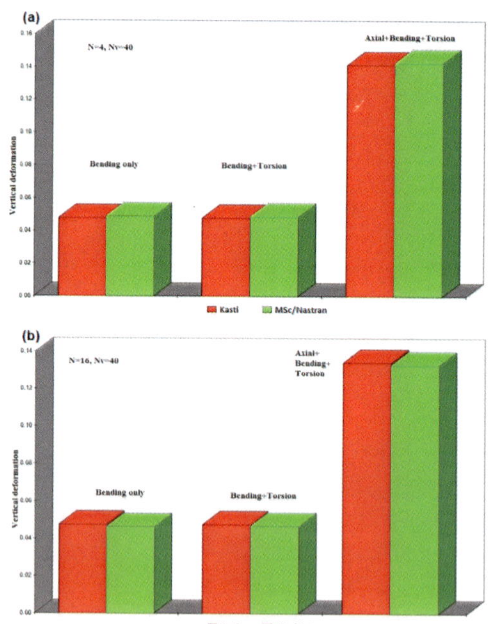

Figure 4. Vertical deformations of armchair carbon nanotubes with lattice translational indices of 4 and 16, respectively. Three cases are considered: 1) bending stiffness only, 2) bending + torsion and 3) axial+bending+torsional.

Zigzag CNTs under simple torsion

When the torsional stiffness of 3D frame elements is included, in addition to the axial and bending deformations, the tangential deformation Δ_B of a zigzag CNT under simple torsion takes the following form [2]:

$$\Delta_B = \frac{N^T KN}{D^T KD} \tag{27}$$

where *N* and *D* are 6x1 vectors function of *n*, the lattice translational index, and *a*, the bond length. *KN* and *KD* are 6x1 vectors of structural stiffnesses. These vectors could be expressed as:

$KN = [\ (EI/L)^3\ (EI/L)^2(GJ/L)\ (EA/L)(GJ/L)^2\ (EA/L)(EI/L)(GJ/L)\ (EI/L)(GJ/L)^2\ (EA/L)(EI/L)^2]^T$

$KD = [\ (EI/L)^4\ (GJ/L)(EI/L)^3\ (EA/L)(GJ/L)^2(EI/L)\ (EA/L)(EI/L)^2(GJ/L)\ (EI/L)^2(GJ/L)^2\ (EA/L)(EI/L)^3]^T$

$N = [\ N1\ N2\ N3\ N4\ N5\ N6\ ]^T$, $D = [\ D1\ D2\ D3\ D4\ D5\ D6]^T$

For example, for n=4,

$N = [\ -21.1721\ -24.53\ -0.0015\ -0.0969\ -0.4074\ -0.0832\ ]^T$

$D = [\ 0.\ -2.3062\text{x}10^3\ -0.4408\ -76.1719\ 497.0279\ -62.4643\ ]^T$

One point worth mentioning is that when the torsional stiffness is neglected, i.e. $GJ / L \to 0$, Δ_B from Eq. (27) takes the form:

$$\Delta_B = \frac{N1 * (EI/L)^3 + N6 * (EA/L) * (EI/L)^2}{D6 * (EA/L) * (EI/L)^3} = \frac{N1/D6}{(EA/L)} + \frac{N6/D6}{(EI/L)}.$$

Thus, in this case of negligible torsional stiffness, the axial and bending deformations are de-coupled.

6.2. Elastic Modulus of an Armchair CNT

Similar to the previous work by Kasti [1], an elastic modulus Y_s (Tpa.nm) could be defined that doesn't include the thickness of CNTs, and is expressed as:

$$Y_s = \frac{F_t}{2\pi R(\Delta / L)} = \frac{2L}{3a\Delta} \tag{28}$$

where F_t (equal to 2nF with F=1) is the total load applied, *L* and *R* are the length and radius of the tube, respectively.

For a finite length cylinder, the elastic modulus obtained from the closed form expressions of Eq. (28) and ABAQUS are compared in Table 5 for a lattice translational index of 4 and variable tube length. The following values of stiffnesses were used:

$EA/a = 652$ nN.nm^{-1}, $EI/a = 0.875$ nN.nm.rad^{-2} and $GJ/a = 0.278$ nN.nm.rad^{-2}.

The closed form results compare very well with the values from ABAQUS.

Lattice Translational Index, n	n_v=40,60 (Eq. 28)	n_v=40 (ABAQUS)	n_v=60 (ABAQUS)
4	358.12	358.284	358.307

Table 5. Elastic Modulus (nN.nm^{-1}).

Zigzag CNTs under simple torsion

Similar to the definition of an elastic modulus Y_s (Tpa.nm) that doesn't include the thickness of CNTs [1], an elastic shear modulus Gs (Tpa.nm) could be defined as [2]:

$$G_s = \frac{T}{(\frac{\varphi}{L})*2\pi R^3} = \frac{nFR}{(\frac{\varphi}{L})*2\pi R^3} = \frac{n}{(\frac{\varphi}{L})*2\pi R^2} \tag{29}$$

where T is the torsional moment applied to the tube, F is the tangential load applied to a single carbon atom which can be taken as unity, L and R are the length and radius of the tube, respectively.

7. Conclusions

Relations between the structural bending stiffness EI/a and the molecular bond bending stiffness C for carbon nanotubes were derived for the cases of simple tension and torsion. In addition, expressions for the deformations and "Young's moduli" of these nanotubes were presented that account for the axial, bending and torsional effects.

Author details

Najib A. Kasti*

Address all correspondence to: najib01@idm.net.lb

Department of Mechanical Engineering, American University of Beirut, Lebanon

References

[1] Kasti, N. (2007). Zigzag Carbon Nanotubes: Molecular/Structural Mechanics and the Finite Element Method. *Int. J. of Solids and Structures*, 44, 6914-6929.

[2] Kasti, N. (2012). Zigzag Carbon Nanotubes under Simple Torsion- Structural Mechanics Formulation. *Advanced Materials Research*, 452-453, 1139-1143.

[3] Chang, T., & Gao, H. (2003). Size-dependent elastic properties of a single-walled carbon nanotube via a molecular mechanics model. *J. of the Mechanics and Physics of Solids*, 51, 1059-1074.

[4] Ijima, S. (1991). Helical microtubules of graphite Carbon. *Nature*, 354, 56-58.

[5] Li, C., & Chou, T. (2003). A structural mechanics approach for the analysis of carbon nanotubes. *Int. J. of Solids and Structures*, 40, 2487-2499.

[6] To, C. (2006). Bending and shear moduli of single-walled carbon nanotubes. *Finite Elements in Analysis and Design*, 42, 404-413.

[7] Tserpes, K. I., & Papanikos, P. (2005). Finite element modeling of single-walled carbon nanotubes. *Composites B*, 36, 468-477.

[8] Xia, J. R., Gama, B. A., Gillespie, J. W. Jr., & , . (2005). An analytical molecular structural mechanics model for the mechanical properties of carbon nanotubes. *Int. J. of Solids and Structures*, 42, 3075-3092.

[9] Chen, W-H., Cheng, H-C., & Liu, Y-L. (2010). Radial mechanical properties of single-walled carbon nanotubes using modified molecular structure mechanics. *Computational Materials Science*, 47, 985-993.

[10] Li, C., & Chou, T-W. (2004). Elastic Properties of single-walled carbon nanotubes in transverse directions. *Physical Reviews B*, 69(7), 073401.

[11] Wang, X-F., Xu, Z-J., & Zhu, Z-Y. (2007). Reversible mechanical bistability of carbon nanotubes under radial compression. *Chemical Physics*, 334, 144-147.

[12] Chang, T., Geng, J., & Guo, X. (2006). *Proceedings of the Royal Society A*, 462, 2523.

[13] Gallagher, R. (1975). Finite Element Analysis: Prentice Hall.

[14] ABAQUS- Finite Element Analysis Program.

[15] MSC/Nastran- Finite Element Analysis Program.

Section 3

Surface Chemistry

Chapter 8

Small Molecules and Peptides Inside Carbon Nanotubes: Impact of Nanoscale Confinement

Peng Xiu , Zhen Xia and Ruhong Zhou

Additional information is available at the end of the chapter

1. Introduction

Carbon-based nanoparticles and nanostructures, such as carbon nanotubes (CNTs), have drawn great attention in both academia and industry due to their wide potential applications. Owing to their well-defined one-dimensional (1D) interior, CNTs serve as desirable materials for encapsulating molecules, such as water [1-4], ionic liquid [5], drug molecules [6], and biomolecules [7]. The nanoscale confinement of CNTs have considerable impact on the inner molecules, including changes in their structure, size distribution, surface area, and dynamics, thus leading to many interesting and striking properties that are quite different from those in bulk [1-5, 7-9]. For example, nanoscale confinement of CNTs can give rise to ordered structure and extra-fast motion of water molecules [1-4], significantly enhanced activity of catalytic particles [8], phase transition of ionic liquids from liquid to high-melting-point crystal [5], and denatured structures of peptide helices [9]. In particular, recent studies [10-13] have shown that these CNT-based nanomaterials can be used as a new paradigm of diagnostic and therapeutic tools, which is beyond the traditional organic chemistry based therapeutics in the current pharmacology. Before their wide applications in the biomedical filed, the effects of CNTs on biomolecules (and drug molecules) need to be understood thoroughly [14-20].

In this book chapter, we review some of our recent works [21-24], with large scale molecular dynamics (MD) simulations using massively parallel supercomputers such as IBM Blue Gene, on the nanoscale confinement of both small molecules and peptides inside the CNT, which demonstrate wide implications in nanoscale signal processing, single-file transportation, drug delivery, and even cytotoxicity. The structure of this chapter will be organized as following. First, we show that water molecules confined within a Y-shaped CNT can realize the molecular signal conversion and multiplication, due to the surprisingly strong dipole-induced orientation ordering of confined water wires [25]. Second, we find a striking phenomen that urea can induce the drying of CNTs and result in single-file urea wires. The unique properties of a urea wire as well as its biological and technological implications are discussed [22, 23]. Third, we show that nanoscale confinement can catalyze the chiral transition of chiral molecules. We further explore the molecular mechanism of CNT-catalyzed enantiomerization and provide some implications for drug delivery [24]. Last, we investigate the effect of confinement of CNT on three important secondary structural motifs of proteins – a hairpin turn, a helix, and a beta-sheet.

2. Results

2.1. Water-mediated signal multiplication with Y-shaped nanotubes

Uunderstanding the molecular-scale signal transmission (amplification, shunting, etc) has attracted intensive attentions in recent years because it is of particular importance in many physical, chemical, and biological applications, such as molecular switches, nano-gates, and biosensors [26-29]. However, due to the intrinsic complexity of these nano-systems and the significant noises coming from thermal fluctuations as well as interferences between branch signals, the molecular details are far from well understood. On the other hand, water molecules confined within nanochannels exhibit structures and dynamics quite different from bulk [3], which might provide a medium for molecular signal transmission. Water molecules inside CNT with a suitable diameter can form a single-file hydrogen-bonded molecular wire, with the concerted water dipole orientations, i.e., either parallel or antiparallel to the CNT axis [1, 30, 31]. The characteristic time for reorientation of the dipole orientation of water wire is in the range of 2–3 ns for CNT with a length of 1.34 nm [1], and the water wire inside a nanochannel can remain dipole-orientation-ordered up to macroscopic lengths of ~ 0.1 mm, with durations up to ~ 0.1 s [30]. If we can "tune" the orientation of a water molecule at one end, we might be able to control the orientations of all water molecules in the molecular wire and even amplify and shunt the orientation signal.

Recently, Y-shaped nanotubes have been successfully fabricated by means of many different methods [32-34]. These nanotubes have been found to exhibit both electrical switching and logic behaviour [27, 35]. In the following, we will show that single-file water wires confined within a Y-shaped single-walled CNT (hereafter referred to it as Y-SWNT, see Fig. 1) can perform both signal amplification and shunting, ignited by a single electron, because of the surprisingly strong interactions between water molecules at the Y-junction. We construct Y-SWNT by jointing three (6, 6) uncapped armchair single-wall CNTs (SWNTs) together symmetrically along three directions neighbouring 120° one another. An external charge, q, is positioned at the centre of a second carbon ring of the main nanotube (see Fig. 1) to monitor the dipole orientation of water wire inside the tube. All carbon atoms were fixed and an opposite charge was assigned at the edge of simulated boxes to keep the whole system charge-neutral. MD simulations were carried out in NVT ensemble (300K, 1atm) with Gromacs 3.3.3 [36]. The TIP3P [37] water model was used.

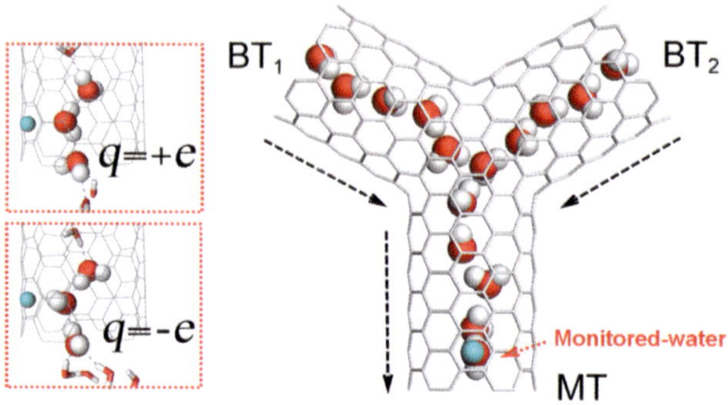

Figure 1. Schematic snapshot of the simulation system in side-view. The Y-SWNT consists of a main tube (MT) and two branch tubes (BT_1, BT_2) positioned in the same plane. Water molecules outside the nanotubes are omitted. The light blue sphere represents the imposed charge. The water molecule facing the external charge is referred to as "Monitored-water". The lengths of MT, BT_1 and BT_2 are 1.44 nm, 1.21 nm, and 1.21 nm, respectively. Insets: Enlarged part for the typical configurations: upper for q = -e and lower for q = +e. This figure is reproduced from ref. [21] with permission.

The simulations show that water molecules in the Y-SWNT form single-file hydrogen-bonded molecular wires. Although the water wires in different tubes interact at the Y-junction, all water's orientations are either parallel or anti-parallel to the nanotube axis, similar as the case of water wire in conventional SWNT [1]. To describe quantitatively the confined water's dipole orientation, we choose an angle ϕ_i between the dipole orientation of ith water molecule and the SWNT axis, and the average angle $\bar{\varphi}(t)$, which the average over all the water molecules inside a nanotube at some time t. The outward direction of the main tube and inward directions of the branch tubes are set as positive directions. The results are displayed in Fig. 2(A). It is clear that $\bar{\varphi}$ dominantly falls in two ranges for each nanotube, $10^\circ < \bar{\varphi} < 70^\circ$ and $110^\circ < \bar{\varphi} < 170^\circ$, indicating that the water molecules within each nanotube are nearly aligned. Furthermore, we have noticed that $\bar{\varphi}(t)$ for all tubes falls in the range from 10° to 70° when q = -e, with few fluctuations to larger values. In contrast, when q = +e, $\bar{\varphi}(t)$ for the main tube primarily falls into the range from 110° to 170°. For the branch tubes, $\bar{\varphi}(t)$ jumps between the two ranges. From the water orientations in each branch tube, we can easily identify the sign of the imposed charge, i.e., the charge signal at the main tube correctly transmits and is amplified/shunted to the two branch tubes.

To further characterize the molecular signal transmission, we define an integer $s(t)$: $s(t)$ = +1 when $10^\circ < \bar{\varphi} < 70^\circ$, and $s(t)$ = -1 when $110^\circ < \bar{\varphi} < 170^\circ$. We calculate the $P(t)$, defined as the occurrence probability of $s(t)$ = +1 from the start of the simulation until the time t in each tube. For a sufficiently long time, $P(t)$ in both branch tubes will approach 1.0 when q=-e, and approach 0.5 when q =+e since $\bar{\varphi}(t)$ falls in the two different ranges with an equal probability. Here, we set P_C = 0.8 as the threshold value to determine the charge. It is expected that $P > P_C$ indicates q = -e, and that $P < P_C$ indicates q = +e. From Fig. 2(B) we can see that, for both branch tubes, when q = -e, $P > P_C$ for $t >$ 1 ns; when q = +e, $P < P_C$ for $t >$ 8 ns. Consequently, the charge signal at the main tube can be readily distinguished from the value of $P(t)$ in each branch tube within a time interval of ~8 ns.

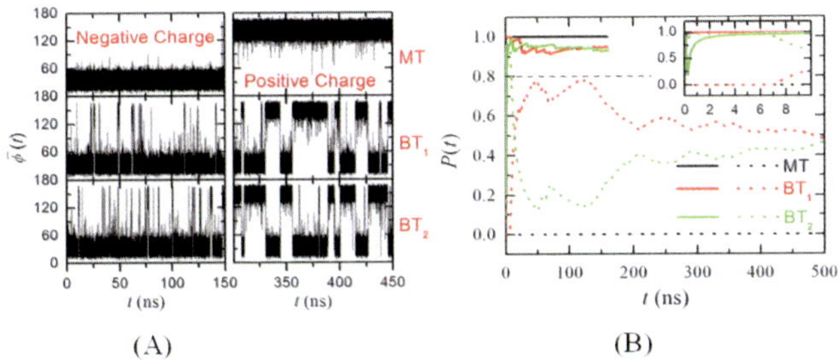

Figure 2. Trajectory of average dipole angle $\bar{\varphi}(t)$ of the water orientation and the probability of dipole orientation $P(t)$ in each tube in a Y-SWNT. (A) Average dipole angle in the main tube (MT), first branch tube (BT_1) and second branch tube (BT_2) for a negative charge (left) and a positive charge (right) in the main tube. (B) $P(t)$ in different tubes for a negative charge (solid lines) and a positive charge (dashed lines). $P(t)$ for a negative charge converges to about 1.0 within a few nanoseconds. This figure is reproduced from ref. [21] with permission.

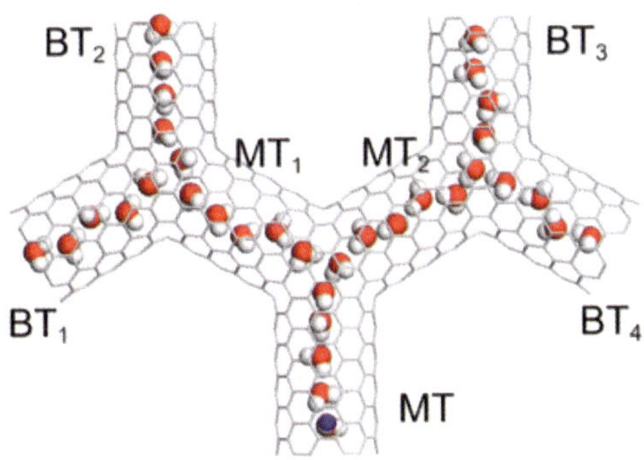

Figure 3. Snapshot of a three Y-junction (3Y-SWNT) system (side view). Colours match those in Fig. 1. The angle between any two neighbouring tubes at each Y-junction is 120°. The lengths of the main tube (MT), two middle tubes denoted by MT_1 and MT_2, and four branch tubes denoted by BT_1, BT_2, BT_3 and BT_4 are 1.44 nm, 1.44 nm, and 1.21 nm, respectively. This figure is reproduced from ref. [21] with permission.

Careful examinations reveal that the external charge "monitors" the water molecule facing this charge (referred to as the "Monitored-water"); the Monitored-water determines the water orientations in the main tube; the uppermost water molecule in the main tube governs the dipole orientations of the bottommost water molecules in branch tubes and hence the water dipole orientations within both branch tubes (see ref. [21] for more discussions). In addition, we find that the response to the switching of the charge signal is very rapid, from a few nanoseconds to a few hundred nanoseconds: In response to $-e \rightarrow +e$ signal switching, the time delay for the branch tubes is 40 ns on average with a maximal duration of 150 ns; in response to $+e \rightarrow -e$ polarity flip, it is only around 4ns.

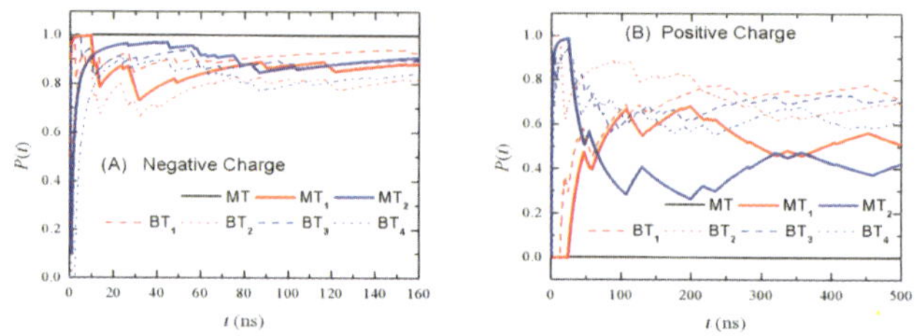

Figure 4. Probability *P*(*t*) in the main tube (black line), two middle tubes (blue and red solid lines), and four branch tubes (dashed lines) in response to a negative (A) and a positive (B) imposed charge signal. This figure is reproduced from ref. [21] with permission.

The charge signal can also be transmitted and amplified/shunted through additional channels. We have simulated a system with three Y-junctions where each of the outlet branch tubes forms a Y-junction connecting two more tubes (see Fig. 3). We refer the two middle tubes as MT_1 and MT_2, and the four branch tubes as BT_1, BT_2, BT_3 and BT_4. Fig. 4 shows the $P(t)$ for different branch tubes. It is found that when t > 200 ns, $P(t) > P_C$ when q = -e, and $P(t) < P_C$ when q = +e, for all branch tubes. As a consequence, the charge signal at the main tube transmits to four branch tubes with a temporal resolution time of ~200 ns.

To summarize, by using MD simulations we show that a signal at the single-electron level can be converted and multiplied into two or more signals by water wires confined within a narrow Y-shaped CNT. This remarkable capability of signal transduction by Y-SWNT derives from the surprisingly strong dipole-induced ordering of such water wires, so that the concerted water orientations in the two branches of the Y-SWNT can be modulated by the orientation of water wire in the main channel. The response to the switching of the charge signal is found to be very rapid, from a few nanoseconds to a few hundred nanoseconds. To our knowledge, this is the first observation of the remarkable signal amplification and shunting with a Y-shaped nanotube at the atomic level and this observation may have significance for future applications in molecular-scale electronic devices. In addition, it is noteworthy that there are Y-shaped biological channels [38, 39], therefore, our findings might also provide useful insight into the molecular signal transmission in biological systems.

2.2. Molecular wire of urea and induced drying in carbon nanotubes

2.2.1. Molecular wire of urea inside narrow carbon nanotube

Molecules confined inside nanoscale space such as narrow nanotubes or membrane proteins can form one-dimensional (1D) molecular wires, which have attracted intense interest recently because of their scientific importance and potential applications in nanotechnology [1, 21, 40-56]. Among them, it is of particular interest in determining the structure and dynamical behavior of water wires [1, 21, 40-49] which have been found to exist in narrow nanotubes[1, 21, 40-42, 46-48] and biological channels [43-45]. Water wires have many interesting properties, such as wavelike density distributions [1, 46], rapid and concerted motions [1, 40, 43], orientation-ordered structures and collective flips [1, 21, 41, 48], and excellent on-off gating behaviors [46, 47]. In addition, it has been observed that the methane [56], methanol [54], and gas molecules (O_2, H_2, and CO_2) [55] preferentially bind to the interiors of narrow SWNT over water and form 1D molecular wires. Despite the above progress, the properties of molecular wires have not been fully understood, particularly for the molecular wires formed by larger polar organic molecules.

Urea plays an important role in the metabolism of nitrogen-containing compounds by animals [57, 58], and serves as a common protein chemical denaturant and an important raw material in chemical industry. It is important to note that the biological urea channel dvUT (a urea transporter from the bacterium *Desulfovibrio vulgaris*) has a long (~ 16 Å) and narrow selectivity filter; this filter consists of closely spaced hydrophobic residues which allows dehydrated urea to permeate in single-file [58]. The hydrophobic SWNTs with appropriate diameters might serve as useful model systems for studying biological urea channel. The current simulations were based on TIP3P water model [37] and two commonly used urea models, namely, KBFF [59] and OPLS [60, 61] models. Below we mainly present the results for the KBFF case; the results for OPLS case are similar, and some of them are also shown as comparison. The simulation were performed using Gromacs 4.0.7 [62] in an NPT (300K, 1 atm) ensemble.

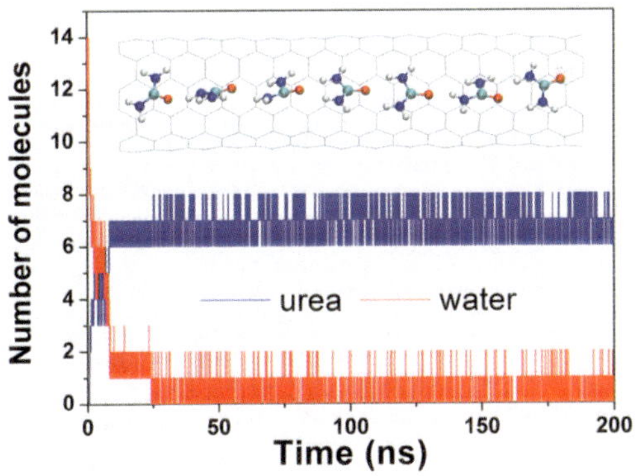

Figure 5. Number of urea (in blue; KBFF urea model is used) and water (in red) molecules within the 336-carbon (6, 6) SWNT as a function of simulation time, at 1 M urea concentration. Inset: Snapshot of a "perfect" urea wire.

We have performed MD simulation of 336-carbon (6, 6) SWNT (3.32 nm in length), solvated in aqueous urea with various urea concentrations (8M, 1 M and 0.5 M, with the simulation lengths 100 ns, 200 ns, and 200 ns, respectively). Fig. 5 shows the number of solvent (water/urea) molecules inside the SWNT in case of 1 M urea concentration during the course of simulation. Almost all water molecules inside the SWNT are replaced by urea within the first 25 ns. The confined urea molecules form a 1D "perfect" urea wire with a contiguous hydrogen-bonded network in most of the simulation time, or occasionally forms a "defective" urea wire [with a very small number of "water defect(s)", commonly near the SWNT edge].

Table 1 summarizes the average number of urea ($\bar{N}_{urea}$) and water molecules ($\bar{N}_{water}$) inside the SWNT after the systems have reached equilibrium with various urea concentrations. Regardless of urea concentration and urea model used, finally, the SWNTs are nearly completely filled with urea molecules. Table 1 also shows the occurrence probability for "perfect" urea wire, $P_{perfect}$, which is high for most cases. These results indicate that urea has a robust capability to form uninterrupted molecular wire.

C_{urea}	KBFF			OPLS		
	$\bar{N}_{urea}$	$\bar{N}_{water}$	$P_{perfect}$	$\bar{N}_{urea}$	$\bar{N}_{water}$	$P_{perfect}$
0.5 M	7.03	0.29	72.6%	6.23	1.67	11.8%
1 M	7.09	0.16	84.0%	6.68	0.85	42.9%
8 M	7.10	0.03	96.7%	7.00	0.14	86.8%

Table 1. Average number of urea and water molecules ($f_{drying}=R_{SWNT}/R_{bulk}$ and $\bar{N}_{urea}$, respectively) inside the 336-carbon (6, 6) SWNT in equilibrium, together with occurrence probabilities for "perfect wire" ($P_{perfect}$ [a]), with various urea concentrations (C_{urea}) and with different urea models.

Next, we explore the structure of the confined urea wire. We use the case of the 336-carbon (6, 6) SWNT in 8 M KBFF urea for illustration because $P_{perfect}$ in this case is very high (see Table 1). We performed two independent 100 ns simulations under same conditions, denoted by case 1 and case 2, respectively. As shown in the inset of Fig. 5, urea molecules inside (6, 6) SWNT form a single-file structure with a contiguous hydrogen-bonded network and concerted dipole orientations [urea's dipole orientation approximates the dipole orientation of its carbonyl (-CO-) group]. Quantitatively, we have computed ϕ (the angle between a urea dipole and the nanotube axis). ϕ is found to fall in two ranges: the angle around 20º (case 1) and around 160º (case 2). No event of flipping between these two ranges is observed during the time period of 100 ns. Even for urea wire in 144-carbon (6, 6) SWNT, no flipping event is observed for KBFF urea, and 1~2 flipping events is observed for OPLS urea, during several independent 100 ns simulations. In contrast, the flipping of water wire inside 144-carbon (6, 6) SWNT occurs every 2~3 ns on average [1, 48]. Further analysis reveal that the lower flipping frequency of urea wire compared with water mainly comes from the larger physical dimension and higher polarity of urea [23].

The above findings have technological implications. Our previous reports [21, 25] have demonstrated water wires can mediate the signal conversion and multiplication because of their ordered 1D structure and collective flipping behavior. However, the very small size of the water and fast flipping of water wire make the experimental realization very difficult [25]. Urea wire has similar ordered 1D structure and flipping behavior as water wire but has a lower flipping frequency and a high molecular polarity which can facilitate the signal detection in practice (urea wire has longer response time [21] to switch its dipole orientation under the influence of a change in charge signal). We therefore expect that urea wire can serve as a better candidate for signal transduction and multiplication.

Next, we have calculated the position distribution of urea along the nanotube axis. There are seven distinct, sharp peaks (with an average peak-to-peak value of ~ 4.6 Å), indicating that the urea wires are translationally ordered along the SWNT axis. The position distribution is found to be much sharper than water wire owing to the larger molecular size of urea (see ref. [23] for details).

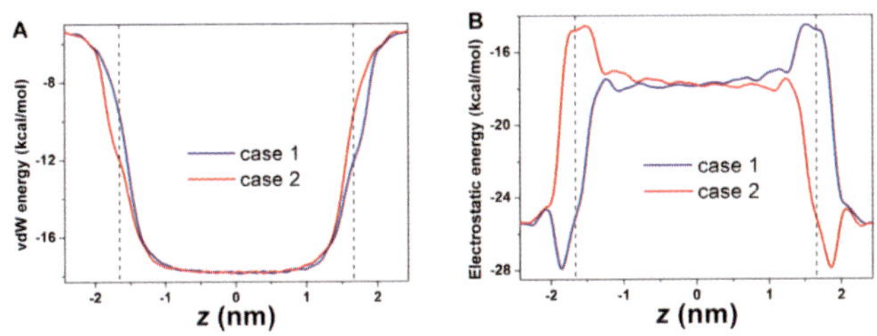

Figure 6. Potential energy profiles of urea along the axis of 336-carbon (6, 6) SWNT (8 M urea, KBFF urea model is used). (A) and (B) show the van der Waals (vdW) and electrostatic potentials, respectively. Case 1 and case 2 denote independent simulation under same conditions. The positions of SWNT inlet/outlet are indicated with dashed lines.

We have also calculated the interaction energies with the rest of the system for a urea molecule with respect to its axial distance from the geometrical center of SWNT (see Fig. 6). Interestingly, the vdW potential curves are approximately symmetric; whereas electrostatic potential curves are observably asymmetric, i.e., correlate to the inner urea's dipole orientations. Urea's asymmetric molecular partial charge distribution together with the extremely confined space result in the orientationally ordered structure (concerted dipole orientations) of molecular wire, thus breaking the symmetry of the system within a finite time period (more than 100 ns for the present case) and causing an asymmetric electrostatic potential.

Although single-file transport of water through SWNT has been intensively investigated in recent years [1, 40, 46-48], much less is known about the single-file transportation for organic small molecules. Here we explore the transport properties of urea wire and make a comparison with water wire. We have calculated the urea flow, defined as the total number of urea molecules per nanosecond that have entered from one end and leave the SWNT from the opposite side. Given that the biological urea channel dvUT [52] has a length of ~ 16 Å (the number of urea molecules accommodated in the selectivity filter is about 3), we chose the 144-carbon (6, 6) SWNT (13.5 Å in length) as the nanochannel, wherein the resulting urea wire also consists of ~ 3 urea molecules. To facilitate a direct comparison with water wire, we performed additional simulations for the SWNT immersed in pure water. The calculated average flows (averaged over three independent 100 ns simulations) are 0.73 ns^{-1} and 0.79 ns^{-1}, for KBFF and OPLS urea, respectively, and it is 16.2 ns^{-1} for water. Transportation of urea seems to be 20+ times slower than water. Fig. 7(A) displays the time evolution of urea flow from a typical simulation trajectory. The urea flow is low, with a maximal value of only 4 ns^{-1}; it vanishes frequently, and the duration time of zero value can be up to 6 ns (e.g., t = 11 ns ~ 17 ns). In contrast, for water wire, its minimal flow is up to 7 ns^{-1}, and its maximal flow reaches a value of 32 ns^{-1}. Furthermore, we have studied the influence of urea concentrations (1 M $\leq C_{urea} \leq$ 10 M) of the surrounding bath on urea's permeability through SWNT and find a maximal urea flow (~0.87 ns^{-1}) around a concentration of 5 M (see ref. [23] for more details).

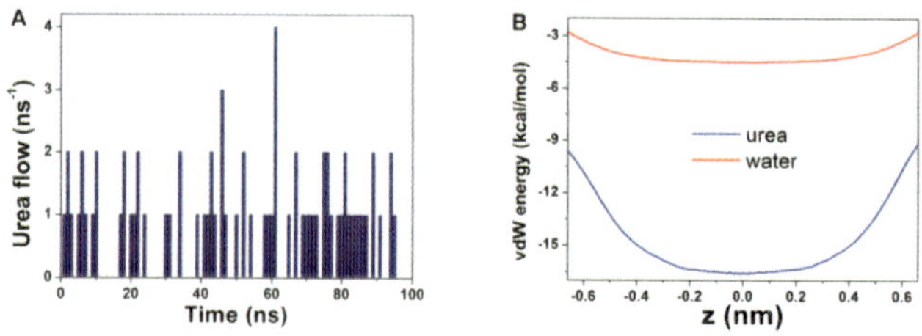

Figure 7. Single-file transport of urea through 144-carbon (6, 6) SWNT (8 M urea, using KBFF urea for demonstration) and the underlying physics. (A) Urea flow versus time from a typical trajectory. (B) The potential energy profiles along the SWNT axis for the urea wire (blue) and the water wire (red), respectively. The data for water derives from the control runs of SWNT immersed in pure water.

To understand the physical mechanism behind the enormously lower permeability of urea relative to water for SWNT, we have calculated the interaction energies of a inner urea/water molecule with the SWNT (the data for water derives from the control runs of SWNT immersed in pure water). Because the carbon atoms of SWNT are modeled as uncharged Lennard-Jones particles, there are only vdW interactions between urea/water and SWNT. As displayed in Fig 7(B), the potential valley for urea is much deeper than that for water, because urea has a stronger dispersion interaction with SWNT than water, which in turn leads to a much lower permeability of urea than water.

In this section, we have investigated the structure and dynamical behavior of urea wire inside the narrow SWNT. Even at relatively low urea concentration (e.g., 0.5 M), we have observed spontaneous and continuous filling of SWNT with a 1D urea wire. The resulting urea wire is translationally and orientationally ordered, with a contiguous hydrogen-bonded network and concerted dipole orientations of urea molecules. Despite the symmetric nature of SWNT, the urea's potential energy profile along SWNT is asymmetric, coming from asymmetric molecular partial charge distribution (or dipole moment) and the ordering of urea's dipole orientation under extremely confinement. Furthermore, we have studied the single-file transportation of confined urea, and find that urea flow decreases significantly (by a factor of ~ 20) compared to that of water, due to the fact that urea has a stronger dispersion interaction with SWNT than water. We also find a maximum in urea permeation around a concentration of 5 M. The studies on the urea wire confined inside SWNT not only help our understanding of the unique properties of confined polar organic molecules, but also present biological (biological urea channel) and technological (e.g., electronic devices for signal transduction and multiplication at nanoscale) implications.

2.2.2. *Urea-induced drying of carbon nanotubes*

In the previous section, we have demonstrated that urea can expel water inside a narrow SWNT [(6, 6) SWNT]. One may wonder if this phenomenon can persist in wider SWNT. To answer this, we performed MD simulations of (17, 8) SWNT (1.73 nm in diameter, it can accommodate several layers of urea and water) immersed in 8 M urea solution. Considering that there are some urea models commonly used in literature whose charge distributions are quite different [22], herein we have used five different urea models to test if the drying phenomenon is sensitive to force fields used.

The five urea models used in the current study are the OPLS [60, 61], KBFF [59], CHARMM (parameters derived from the CHARMM22 force field [63]), AMBER* [64], and AMBER [parameters derived from the file embedded in the AMBER 10 simulation package (University of California at San Francisco)] urea models. The simulation were performed using Gromacs 4.0.7 [62] in an NPT (300K, 1 atm) ensemble with the simulation lengths of 100 ns for all systems. In all cases, we observe that most of water molecules initially inside the SWNT (C_{urea} inside the SWNT is approximately 8 M from the initial solvation setup) are repelled from the SWNT within the first 10 ns; after that, the hydrophobic nanopores are dominantly occupied by urea. Table 2 lists the average number of urea and water molecules inside (17, 8) SWNT with different urea models. To quantitatively characterize the drying effect, we have calculated the "drying factor", f_{drying}, defined as following:

$$f_{\text{drying}} = R_{\text{SWNT}} / R_{\text{bulk}} \quad (1)$$

where R_{SWNT} and R_{bulk} are the ratios of the average number of urea to water molecules inside SWNT and in the bulk region, respectively. A larger f_{drying} means a stronger urea-induced drying effect. f_{drying} for different urea models are also shown in Table 2. In all cases, f_{drying} is very high, indicating that strong drying phenomena occur in all cases.

Model	$\bar{N}_{urea}$	$\bar{N}_{water}$	f_{drying}
OPLS	43.6	12.8	14.8
KBFF	37.9	25.7	6.40
CHARMM	45.5	6.6	29.9
AMBER*	49.6	5.9	36.8
AMBER	47.8	2.9	71.7

Table 2. Average number of urea ($\bar{N}_{urea}$) and water molecules ($\bar{N}_{water}$) inside (17, 8) SWNT together with the drying factors, f_{drying} (see text for the definition) with different urea models. These data were averaged over the time region wherein the systems have reached equilibrium (t ≥ 90 ns).

To understand the observed phenomenon of urea-induced drying of SWNTs, we have calculated the difference in average interaction energies for a solvent (urea/water) in bulk and in (17, 8) SWNT with the rest of the system. As the solvent molecules move from bulk into the (17, 8) SWNT, both urea and water lose electrostatic interaction energies, but urea gains more vdW energy than water (about 3~4 times larger than water), which mainly comes from the stronger dispersion interaction of urea than water with nanotube. As a consequence, after a solvent penetrates the SWNT, on average each urea gains 2.55~4.58 kcal/mol whereas each water loses 0.12~1.64 kcal/mol. It is noteworthy that the replacement of structurally confined water by larger urea (on average each urea molecule can replace ~2.5 water molecules) is also favorable in overall free energy due to an overall solvent entropy gain. In addition, the free energy analysis [by calculating the potential of mean force (PMF)] also support that the phenomenon of urea-induced drying of SWNT derives from the stronger dispersion interaction of urea with SWNT than water (see ref. [22] for details).

In conclusion, by using MD simulation we have observed a striking phenomenon of urea-induced drying of hydrophobic nanotubes and demonstrated the robustness of this phenomenon by using five different urea models. By decomposing the interaction energies for a solvent molecule into electrostatic and vdW components, we find that the drying phenomenon results from the stronger dispersion interaction of urea than water with nanotube. These results also have implications on understanding the urea-induced denaturation of proteins by providing further evidence of the potential existence of a "dry globule"-like transient state [65] during early stage of protein unfolding and the "direct interaction mechanism" whereby urea attacks protein directly via favorable dispersion interaction, rather than disrupts water structure as a "water breaker". In addition, this study points out the crucial role of dispersion interaction in the selective absorption of molecules inside hydrophobic nanopores [54-56], which might be important for nanoscience and nanotechnology.

2.3. Chirality switch of drug-like molecules inside boron-nitride nanotubes

Many basic building materials of organism, such as amino acids and saccharides, are chiral in nature. Understanding the molecular chirality is very important for pharmaceutical products because the biological systems have stereoselectivity [66]. Some molecules chiral stable in bulk systems may undergo conformational transitions in human body [67]. For example, in late 1950s and early 1960s, thalidomide caused serious damages to the fetal growth, known as the "thalidomide tragedy" [67, 68], which correlates to a chiral transition of thalidomide occurred in human body. Hence, good conformational stability is an important requirement for chiral molecules used in pharmaceutical products and drug delivery.

It is well-known that there are various nanoscale confinement environments in human body, but the effect of nano-confinement on molecular chirality is still poorly understood so far. Here we use MD simulations (employing Gromacs 3.3.1 [36]) to study the chiral transition of difluorobenzo[c]phenanthrene molecules ($C_{18}H_{12}F_2$, referred to as "D molecule") in single-walled boron-nitride nanotubes (SWBNNTs). Molecular systems can be chiral by asymmetrically arranging atoms in space around a center, axis, or plane, which are called point, axial, and planar chirality, respectively [69]. It has been reported using infrared laser pulses that D molecule show the planar chirality transition between *P*-enantiomer and *M*-enantiomer, and the energy barrier for this transition in bulk was estimated to be only 6.7-8.0 kcal/mol [70]. The chiral character of enantiomers can be characterized by dihedral angle of four atoms (a-b-c-d) shown in Fig. 8(a). When the dihedral angle is averaged over a certain time period (0.1 ns is used), the value of the chiral character is positive for *P*-enantiomer and negative for *M*-enantiomer.

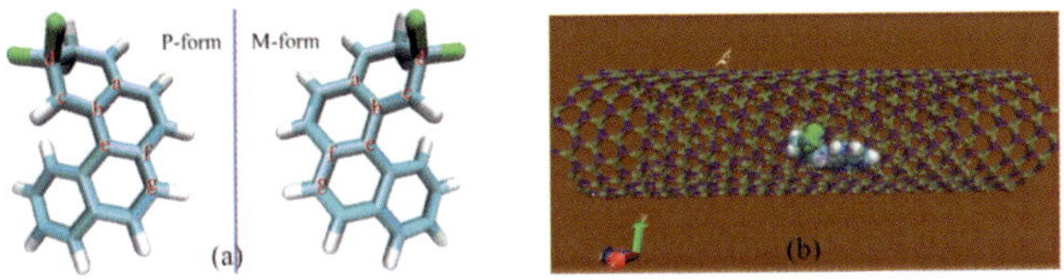

Figure 8. a) *P*- and *M*-form enantiomers of D molecule. The dihedral angle of four atoms (a-b-c-d) is used to identify the chiral geometry of different enantiomers. The e, f and g atoms are used to determine a plane of the D molecule. (b) Snapshot of D molecule inside a (15, 6) SWBNNT to illustrate the simulation system. This figure is reproduced from ref. [24] with permission.

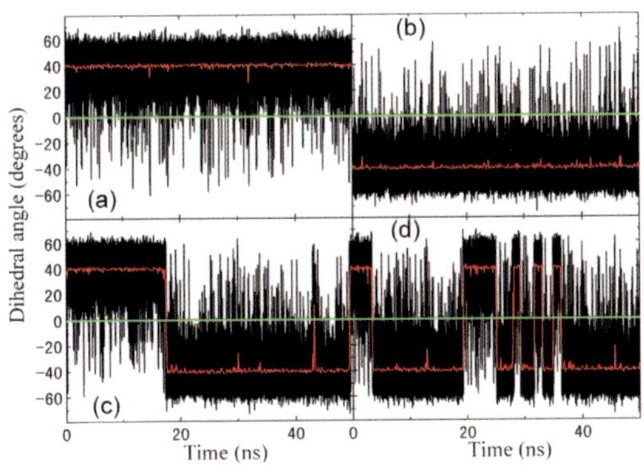

Figure 9. Time evolution of dihedral angle of the D molecule in a (15, 6) SWBNNT at different temperatures. (a) *P*-form at 420 K. (b) *M*-form at 420 K. (c) *P*-form at 440 K, showing chiral transition. (d) *P*-form at 460 K, showing chiral transition. This figure is reproduced from ref. [24] with permission.

Figs. 9(a) and (b) show the chiral character of *P*- and *M*-enantiomers inside a (15, 6) SWBNNT at 420 K. In all of 50 ns simulation times, the averaged values of dihedral angle keep their original signs, indicating that both *P*- and *M*-enantiomers are chiral stable at (and below) 420 K. When the temperature increases to 440 K, the chiral transitions occur, as shown in Figs. 9(c) and (d). Similar phenomena have been observed in other SWBNNTs systems in which the transition occurs at different temperature thresholds.

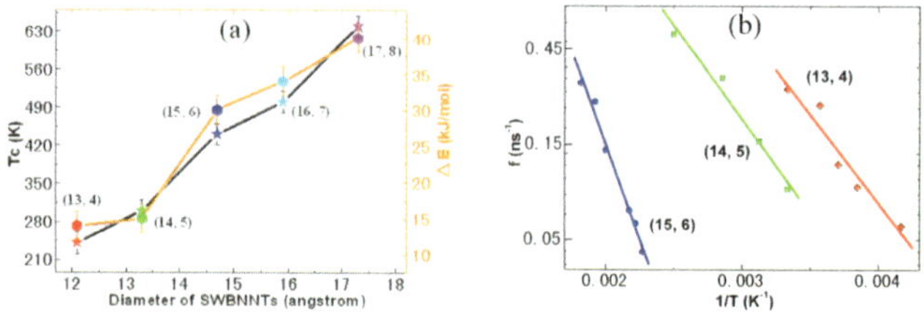

Figure 10. a) Transition critical temperature T_C (star representation, left axis) and corresponding interaction energy barrier ΔE between SWBNNT and D molecules in the chiral transition process (● representation, right axis). Symbols of the same color denote the data for the same SWBNNT. (b) The dependence of chiral transition frequency f on temperature T. Solid lines are fitted with the exponential functions $f = f_0 \exp(-E_a/k_BT)$ for different SWBNNTs. This figure is reproduced from ref. [24] with permission.

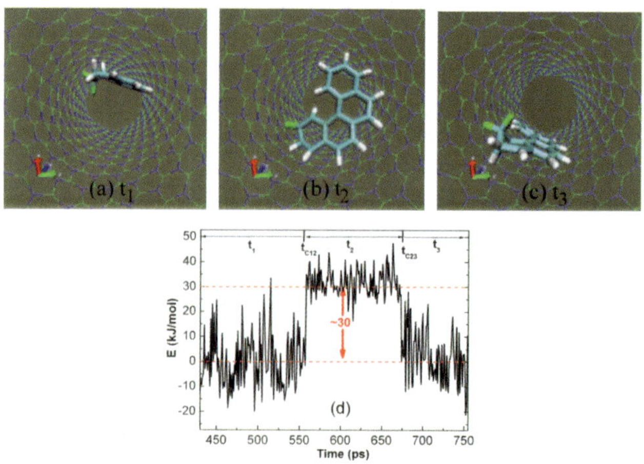

Figure 11. Typical configurations of D molecule and the corresponding interaction energies E inside (15, 6) SWBNNT in different time periods. t_1 and t_3 denote the time periods wherein the enantiomers are stable; t_2 denotes the time periods wherein the chiral transition occurs. This figure is reproduced from ref. [24] with permission.

We have computed the critical temperature T_C for chiral transitions for different SWBNNTs. Here T_C is defined as the temperature at which the enantiomers can transform within 30 ns, and meanwhile, the enantiomer keeps intact at T_C -20 K for 30 ns, for a large number of trajectories starting from different initial configurations, with the error bars of T_C approximately 20 K by this definition. As displayed in Fig. 10(a), T_C increases monotonically with the diameter of SWBNNT. We have also calculated the frequencies of chiral transition, f, for different temperatures inside various SWBNNTs, as shown in Fig. 10(b). The data can be fitted with the Arrhenius activation energy function ($f = f_0 \exp(-E_a/k_B T)$) very well, where E_a is the activation energy, k_B is the Boltzmann constant. For the current cases, f_0 = 937, 139, 276 ns^{-1}, and E_a = 36, 18, 17 kJ/mol, for (15, 6), (14, 5), and (13, 4) SWBNNTs, respectively.

Now we focus on how enantiomerization occurs in nanotubes and the mechanism behind those observations. The D molecule consists of four six-membered rings, with a nearly planar structure. At low temperatures, the D molecule prefers to cling to the inner surface of SWBNNT [with its rings parallel to the SWBNNT axis, see Fig. 11(a)]. It is observed that when chiral transition occurs, the D molecule changes its orientation first so that the angle between plane of D molecule [determined by atoms e, f and g, see Fig. 8(a)] and the axis of SWBNNT increases considerably, even reaches 90° in a SWBNNT with a large diameter, e.g., the (15, 6) SWBNNT [see Fig. 11(b)]. This observation is quite different from the chiral transition in bulk systems. When the D molecule clings to the SWBNNT surface again, its chirality may be changed [see Fig. 11(c)]. We have computed the interaction energies between (15, 6) SWBNNT and D molecule, and find that when chiral transition occurs (at this time, D molecule is almost perpendicular to the nanotube axis), D molecule loses interaction energies [~30 kJ/mol, see Fig. 11(d)], mainly comes from the lost in vdW interactions (the electrostatic interactions between D molecules and nanotube is very small, in the order of 0.1 kJ/mol).

We have also obtained the interaction energy barrier ΔE for the chiral transitions inside different SWBNNTs. The results are displayed in Fig. 10(a) (●, right axis). ΔE is defined as the average interaction energy in the t_2 period, minus the average interaction energy in the t_1 and t_3 periods [see Fig. 11(d)]. It is found that ΔE gradually increases with the diameter of SWBNNTs and the tendency is quite similar to that of the threshold temperature T_C. It appears that the T_C for the D molecule is mainly determined by the transition barrier from a parallel conformation to a perpendicular conformation relative to the nanotube axis. Therefore, we can control the transition temperature by using SWBNNTs with appropriate diameters. To further characterize the effect of confined environments on the chiral transition, we have calculated the free energy of chiral transition for isolated D molecule, and a D molecule inside (13, 4) and (14, 5) SWBNNTs, at the room temperature (300 K). Compared to that of isolated D molecule, the free energy barriers for (13, 4) and (14, 5) SWBNNTs decrease by ~5 kJ/mol and ~3 kJ/mol, respectively (see ref. [24] for more details), indicating that the confined environment can indeed catalyze the enantiomerization of molecules with planar chirality.

In summary, we have performed MD simulations of chiral transition of D molecule (with planar chirality) in SWBNNTs and revealed remarkable effects of nanoscale confinement on molecular chirality. The critical temperature, above which the enantiomerization occurs, increases considerably with the diameter of nanotube, and the frequency of chiral transition decreases exponentially with respect to the reciprocal of temperature. The chiral transitions are found to closely correlate with the orientational transformations of D molecule. Furthermore, the barriers of interaction energies between D molecule and SWBNNT for different orientational states can characterize the chiral transition, implying that the temperature thresholds of chiral transitions can be controlled by nanotubes with appropriate diameters.

These findings provide new insights to the effect of nano-confinement on molecular chirality, and offer some guidance for the safe delivery of the chiral drugs since an unexpected chiral transition may cause serious cytotoxicity.

2.4. Conformational change of small peptides in carbon nanotubes

How proteins fold and unfold in nanoscale confinement has been an open question to the society. Currently, most of the experimental and theoretical studies on protein folding are performed in dilute solutions [71-73]. However, in *vivo*, proteins fold in a heterogeneous, crowded, and confined space, in which the energy landscapes, the folding thermodynamics and kinetics may alter from that in bulk [74-92]. Interestingly in some situations, the confined environment could facilitate the proteins folding to their desired native structures, such as the confinement in chaperonin-assisted folding cavity [93-96], or the exit tunnel of the ribosome [97, 98].

Previous studies using polymer physics models have proposed an entropic stabilization theory, pointing out that the stability of folded protein can be enhanced in confined space because of the reduction of conformational entropy to the unfolded structural ensemble [80, 85, 92, 94]. On the other hand, the additional hydrophobic interaction between the protein and the confined boundary may destabilize the folded state [76-78, 81]. Both the stabilization and destabilization effects due to the confinement were then examined in amino acid side chain level using molecular dynamics simulations by Vaitheeswaran and Thirumalai [99]. In their work, three types of side chain interactions, hydrophobic (Ala:Phe), polar (Ser:Asn) and charged (Lys:Glu), were simulated in a cylinder nanopore confinement with different lengths and diameters, showing that the hydrophobic side chain pair was strongly destabilized and then separated in the confined environment, while both the interactions of polar side chain pair and charged side chain pair were enhanced in the cylindrical confinement [99].

Later, the effect of different confining geometries on protein-folding thermodynamics and kinetics were studied by Mittal and Best [100], in which two proteins, a 3-helix bundle protein prb and protein G, were tested in a coarse-grained model. A quantitative exponential relationship ($R^{-\gamma c}$, where $\gamma c \approx 5/3$) was found between the characteristic size R of the confining boundary and its stabilization effect on the folded state. Surprisingly, the stabilization effect was not relevant to the dimension of the confinement (*e.g.*, planar, cylindrical, or spherical) [100]. The dominant effect of stability and kinetics by confinement was due to the free energy change of the unfolded state in proteins, in which the diffusion coefficients only show difference in the unfolded state basin.

The role of solvent in protein folding kinetics and thermodynamics in confined environment was investigated by Pande's group [81]. In a small representative protein (villin) system, Pande and co-workers found that the protein was promoted to folded state and more unlikely to change to the unfolded state when only the protein was confined [81]. However, the folded state was destabilized when both the protein and waters were confined. Comparing to the bulk, a compact unfolded state was promoted instead of native state, which points out the confined solvent may be another crucial aspect to the protein folding under nanoscale confinement.

Carbon nanotubes (CNTs) are good cylindrical condiment carriers with hydrophobic surface [9]. CNTs are recognized as promising candidates to be biocompatible cargos for drugs, nucleic acids, and proteins because they can spontaneously penetrate mammalian cells [101, 102]. Towards this goal, lots of efforts have been put on studying the biosafety of using CNTs *in vivo*, where the potential influence of CNTs to the biomolecules need to be carefully investigated [10-19]. Our recent work indicates that four main types of interactions -- hydrophobic interaction, π-π stacking interaction, electrostatic interaction, and cation-π interaction -- could affect the structure and function of protein [103, 104]. However, the interactions of proteins with inner side of CNTs are not fully studied yet. The hydrophobic wall of CNT could drastically change the original strong-polar environment (*e.g.*, water) around proteins. In addition, the CNT confinement could affect the solvent by decreasing its entropy. For example, a 23-residue helical peptide was found unstable in CNT by Ponder's group, in which the change of solvent entropy was considered to be the main reason alter the protein stability [9].

In this section, the stability of protein motifs are systematically investigated in CNT confinement with various secondary structures, including a helix, a beta-sheet, and a hairpin turn. Our simulations show that the stability of tested peptides is mainly dependent on their secondary structural types. Interestingly, the stability of beta-sheet peptides is enhanced by the CNTs confinement, but those stabilized beta-sheets can become totally unfolded when a hairpin turn is added to connect these two beta-sheets. The helical structure was bended inside the CNTs in order to adapt to the curved surface, forming stable coil-coil structures (see Table 3).

System	CNT (15, 15) D = 20 Å[a]	CNT (22, 22) D = 30 Å	Bulk water
Hairpin turn (GB1)	unfolded	unfolded	stable
Single-strand beta Ac-KLVFFAE-NH_2	stable	stable	unstable
Double-strand antiparallel beta Ac-KLVFFAE-NH_2	stabilized	stabilized	unstable
Alpha-helix (26-mer poly-alanine)	coil-coil	coil-coil	stable

a D refers to the diameter of the CNTs

Table 3. Comparison the stability of peptide with various secondary structures in bulk water and under CNTs confinement.

The structure of hairpin turn in CNT confinement was investigated by all-atom MD simulations with explicit solvent. The GB1 hairpin turn (PDB entry 2GB1, residue index 41 to 56) was put into CNTs with diameters of D=20 Å (D20) and 30 Å (D30), respectively [105]. We found both hairpin turns were unfolded to random coils after 30 ns simulations in CNTs of different sizes [see Figs. 12 and 13(a)]. The hairpin turn was unfolded to a more relaxed form in the larger size CNT, with radius of gyrations (Rg) 10.1 Å in D30 CNT and 6.7 Å in D20 CNT. Both unfolding processes were started at the turn segment, where the hydrogen bonds formed in the beta region were broken gradually (Fig. 13b). Meanwhile, the aromatic side-chains of Trp_{43}, Tyr_{45}, and Phe_{52} in the beta-region were tightly stuck to the inner wall of CNT by their strong π-π stacking interactions. A helix-like structure was formed in the turn segment [Figs. 12(b) and (d)]. The φ/ψ backbone dihedral angle distributions indicated the alpha-helix and poly-Pro II were the dominant conformations in the CNT confinements for hairpin turns [Figs.13 (c) and (d)].

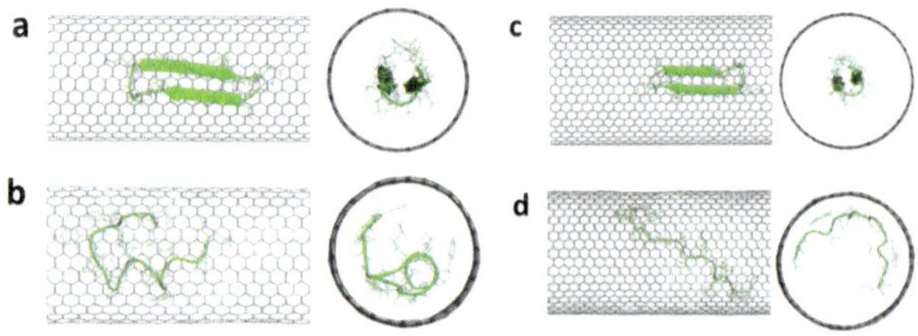

Figure 12. Conformational changes of hairpin turn GB1 inside CNTs. (**a**) and (**b**) The starting structures and the final snapshots of hairpin turn in CNTs with D = 20 Å. (**c**) and (**d**) The starting structures and the final snapshots of hairpin turn in CNTs with D = 30 Å. The final snapshots were obtained from 100 ns MD simulations.

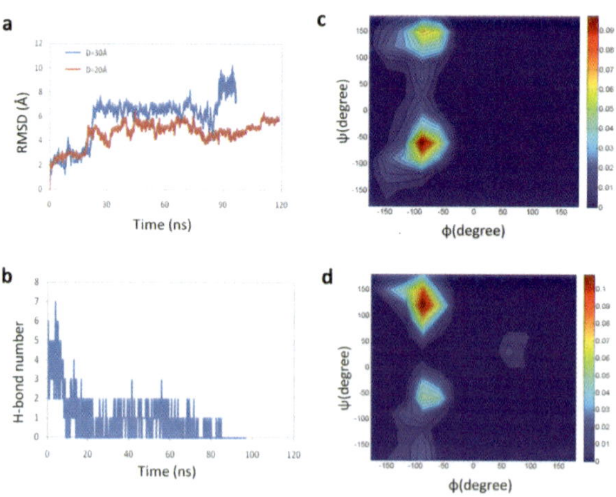

Figure 13. Conformational change of hairpin backbone. (**a**) The RMSD values of hairpin backbone by comparing each snapshot to the starting native structure during the simulations. (**b**) The number of hydrogen bonds formed in the zip

region between backbone atoms. (**c**) and (**d**) Distribution of backbone dihedral angles (φ and ψ) of hairpin turn in D20 and D30 CNTs.

The polyalanine chain was then utilized as the model system to study the stability of helix in CNT confinement. A 26-residue alanine chain was started from alpha-helix form. At the beginning of the simulations, the alanine chain was put in the middle of the CNT along the tube direction [Figs. 14(a) and (c)]. To our surprise, in just a few nanosceonds of the simulations, the entire alanine chain was quickly stuck to the inner side of CNT wall for all sizes of CNTs. Then the helix was bent to adapt the curved surface of CNT and extended along the unit vector, and finally the alpha-hliex turned to the coil-coil superhelix structure [Figs. 14(b) and (d)]. We performed 3 extra independent simulations for each size of CNT systems to conform the fast conformational changes and the final coil-coil superhelix structure for all the alanine chains. The superhelix conformation is an important feature to design proteins that can wrap CNTs, which has been successfully applied to virus-like protein assemblies on CNT surfaces in DeGrado's group [106]. Our simulations indicate that similar strategy could be applied to wrap inner side of CNTs with preferred of coil-coil superhelix structure.

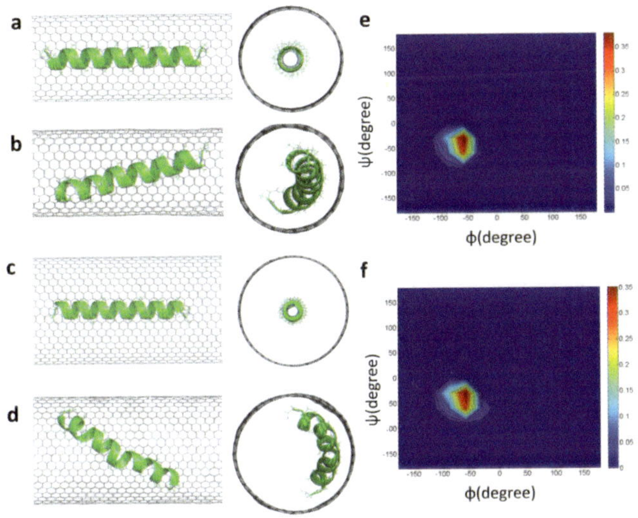

Figure 14. Conformational changes of helical polyalanine inside CNTs. (**a**) and (**b**) The starting structures and the final snapshots of polyalanine in CNTs with D=20 Å. (**c**) and (**d**) The starting structures and the final snapshots of polyalanine in CNTs with D=30 Å. The final snapshots were obtained from 100 ns MD simulations. (**e**) and (**f**) Distribution of backbone dihedral angles (φ and ψ) in polyalanine in D20 and D30 CNTs.

For beta-strand structure, we used Alzheimer amyloid-$\beta_{16\text{-}22}$ peptides (Ace-KLVFFAE-NH_2) as an example. Both single- and double-strand beta were put into the center of CNT (15, 15) [Figs. 15(a) and (c)]. The anti-parallel double-strand sheet was stable inside the CNT during the simulation; in each strand, two phenylalanine were stuck to the inside wall of CNT, and the backbone-backbone hydrogen bonds between two strands were well kept [Figs. 15(d) and (f)]. For single beta strand, large fluctuations can be seen at two charged terminals. However, the middle 4-residue (with sequence "LVFFA") still remained the beta shape [Figs. 15(b) and (e)], which was much more stable than single strand in bulk water. Our recently theoretical investigation has shown that the hydrophobic effect plays a significant role in protein self-assembly in water, in which the "dewetting transition" can be induced by the hydrophobic interaction between two strands in both amyloid-β peptides (KLVFFAE) and $hIAPP_{22\text{-}27}$ peptides (NFGAIL) [107, 108]. Our simulations confirm that beta-strand conformation can be stabilized in hydrophobic environment, which could further promote the formation of protofilaments and form amyloid fibrils. Further study is needed to confirm the role of hydrophobic confinement in facilitating the formation of amyloid fibrils.

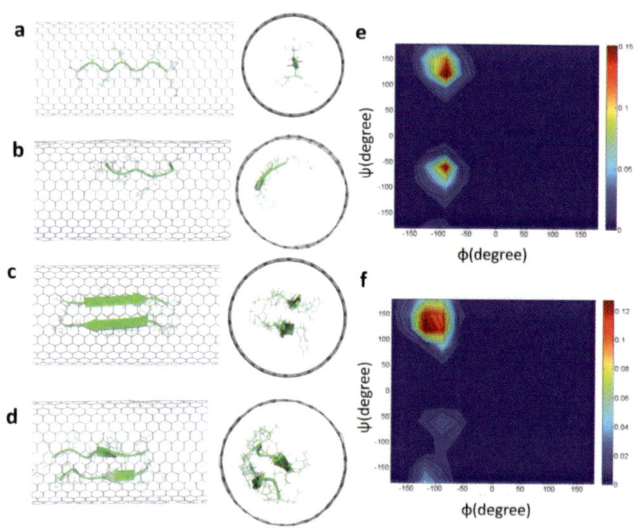

Figure 15. Conformational changes of beta-sheet(s) inside CNTs. (**a**) and (**b**) The starting structures and the final snapshots of single-strand amyloid-beta in CNTs with D=20 Å. (**c**) and (**d**) The starting structures and the final snapshots of double-strand antiparallel Amyloid-beta sheets in CNTs with D=20 Å. The final snapshots were obtained from 100 ns MD simulations. (**e**) and (**f**) Distribution of backbone dihedral angles (φ and ψ) in single-strand and double-strand amyloid-beta sheet(s).

In conclusion, we have investigated three important secondary structural motifs in protein – hairpin turn, helix, and beta-sheet – with CNT confinements by all-atom MD simulations. We find only beta-strand conformation is stabilized in the CNTs. The alpha-helical polyalanine is turned to form coil-coil superhelix structure in order to adapt the curved surface of CNTs. The hairpin turn becomes the most unstable structure in the CNT which totally unfolds to random coil structure and sticks to the CNT walls. Therefore, it is hard to make simple conclusions that CNT confinement could stabilize or destabilize the protein structures.

The conformation of protein in the CNT confinement could be largely dependent on its residue types and building motifs.

3. Conclusion

In this book chapter, we review some of our recent computational works, including: i) the water-mediated signal conversion and multiplication with Y-SWNT; ii) structure, dynamics, and transportation of urea wire and the phenomenon of urea-induced drying inside SWNT; iii) remarkable effect of nanoscale confinement on molecular chirality; and iv) conformational changes of various peptides under nanoscale confinement. These studies provide a deeper understanding towards the unique structure and behaviors of small molecules (water and small organic molecules) and peptides under nanoscale confinement, and demonstrate potential wide implications in nanoscale signal processing, single-file transportation, drug delivery, and even cytotoxicity.

Acknowledgements

We thank Prof. Zhigang Wang, and Dr. Yusong Tu for helpful discussions. This research is supported in part by grants from Zhejiang Provincial Natural Science Foundation of China (Grant No. LY12A04007), the China Postdoctoral Science Foundation (Grant No. 201104738), and the Fundamental Research Funds for the Central Universities. RZ acknowledges the support from the IBM BlueGene Science Program.

Author details

Peng Xiu[3], Zhen Xia[1,2] and Ruhong Zhou[1,4]

1 Computational Biology Center, IBM Thomas J. Watson Research Center, Yorktown Heights, NY 10598

2 Department of Biomedical Engineering, The University of Texas at Austin, Austin , TX 78712

3 Department of Engineering Mechanics, and Soft Matter Research Center, Zhejiang University, Hangzhou , 310027, China

4 Department of Chemistry, Columbia University , New York, NY 10027

References

[1] Hummer, G., Rasaiah, J. C., & Noworyta, J. P. (2001). Water conduction through the hydrophobic channel of a carbon nanotube. *Nature* [414], 188-190.

[2] Koga, K., Gao, G. T., Tanaka, H., et al. (2001). Formation of ordered ice nanotubes inside carbon nanotubes. *Nature* [412], 802-805.

[3] Ball, P. (2008). Water as an active constituent in cell biology. *Chem. Rev.* [108], 74-108.

[4] Holt, J. K., Park, H. G., Wang, Y. M., et al. (2006). Fast mass transport through sub-2-nanometer carbon nanotubes. *Science* [312], 1034-1037.

[5] Chen, S. M., Wu, G. Z., Sha, M. L., et al. (2007). Transition of ionic liquid [bmim][PF6] from liquid to high-melting-point crystal when confined in multiwalled carbon nanotubes. *J. Am. Chem. Soc.* [129], 2416.

[6] Su, Z., Zhu, S., Donkor, A. D., et al. (2011). Controllable Delivery of Small-Molecule Compounds to Targeted Cells Utilizing Carbon Nanotubes. *J. Am. Chem. Soc.* [133], 6874-6877.

[7] Tasis, D., Tagmatarchis, N., Bianco, A., et al. (2006). Chemistry of carbon nanotubes. *Chem. Rev.* [106], 1105-1136.

[8] Pan, X. L., Fan, Z. L., Chen, W., et al. (2007). Enhanced ethanol production inside carbon-nanotube reactors containing catalytic particles. *Nat. Mater.* [6], 507-511.

[9] Sorin, E. J., & Pande, V. S. (2006). Nanotube confinement denatures protein helices. *J. Am. Chem. Soc.* [128], 6316-6317.

[10] Zanello, L. P., Zhao, B., Hu, H., et al. (2006). Bone cell proliferation on carbon nanotubes. *Nano Lett.* [6], 562-567.

[11] Prato, M., Kostarelos, K., & Bianco, A. (2008). Functionalized carbon nanotubes in drug design and discovery. *Acc. Chem. Res.* [41], 60-68.

[12] Bhirde, A. A., Patel, V., Gavard, J., et al. (2009). Targeted Killing of Cancer Cells in Vivo and in Vitro with EGF-Directed Carbon Nanotube-Based Drug Delivery. *ACS Nano* [3], 307-316.

[13] Thakare, V. S., Das, M., Jain, A. K., et al. (2010). Carbon nanotubes in cancer theragnosis. *Nanomedicine* [5], 1277-1301.

[14] Bi, S. P., Zhang, J., & Cheng, J. J. (2009). Call from China for joint nanotech toxicity-testing effort. *Nature* [461], 593.

[15] Donaldson, K., & Poland, C. A. (2009). NANOTOXICOLOGY New insights into nanotubes. *Nat. Nanotechnol.* [4], 708-710.

[16] Gilbert, N. (2009). Nanoparticle safety in doubt. *Nature* [460], 937.

[17] Nel, A., Xia, T., Madler, L., et al. (2006). Toxic potential of materials at the nanolevel. *Science*, 2006(311), 622-627.

[18] Service, R. F. (2000). Is nanotechnology dangerous? *Science*, 2000(290), 1526-1527.

[19] Zhao, Y. L., Xing, G. M., & Chai, Z.F. (2008). Nanotoxicology: Are carbon nanotubes safe? *Nat. Nanotechnol.* , 2008(3), 191-192.

[20] Ge, C. C., Du, J. F., Zhao, L. N., et al. (2011). Binding of blood proteins to carbon nanotubes reduces cytotoxicity. *Proc. Natl. Acad. Sci. U. S. A.* , 2011(108), 16968-16973.

[21] Tu, Y. S., Xiu, P., Wan, R. Z., et al. (2009). Water-mediated signal multiplication with Y-shaped carbon nanotubes. *Proc. Natl. Acad. Sci. U. S. A.* , 2009(106), 18120-18124.

[22] Xiu, P., Yang, Z. X., Zhou, B., & et, al. . (2011). Urea-Induced Drying of Hydrophobic Nanotubes: Comparison of Different Urea Models. *J. Phys. Chem. B* , 2011(115), 2988-2994.

[23] Xiu, P., Tu, Y., Tian, X., et al. (2012). Molecular wire of urea in carbon nanotube: a molecular dynamics study. *Nanoscale*, 2012(4), 652-658.

[24] Zhang, R. Q., Wang, Z. G., Wang, C. L., et al. (2010). Size Dependence of Nanoscale Confinement on Chiral Transformation. *Chem. Eur. J.* , 2010(16), 6482-6487.

[25] Tu, Y. S., Zhou, R. H., & Fang, H. P. (2010). Signal transmission, conversion and multiplication by polar molecules confined in nanochannels. *Nanoscale*, 2010(2), 1976-1983.

[26] Kwok, K., & Ellenbogen, J. (2002). Moletronics: future electronics. *Materials Today* , 2002(5), 28-37.

[27] Xu, H. Q. (2005). Nanotubes: The logical choice for electronics? *Nat. Mater.*, 2005(4), 649-650.

[28] Koenig, D. R., Weig, E. M., & Kotthaus, J. P. (2008). Ultrasonically driven nanomechanical single-electron shuttle. *Nat Nano*, 2008(3), 482-485.

[29] Litvinchuk, S., Tanaka, H., Miyatake, T., et al. (2007). Synthetic pores with reactive signal amplifiers as artificial tongues. *Nat. Mater.* [6], 576-580.

[30] Köfinger, J., Hummer, G., & Dellago, C. (2008). Macroscopically ordered water in nanopores. *Proc. Natl. Acad. Sci. U. S. A.* [105], 13218-13222.

[31] Li, J. Y., Gong, X. J., Lu, H. J., et al. (2007). Electrostatic gating of a nanometer water channel. *Proc. Natl. Acad. Sci. U. S. A.* [104], 3687-3692.

[32] Papadopoulos, C., Rakitin, A., Li, J., et al. (2000). Electronic Transport in Y-Junction Carbon Nanotubes. *Phys. Rev. Lett.* , 85, 3476 -3479 .

[33] Terrones, M., Banhart, F., Grobert, N., et al. (2002). Molecular Junctions by Joining Single-Walled Carbon Nanotubes. *Phys. Rev. Lett.* , 89, 075505 .

[34] Gothard, N., Daraio, C., Gaillard, J., et al. (2004). Controlled Growth of Y-Junction Nanotubes Using Ti-Doped Vapor Catalyst. *Nano Lett.* [4], 213-217.

[35] Bandaru, P. R., Daraio, C., Jin, S., et al. (2005). Novel electrical switching behaviour and logic in carbon nanotube Y-junctions. *Nat. Mater.* [4], 663-666.

[36] Lindahl, E., Hess, B., & van der Spoel, D. (2001). GROMACS 3.0: a package for molecular simulation and trajectory analysis. *J. Mol. Modeling* [7], 306-317.

[37] Jorgensen, W. L., Chandrasekhar, J., Madura, J. D., et al. (1983). Comparison of Simple Potential Functions for Simulating Liquid Water. *J. Chem. Phys.* [79], 926-935.

[38] Luna, V. M., Chen, Y., Fee, J. A., et al. (2008). Crystallographic Studies of Xe and Kr Binding within the Large Internal Cavity of Cytochrome ba3 from Thermus thermophilus: Structural Analysis and Role of Oxygen Transport Channels in the Heme-Cu Oxidases. *Biochemistry* , 47(47), 4657-4665.

[39] Moustafa, I. M., Foster, S., Lyubimov, A. Y., et al. (2006). Crystal Structure of LAAO from Calloselasma rhodostoma with an l-Phenylalanine Substrate: Insights into Structure and Mechanism. *J. Mol. Biol.* , 2006(364), 991-1002.

[40] Berezhkovskii, A., & Hummer, G. (2002). Single-file transport of water molecules through a carbon nanotube. *Phys. Rev. Lett.* , 89, 064503 .

[41] Best, R. B., & Hummer, G. (2005). Reaction coordinates and rates from transition paths. *Proc. Natl. Acad. Sci. U. S. A.* , 2005(102), 6732-6737.

[42] Kofinger, J., Hummer, G., & Dellago, C. (2008). Macroscopically ordered water in nanopores. *Proc. Natl. Acad. Sci. U. S. A.* , 2008(105), 13218-13222.

[43] de Groot, B. L., & Grubmuller, H. (2001). Water permeation across biological membranes: Mechanism and dynamics of aquaporin-1 and GlpF. *Science* [294], 2353-2357.

[44] Tajkhorshid, E., Nollert, P., Jensen, M. O., et al. (2002). Control of the selectivity of the aquaporin water channel family by global orientational tuning. *Science* [296], 525-530.

[45] Raghavender, U. S., Kantharaju, , Aravinda, S., et al. (2010). Hydrophobic Peptide Channels and Encapsulated Water Wires. *J. Am. Chem. Soc.* [132], 1075-1086.

[46] Wan, R. Z., Li, J. Y., Lu, H. J., et al. (2005). Controllable water channel gating of nanometer dimensions. *J. Am. Chem. Soc.* [127], 7166-7170.

[47] Li, J. Y., Gong, X. J., Lu, H. J., et al. (2007). Electrostatic gating of a nanometer water channel. *Proc. Natl. Acad. Sci. U. S. A.* [104], 3687-3692.

[48] Wan, R. Z., Lu, H. J., Li, J. Y., et al. (2009). Concerted orientation induced unidirectional water transport through nanochannels. *Phys. Chem. Chem. Phys.* [11], 9898-9902.

[49] Reddy, G., Straub, J. E., & Thirumalai, D. (2010). Dry amyloid fibril assembly in a yeast prion peptide is mediated by long-lived structures containing water wires. *Proc. Natl. Acad. Sci. U. S. A.* [107], 21459-21464.

[50] Cao, Z., Peng, Y. X., Yan, T. Y., et al. (2010). Mechanism of Fast Proton Transport along One-Dimensional Water Chains Confined in Carbon Nanotubes. *J. Am. Chem. Soc.* [132], 11395-11397.

[51] Jensen, M. O., Borhani, D. W., Lindorff-Larsen, K., et al. (2010). Principles of conduction and hydrophobic gating in K+ channels. *Proc. Natl. Acad. Sci. U. S. A.* [107], 5833-5838.

[52] Dong, K., Zhou, G. H., Liu, X. M., et al. (2009). Structural Evidence for the Ordered Crystallites of Ionic Liquid in Confined Carbon Nanotubes. *J. Phys. Chem. C* [113], 10013-10020.

[53] Chaban, V. (2010). Filling carbon nanotubes with liquid acetonitrile. *Chem. Phys. Lett.* [496], 50-55.

[54] Liu, Y., Consta, S., & Goddard, W. A. (2010). Nanoimmiscibility: Selective Absorption of Liquid Methanol-Water Mixtures in Carbon Nanotubes. *J. Nanosci. Nanotechnol.* , 2010(10), 3834-3843.

[55] Lee, J., & Aluru, N. R. (2010). Separation of gases from gas-water mixtures using carbon nanotubes. *Appl. Phys. Lett.* , 96, 133108 .

[56] Kalra, A., Hummer, G., & Garde, S. (2004). Methane partitioning and transport in hydrated carbon nanotubes. *J. Phys. Chem. B* [108], 544-549.

[57] Knepper, M. A., & Mindell, J. A. (2009). STRUCTURAL BIOLOGY Molecular coin slots for urea. *Nature* [462], 733-734.

[58] Levin, E. J., Quick, M., & Zhou, M. (2009). Crystal structure of a bacterial homologue of the kidney urea transporter. *Nature*, 2009(462), 757-761.

[59] Weerasinghe, S., & Smith, P. E. (2003). A Kirkwood-Buff derived force field for mixtures of urea and water. *J. Phys. Chem. B* [107], 3891-3898.

[60] Duffy, E. M., Severance, D. L., & Jorgensen, W. L. (1993). UREA- POTENTIAL FUNCTIONS, LOG-P, AND FREE-ENERGY OF HYDRATION. *Isr. J. Chem.* , 1993(33), 323-330.

[61] Smith, L. J., Berendsen, H. J. C., & van Gunsteren, W. F. (2004). Computer simulation of urea-water mixtures: A test of force field parameters for use in biomolecular simulation. *J. Phys. Chem. B* [108], 1065-1071.

[62] Hess, B., Kutzner, C., van der Spoel, D., et al. (2008). GROMACS 4: Algorithms for highly efficient, load-balanced, and scalable molecular simulation. *J. Chem. Theory Comput.* , 2008(4), 435-447.

[63] Mac, Kerell. A. D., Bashford, D., Bellott, M., et al. (1998). All-atom empirical potential for molecular modeling and dynamics studies of proteins. *J. Phys. Chem. B* [102], 3586-3616.

[64] Sorin, E. J., & Pande, V. S. (2005). Exploring the helix-coil transition via all-atom equilibrium ensemble simulations. *Biophys. J.* [88], 2472-2493.

[65] Hua, L., Zhou, R. H., Thirumalai, D., et al. (2008). Urea denaturation by stronger dispersion interactions with proteins than water implies a 2-stage unfolding. *Proc. Natl. Acad. Sci. U. S. A.* [105], 16928-16933.

[66] Agranat, I., Caner, H., & Caldwell, A. (2002). Putting chirality to work: The strategy of chiral switches. *Nat. Rev. Drug Discovery* [1], 753-768.

[67] Eriksson, T., Björkman, S., Roth, B., et al. (1998). Enantiomers of thalidomide: blood distribution and the influence of serum albumin on chiral inversion and hydrolysis. *Chirality* [10], 223-228.

[68] Blaschke, G., Kraft, H., , P., & Markgraf, H. (1980). Chromatographische Racemattrennungen, X. Racemattrennung des Thalidomids und anderer Glutarimid-Derivate. *Chem. Ber.* [113], 2318-2322.

[69] Hembury, G. A., Borovkov, V. V., & Inoue, Y. (2008). Chirality-sensing supramolecular systems. *Chem. Rev.* [108], 1-73.

[70] Umeda, H., Takagi, M., Yamada, S., et al. (2002). Quantum control of molecular chirality: Optical isomerization of difluorobenzo c phenanthrene. *J. Am. Chem. Soc.* [124], 9265-9271.

[71] Fersht, A. R., Matouschek, A., & Serrano, L. (1992). The folding of an enzyme.1. Theory of protein engineering analysis of stability and pathway of protein folding. *J. Mol. Biol.* [224], 771-782.

[72] Snow, C. D., Sorin, E. J., Rhee, Y. M., et al. (2005). How well can simulation predict protein folding kinetics and thermodynamics? *Annu. Rev. Biophys. Biomol. Struct.*, 43-69.

[73] Zhou, R., Huang, X., Margulis, C. J., et al. (2004). Hydrophobic collapse in multidomain protein folding. *Science* [305], 1605-9.

[74] Betancourt, M. R., & Thirumalai, D. (1999). Exploring the kinetic requirements for enhancement of protein folding rates in the GroEL cavity. *J. Mol. Biol.* [287], 627-644.

[75] Bolis, D., Politou, A. S., Kelly, G., et al. (2004). Protein stability in nanocages: A novel approach for influencing protein stability by molecular confinement. *J. Mol. Biol.* [336], 203-212.

[76] Cheung, M. S., & Thirumalai, D. (2006). Nanopore-protein interactions dramatically alter stability and yield of the native state in restricted spaces. *J. Mol. Biol.* [357], 632-643.

[77] Eggers, D. K., & Valentine, J. S. (2001). Crowding and hydration effects on protein conformation: A study with sol-gel encapsulated proteins. *J. Mol. Biol.* [314], 911-922.

[78] Eggers, D. K., & Valentine, J. S. (2001). Molecular confinement influences protein structure and enhances thermal protein stability. *Protein Sci.* [10], 250-261.

[79] Ellis, R. J., & Minton, A. P. (2003). Cell biology-. *Join the crowd. Nature* [425], 27-28.

[80] Klimov, D. K., Newfield, D., & Thirumalai, D. (2002). Simulations of beta-hairpin folding confined to spherical pores using distributed computing. *Proc. Natl. Acad. Sci. U. S. A.* [99], 8019-8024.

[81] Lucent, D., Vishal, V., & Pande, V. S. (2007). Protein folding under confinement: A role for solvent. *Proc. Natl. Acad. Sci. U. S. A.* [104], 10430-10434.

[82] Minton, A. P. (2000). Implications of macromolecular crowding for protein assembly. *Curr. Opin. Struct. Biol.* [10], 34-39.

[83] Ravindra, R., Shuang, Z., Gies, H., et al. (2004). Protein encapsulation in mesoporous silicate: The effects of confinement on protein stability, hydration, and volumetric properties. *J. Am. Chem. Soc.* [126], 12224-12225.

[84] Zhou, H.X. (2007). Helix formation inside a nanotube: Possible influence of backbone-water hydrogen bonding by the confining surface through modulation of water activity. *J. Chem. Phys.*, 127, 245101.

[85] Zhou, H. X., & Dill, K. A. (2001). Stabilization of proteins in confined spaces. *Biochemistry* [40], 11289-11293.

[86] Ellis, R. J. (2001). Macromolecular crowding: obvious but underappreciated. *Trends Biochem. Sci.* [26], 597-604.

[87] Friedel, M., Sheeler, D. J., & Shea, J. E. (2003). Effects of confinement and crowding on the thermodynamics and kinetics of folding of a minimalist beta-barrel protein. *J. Chem. Phys.* [118], 8106-8113.

[88] Stagg, L., Zhang, S., , Q., Cheung, M. S., et al. (2007). Molecular crowding enhances native structure and stability of alpha/beta protein flavodoxin. *Proc. Natl. Acad. Sci. U. S. A.* [104], 18976-18981.

[89] van den, Berg. B., Ellis, R. J., & Dobson, C. M. (1999). Effects of macromolecular crowding on protein folding and aggregation. *EMBO J.* [18], 6927-6933.

[90] Zhou, H., & , X. (2008). Protein folding in confined and crowded environments. *Arch. Biochem. Biophys.* [469], 76-82.

[91] Zhou, H., , X., Rivas, G., & Minton, A. P. (2008). Macromolecular crowding and confinement: Biochemical, biophysical, and potential physiological consequences. *In Annual Review of Biophysics*, 375-397.

[92] Thirumalai, D., Klimov, D. K., & Lorimer, G. H. (2003). Caging helps proteins fold. *Proc. Natl. Acad. Sci. U. S. A.* [100], 11195-11197.

[93] Thirumalai, D., & Lorimer, G. H. (2001). Chaperonin-mediated protein folding. *Annu. Rev. Biophys. Biomol. Struct.* [30], 245-269.

[94] Takagi, F., Koga, N., & Takada, S. (2003). How protein thermodynamics and folding mechanisms are altered by the chaperonin cage: Molecular simulations. *Proc. Natl. Acad. Sci. U. S. A.* [100], 11367-11372.

[95] Jewett, A. I., Baumketner, A., & Shea, J. E. (2004). Accelerated folding in the weak hydrophobic environment of a chaperonin cavity: Creation of an alternate fast folding pathway. *Proc. Natl. Acad. Sci. U. S. A.* [101], 13192-13197.

[96] Baumketner, A., Jewett, A., & Shea, J. E. (2003). Effects of confinement in chaperonin assisted protein folding: Rate enhancement by decreasing the roughness of the folding energy landscape. *J. Mol. Biol.* [332], 701-713.

[97] Woolhead, C. A., Mc Cormick, P. J., & Johnson, A. E. (2004). Nascent membrane and secretory proteins differ in FRET-detected folding far inside the ribosome and in their exposure to ribosomal proteins. *Cell*, 2004(116), 725-736.

[98] Ziv, G., Haran, G., & Thirumalai, D. (2005). Ribosome exit tunnel can entropically stabilize alpha-helices. *Proc. Natl. Acad. Sci. U. S. A.* [102], 18956-18961.

[99] Vaitheeswaran, S., & Thirumalai, D. (2008). Interactions between amino acid side chains in cylindrical hydrophobic nanopores with applications to peptide stability. *Proc. Natl. Acad. Sci. U. S. A.* [105], 17636-17641.

[100] Mittal, J., & Best, R. B. (2008). Thermodynamics and kinetics of protein folding under confinement. *Proc. Natl. Acad. Sci. U. S. A.* [105], 20233-20238.

[101] Kam, N. W. S., Liu, Z., & Dai, H. J. (2005). Functionalization of carbon nanotubes via cleavable disulfide bonds for efficient intracellular delivery of siRNA and potent gene silencing. *J. Am. Chem. Soc.* [127], 12492-12493.

[102] Kam, N. W. S., Liu, Z. A., & Dai, H. J. (2006). Carbon nanotubes as intracellular transporters for proteins and DNA: An investigation of the uptake mechanism and pathway. *Angew. Chem. Int. Ed.* [45], 577-581.

[103] Zuo, G. H., Huang, Q., Wei, G. H., et al. (2010). Plugging into Proteins: Poisoning Protein Function by a Hydrophobic Nanoparticle. *ACS Nano* [4], 7508-7514.

[104] Yang, Z., Wang, Z., Tian, X., et al. (2012). Amino acid analogues bind to carbon nanotube via pi-pi interactions: Comparison of molecular mechanical and quantum mechanical calculations. *J. Chem. Phys*, 136, 025103.

[105] Gronenborn, A. M., Filpula, D. R., Essig, N. Z., & et, al. . (1991). A Novel, Highly Stable Fold of the Immunoglobulin Binding Domain of Streptococcal Protein-G. *Science* [253], 657-661.

[106] Grigoryan, G., Kim, Y. H., Acharya, R., et al. (2011). Computational Design of Virus-Like Protein Assemblies on Carbon Nanotube Surfaces. *Science*, 2011(332), 1071-1076.

[107] Krone, M. G., Hua, L., Soto, P., et al. (2008). Role of water in mediating the assembly of Alzheimer amyloid-beta a beta 16-22 protofilaments. *J. Am. Chem. Soc.* [130], 11066-11072.

[108] Yang, Z. X., Shi, B. Y., Lu, H. J., et al. (2011). Dewetting Transitions in the Self-Assembly of Two Amyloidogenic beta-Sheets and the Importance of Matching Surfaces. *J. Phys. Chem. B* [115], 11137-11144.

Chapter 9

Preparation, Characterization and Applicability of Covalently Functionalized MWNT

Eun-Soo Park

Additional information is available at the end of the chapter

1. Introduction

Over the last twenty years, carbon nanotubes (CNTs) have received much attention for their unique structural, mechanical, and electronic properties as well as their broad range of potential applications [Lee et. al., 2012; Kim and Park, 2008; Kang et. al., 2008; Xu et. al., 2008; Meyyappan et al., 2005; Kumar, 2002; Wong et al., 1998]. CNTs are cylinder-shaped macromolecules with a radius as small as a few nanometers, which can be grown up to 20 cm in length [Zhu et. al., 2002]. Their properties depend on the atomic arrangement, chirality, diameter, and length of the tube and the overall morphology. They exist in one of two structural forms, single-walled CNT (SWNT) or multi-walled CNT (MWNT). SWNTs are best described as a 2-D graphene sheet rolled into a tube with pentagonal rings as end caps [Harris, 2004]. SWNTs have aspect ratios of 1000 or more and an approximate diameter of 1 nm. Similarly, MWNTs can be described as multiple layers of concentric graphene cylinders also with pentagonal ring end caps. Conventional MWNT diameters range from 2-50 microns [Harris, 2004]. Measurements using in situ transmission electron microscopy (TEM) and atomic force microscopy (AFM) have produced estimates that Young's modulus of CNTs is approximately 1 TPa [Treacy et. al., 1996; Wong et. al., 1997]. For comparison, the stiffest conventional glass fibers have Young's modulus of approximately 70 GPa, while carbon fibers typically have modulus of about 800 GPa. CNTs can accommodate extreme deformations without fracturing and also have the extraordinary capability of returning to their original, straight, structure following deformation [Harris, 2004]. In addition, they are excellent electrical conductors and have very high thermal conductivities. Possible applications for CNTs range from nanoelectronics, quantum wire interconnects, sensors and field emitters to nanocomposites [Meyyappan et. al., 2005].

Despite these great promises, many real applications of CNTs have been impeded by difficulties associated with their processing and manipulation. As produced CNTs have the tendency to exist in bundles rather than as individual tubes, because of strong van der Waals interactions, leading to insolubility in most organic media, and therefore limiting the range of applications [Neelgund and Oki, 2011]. To make CNTs more easily dispersible in various media, it is necessary to physically or chemically attach certain molecules, or functional groups, to their smooth sidewalls without significantly changing the CNTs desirable properties. This process is called functionalization. Various functionalization methods such as chopping, oxidation, wrapping and irradiation of the CNTs can be created more active bonding sites on the surface of the nanotubes. Among them, electron beam (EB) irradiation is potent to induce the uniform and consistent modification of the nanotubes because of the high amount of energy, it imparts to the atoms via the primary knock-on atom mechanism. This chapter describes a novel method to covalently functionalized nanotubes that bear terminated isocyanate, hydroxyl, amine and epoxy group, which then react covalently with other molecules. The first step is preparation of COOH-terminated MWNT by EB irradiation of unmodified nanotubes. These carboxylic groups were used as reaction precursors in the covalent functionalization. The MWNTs attached to the organofunctional moieties have greater versatility for further utilization in different application fields such as macroinitiator, electroconductive nanocomposite, biology, water treatment, and starting material for another cycle of functionalization. Moreover covalently functionalized nanotubes can extend the field of application in nanoelectronics, sensorics, hydrogen power engineering, bioengineering, and medicine [Dresselhaus and Dresselhaus, 2001; Burghard, 2005].

2. Preparation and characterization of covalently functionalized MWNT

The non-reactive nature of the CNT surface appears as a constraint in several technological applications. To manipulate and process CNTs, it is desirable to functionalize the sidewall of CNTs, thereby generating CNT-derivatives that are compatible with solvent as well as organic matrix materials. Modification of the CNT surface by changing its chemical composition has proved to be efficient to overcome this problem. Several methods such as chemical functionalization, non-covalent wrapping and high energy beam irradiation have been used to modify the chemical composition of the CNT surface by grafting functional groups to it.

Chemical functionalization of CNT has been performed mainly on the basis of oxidative treatments [Abuilaiwi et. al., 2010; Basiuk et. al., 2004; Lee et. al. 2005; Balasubramanian and Burghard, 2004]. Typically this is achieved by the oxidative process of CNTs using strong inorganic acids or oxidizing agents. This is a lengthy process that also generates a lot of waste and that can damage the CNT structure. Moreover these conventional surface treatment methods utilize a reaction between a liquid and a solid since an oxidizing agent contacts with the entire surface of CNT and the surface is uniformly reacted or physically treated, it is difficult to control the surface state.

The non-covalent method to functionalize CNTs involves using surfactants, oligomers, biomolecules, and polymers to wrap CNTs to enhance their solubility [Hirsch, 2002]. Using polymer chains to wrap CNTs is a versatile and effective way for CNT functionalization. Block copolymers may provide a series of attractive non-covalent wrapping and decoration for the functionalization of CNTs [Chen et. al., 2011]. These approaches can be driven by distinct interactions between nanotubes and polymers including p-stacking, electrostatic interactions, and decoration of CNTs with micelles [Zou et. al., 2008]. One block of the block copolymers forms a close interaction with CNTs, while the other block provide the dispersibility and chemical compatibility to the CNTs [Szleifer and Yerushalmi-Rozen, 2005]. However, the non-covalent interaction between the wrapping molecules and the CNTs is not as strong as the covalent bonding formed in the chemical functionalization methods [Hirsch, 2002].

Electron and ion irradiation is generally used nowadays for modifying properties of semiconductors; beams of energetic particles are also expected to be widely employed for nanotube-based materials processing [Krasheninnikov and Nordlund, 2004]. EB irradiation is a form of ionizing energy that is generally characterized by its low penetration and high dosage rates. The beam with a concentrated and highly charged stream of electrons is generated by the acceleration and conversion of electricity. The electrons are generated by equipment referred to as accelerators which are capable of producing beams that are either pulsed or continuous. When an electron hits the target, different mechanisms of damage creation can work. Depending on the target material, the main mechanism can be the kinetic energy transfer, electronic excitations and ionization [Krasheninnikov and Nordlund, 2004]. For CNTs, the most important mechanism is the knock-on atom displacements due to kinetic energy transfer for electrons [Dresselhaus and Avouris, 2001]. Electronic excitations and ionization effects seem to be less important due to a high thermal and electrical conductivity of graphene shells [Banhar, 1999].

2.1. Functionalization of MWNT by electron-beam irradiation

The MWNT (purity = 95 wt %, average diameter = 15 nm, average length = 20 μm, specific gravity = 1.8) was received from the Iljin Nanotech Co., Ltd., Korea. Fig. 1 shows the TEM image and energy-dispersive X-ray spectroscope analysis (EDX) result of MWNT produced by a chemical vapour deposition (CVD) process without any purification. The TEM measurements were performed with a Philips CM200 operated at 200 kV. Scanning electron microscopy (SEM) observations of the MWNT samples were performed on a Hitachi model S-4300, Japan. The morphology was determined at an accelerating voltage of 15 kV. The surface sample composition was evaluated with SEM equipped with an EDX spectroscope.

As-received MWNT contain some impurities and entangle into a bulk piece (Fig. 1a). EDX results of the pristine MWNT show small peaks which are corresponding to Fe, Si and S. The Si peak has its origin in silicon substrate whereas the other peaks are due to the precursor gases present in the gas mixture and catalyst. The Pt peaks was due to the platinum sputtering process during sample preparation. CNTs are often formed in entangled ropes with 10–100 CNTs per bundle depending on the method of synthesis. They can be produced by a number of methods: direct-current arc discharge, laser ablation, thermal and plasma enhanced CVD process [Lau and Hui, 2002]. The method of production affects the level of purity of the sample and whether SWNTs or MWNTs are formed. Impurities exist as catalysis particles, amorphous carbons and non-tubular fullerenes [Thostenson et. al., 2001].

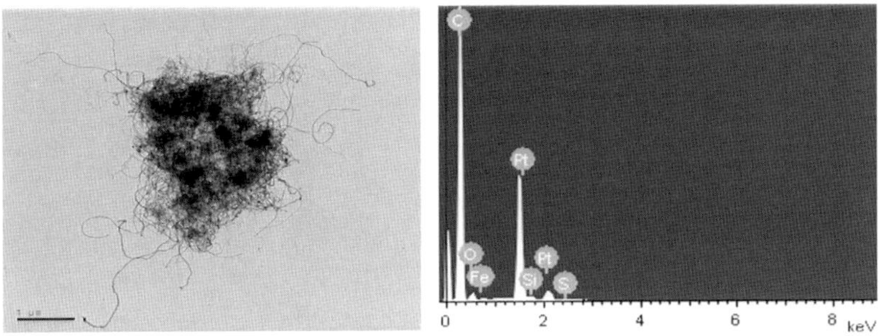

(a) (b)

Figure 1. TEM image (a) and EDX analysis (b) result of the pristine MWNT.

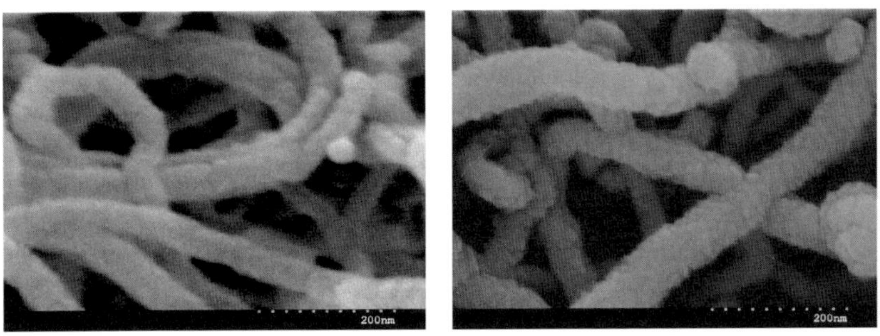

(a) (b)

Figure 2. SEM image of the MWNT before (a) and after (b) EB irradiation at 1200 kGy.

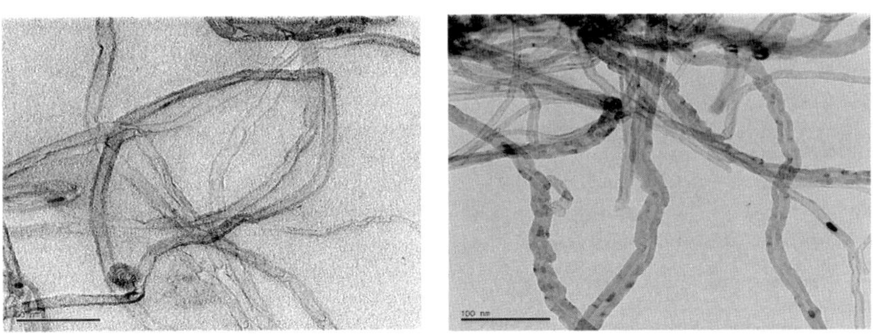

(a) (b)

Figure 3. TEM image of the MWNT before (a) and after (b) EB irradiation at 1200 kGy.

The MWNT were EB-irradiated in air at room temperature using a 1.5MeV electrostatic accelerator (ELV-4, EB Tech Co., Ltd., Korea). Irradiation dose of 800, 1000, and 1200 kGy were used, respectively. The specifications of the ELV-4 are presented Table 1.

Model	Energy (MeV)	Maximum Current (mA)	Output (kW)	Window length (mm)	Height (mm)
ELV-4	0.8 ~ 1.5	50	50	980	4330

Table 1. Specifications of the ELV-4 EB accelerator.

Fig. 2 demonstrates higher magnification SEM micrographs of MWNT before and after treatment with the EB irradiation. The pristine MWNT has relatively smooth surface without extra phase or stain attached on its sidewall. Although the EB irradiation increased up to 1000 kGy, the surface appearance little changed compare to the pristine MWNT. After the 1200 kGy EB irradiation, the smooth surface was disappeared, many wrinkled structure were formed, and the surface roughness increased. Additional sample characterization is carried out using TEM. From the Fig. 3, the presence of dark spots on the outer wall of the MWNT1200 suggests that damage and formation change of MWNT induced by high-dose irradiation.

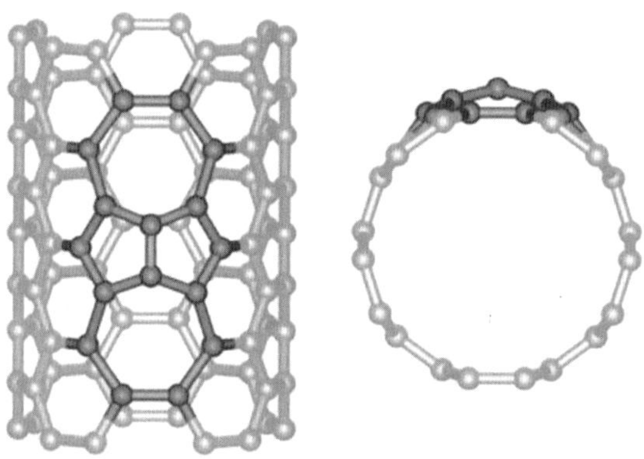

Figure 4. Stone–Wales defect on the sidewall of a nanotube [Burghard and Balasubramanian, 2005].

In general, the surface of the synthesized CNT is smooth and relatively defects free. However, stresses can induce Stone-Wales transformations, resulting in the formation of heptagons and concave areas of deformation on the nanotubes [Thostenson et. al., 2001; Burghard and Balasubramanian, 2005]. Moreover EB irradiation of MWNTs resulted in forming vacancies on their walls and eventual amorphization upon high-dose irradiation [Banhart, 1999]. The irradiation induced damage manifested itself in the deterioration of mechanical properties of MWNTs exposed to prolonged 2-MeV electron irradiation [Salvetat et. al., 1999].

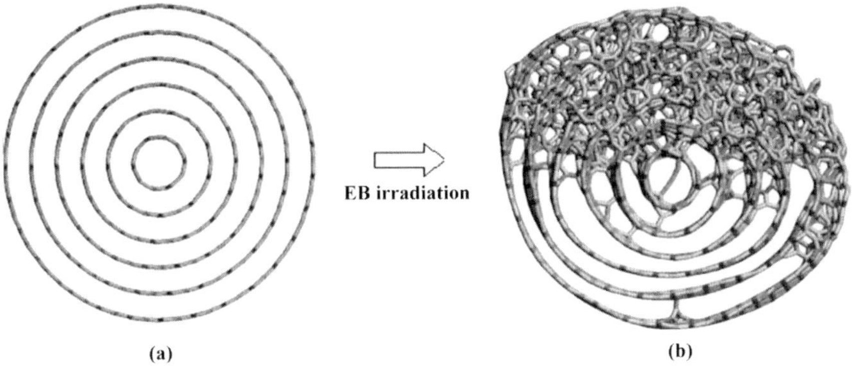

Figure 5. Molecular model of MWNT before (a) and after (b) 300-eV Ar ion irradiation with a dose of $2\times10^{16}/cm^2$ [Krasheninnikov and Nordlund, 2004].

2.2. Characterization of EB-MWNT

The pristine MWNT and EB-irradiated MWNT were characterized by Fourier transform infrared (FTIR) spectroscopy. FTIR spectra of the KBr pelleted samples were measured with a PerkinElmer infrared spectrometer (Spectrum 2000) in the wave-number range from 4000 to 400 cm^{-1} and were analyzed with commercial software.

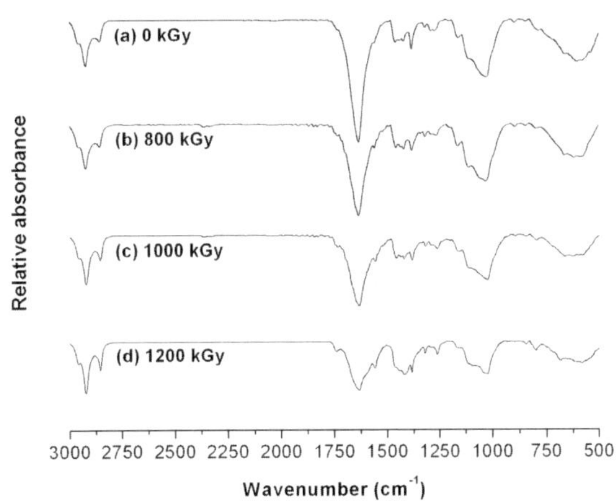

Figure 6. FTIR spectra of the EB-irradiated MWNT.

From Fig. 6, the strong bands at 2920 and 2852 cm^{-1} on the curve are well known, due to asymmetrical and symmetrical stretching of $-CH_2$, respectively. The band at 2958 cm^{-1} is assigned to the asymmetrical stretching of $-CH_3$. The peak at 1635 cm^{-1} can be associated with the stretching of the MWNT backbone. FTIR spectra of MWNT after EB irradiation more than 1000 kGy showed new peaks at 1782-1720 cm^{-1} due to the C=O bond resulting from the stretch mode of carboxylic groups (Fig. 6). These groups can then be used to link molecules via covalent bond formation.

EDX results also confirmed that the oxygen content in the MWNTs increased significantly after irradiation at 1000 kGy. The abbreviation of the sample code in Table 2, MWNT800, for example, means that the MWNT was EB-irradiated at radiation dose of 800 kGy. Oxygen atom on the surfaces of pristine MWNT may be due to the partial oxidation of the surfaces of MWNTs during manufacturing or purification by the manufacturer.

Element	Composition (atomic%)			
	MWNT	MWNT800	MWNT1000	MWNT1200
C	94.63	91.25	88.32	83.47
O	5.11	7.62	9.81	15.32
Si	0.14	1.05	1.70	1.21
S	0.03	-	-	-
Fe	0.09	0.08	0.17	-

Table 2. EDX analysis result of the pristine MWNT and EB-MWNT.

Elemental analyses (EA) results of the MWNT and EB-MWNT are shown in Table 3. EA was performed in a Thermo EA1112 apparatus. The results presented a decrease in the hydrogen content up to 1000 kGy. After the 1200 kGy irradiation, the hydrogen content was significantly increased. This indicated that the low irradiation dose cleaned the MWNT surface of impurities, according to the SEM, EDX and EA results, but the increase in the irradiation doses could have affected the surface roughness and chemical composition [Lee et. al., 2012].

Element	Composition (%)			
	MWNT	MWNT800	MWNT1000	MWNT1200
C	99.50	99.52	99.52	99.35
H	0.50	0.48	0.48	0.57
N	-	-	-	0.08

Table 3. EA results of the pristine MWNT and EB-MWNT.

2.3. Properties of EB-MWNT

2.3.1. Thermal stability

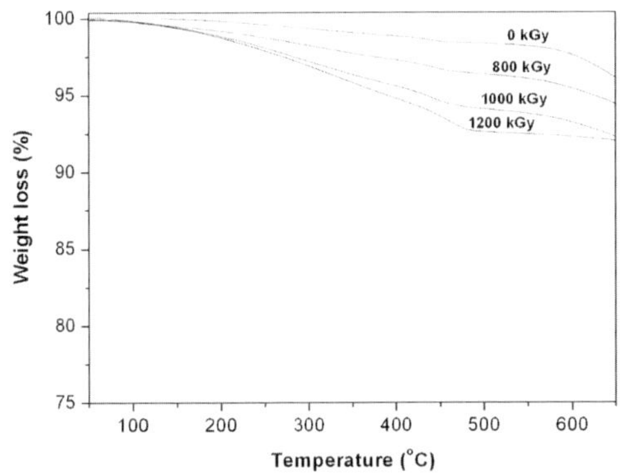

Figure 7. TGA thermograms of the pristine MWNT and EB–MWNTs (Lee et. al., 2012).

Fig. 7 provides quantitative information on the EB-MWNT by using thermogravimetry (TGA, PerkinElmer TGS-2) results. The TGA curves were obtained under an N_2 atmosphere and scanned from 20 to 800°C at a heating rate of 20°C/min. As shown in Fig. 7, the pristine MWNT did not show any discernible thermal degradation, with only 2 wt% degradation at 600°C. On the contrary, the weight loss of the EB–MWNTs significantly increased with increasing irradiation dose because of the possible destruction of the CNT structure. The damage formation in CNTs is quite different from that observed in most other solids [Krasheninnikov and Nordlund, 2004]. Nitric acid or other oxidizing media, such as ozone or oxygen plasma, have been reported to be effective for the partial surface oxidation of CNTs [Banerjee et. al., 2005]. It has been shown that the basal planes of graphite are attacked by molecular oxygen only at their periphery or at defect sites, such as edge planes and vacancies [Radovic, 2003]. Along with the simple defects, a number of more complex defects can be formed such like the Stone–Wales defects [Burghard and Balasubramanian, 2005] associated with a rotation of a bond in the CNT atom network, other topological defects in the graphitic network, and amorphous complexes. Besides this, defect-mediated covalent bonds between adjacent SWNTs in the bundle can appear. Likewise, similar links between shells can appear in MWNTs [Krasheninnikov and Nordlund, 2004].

2.3.2. Electrical resistivity

Table 4 shows a rapid decrease in volume resistivity of the poly(ethylene-*co*-vinyl acetate (EVA, vinyl acetate content = 28%)/EB–MWNT nanocomposites with increasing nanotube content. The surface electrical resistance of specimens (80 mm × 10 mm) was detected by a megohmmeter (TeraOhm 5 kV, Metrel) according to ASTM D 257. The charge time was 30 s, and the current stress of the measurements was 2500 V at 20 ± 1°C. Volume resitivity values of the prepared films were calculated with the following equation:

$$\varrho_V = \frac{AR_V}{L} \tag{1}$$

Where ρ_v, A, R_v and L represent the area of the volume resistivity (Ω-cm), effective electrode (cm^2), measured resistance (Ω), and distance between electrodes (cm), respectively.

The percolation threshold of the EVA/MWNT nanocomposites formed by solution mixing was approximately ~ 5 wt %; this was due to the advantageous effect of composites with higher aspect ratios compared with spherical or elliptical fillers in the formation of conducting networks in the polymer matrix [Lee et. al., 2012]. However, volume resistivity of nanocomposites was not significantly changed with irradiation dose indicated that EB irradiation did not affect the electroconductivity of MWNT.

Nanotube content (wt%)	Volume resistivity (Ω-cm x 10^{-5})			
	MWNT	MWNT800	MWNT1000	MWNT1200
0.0	474,600	474,600	474,600	474,600
2.5	107,083	104,900	100,983	100,271
5.0	2.48	2.46	2.10	2.10
10	0.015	0.015	0.014	0.014

Table 4. Volume resistivity changes of the EVA/EB-MWNT nanocomposites (Lee et. al., 2012).

2.3.3. *Biological activity*

The biological activity of the pristine MWNT and EB-MWNT was compared against *Staphylococcus aureus* (*S. aureus*, ATCC 25923) and *Escherichia coli* (*E. coli*, ATCC 25922) with the shake flask method. The bacteria cell were subcultured on nutrient broth and incubated for 20 h at 37°C. The cells were suspended in 50 ml of phosphate-buffered saline (PBS) to yield a bacterial suspension of $2.32{\times}10^9$ – $2.49{\times}10^9$ colony forming units/ml (cfu/ml). The nanotube (0.5 g) was weighed and shaken in 20 ml of a bacterial suspension for 24 h. The suspension (25 wt/vol%) was serially diluted in PBS and cultured on nutrient broth at 37°C for 24 h. The number of viable organisms in the suspension was determined by multiplication of the number of colonies with the dilution factor, and the percentage reduction was calculated on the basis of the initial count. *S. aureus* and *E. coli* are two of the most common nosocomial pathogens and they represent Gram-positive and Gram-negative bacteria, respectively. The number of viable bacteria and the percentage reduction of the number of bacteria are summarized in Table 5.

Sample	Antibacterial activity			
	S. aureus (+)		*E. coil (-)*	
	cfu/ml (×10⁻⁹)	Reduction (%)	cfu/ml (×10⁻⁹)	Reduction (%)
Blank	2.32	-	2.49	-
MWNT	2.13	8.2	2.08	10.3
MWNT800	1.85	20.3	1.93	16.8
MWNT1000	1.72	25.9	1.78	23.3
MWNT1200	1.65	28.9	1.55	33.2

Table 5. Shake flask test results for the pristine MWNT and EB-MWNT.

After 24 h of bacterial contact, pristine MWNT extirpated 8.2 and 10.3 % of the viable cells of *S. aureus* and *E. coil*, respectively. This indicated that pristine MWNT has some interesting biological activities. Harmful effect of nanoparticles arises due to high surface area and intrinsic toxicity of the surface. The nano-scale dimensions of CNT make quantities of milligrams possess a large number of cylindrical particles with a concurrent very high total surface area. The intrinsic toxicity of CNT depends on the degree of surface functionalization and the different toxicity of functional groups. Batches of pristine CNT readily after synthesis contain impurities such as amorphous carbon and metallic catalysts which can also be the source of toxic effects [Singh et. al., 2010]. Kang and co-workers [Kang et. al., 2008] showed that the size of CNTs is a key factor governing their antibacterial effects and that the likely main CNT-cytotoxicity mechanism is cell membrane damage by direct contact with CNTs. As the size of CNTs decreases, the specific surface area increases, leading to increased opportunity for interaction and uptake by living cells. This characteristic could result in adverse biological effects that otherwise would not be possible with the same material in a larger form [Donaldson et. al., 2004; Nel et. al., 2006; Jia et. al., 2005]. Several studies have shown that SWNTs exhibit significant cytotoxicity to human and animal cells, whereas MWNTs exhibit a milder toxicity [Jia et. al., 2005].

With the EB irradiation dose the biological activity of MWNT against both the *S. aureus* and *E. coil* was gradually increased. It is noteworthy that 1200 kGy irradiated MWNT exhibits highest antibacterial activity against *S. aureus*. After 24h of shaking, MWNT1200 showed 33.2 % inhibition of the growth of *S. aureus*. In order to inactivate or kill microbes, the nanocomposite particles must come close to or touch the microbes. Such interactions are either attraction or repulsion. As most bacteria carry a net negative surface charge [Jucker et. al., 1996], adhesion of bacteria is discouraged on negatively charged surfaces, while it is promoted on positively charged surfaces [Hogt et. al., 1986]. The increase in polarity of MWNT after EB irradiation is reflected in the relative polar surface area, hydrogen bond donor, and hydrogen bond acceptor numbers, all of which increase substantially for biological activity [Lee et. al., 2011].

2.4. Covalent functionalization of EB-MWNT

The sites of highest chemical reactivity within CNTs are the caps, which have a fullerene like structure [Balasubramanian and Burghard, 2008]. CNTs are not ideal structures, but rather contain defects formed during synthesis. Typically around 1–3 % of the carbon atoms of a CNT are located at a defect site [Hu et. al., 2001]. A frequently encountered type of defect is so-called Stone–Wales defect, which is comprised of two pairs of 5- and 7-membered rings, and is hence referred to as a 7-5-5-7 defect. A Stone–Wales defect leads to a local deformation of the graphitic sidewall and thereby introduces an increased curvature in this region. The strongest curvature exists at the interface between the 2 5-membered rings; as a result of this curvature, addition reactions are most favored at the carbon–carbon double bonds in these positions [Zhao et. al., 2004].

The EB irradiation procedure results in the formation of carboxylic moieties, preferentially on the end caps of the CNT, since the regions where pentagons are located suffer more strain compared with that of purely hexagonal lattice. Under these conditions, the end caps of the nanotubes are opened and acidic functionalities are formed at these defect sites and at the side walls. The carboxyl groups represent useful sites for further modifications, as they enable the covalent coupling of molecules through the creation of amine or ester bonds (Fig. 8). Isocyanate groups are highly unsaturated organic compounds. They can react readily with many diverse compounds containing active protons such as alcohol, amines and carboxylic acids. Thus, amine-functionalized MWNTs can be obtained where surface-bound isocyanate groups are subsequently reacted with H_2O and converted into amine. The detail of these reactions was summarized as follows. The bisphenol A type low molecular weight epoxy resin (DGEBA, epoxide equivalent weight 121 g/equiv) and PDMS (Mw ≈ 5000) were donated by Huenvi Co., Ltd., Korea and used without further purification. Other chemical compounds were reagent grade and were used as received.

The EB-MWNT and toluene were fed to a glass reactor and the mixture was dispersed for 30 min in an ultrasonic bath at 60°C. Methylene diphenyl diisocyanate (MDI) [diisocyanate terminated polydimethylsiloxane (PDMS) or 3-(triethoxysily)propyl isocyanate (TEPI)], and 1,4-diazabicyclo [2.2.2] octane (DABCO) were added into reactor and the mixture was sonicated for 2 h. Upon completion of the reaction, the mixture was filtered through a filter paper. The filtrate was then washed with toluene and the functionalized MWNT was dried in an oven at 60°C. EB-MWNTs are also reacted with an epoxide to produce an epoxide terminated MWNT. A reaction mixture consisting of toluene, MWNT1200, and DGEBA were charged into the reactor. The mixture was dispersed for 1 h in an ultrasonic bath at 60°C followed by addition of a trace amount of triethylamine as the catalyst. The reaction was carried out for 2 h; then, the mixture was filtered, and the filtrate was washed with toluene and methanol. The filtrate was dried in an oven at 60°C. On the other hand, the covalently functionalized MWNT with PTMG was prepared by using Fischer esterification method [Abuilaiwi et. al, 2010].

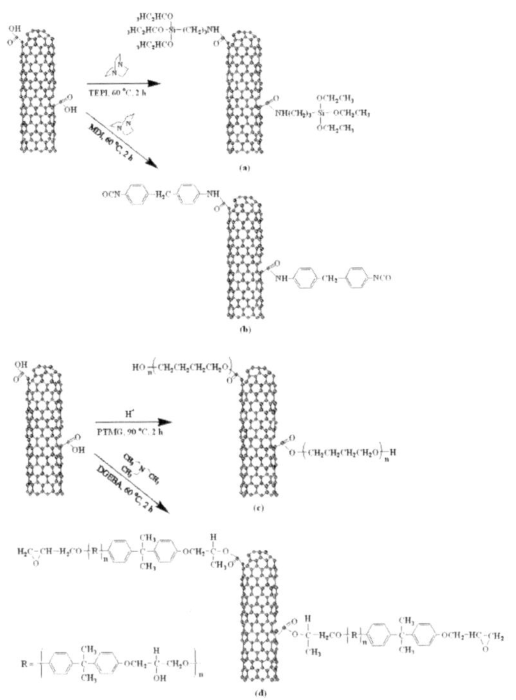

Figure 8. Covalent functionalization EB-MWNT through reaction with TEPI (a), MDI (b), PTMG (c) and DGEBA (d).

Fig. 9 demonstrates the FTIR spectra of the MWNT1200 and covalently functionalized MWNT1200. In the MWNT-TEPI (Fig. 9b), the new bands at 2878 cm^{-1} associated with the stretching of the methylene groups from the TEPI molecules appeared. In addition, the functionalized MWNT shows the peaks of $-Si-O-CH_2CH_3$ groups at 1175, 1100, 1075, and 970–940 cm^{-1}. As depicted in Fig. 9c, the small peak at 2281 cm^{-1} is from the N–C=O asymmetric vibration, while a new peak is observed at 1544 cm^{-1} which is attributed to the overlapping of a signal from the N–H, N–C bands and N–C=O group [Abuilaiwi et. al., 2010]. The peaks at 3000–2800 cm^{-1} in Fig. 9d are due to C–H antisymmetric and symmetric stretching vibrations of methylene groups of PTMG. The peak at 1460 cm^{-1} originates from the C–H bend of the alkyl chain and the peak at 1108 cm^{-1} arises from the C–O stretch of the ester group. The characteristic epoxy group in synthesized MWNT-DGEBA can be identified on the basis of band presence in the FTIR spectrum 3060-3000 cm^{-1} from -CH vibration of the epoxy ring, 1250 cm^{-1}, from -CO vibrations of the epoxy ring and at 960-815 cm^{-1} from the deformed -CH vibrations of the epoxy ring (Fig. 9e). These results confirmed the attachment of functional molecules onto the EB-MWNT surface.

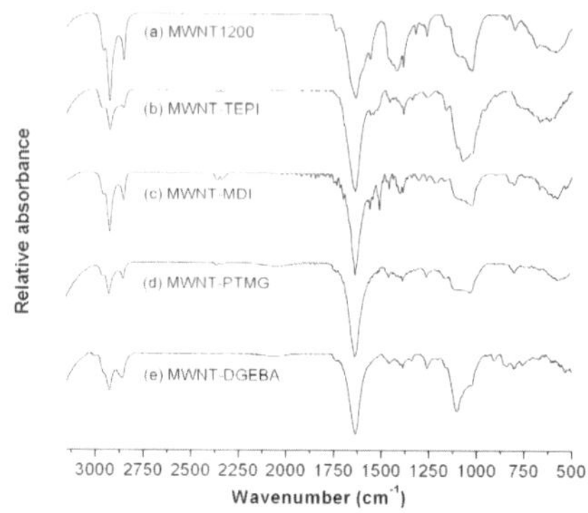

Figure 9. FTIR spectra of the EB-MWNT and covalently functionalized EB-MWNT.

More direct evidence for the covalent functionalization of EB-MWNT is manifested by SEM images. In Fig. 10, SEM images of MWNT1200, MWNT-MDI, MWNT-TEPI, and MWNT-DGEBA are shown. It indicates that the MWNT1200 (Fig. 10a) has a wrinkled and rough surface. However, after covalent functionalization, the wrinkled structures of MWNT1200 were almost disappeared and the surface roughness decreased. These changes in the morphology for MWNT-DGEBA (Fig. 10d) are remarkable. A uniform tubular layer due to covalently bonded DGEBA on the surface of the MWNT1200 is observable. It seems that the average diameters of MWNT-DGEBA are slightly increased in comparison to MWNT1200.

Figure 10. SEM images of the MWNT1200 (a) and covalently functionalized MWNT1200 with MDI (b), TEPI (c), and DGEBA (d).

TGA measurements were conducted on the covalently functionalized EB-MWNT to elucidate their thermal degradation behaviors. Some typical weight-loss curves as a function of temperature are shown in Fig. 11. Several weight loss steps were observed below ~150°C which are due to the release of moisture and the decomposition of the associated organic groups. Fig. 11a, the initial degradation of –COOH group for MWNT1200 starts at approximately 170°C and completes at about 480°C. It also showed no significant weight loss at 480-700°C. Assuming, the portion of the weight loss of functional group at 600°C is the same as that of in the MWNT1200. When the weight loss of the MWNT1200 at 600°C (6.5 %) is used as the reference, the weight loss of covalently functionalized MWNT by TEPI, PDMS, PTMG, and MDI of MWNT-TEPI, MWNT-PDMS, MWNT-PTMG and MWNT-MDI at 600°C is about 4.2, 6.3, 18.8 and 19.8 %, respectively.

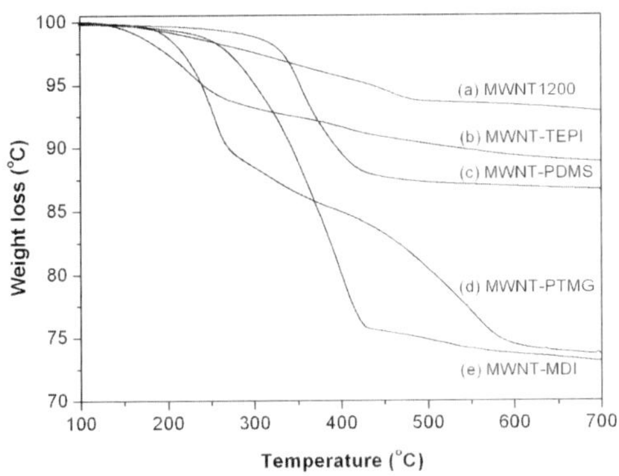

Figure 11. TGA traces of the covalently functionalized EB-MWNT.

3. Applicability of EB-MWNT and covalently functionalized MWNT

The covalent functionalization of CNT has been more interesting because it allows the modification of CNTs surface for subsequent alignment [Tahermansouri et. al., 2010]. The recently developed methods for functionalization of CNTs have opened up a broad range of novel application perspectives. These surface modifications play an important role for application of nanotubes in composite, sensors and many other fields [Chiu and Chang, 2007]. In this section, eight specific applications of EB irradiation and covalently modified MWNTs are described in detail.

3.1. Mechanically reinforced nanocomposites

CNT's polymeric nanocomposites are aimed at the exploitation of the high electrical conductivity of CNTs coupled to high mechanical properties, thermal properties and others unique properties. However, high molecular weight and strong inter-tube forces keep CNT together in bundles, making their manipulation, characterization and analytical investigation very difficult. Therefore the functionalization of CNTs offers the great advantages of producing soluble and easy-to-handle CNT [Yoon et. al., 2004]. Consequently, compatibility and reactivity of CNT with other material, such as polymer should be strongly improved. Recent work has demonstrated superior dispersion of MWNTs in polymers by functionalization of the nanotubes to compatibilize them with solvents and the matrix polymers [Chiu and Chang, 2007; Wu et. al., 2006; Balasubramanian and Burghard, 2005]. The improved dispersion of nanotubes with functional groups has been accompanied by increased mechanical properties of the nanocomposite. Fig. 12 shows SEM micrographs of the fractured surface of EVA/MWNT and EVA/EB-MWNT nanocomposites after saponification for 6 h and heterogeneous solution reaction in the presence of tetrabutyl titanate (TNBT) catalyst for 20 min. For EVOH/MWNT1200–TNBT [Fig. 12b], the MWNTs dispersed well in the EVOH matrix and most of the MWNTs were broken in the interface rather than pulled out from the polymer matrix. However, the EVOH/MWNT specimen showed a different morphology [Fig. 12a]. Most of the MWNT fibers were pulled out from the EVOH matrix. Such a discrepancy demonstrated that a stronger interfacial adhesion existed between the MWNTs and the EVOH matrix. The presence of the chemical bonding leads to improvement in the mechanical properties of final composite. Consequentially there was about 17% increase in tensile strength for the MWNT1200 nanocomposite compared to pristine MWNT one.

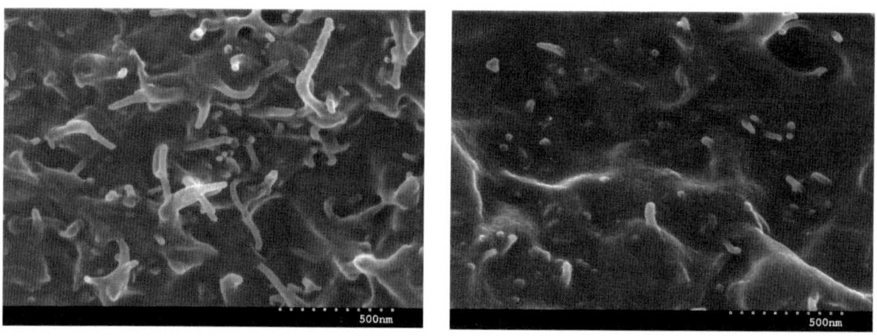

(a) (b)

Figure 12. SEM micrographs of the fractured surface of EVA/MWNT and EVA/MWNT1200 nanocomposites (nanotube content = 10 wt%) after saponification for 6 h and heterogeneous solution reaction in the presence of TNBT.

3.2. Inorganic initiator and catalysts

Functionalized nanotubes with different initiator moieties can be used in in-situ polymerizations to produce composites that have covalent bonds between the filler and the polymer chains. They can be produced by incorporating various reactive groups onto the convex surfaces of nanotubes by adjusting the feed ratio of azide compounds to CNTs and were used in surface-initiated polymerizations, amidation, and reduction of metal ions affording various CNT-polymer and CNT-Pt nanohybrids [Liu and Chen, 2007; Sahoo et. al., 2010]. For example, CNTs functionalized by butyl lithium act as an initiator for polymerization to produce grafted CNT polymer nanocomposites where the anions disperse the nanotubes due to electrostatic repulsive force between the tubes [Abuilaiwi et al., 2010]. When CNTs are functionalized by oxy radicals, they also act as initiators for polymerization of polyethylene chains grafted on the CNTs [Gao et. al., 2009]. MWNT with hydroxyl groups can be used as co-initiators to polymerize poly(ε-caprolactone) or poly(α-chloro-ε-caprolactone) by surface-initiated ring-opening polymerization. Pendent chlorides were converted into azides by the reaction with sodium azides. Finally, various types of terminal alkynes were reacted with pendent azides by copper-catalyzed Huisgen's 1,3-dipolar cycloaddition [Lee, et. al., 2011]. Moreover, for a long time, active carbon has found wide spread application as a support material in heterogeneous catalysis. Compared to this form of carbon, CNTs offer the advantage of a more defined morphology and chemical composition as well as the possibility to attach catalysts onto their surface through covalent bonds [Balasubramanian and Burghard, 2008]. The applicability of CNTs as carriers for catalytically active molecular functional units has recently been demonstrated through the covalent coupling of an organic vanadyl complex [Baleizão et. al., 2004].

3.3. Drug delivery and gene therapy

A drug delivery systems are continuously being developed to improve the pharmacological profile and the therapeutic properties of administered drugs [Allen and Cullis, 2004; Shohet et. al., 2000; Davis, 1997]. They have been developed according to the different classes of bioactive molecules to be delivered and the characteristics of the target tissues. Liposomes, emulsions, cationic polymers, micro and nanoparticles are the most commonly studied vehicles [Singh, et. al., 2010]. MWNTs can be functionalized by attaching biological molecules such as different peptides, proteins, nucleic acids and small organic molecules are able to deliver their cargos into cells, thus opening the path for their facile manipulation and processing in physiological environments. For instance, enzymes which contain many amine and hydroxyl groups react with isocyanate groups in MWNT-NCO thereby forming covalently bound enzymes. Then they can usefully mimic certain biological functions, such as drug delivery and gene therapy. Because large part of the human body consists of carbon, CNTs are generally thought of as a very biocompatible material [Singh, et. al., 2010]. Cells have been shown to grow on CNTs so they appear to have no toxic effect. The cells also do not adhere to the CNTs, potentially giving rise to application such as coatings for prosthetics. The molecular targeting of CNT delivery systems derivatised with a therapeutic agent is possible if an active recognition moiety is simultaneously present at the surface of the carrier [Kam et. al., 2005]. Moreover attachment of a fluorescent molecule would provide optical signals for imaging and localisation of the CNT–drug conjugates [Pastorin et. al., 2006]. The ability to functionalize the sidewalls of CNTs also leads to biomedical application such as vascular stents, neuron growth and regeneration [Guzman et. al., 1996].

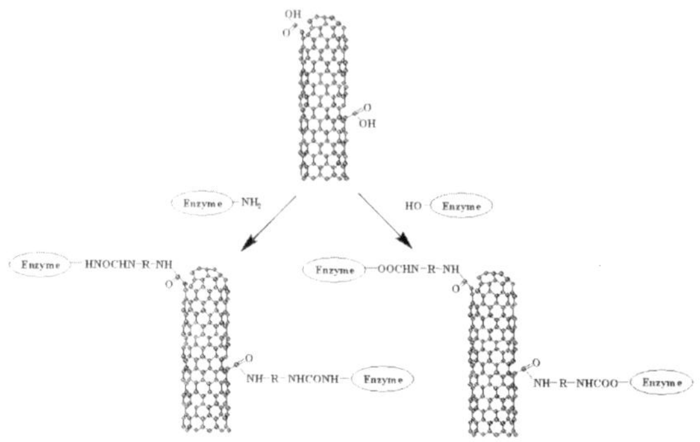

Figure 13. Immobilization of enzymes onto isocyanate functionalized MWNT.

3.4. Energy storage and conversion device

Self-assembly is a process that occurs due to the spontaneous and uninstructed structural reorganization that forms from a disordered system. Such processes are reversible and held together by non-covalent intermolecular forces. Layer-by-layer (LBL) assembly is a versatile method to form dense thin films from dispersed solutions containing functionalized nanomaterials. This technology takes advantage of the charge-charge interaction between substrate and monolayers to create multiple layers held together by electrostatic forces. The LBL approach consists of the repeated, sequential immersion of a substrate into aqueous solutions of functionalized materials having complementary charge, thereby producing conformal ultra-thin films and controllable surface morphology using various nanomaterials. Electrostatic LBL assembly, in which CNTs have been alternated with polymers for various energy storage and conversion devices, has exhibited improved networks for electrochemical energy applications. Recently, Lee and co-workers [Lee, et. al., 2009] reported the preparation of MWNT thin films based on the LBL assembly method. The authors prepared negatively and positively charged MWNTs by surface functionalization using carboxylic acid and amine groups, respectively. The complementary charged and functionalized MWNT dispersions enable the incorporation of MWNTs into highly controlled thin films using LBL assembly. The prepared MWNT exhibited high electronic conductivity in comparison with polymer composites with SWNTs, and high capacitive behaviour with the ability to precisely control the capacity [Lee, et. al., 2009]. The average capacitance was considerably higher than those of vertically aligned or conventional CNT electrodes due to the high nanotube densities and well developed nanopores in the LBL MWNT thin films. This study indicates the potential to precisely control the charge and energy storage parameters in MWNT thin films by controlling the number of bilayers and film thickness in the LBL assembly. Therefore precise control of the LBL system can be used to design ideal electrode materials for fuel cells, photoelectrochemical cells, batteries, supercapacitors [Lee, et. al., 2009], and strain and corrosion sensing [Loh, et. al, 2007].

3.5. Nanofiltration membrane for purification and separation

Nanofiltration is a relatively recent membrane filtration process used most often with low total dissolved solids water such as surface water and fresh groundwater, with the purpose of softening and removal of disinfection by-product precursors [Raymond, 1999]. The development of advanced membrane technologies with controlled and novel pore architectures is important for the achievement of more efficient and cost effective purification. Present polymeric membranes are well known to suffer from a trade off between selectivity and permeability, and in some cases are also susceptible to fouling or exhibit low chemical resistance [Sears, et. al., 2010]. Modification of membrane surfaces through the introduction of nano materials is another strategy to reduce fouling, or to alter the affinity of the membrane for organic solutes. One premise of the current effort if that these properties might be exploited to create membrane materials of very high strength. MWNTs exhibit characteristics such as strength, and thermal and electrical conductivity that make them promising materials for use in developing new nanocomposite materials. Moreover their toxicity might be exploited when inserted in membranes as a basis for inhibiting bacterial growth and therefore reducing biofouling. Such toxic effects towards bacteria were detected MWNT immobilized within the membrane skin might serve as a basis for fouling produced by microbial growth.

Recently, our group has developed a process of simple saponification to make highly porous nanocomposites [Lee et. al, 2012]. In this process, at least one vinyl acetate (VAc) containing polymer or blend is dissolved in an appropriate solvent and a suitable viscosity of the solution is achieved. EB-irradiated MWNT was dispersed in polymer solution and then the polymer suspension was precipitated and saponified in alkaline non-solvent. After rinsing off the coagulant and drying, sponge-like structure of connected matrix polymer and nanotube were obtained. Production parameters that affect the pore structure and properties include polymer and nanotube concentration, VAc content in polymer, saonification time and temperature, and precipitation media. These factors can be varied to produce porous structure with a large range of pore sizes, and altering chemical, thermal and mechanical properties. These nanocomposites with highly porous and excellent antibacterial activity have potential use of nanofiltration membranes in treatment of industrial wastewater and removal of disinfection (Fig. 14).

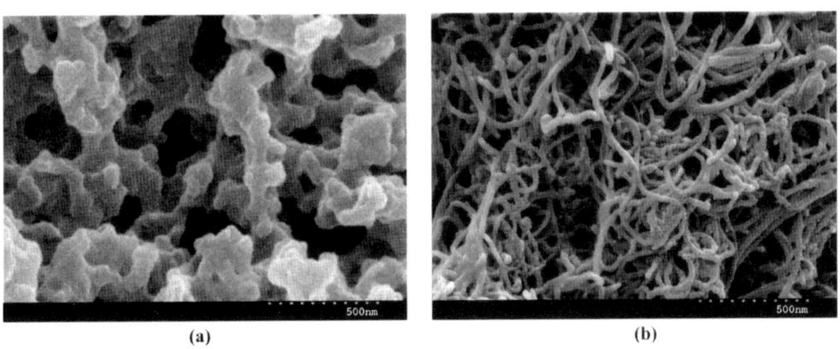

Figure 14. SEM micrographs of the PVA (a) and PVA/MWNT1200 (b) nanocomposite particle prepared by simple saponification method using ethanol/NaOH solution.

3.6. Antibacterial agents

The microbial contaminated materials could serve as important sources of cross-infections, causing a variety of serious consequences in medical devices, hospital equipment, water purification and delivery systems, bio-protective equipment. Even though they cannot be directly assimilated by microorganisms, microbes can grow and propagate using bioassimilable contaminants on the surface of the polymeric materials. One possible way to avoid microbial contamination is to develop materials that possess antimicrobial activities [Park et. al., 2001; Park et. al., 2001; Moon et. al., 2003, Moon et. al., 2003; Park et. al., 2004; Park et. al., 2004; Kim et. al., 2004; Yang and Park, 2006; Park et. al., 2008]. Moreover, increased efficiency, selectivity, and handling safety are additional benefits which may be realized [Kim et. al., 2004]. Antimicrobial agents used in antimicrobial-processed products are classified into organic, inorganic and natural organic compounds. Organic antimicrobial agents raise health concerns and many of them do not have sufficient antimicrobial activity. Polymeric biocides can significantly reduce loss of antimicrobial activity associated with volatilization, photolytic decomposition, dissolution, and permeation [Kim et. al., 2004]. On the other hand, inorganic antimicrobial agents employ Ag, Cu, and Zn compounds and are excellent in safety and antimicrobial activity. These metalic compounds are used in many types of household and medical products due to their good balance between antimicrobial activity and endurance. However, patients with metal allergy due to Cu or Zn have been reported [Nakashima et. al., 2008].

As previously mentioned, CNTs possess antimicrobial properties themselves, and their relevant activities were ascribed to the behavior of 'nanodart' with the proposed physical damage mechanism [Kang et. al., 2008]. The intrinsic toxicity of CNT depends on the degree of surface functionalization and the different toxicity of functional groups. In the presence of water, the isocyanate groups in MWNT can react with water. The reaction is a three step process (Fig. 15). A water molecule reacts with an isocyanate group to form cabamic acid. Carbamic acids are unstable, and decompose forming CO_2 and an amine. Existence of high abundance of amine groups on the surface of functionalized MWNTs provided sites for binding of various antibiotics such like norfloxacin, and for formation of silver nanoparticles by the reduction of aqueous solution of $AgNO_3$.

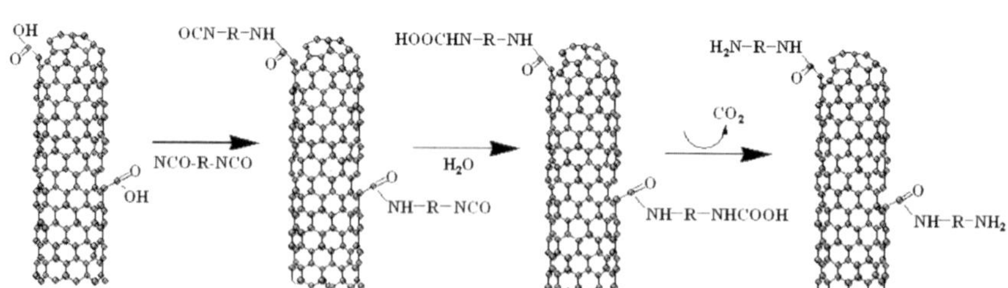

Figure 15. Synthesis of amine covalently functionalized MWNT.

Recently, Neelgund and Oki synthesized the nanohybrids composed of silver nanoparticles and aromatic polyamide functionalized MWNTs. Prior to deposition of silver nanoparticles, acid treated MWNTs were successively reacted with p-phenylenediamine and methylmethacrylate to form series of NH_2-terminated aromatic polyamide dendrimers on the surface of MWNTs through Michael addition and amidation. The antimicrobial activity of MWNT-Ar-NH_2/Ag nanohybrids were measured against *E. coli*, *P. aeruginosa* and *S. aureu* and compared with MWNT-COOH and MWNT-Ar-NH_2. The results showed that functionalization of MWNTs with aromatic polyamide dendrimers and successive deposition of Ag nanoparticles could play an important role in the enhancement of antimicrobial activity [Neelgund and Oki, 2011].

3.7. Environmental friendly aqueous coating system

For coating applications a uniform and stable dispersion of particulate matter plays an important role. This requirement is especially critical when submicron or nano-sized particles are involved. CNTs tend to cohere in aqueous dispersion due to their high surface energy and lack of chemical affinity with the dispersing medium [Park et. al., 2002]. Surfactant adsorption on nanotube surfaces and chemical functionalization of nanotube sidewalls are two of the most widely-used methods for solubilization of nanotubes. Non-covalent surface treatment by surfactants or polymers has been used in the preparation of both aqueous and organic solutions to obtain high weight fraction of individually dispersed nanotubes [Barraza et. al., 2002; Jiang et. al., 2003; Yurekli et. al., 2004]. When surfactants are employed in CNT dispersions, surfactant molecules work by adsorption at the interface and self-accumulation into supra-molecular structures, which help their dispersion retain a stable colloidal state. However, surfactants adsorbed on nanotubes create a physical barrier between the nanotubes and the environment [Hobbie et. al., 2006; Lee et. al., 2007]. Chemical methods use surface functionalization of CNT to improve their chemical compatibility with the target medium that is to enhance wetting and reduce their tendency to agglomerate [Vaisman et. al., 2006]. Systematic investigation of the effects of nanotube length and functionalization for MWNT has revealed that the introduction of carboxylic or thiol groups on the surface of shortened nanotubes increases the stability of MWNT dispersions up to 0.24 mg/ml. Octadecyl-amide functionalized MWNT were reported to exhibit good solubility in polar solvents [Qin et. al., 2003]. These long chain alkylamide-functionalized nanotubes were obtained where surface-bound COOH groups are converted into thionyl chloride groups and subsequently reacted with amine. The introduction of surface charge on MWNT also has contrasting effects on stabilising their dispersions. MWNTs are modified with carboxylic anion groups; the dispersion stability in water was significantly enhanced due to the combination of polar–polar affinity and electrostatic repulsion [Lee et. al., 2007]. MWNT uniformly dispersed in water can be utilized in environmental friendly aqueous coating, direct conductive coating, further sol–sol process and aqueous nanocomposite system.

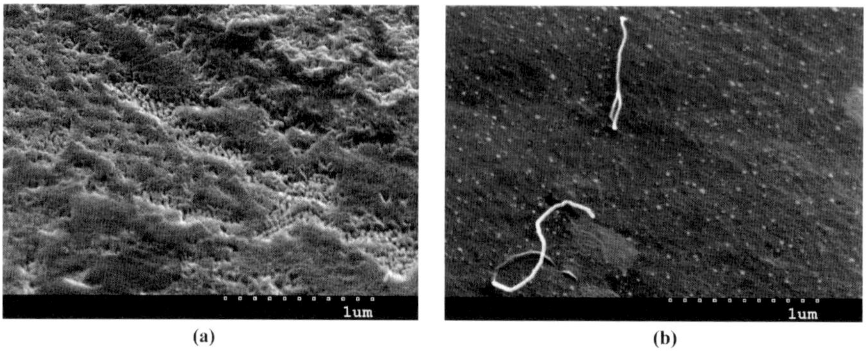

Figure 16. SEM image of the sodium silicate (a) and sodium silicate/functionalized MWNT composites prepared from aqueous coating system (b).

3.8. Electromagnetic interference (EMI) shielding materials

As electromagnetic radiation, particularly that at high frequencies tend to interfere with electronics, EMI shielding of both electronics and radiation source is needed and is increasingly required by governments around the world [Chung, 2001]. The radiation may be either electromagnetic in nature, such as X-rays and gamma rays, or charged particles, such as beta particles and electrons. The lifetime and efficiency of them can be increased by the effective shielding. Generally, highly electroconductive materials such like metals are used for shielding application. However, metals have their own shortcomings like heavy weight, susceptibility to corrosion, wear, and physical rigidity [Wu et al., 2006]. Many researches have been conducted to improve the EMI shielding of polymer materials by coating an electroconductive layer on the surface, incorporating electroconductive fillers, or utilizing electroconductive polymers. Among various electroconductive fillers that have been utilized, covalently functionalized MWNT is one of the most promising candidates, not only because of its good electrical conductivity but also because of its ability to improve mechanical properties. Recently, the mass production of MWNTs causes price reduction. They are more affordable for application in nanocomposites [Wu et. al., 2006].

4. Conclusion

Current interest in CNTs has been generated and maintained because nanotubes exhibit unique properties include high modulus, high aspect ratios, excellent thermal and electrical conductivities, and magnetic properties not achievable with traditional filler. In this chapter, MWNTs were subjected to EB irradiation at various doses to determine the incidence of surface modification and, resultantly, deformation or destruction to the otherwise pristine graphitic structure. FTIR spectra obtained from EB-MWNT samples provide insight into the level of surface modification. The introduced carboxyl groups represent useful sites for further modifications, as they enable the covalent coupling of molecules through the creation of ester or amide bonds. Consequently, EB-MWNT was covalently functionalized with isocyanate, epoxy, and hydroxyl compounds and their applicability was investigated. As has been shown in this study, the possible applications of them range widely, from nanocomposite materials to antibacterial agents. Functional groups such like isocyanate group on MWNT surface can interact with -OH group in polymers by urethane bonding and result in a better dispersion of MWNT in polyurethane matrix. Afterward we carry out extensive studies to investigate the properties and applicability for urethane link polydimethylsiloxane/MWNT nanocomposites using various isocyanate functionalized MWNT.

Acknowledgements

We are grateful to the Small and Medium Enterprises (SMEs) Technology Innovation Program, Republic of Korea, for financial support of this experimental work.

Author details

Eun-Soo Park[1*]

Address all correspondence to: t2phage@hitel.net

1 Youngchang Silicone Co., Ltd., Korea

References

[1] Abuilaiwi, F. A., Laoui, T., Al-Harthi, M., & Atieh, M. A. (2010). Modification and functionalization of multiwalled carbon nanotube (MWCNT) via Fischer esterification. *The Arabian Journal for Science and Engineering*, 35, 38-48.

[2] Ajayan, P., Stephan, O., Colliex, C., & Trauth, D. (1994). Aligned carbon nanotube arrays formed by cutting a polymer resin-nanotube composite. *Science*, 265(5176), 1212-1214.

[3] Allen, T. M., & Cullis, P. R. (2004). Drug delivery systems: entering the mainstream. *Science*, 303, 1818-1822.

[4] Balasubramanian, K., & Burghard, M. (2008). Electrochemically functionalized carbon nanotubes for device applications. *Journal of Materials Chemistry*, 18, 3071-3083.

[5] Baleizão, C., Gigante, B., Garcia, H., & Corma, A. (2004). Vanadyl salen complexes covalently anchored to single wall carbon nanotubes as heterogeneous catalyst for the cyanosilylation of aldehydes. *Journal of Catalysis*, 221, 77-84.

[6] Banerjee, S., Hemraj-Benny, T., & Wong, S. S. (2005). Covalent surface chemistry of single-walled carbon nanotubes. *Advanced Materials*, 17, 17-29.

[7] Banhart, F. (1999). Irradiation Effects in Carbon Nanostructures. *Reports on Progress in Physics*, 62, 1181-1221.

[8] Barraza, H. J., Pompeo, F., O'Rear, E. A., & Resasco, D. E. (2002). SWNT-filled thermoplastic and elastomeric composites prepared by miniemulsion polymerization. *Nano Letters*, 2, 797-802.

[9] Basiuk, E. V., Monroy-Pelaez, M., Puente-Lee, I., & Basiuk, V. A. (2004). Direct solvent-free amination of closed-cap carbon nanotubes: A link to fullerene chemistry. *Nano Letters*, 4, 863-866.

[10] Burghard, M., & Balasubramanian, K. (2005). Chemically Functionalized Carbon Nanotubes. *Small*, 1(2), 180-192, DOI:10.1002/smll.200400118.

[11] Chen, L., Xie, H., & Yu, W. (2011). *Functionalization Methods of Carbon Nanotubes and Its Applications In Carbon Nanotubes Applications on Electron Devices*, Mauricio, J. M., Eds, 213-232, 978-9-53307-496-2.

[12] Chiu, W. M., & Chang, Y. A. (2007). Chemical modification of multiwalled carbon nanotube with the liquid phase method. *Journal of Applied Polymer Science*, 107, 1655-1660, DOI:10.1002/app.26633.

[13] Chung, D. D. L. (2001). Electromagnetic interference shielding effectiveness of carbon. *Carbon*, 39, 279-285.

[14] Davis, S. S. (1997). Biomedical applications of nanotechnology-implications for drug targeting and gene therapy. *Ophthalmic Genetics*, 15, 217-224.

[15] Donaldson, K., Stone, V., Tran, C. L., Kreyling, W., & Borm, P. J. A. (2009). Nanotoxicology. *Occupational and Environmental Medicine*, 61, 727-728.

[16] Dresselhaus, G., & Avouris, P. (2001). Eds, *Carbon Nanotubes: Synthesis, Structure, Properties and Applications*, 80, Springer-Verlag, Berlin, 81-112.

[17] Gao, C., He, H., Zhou, L., Zheng, X., & Zhang, Y. (2009). Scalable Functional Group Engineering of Carbon Nanotubes by Improved One-Step Nitrene Chemistry. *Chemistry of Materials*, 21, 360-370.

[18] George, J. J., & Bhowmick, A. K. (2009). Influence of Matrix Polarity on the Properties of Ethylene Vinyl acetate-carbon nanofiller nanocomposites. *Nanoscale Research Letters*, 4, 655-664, DOI:10.1007/s11671-009-9296-8.

[19] Guzman, L. A., Labhasetwar, V., Song, C., Jang, Y., Lincoff, A. M., Levy, R., & Topol, E. J. (1996). Local intraluminal infusion of biodegradable polymeric nanoparticles. A novel approach for prolonged drug delivery after balloon angioplasty. *Circulation*, 94, 1441-1448.

[20] Harris, P. J. F. (2004). Carbon nanotube composites. *International Materials*, 49(1), 31-43, DOI: 10.1179/095066004225010505.

[21] Hirsch, A. (2002). Functionalization of Single-Walled Carbon Nanotubes. *Angewandte Chemie International Edition*, 41, 1853-1859.

[22] Hobbie, E. K., Obrzut, J., Kharchenko, S. B., & Grulke, E. A. (2006). Charge transport in melt-dispersed carbon nanotubes. *Journal of Chemical Physics*, 125, 044712-1-044712-3.

[23] Hogt, A. H., Dankert, J., & Feijen, J. (1986). Adhesion of coagulase-negative staphylococci to methacrylate polymers and copolymers. *Journal of Biomedical Materials Research*, 20, 533-545.

[24] Hu, H., Bhowmik, P., Zhao, B., Hamon, M. A., Itkis, M. E., & Haddon, R. C. (2001). Determination of the acidic sites of purified single-walled carbon nanotubes by acid-base titration. *Chemical Physics Letters*, 345, 25-28.

[25] Jia, G., Wang, H. F., Yan, L., Wang, X., Pei, R. J., Yan, T., Zhao, Y. L., & Guo, X. B. (2005). Cytotoxicity of Carbon Nanomaterials: Single-Wall. Nanotube, Multi-Wall Nanotube, and Fullerene. *Environmental Science & Technology*, 39, 1378-1383.

[26] Jiang, L., Gao, L., & Sun, J. (2003). Production of aqueous colloidal dispersions of carbon nanotubes. *Journal of Colloid and Interface Science*, 260, 89-94.

[27] Jucker, B. A., Harms, H., & Zehnder, A. B. (1996). Adhesion of the positively charged bacterium Stenotrophomonas (Xanthomonas) maltophilia 70401 to glass and Teflon. *Journal of Bacteriology*, 178, 5472-5479.

[28] Kam, N. W. S., Connel, M. O'., Wisdom, J. A., & Dai, H. (2005). Carbon nanotubes as multifunctional biological transporters and near-infrared agents for selective cancer cell destruction. *PNAS*, 102, 11600-11605.

[29] Kang, S., Herzberg, M., Rodrigues, D. F., & Elimelech, M. (2008). Antibacterial Effects of Carbon Nanotubes: Size Does Matter! *Langmuir*, 24, 6409-6413, DOI: 10.1021/la800951v.

[30] Kanny, K., Muhfuz, H., Carlsson, L. A., Thomas, T., & Jeelani, S. (2002). Dynamic mechanicalanalyses and flexural fatigue of PVC foams. *Composite Structures*, 58, 175-183.

[31] Kim, J. H., Park, E. S., Shim, J. H., Kim, M. N., Moon, W. S., Chung, K. H., & Yoon, J. S. (2004). Antimicrobial activity of p-hydroxyphenyl acrylate derivatives. *Journal of Agricultural and Food Chemistry*, 52, 7480-7483.

[32] Kim, G. T., & Park, E. S. (2008). Thermal Reproducibility and Voltage Stability of Carbon Black/Multiwalled Carbon Nanotube and Carbon Black/SnO_2-Sb Coated Titanium Dioxide Filled Silicone Rubber Heaters. *Journal of Applied Polymer Science*, 109, 1381-1387, DOI:10.1002/app.28200.

[33] Krasheninnikov, A. V., & Nordlund, K. (2004). Irradiation effects in carbon nanotubes. *Nuclear Instruments and Methods in Physics Research*, B216, 355-366.

[34] Krijgsman, J. (1992). *Product Recovery in Bioprocess Technology*, Butterworth-Heinemann, 0-75061-510-9.

[35] Kumar, S., Doshi, H., Srinivasarao, M., Park, J. O., & Schiraldi, D. A. (2002). Fibers from polypropylene/nano carbon fiber composites. *Polymer*, 43, 1701-1703.

[36] Lau, A. K.-T., & Hui, D. (2002). The Revolutionary Creation of New Advanced Materials: Carbon Nanotube Composites. *Composites Part B: Engineering*, 33(2), 263-267.

[37] Li, Q. W., Li, Y., Zhang, X. F., Chikkannanavar, S. B., Zhao, Y. H., Dangelewicz, A. M., Zheng, L. X., Doorn, S. K., Jia, Q. X., Peterson, D. E., Arendt, P. N., & Zhu, Y. T. (2007). Structure-Dependent Electrical Properties of Carbon Nanotube Fibers. *Advanced Materials*, 19, 3358-3363, DOI: 10.1002/adma.200602966.

[38] Liu, Y.-L., & Chen, W.-H. (2007). Modification of Multiwall Carbon Nanotubes With Initiators and Macroinitiators of Atom Transfer Radical Polymerization. *Macromolecules*, 40, 8881-8886.

[39] Lee, E. J., Yoon, J. S., & Park, E. S. (2011). Morphology, Resistivity, and Thermal Behavior of EVOH/Carbon Black and EVOH/Graphite Composites Prepared by Simple Saponification Method. *Polymer Composites*, 32, 714-726.

[40] Lee, E. J., Yoon, J. S., & Park, E. S. (2012). Preparation and Properties of the Highly Porous Poly(ethylene-co-vinyl alcohol)/Multiwalled Carbon Nanotube Nanocomposites Prepared by a Simple Saponification Method. *Journal of Applied Polymer Science*, 125, E691-E704, DOI: 10.1002/app.36537.

[41] Lee, J. W., Kim, M. H., Hong, C. K., & Shim, S. E. (2007). Measurement of the dispersion stability of pristine and surface-modified multiwalled carbon nanotubes in various nonpolar and polar solvents. *Measurement Science and Technology*, 18, 3707-3712, doi:10.1088/0957-0233/18/12/005.

[42] Lee, K., Li, L., & Dai, L. (2005). Asymmetric end-functionalization of multiwalled carbon nanotubes. *Journal of the American Chemical Society*, 127, 4122-4123.

[43] Lee, R. S., Chen, W. H., & Lin, J. H. (2011). Polymer-grafted multi-walled carbon nanotubes through surface-initiated ring-opening polymerization and click reaction. *Polymer*, 52, 2180-2188.

[44] Lee, S. W., Kim, B. S., Chen, S., Shao-Horn, Y., & Hammond, P. T. (2009). Layer-by-Layer Assembly of All Carbon Nanotube Ultrathin Films for Electrochemical Applications. *Journal of the American Chemical Society*, 131, 671-679, DOI: 10.1021/ja807059k.

[45] Lewinski, N., Colvin, V., & Drezek, R. (2008). Cytotoxicity of Nanoparticles. *Small*, 4(1), 26-49.

[46] Loh, K. J., Kim, J., Lynch, J. P., Kam, N. S., & Kotov, N. A. (2007). Multifunctional layer-by-layer carbon nanotube-polyelectrolyte thin films for strain and corrosion sensing. *Smart Materials and Structures*, 16, 429-438.

[47] Moon, W. S., Chung, K. H., Seol, D. J., Park, E. S., Shim, J. H., Kim, M. N., & Yoon, J. S. (2003). Antimicrobial effect of monomers and polymers with azole moieties. *Journal of Applied Polymer Science*, 90, 2933-2937.

[48] Moon, W. S., Kim, J. C., Chung, K. H., Park, E. S., Kim, M. N., & Yoon, J. S. (2003). Antimicrobial activity of a monomer and its polymer based on quinolone. *Journal of Applied Polymer Science*, 90, 1797-1801.

[49] Meyyappan, M., Yamada, T., Sarrazin, P., & Li, J. (2005). *Carbon Nanotubes: Science and Applications*, CRC Press.

[50] Nakashima, H., Miyano, N., & Takatuka, T. (2008). Elution of metals with artificial sweat/saliva from inorganic antimicrobials/processed cloths and evaluation of antimicrobial activity of cloths. *Journal of Health Science*, 54, 390-399.

[51] Neelgund, G. M., & Oki, A. (2011). Deposition of Silver Nanoparticles on Dendrimer Functionalized Multiwalled Carbon Nanotubes: Synthesis, Characterization and Antimicrobial Activity. *Journal of Nanoscience and Nanotechnology*, 11(4), 3621-3629.

[52] Nel, A., Xia, T., Madler, L., & Li, N. (2006). Toxic Potential of Materials at the Nanolevel. *Science*, 311, 622-627.

[53] Okaya, T., & Ikari, K. (1997). *Polyvinylalcohol-Development*, Finch, C. A. (Ed.), John Wiley & Sons, 978-0-47199-850-1, Chichester (Chapter 8).

[54] Park, C., Oundaies, Z., Watson, K. A., Crooks, R. E., Smith, J. Jr., Lowther, S. E., Connell, J. W., Siochi, E. J., Harrison, J. S., & St Clair, T. L. (2002). Dispersion of single wall carbon nanotubes by in-situ polymerization under sonication. *Chemical Physics Letters*, 364, 303-308.

[55] Park, E. S. (2008). Mechanical properties and antibacterial activity of peroxide-cured silicone rubber foams. *Journal of Applied Polymer Science*, 110, 1723-1729.

[56] Park, E. S., Kim, H. K., Shim, J. H., Kim, M. N., & Yoon, J. S. (2004). Synthesis and properties of polymeric biocides based on poly ethylene-co-vinyl alcohol. *Journal of Applied Polymer Science*, 93, 765-770.

[57] Park, E. S., Kim, H. S., Kim, M. N., & Yoon, J. S. (2004). Antibacterial activities of polystyrene-block-poly(4-vinyl pyridine) and poly(styrene-random-4-vinyl pyridine). *European Polymer Journal*, 40, 2819-2822.

[58] Park, E. S., Lee, H. J., Park, H. Y., Kim, M. N., Chung, K. H., & Yoon, J. S. (2001). Antifungal effect of carbendazim supported on poly(ethylene-co-vinyl alcohol) and epoxy resin. *Journal of Applied Polymer Science*, 80, 728-736.

[59] Park, E. S., Moon, W. S., Song, M. J., Kim, M. N., Chung, K. H., & Yoon, J. S. (2001). Antimicrobial activity of phenol and benzoic acid derivatives. *International Biodeterioration & Biodegradation*, 47, 209-214.

[60] Pastorin, G., Wu, W., Wieckowski, S., Briand, J. P., Kostarelos, K., Prato, M., & Bianco, A. (2006). Double functionalisation of carbon nanotubes for multimodal drug delivery. *Chemical Communications*, 1182-1184.

[61] Petrova, S., Miloshev, S., Mateva, R., & Iliev, I. (2008). Synthesis of Amphiphilic PEG-PCL-PEG Triblock Copolymers. *Journal of the University of Chemical Technology and Metallurgy*, 43, 199-204.

[62] Qin, Y., Liu, L., Shi, J., Wu, W., Zhang, J., Guo, Z. X., Li, Y., & Zhu, D. (2003). Large Scale Preparation of Solubilized Carbon Nanotubes. *Chemistry of Materials*, 15, 3256-3260.

[63] Ramanathan, T., Stankovich, S., Dikin, D. A., Liu, H., Shen, H., & Nguyen, S. T. (2007). Graphitic nanofillers in PMMA nanocomposites-An investigation of particle size and dispersion and their influence on nanocomposite properties. *Journal of Polymer Science Part B: Polymer Physics*, 45, 2097-2112.

[64] Radovic, L. R. (2003). *Chemistry and Physics of Carbon*, Marcel Dekker, New York, USA.

[65] Raymond, D. L. (1999). *Water Quality and Treatment*, 5th, American Water Works Association and McGraw-Hill, New York, 0-07001-659-3.

[66] Sahoo, N. G., Rana, S., Cho, J. W., Li, L., & Chan, S. H. (2010). Polymer Nanocomposites Based on Functionalized Carbon Nanotubes. *Prog. Polym. Sci.*, 35, 837-867.

[67] Salvetat, J. P., Bonard, J. M., Thomson, N. H., Kulik, A. J., Forro, L., Benoit, W., & Zuppiroli, L. (1999). Mechanical properties of carbon nanotubes. *Applied Physics A*, 69, 255-260.

[68] Sears, K., Dumée, L., Schütz, J., She, M., Huynh, C., Hawkins, S., Mikel, M., & Gray, S. (2010). Recent Developments in Carbon Nanotube Membranes for Water Purification and Gas Separation. *Materials*, 3, 127-149, DOI:10.3390/ma3010127.

[69] Shohet, R. V., Chen, S., Zhou, Y. T., Wang, Z., Meidell, R. S., Unger, R. H., & Grayburn, P. A. (2000). Echocardiographic destruction of albumin microbubbles directs gene delivery to the myocardium. *Circulation*, 101, 2554-2556.

[70] Singh, P., Tripathi, R. M., & Saxena, A. (2010). Synthesis of carbon nanotubes and their biomedical applications. *Journal of Optoelectronics and Biomedical Materials*, 2, 91-98.

[71] Szleifer, I., & Yerushalmi-Rozen, R. (2005). Polymers and carbon nanotubes- dimensionality, interactions and nanotechnology. *Polymer*, 46(19), 7803-7818, 0032-3861.

[72] Tahermansouri, H., Chobfrosh khoei, D., & Meskinfam, M. (2010). Functionalization of Carboxylated Multi-wall Nanotubes with 1,2-phenylenediamine. *International Journal of Nano Dimension*, 1, 153-158, 2008-8868.

[73] Thostenson, Erik T., Zhifeng, Ren., & Tsu-Wei, Chou. (2001). Advances in the science and technology of carbon nanotubes and their composites: a review. *Composite Science and Technology*, 61, 1899-1912.

[74] Thostenson, E. T., & Chou, T. W. (2002). Aligned multi-walled carbon nanotube-reinforced composites: processing and mechanical characterization. *Journal of Physics D: Applied Physics*, 35(16), L77-L80.

[75] Treacy, M. M. J., Ebbesen, T. W., & Gibson, J. M. (1996). Exceptionally high Young's modulus observed for individual carbon nanotubes. *Nature*, 381(6584), 678-680.

[76] Vaisman, L., Wagner, H. D., & Marom, G. (2006). The role of surfactants in dispersion of carbon nanotubes. *Advances in Colloid and Interface Science*, 128-130, 37-46.

[77] Vaisman, L., Wachtel, E., Wagner, H. D., & Marom, G. (2007). Polymer-nanoinclusion interactions in carbon nanotube based polyacrylonitrile extruded and electrospun fibers. *Polymer*, 48, 6843-6854.

[78] Wong, S. S., Joselevich, E., Wooley, A. T., Cheung, C. L., & Lieber, C. M. (1998). Covalently functionalized nanotubes as nanometre-sized probes in chemistry and biology. *Nature*, 394, 52-55.

[79] Wong, E. W., Sheehan, P. E., & Lieber, C. M. (1997). Nanobeam Mechanics: Elasticity, Strength, and Toughness of Nanorods and Nanotubes. *Science*, 277(5334), 1971-1975.

[80] Wu, H. L., Ma, C. C. M., Yang, Y. T., Kuan, H. C., Yang, C.-C., & Chiang, C. L. (2006). Morphology, Electrical Resistance, Electromagnetic Interference Shielding and Mechanical Properties of Functionalized MWNT and Poly(urea urethane) Nanocomposites. *Journal of Polymer Science: Part B: Polymer Physics*, 44, 1096-1105, DOI:10.1002/polb.20766.

[81] Yang, H. S., & Park, E. S. (2006). Mechanical properties and antimicrobial activity of silicone rubber compounds containing acrylated norfloxacin and its polymer. *Macromolecular Materials and Engineering*, 291, 621-628.

[82] Yoon, K. R., Kim, W. J., & Choi, I. S. (2004). Functionalization of Shortened Single-Walled Carbon Nanotubes with Poly(p-dioxanone) by "Grafting-From" Approach. *Macromolecular Chemistry and Physics*, 205, 1218-1221.

[83] Yurekli, K., Mitchell, C. A., & Krishnamootri, R. (2004). Small-angle neutron scattering from surfactant-assisted aqueous dispersions of carbon nanotubes. *Journal of the American Chemical Society*, 126, 9902-9903.

[84] Zhao, J. J., Park, H. K., Han, J., & Lu, J. P. (2004). Electronic properties of carbon nanotubes with covalent sidewall functionalization. *J. Phys. Chem. B*, 108, 4227-4230.

[85] Zeleniakiene, D. (2006). The influence of microstructural stiffness changes on the stress concentration factor of porous polymer materials. *Proceedings of the Estonian Academy of Sciences Engineering*, 12(2), 147-155, 1406-0175.

[86] Zhang, M., Atkinson, K. R., & Baughman, R. H. (2004). Multifunctional Carbon Nanotube Yarns by Downsizing an Ancient Technology. *Science*, 306(5700), 1358-1361.

[87] Zhu, L., Xu, J., Xiu, Y., Hess, W. H., & Wong, C. P. (2006). Controlled Growth and Electrical Characterization of High-aspect ratio Carbon Nanotube Arrays. *Carbon*, 4, 253-258.

[88] Zhu, H. W., Xu, C. L., Wu, D. H., Wei, B. Q., Vajtai, R., & Ajayan, P. M. (2002). Direct synthesis of long nanotube strands. *Science*, 296, 884-886.

[89] Zou, J. H., Liu, L. W., Chen, H., Khondaker, S. I., McCullough, R. D., & Huo, Q. (2008). Dispersion of pristine carbon nanotubes using conjugated block copolymers. *Advanced Materials*, 20, 2055-2060, 1521-4095.

Chapter 10

Aqueous Solution Surface Chemistry of Carbon Nanotubes

Anup K. Deb and Charles C. Chusuei

Additional information is available at the end of the chapter

1. Introduction

Since the rediscovery of carbon nanotubes (CNTs) by Iijima in 1991, a plethora of applications have been developed in the fields of biomolecular science, catalysis, environmental chemistry and medicine. Relevant to the development of these new technologies, it is important to effectively characterize and tune the chemical and electronic structures of these materials for desired properties. Within the last 15 years, an array of surface characterization methods have been developed to assay the surface structures of single- (SWNTs) and multi-walled (MWNTs) carbon nanotubes, in particular as organic moieties and catalytically active metal nanoparticles are tethered to them. Distinctive physical, chemical, electrical and high thermal properties of CNTs make these materials suitable for widespread applications, such as fuel cells, semiconducting materials in electronics, atomic force microscopy probes, microelectrodes, adsorbents to remove pollutants from waste water, electrochemical sensing and drug carriers. Aqueous surface chemistry plays a vital role in determining the fate and transport of CNTs. A large fraction of the atoms in CNTs reside at or near the surface (Sayes et al., 2006; Bottini et al., 2006). Pristine carbon nanotubes are barely soluble in liquids. To introduce nanotubes in more easily dispersible forms, they require functionalization. These processes entail attaching various organic moities to the sidewalls, which can be used to tether catalytically reactive nanoparticles. Biomolecules require electron mediators to promote electron transfer needed for effective biosensing (Sampath et al., 1998). Electrochemical metal ion sensors require certain functional groups which show potential affinity towards particular metal ions (Mojica et al., 2007). Surface electrostatic interactions in solution also influence the sorption properties of these materials to entrain environmental contaminants on the CNT sidewalls (Tavallai et al., 2012).

Historically, the synthesis, fabrication and characterization of carbon nanomaterials have been carried out in vacuum environments. As these materials proliferate in use, knowledge pertaining to their environmental impact (i.e., involving fate and transport in aqueous systems) becomes increasingly important (Cho et al., 2008). Furthermore, preparation and synthesis of these materials in non-vacuum conditions makes these processes more amenable for industrial scale up. Recent attention has focused on modifying SWNTs and MWNTs in solution media.

A review of recent advancements to modify CNT surfaces in aqueous media is described in this chapter. Changes in the material properties are often observed concomitant to alterations in surface structure, such as colloidal dispersion and electrocatalytic activity. In the introductory section, the strengths and weaknesses of various traditional CNT surface chemistry probes are presented. Following this, nanotubes that have been chemically modified via chemical oxidation and organic derivatization are discussed. The technique of electrochemical functionalization using carbon nanotubes as the working electrode surface is presented. The next section describes the applications of derivatized carbon nanotubes as it applies to catalysis (involving noble metal nanoparticles), sensing, and selective cancer cell destruction, in which the nanotube sidewall structure plays a key role. The use of transmission electron microscopy (TEM) in conjunction with point-of-zero charge (PZC) measurements for exploring structure-property relationships is shown. The final sections present the effects of CNT functionalization on properties pertaining to colloidal stability and isoelectric points relevant for applications in environmental chemistry, catalyst synthesis, and designing materials for the remediation of contaminated ground water.

2. Overview of analytical techniques: strengths and weaknesses

Traditional analysis methods of carbon nanotubes include Boehm titrations, settling speed measurements, atomic force microscopy (AFM), and quartz crystal microbalance (QCM) measurements, X-ray photoelectron spectroscopy (XPS), attenuated total reflection infrared spectroscopy (ATR-IR), transmission electron microscopy (TEM), Raman spectroscopy, thermogravimetric analysis (TGA) and temperature programmed desorption (TPD). Each of these techniques has its own advantages in the chemical/structural information that they can provide as well as drawbacks.

Wet chemical characterization methods provide a rapid means of characterizing the CNT surface structure. Settling speed measurements is a crude, but rapid technique for measuring the extent of CNT sidewall oxidation containing protic groups with which the solvent can undergo hydrogen bonding (Xing et al., 2005). In this simple experimental setup, the rate at which CNTs fall in a buret (by gravity) is measured and correlated with the extent of surface functionalization. However, no qualitative information regarding the identity of the surface groups is available using this approach. Boehm titrations can be used to quantify the number of proton-containing functional groups (carboxylic acids, hydroxyl groups, lactones, etc.) on the CNT sidewall surface (Boehm et al., 1964). The titrant typically involves various bases ideal for each protic group, e.g., Na_2CO_3, NaOH, $NaHCO_3$, etc.), the acidity constants (pK_a) of which differ by orders of magnitude, rendering the analysis selective to the functional group of interest. But, this technique is ineffective for characterizing CNTs functionalized with aprotic moieties.

ATR-IR, AFM, QCM and Raman spectroscopy can be performed in ambient environments (i.e., not requiring vacuum conditions for analysis). ATR-IR is useful for qualitative identification of CNT surface moieties; however, quantitation is not available and some modes are too small to be observed relative to background (Brundle et al., 1992). AFM offers the capability of probing changes in CNT surface morphology, sidewall surface coverage and CNT lengths. However, the technique is not amenable to subnanometric scales as thermal noise becomes a major interference at this lengthscale (Magonov et al., 1996). QCM provides a means of monitoring mass changes during the assembly process as CNTs undergo functionalization, but accurate mass measurements are readily hampered by changes in temperature or cavitation (i.e., during ultrasonication) (Brown and Gallagher, 2007). The "diamond" D and G band shifts observed in Raman spectroscopy at ~1300 and ~1600 cm^{-1}, respectively, is a useful tool for assessing the degree of sidewall surface damage encountered in some functionalization methods (i.e., ultrasonication) as well as CNT purity and composition. D and G bands emanate from disordered and ordered sp^2-hybridized carbon from the graphene sheets, respectively, and are commonly used markers for elucidating covalent bond formation (Dresselhaus et al., 2001). However, spectral interpretion, involving relative D and G band intensity determinations can be complex (Brundle et al., 1992).

Vacuum-based characterization tools (the most cost-prohibitive class of these analytical methods), include electron spectroscopy, microscopy and mass analysis. XPS is an excellent tool for monitoring analyte surface oxidation states, and useful for elemental quantification and qualitative identification of surface functional groups. However, large amounts of sample (~ 5 mg) are needed for analysis and peakfitted interpretation can be complex. TEM offers powerful imaging capabilities of the CNT sidewalls to allow for observation of surface roughening that can result from either functionalization or the creation of surface defects. Material length, diameter and dispersion state can also be readily determined by TEM. In addition, spatial elemental analysis is available via energy dispersive X-ray spectroscopy (EDX), as it is often an available technique built into many TEM instruments. However, CNTs are susceptible to beam damage from TEM electrons. Another caveat is that variation in technique involving dispersing samples onto TEM grids and subsequent drying can skew observed results. TGA and TPD can be used to quantify the concentration of moieties tethered to the CNT sidewalls; but, limiting case assumptions, e.g., all of the mass lost (TGA) and bonding modes remain unchanged (TPD), need to be made for assessments, which may not be accurate if the CNT surface chemistry is complex. In addition, large amounts of sample (> 10 mg) are required for TPD and TGA analysis.

3. Functionalizing carbon nanotubes

Both single- and multiwalled carbon nanotubes have a tendency to aggregate into bundles very efficiently via van der Waals interactions in solution. These bundles can be exfoliated by using ultrasonication in combination with suitable surfactants. Typically, the outer walls of pristine carbon nanotubes are chemically inactive. Two major functionalization routes are used to activate CNT sidewalls: (i) endohedral and (ii) exohedral functionalization.

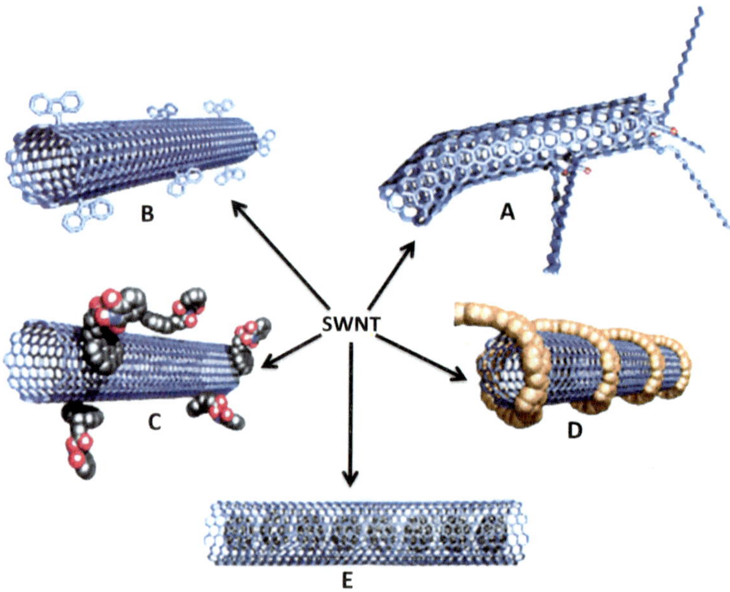

Figure 1. Functionalization pathways of SWNTs: A) defect-group functionalization, B) covalent sidewall functionalization, C) noncovalent exohedral functionalization with surfactants, D) noncovalent exohedral functionalization with polymers, and E) endohedral functionalization with, for example, C_{60}. (Reprinted with permission from [Hirsch., 2002]. Copyright, WILEY-VCH Verlag).

Endohedral functionalization involves insertion of various nanoparticles into the inner walls (Fig. 1E) (Hirsch, 2002). This task can be achieved either by (i) spontaneous penetration with colloidal nanoparticle suspensions filling the inner walls by evaporation of the carrier solvent; or (ii) by wet chemistry, as compounds are introduced into the inner walls of the nanotubes where they are transformed into nanoparticles while maintaining predetermined thermal/chemical conditions. Various pathways for exohedral functionalization is summarized in Figs. 1A-D (Hirsch., 2002). These avenues include defect group functionalization (Fig. 1A), covalent sidewall functionalization (Fig. 1B), and noncovalent exohedral functionalization using surfactants (Fig. 1C) and polymers (Fig. 1D). Covalent functionalization, which typically damages the carbon framework and is an irreversible process, is achieved by attaching functional groups to the nanotube ends or defects (Hirsch, 2002; Banerjee et al.,

2005). Noncovalent exohedral functionalization, on the other hand, is achieved by wrapping nanotubes using polymers or surfactants (Hirsch, 2002). They leave the CNT carbon framework intact and, it is usually a reversible process.

Hu et al. (2005) exohedrally functionalized SWNTs with DNA (noncovalently) by wrapping the outer surface of dispersed SWNTs with single-stranded DNA (ss-DNA). The functionalized ss-DNA-SWNTs have a strong tendency to attach onto glass substrates, forming a uniform film. These behaviors make it possible for electrochemical analysis and sensing. The material is amenable for use as a working electrode, exhibiting good electrochemical voltammetric properties. The electrode has well-defined quasi-reversible voltammetric responses, showing rapid electron transfer properties for $Fe(CN)_6^{3-}/Fe(CN)_6^{4-}$ redox pair systems, important for biosensing as this redox couple has demonstrated the ability to traverse bilayer lipid membranes (Lu et al., 2008).

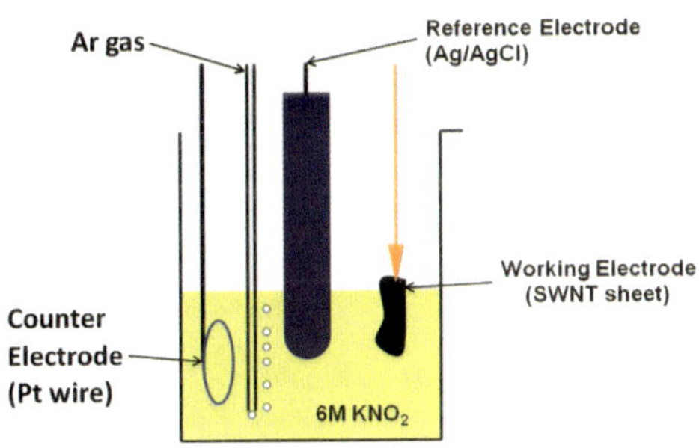

Figure 2. Arrangement of SWNT sheet in an electrochemical cell. The free-standing sheet of SWNTs underwent electrochemical oxidation upon reaction in potassium nitrite (KNO_2) solution (McPhail et al., 2009).

CNTs can also be functionalized electrochemically. Fig. 2 shows the general arrangement of an electrochemical cell where a SWNT sheet is used as a working electrode. This particular set up has been used to functionalize pristine HiPco SWNTs with nitroso (NO) functional groups in which free-standing SWNT sheets were produced via ultrasonication in 1% Triton X-100 solution surfactant. Prior to use as a working electrode for the electrochemical NO group attachment reaction, the Triton X-100 surfactant is removed via thermal decomposition in a tube furnace while flowing inert Ar gas is heated to 800°C. The electrochemical reaction forms a N_2O_4 dimer, which then dissociates into NO groups that attach to the SWNT sidewalls (Piela and Wrona, 2002; McPhail et al., 2009). The mechanistic scheme for the reaction is as follows:

$$NO_2^-(aq) \rightarrow NO_2(g) + e^-$$

$$2\,NO_2(g) \rightleftarrows N_2O_4(g)$$

$$N_2O_4(g) + H_2O \rightleftarrows HNO_2 + NO_3^-(aq) + H^+(aq)$$

$$HNO_2 + H_2O \rightarrow 3H^+(aq) + NO_3^-(aq) + NO$$

Nitric oxide (NO) is formed from nitrite (NO_2^-), in which dimerization occurs and followed by disproportionation. The observed nitrogen dioxide (NO_2) gas is liberated from the free standing SWNT working electrode during electrolysis. It should be noted that the fabrication technique for the free-standing sheet is not effective for homogeneously electrografting large quantitites of SWNTs (with a ~2 μm thickness), hampering industrial scale-up.

This task can be accomplished by using room-temperature ionic liquid (RTIL) to fabricate a supported three-dimensional network of SWNTs as the working electrode (Zhang et al., 2005). In this design, N-succinimidal acrylate (NSA) serves as a monomer dissolved in the supporting RTIL, 1-butyl-3-methylimidazolium hexafluorophosphate ($BMIMPF_6$), electrografted onto the SWNTs. The resulting linear sweep voltammogram (LSV) for the oxidation of glucose is shown in Fig. 3. Voltage is applied to the three dimensional network SWNT electrode from 0 to −2.4 V before and after the electrografting. The passivation peak due to the chemisorption (grafting) of an insulative polymer film on the cathode surface is observed at about −2.0 V in the first scan. After electrografting, the passivation peak disappears, denoting electrografting saturation.

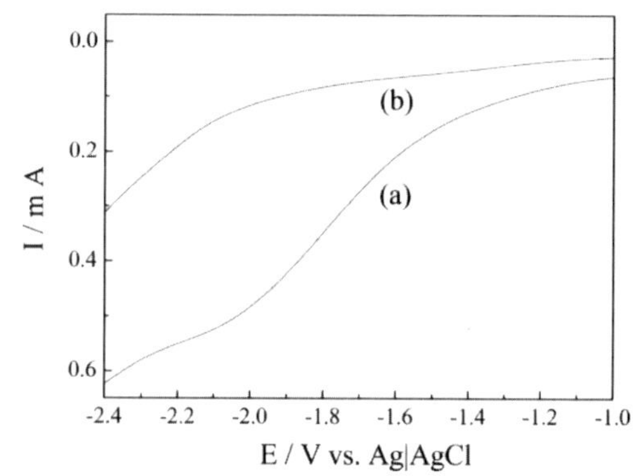

Figure 3. Linear sweep voltammograms of electrografting N-succinimidyl acrylate (NSA) at the three-dimensional network SWNT electrode in $BMIMPF_6$ during the first scan (a) and the second scan after conditioning at the passivation potential for a period of 40 min. (b). Scan rate: 20 mV/s. (Reprinted with permission from [Zhang et al., 2005]. Copyright, American Chemical Society).

Raman bands (Fig. 4) at 1591 cm^{-1} (tangential modes) and at 1278 cm^{-1} (disorder mode) are observed in both pristine SWNTs (Fig. 4a) and the SWNTs tethered to poly-NSA (Fig. 4b), showing direct evidence of covalent electrografting. Raman spectra were collected at several different spots for each of these surfaces; no distinctive differences in spectral features were observed, confirming homogeneous functionalization. Control experiments without NSA addition showed no affect on the structure of pristine SWNTs, as observed by Raman spectroscopy.

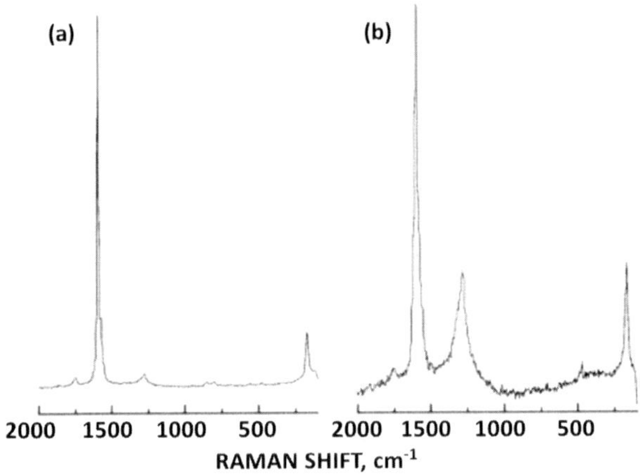

Figure 4. Normalized Raman spectra (the intensity of the strongest tangential modes) of pristine SWNTs (a) and SWNTs-poly-NSA (b). (Reprinted with permission from [Zhang et al., 2005]. Copyright, American Chemical Society).

Ultrasonication has become a standard technique for accelerating surface functionalization, employing the cavitation process from sound waves to facilitate acid oxidation. Defect sites are created during this process to facilitate sidewall functionalization (Fig. 1A). It should be noted that acid oxidized functionalization is more amenable to MWNTs than to SWNTs as robust conditions render the latter more susceptible to material decomposition. A sonochemical treatment method under acidic condition has been carried out to functionalize carbon nanotubes with –C=O, –C-O-C–, –COO–, –C-OH groups, which serve as effective tethering points for attaching catalytically active Pt nanoparticles for improved direct methanol fuel cell performance (Xing et al., 2005; Hull et al., 2006; Chusuei and Wayu, 2011). Raman spectra of the D and G "diamond" bands indicate minimal surface damage of the underlying graphene sheet during the sonication process (applied up to 8 hours). Pt nanoparticles were deposited onto these functionalized surfaces via O-containing moieties resulting in the improved electrocatalytic activity. Hull et al. (2006) demonstrated from ATR-IR data that, specifically, the carboxylate oxygen atoms were responsible for effective tethering of the catalytically active nanoparticles. It should be noted, however, that while sonication improves and facilitates functionalization of the MWNT sidewalls, it is possible to overtreat the MWNT sidewalls using this process. In the study, catalytic activity improved when sonication was performed over a 1-hour period, maximizing after a 2-hour sonication treatment. At a 4-hour sonication treatment, however, performance (for the direct methanol fuel cell reaction) diminished. ATR-IR peaks indicated that carboxylate surface structure was damaged after prolonged treatment. These same surfaces also show signs of roughening in the TEM images (Xing et al., 2005), attributable to defect formation and an increased degree of functionalization. Well-defined MWNT sidewall surface structures of the MWNTs are paramount for effective catalytic performance (*vide infra*).

4. Applications of functionalized carbon nanotubes

Functionalized CNTs have distinctive physicochemical properties, such as ordered structure with high aspect ratio, high mechanical strength, ultra-light weight, high electrical conductivity, high thermal conductivity, metallic or semi-metallic behavior and high surface area, which make them amenable for diverse applications (Ajayan, 1999). For example, Zhang et al. (2006) showed that electrochemically functionalized SWNT with polyaniline (PANI) can be used to fabricate chemical gas sensors. In monitoring ammonia gas with the PANI-SWNT composite, superior sensitivity and detection limits with good reproducibility were observed. Fig. 5 shows gas sensing response to various concentrations of NH_3, ranging from 50 ppm to 15 ppm, relative to initial baseline. It is clear from the graph that, after exposure to NH_3, the resistance of the PANI-SWNT sensor dramatically increased.

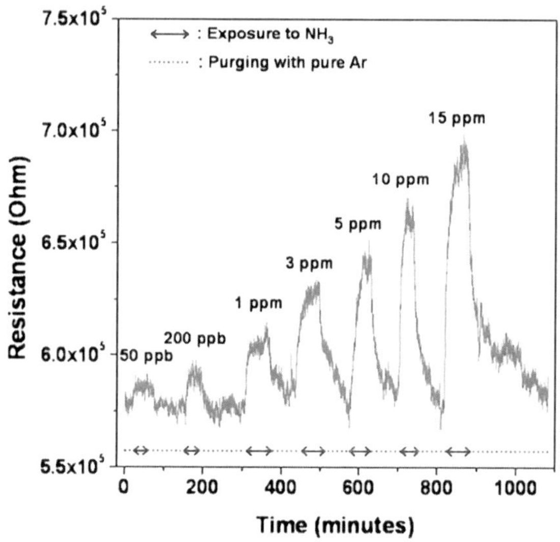

Figure 5. NH_3 gas sensing results using polyaniline coated SWNTs. The arrows (⟷) show exposure times to NH_3. PANI was coated on SWNTs using a two electrode configuration at 0.8 V for 5 minutes. (Reprinted with permission from [Zhang et al., 2006]. Copyright, WILEY-VCH Verlag).

When comparing the performance of the functionalized MWNT surface for the direct methanol fuel cell reaction in the previous section (*vide supra*), catalysts with the functionalized MWNT support exhibited a 48% increase in electrocatalytic activity compared to Pt nanoparticles tethered to the more commercially used Vulcan XC-72 fibrous carbon black support (Xing, 2004). The increased activity was due to the finer dispersion of catalytically active Pt nanoparticles (~3.5 nm in diameter) tethered to the CNT sidewalls (as compared to carbon black) made available by uniform attachment of the Pt nanoparticle precursors to ester-like oxygen atoms (Hull et al., 2006). Hence, sonication in aqueous acid environment has been shown to be effective for creating functional tethering points for practical catalyst synthesis. Overjero et al. (2006), similarly applied acidic (liquid phase) oxidation to MWNTs using nitric acid (HNO_3) to tether catalytically active Pt, Cu, and Ru nanoparticles. The sturdy support provided by functionalized MWNTs (with oxygen-containing moieties) was responsible for the observed, enhanced catalytic activity.

In fact, aqueous solution acid treatments have been found to be more effective than oxygen plasma treatments to functionalize CNTs with oxygen containing moieties. Xia et al. (2007) showed that nitric acid treatment yielded a 60% higher surface oxygen concentration compared to plasma treatment. Fig. 6 shows XPS survey spectra of the intensity of nitric acid-treated and plasma-treated MWNTs were recorded to identify the chemical composition. There was no evidence of metallic impurities (i.e., FeCo used to synthesize the nanotubes) present. In addition, after the nitric acid treatment an N 1s peak was found. It can be clearly seen that the intensity of the O 1s peak increased, whereas the C 1s peak decreased due to the oxidizing treatment with nitric acid and plasma treatments. The atomic percent oxygen-to-carbon ratios (taking into account differences in instrumental atomic sensitivity factors in the XPS) for the as-received, nitric acid-treated and oxygen plasma-treated MWNTs were found to be 0.118, 0.214 and 0.052_6, respectively. The acid-treated MWNTs clearly yielded the higher density of surface oxygen.

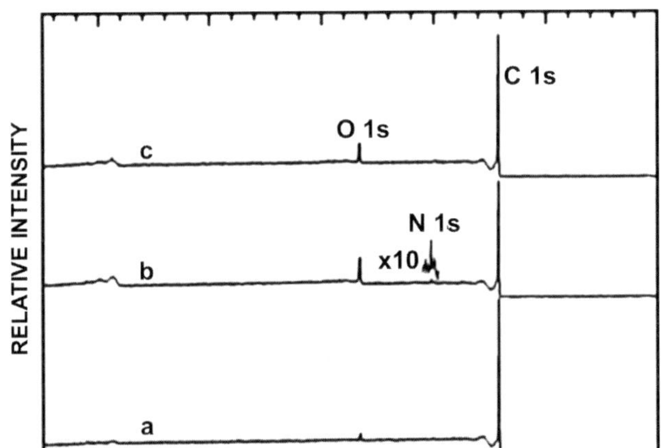

Figure 6. XP survey spectra of the MWNTs: (a) as received; (b) treated with nitric acid; (c) treated with oxygen plasma. The N 1s region at around 400 eV in trace (b) is magnified 10 times. (Reprinted with permission from [Xia et al., 2007]. Copyright, Elsevier B.V.).

Functionalized SWNTs have also received attention for their potential applications in medicine. Carboxylic acid functionalization on SWNTs improves electrocatalytic reactivity towards the oxidation of an array of biomolecules, such as dopamine, ephinephrine and ascorbic acid (Luo et al., 2001). SWNTs functionalized with hydroxyl (–OH) and carboxylic acid (–COOH) exhibit antimicrobial properties, capable of inactivating bacterial pathogens. In a study by Arias et al. (2009), modified SWNTs inactivated both Gram-positive and

Gram-negative bacterial cells in deionized water and 0.9% NaCl solution regardless of cell shape. Antimicrobial activity increased with both increasing concentration of the CNTs (in colloidal suspension) and treatment time (Arias et al., 2009). In either deionized water or 0.9% NaCl aqueous solution, 200-250 μg/mL of either OH-SWNTs or COOH-SWNTs have the ability to inactivate ~10^7 cfu/mL *Salmonella* cells in 15 minutes. The oxygen-containing moieties attached to the cell surface facilitated inactivation. Functionalized SWNTs have also shown promise as near-infrared agents for selective cancer cell destruction (Kam et al., 2005). Engineering SWNTs for this purpose is achieved by tethering pristine SWNTs with folate groups using sonication and centrifugation, in which the HiPco SWNTs are incorporated into a solution of phospholipids with polyethylene glycol moieties and folic acid terminal groups. These folate-SWNTs selectively attach to the inside structures of cancer cells that contain folate receptor tumor markers. Cell death is then triggered using near infrared irradiation that thermally decompose cancer cells without harming normal cells, which are folate receptor-free. Kam et al. (2005) demonstrate that while biological systems are transparent to 700-to-1100-nm near-infrared light, there is a strong absorbance of SWNTs within this wavelength region resulting in selective thermal heating of cancer cells.

In applications pertaining to ground water remediation, –OH, –COOH, and carbonyl (–C=O) functionalized MWNTs have been shown to have high sorption capacities. In fact, carboxyl-carbon sites are 20 times more energetic for zinc sorption than unoxidized carbon sites (Cho et al., 2010). Along with Zn(II) and Cd(II) chemically modified MWNTs have also been used as sorbent material (Tavallai et al., 2012) for separation and preconcentration of trace amounts of Co(II) and Cu(II) in the environmental and biological samples. In this study, MWNTs were modified with thiosemicarbazide and found to be an easily prepared solid and cost effective sorbent. These MWNT materials can be used several times without marked loss in sorption capacity.

In another study by Shamspur and Mostafavi (2009), MWNTs were modified using the reagent, N,N-bis(2-hydroxybenzylidene)-2,2(aminophenylthio)ethane for applications in ground water remediation. The resulting composite (incorporated into column material) was found to be a useful sorbent for simultaneous separation and preconcentration trace amounts of Au(III) and Mn(II). The reagent remained in the column and it's use could be cycled several times. Analytical ions were quantitatively recovered with detection limits and enrichment factors comparable or better than an array of commercially available matrices, such as Mberlite XAD-2000, silica gel/nanometer-sized TiO_2, Cu(II)-9-phenyl-3-fluorone, Kaolinite/5-Br-PADAP, and Penicillum italicum/Sepabeads SP 70 systems.

Furthermore, MWNTs can be modified using electrolysis. Unger et al. (2002) discovered that halogens, such as chlorine or bromine, can electrochemically be bonded to the nanotube lattice. Halogen gases are evolved from the anode and are attached to free-standing MWNT bucky sheets. These chlorine and bromine carbon nanotubes offer a pathway to a wide spectrum of nanotube derivatives. Oxygen-bearing functional groups, such as –OH and –COOH groups, are formed simultaneously, promoting solvation of the nanotubes in water or alcohol without any surfactant. Impurities and low grade modified nanotubes remain insoluble and can be filtered out. Since the functionalized nanotube structure is maintained, soluble material can readily be applied in aqueous solution to solid surfaces for applications, such as electric circuit patterning.

5. Effect of functionalization on colloidal stability

Without the use of sonication, pristine CNTs are generally hydrophobic in nature and cannot be dispersed in most solvents. The disparity of functionalized CNTs in colloidal particle depends on the nature of the functional groups and colloidal particles. Smith et al. (2009a; 2009b) found that the difference in the colloidal stability of the O-MWNTs was due to the effects of surface oxygen. Small changes of surface oxygen concentration, by as little as 1-to-5 percent results in drastic changes in the colloidal stability of O-MWNTs. The amount of oxygen incorporated onto the surface of nanotubes depends on the oxidizing agent used (HNO_3, $KMnO_4$, H_2SO_4/HNO_3, O_3, H_2O_2, etc.). Fig. 7a shows the relation between critical coagulation concentrations (CCC) for each O-MWNT with the surface oxygen concentration. The effect of pH with the above two parameters are also shown in Fig. 7b. These plots confirm that, for the vast majority of the O-MWNTs studied, CCC has a linear dependence on surface oxygen concentration over the pH range of 4-to-8).

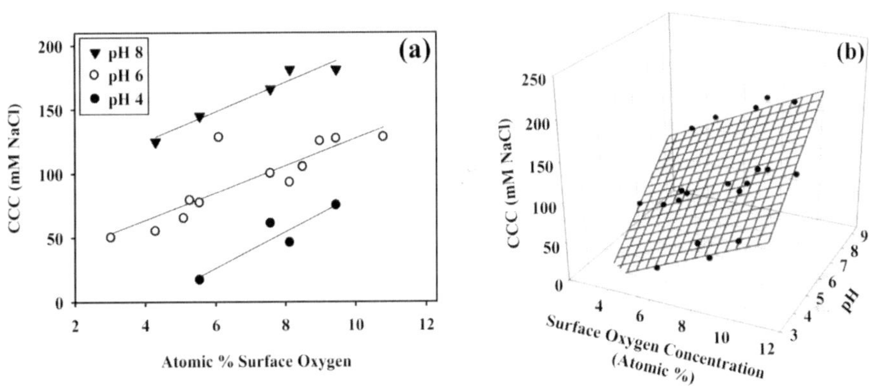

Figure 7. (a) Influence of surface oxygen concentration on the critical coagulation of O-MWNTs at pH = 4, 6 and 8; (b) three dimensional plot showing the functional interdependence of surface oxygen concentration, pH, and CCC of O-MWNTs. (Reprinted with permission from [Smith et al., 2009b]. Copyright, American Chemical Society).

Fig. 7 shows that for a given concentration of surface oxygen, the colloidal stability of O-MWNTs increases with increasing pH (Wepasnick et al., 2011). In the study, chemical derivatization was used in conjunction with XPS to quantify the distribution of oxygen containing functional groups (e.g., –OH, –COOH, –C=O) on the differently functionalized O-MWNTs. At high pH, the carboxylic acid group was the most predominant surface oxide present. This same result was observed by other researchers (Blanchard et al., 2007). Of the various MWNTs studied, the CCC correlated best with carboxylic acid group surface concentration. A significantly poorer correlation was found with both hydroxyl and carbonyl group surface concentration. In terms of MWNT electrophoretic mobility, Smith et al. (2009a) observed that surface oxygen concentration had no measurable affect on electrophoretic mobility. No correlation was observed between colloidal stability of O-MWNTs and its electrophoretic mobility. However, in terms of environmental impact, it is noteworthy that CCC values fell within the range of salinity conditions in estuaries and other fresh water bodies, indicating that O-MWNTs are likely stable and prone to aggregate and/or settle prior to being transported to oceanic environments.

Colloidal stability of oxidized MWNTs also changes with pH and electrolytic composition. Smith et al. (2009a) found that the colloidal stability of O-MWNTs increases with increasing pH, which is consistent with previous UV-vis studies of acid treated CNTs (Shieh et al., 2007). CCC values of O-MWNTs vary with counter ion concentration and valence in a manner consistent with Derjaguin-Landau-Verwey-Overbeek (DLVO) theory (Derjaguin et al., 1941; Verwey and Overbeek, 1948). MWNT surface oxygen density also affected MWNT adsorption properties. For instance, when adsorption of naphthalene onto O-MWNTs were carried out with variable surface oxygen concentrations (Ball et al., 2008), the MWNTs with the most concentrated surface oxygen content had the least adsorption capacity in the series.

The selection of acid oxidant can have markedly different effects on MWNT sidewall oxidation, as shown by Wepasnick et al. (2011). In this study, MWNTs were treated with six commonly used wet chemical oxidants (HNO_3, $KMnO_4$, H_2SO_4/HNO_3, $(NH_4)_2S_2O_8$, H_2O_2 and O_3). Using XPS and EDX to characterize and quantify the extent of surface oxidation, density of –OH, –COOH, –C=O surface groups, and their distribution, these parameters were found to be independent of reaction conditions, but sensitive to identity the oxidant. As MWNTs were treated with $(NH_4)_2S_2O_8$, H_2O_2 and O_3, higher concentrations of carbonyl and hydroxyl functional groups were found to form on the surface. In contrast, as more aggressive oxidant agents (HNO_3, $KMnO_4$) were used, higher fractional concentrations of carboxylic acid groups formed. Fig. 8 shows representative transmission electron micrographs of pristine MWNTs exposed to various oxidants, comparing the effects of equal concentrations of H_2O_2 and H_2SO_4/HNO_3.

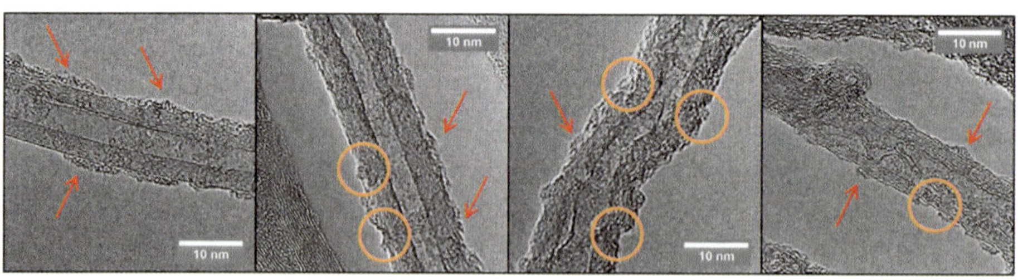

Figure 8. Representative TEM micrographs (left to right): Pristine MWNTs (0.9%), H_2O_2-treated MWNTs (4.5% O), and H_2SO_4/HNO_3-treated MWNTs (5.3% O). Amorphous carbon is indicated with arrows, and sidewall defects are highlighted by circles. (Reprinted with permission from [Wepasnick et al., 2011]. Copyright, Elsevier Ltd.).

Noteworthy are the effects of the oxidants on amorphous carbon and sidewall defects. The long and straight outermost wall of MWNT denotes uniform and largely defect-free sidewall structure. The overall level of amorphous carbon was reduced during H_2O_2 treatment, and few defects were generated on the sidewalls. On the other hand, treatment with H_2SO_4/HNO_3 produced a distortion in the linearity of the MWNT structure. Following $KMNO_4$ treatment, MWNTs exhibited a larger fraction of tethered COOH groups compared to other oxidized MWNTs with a relatively low amount of sidewall damage.

The identity of the CNT surface functional group has a large impact on the surface charge of the sidewalls. While using MWNTs as catalyst supports, the point-of-zero charge [PZC, defined as the pH at which the solid-aqueous solution interface is electrostatically neutral, according to the electrical double layer model described by Gouy-Chapman theory (Brown et al., 1999)] is an important parameter to consider when anchoring metal complex precursors to maximize dispersion and loading on the MWNT sidewalls. Lee et al. (2011) showed that the treatment of nitric acid-oxidized MWNTs by ethanol reduction at 20 atm and 180°C was an efficient method for producing a high surface density of –OH groups, which in turn provided effective tethering points for grafting metal acetylacetone metal complexes to the MWNT surface. Since the tethering of cationic/anionic precursors is Coulombic in nature, the PZC can serve as a guide for electrostatic attachment of precursors to engineer the MWNT sidewalls.

Similarly, when functionalizing CNT sidewalls with specific moieties, the PZC is an important parameter for depositing finely dispersed metal nanoparticles from precursors in solution. McPhail et al. (2009) functionalized HiPco single-walled carbon nanotubes (p-SWNTs) with carboxyl acid (COOH-SWNT), nitroso (NO-SWNT), and maleic anhydride (MA-SWNT) groups. PZC values measured using a method described by Park and Regalbuto (1995) were found to be in the descending order: NO-SWNTs (7.5) > p-SWNTs (3.5) > MA-SWNTs (2.0) > COOH-SWNTs (1.2). The trend in measured PZC values correlated well with the electron withdrawing character of the moieties. Of the functional groups used, those with a greater electron donating character resulted in a higher PZC. By varying only the predetermined selection of the functional groups for sidewall attachment, the PZC of HiPco SWNTs could be tuned within a range of 6.3 pH units. Furthermore, UV-vis-NIR and Raman spectra showed that increasing electron withdrawing character of the functional groups led to greater selectivity for covalent attachment to those SWNTs with greater semiconducting character.

The extent of CNT surface oxidation has also been shown to directly impact catalytic reaction rate. Rocha et al. (2011) modified MWNTs using nitric acid at 100°C (boiling temperature), liquid phase urea at 200°C, and gas-phase nitrogen at 600°C in order to produce materials with different textural and chemical properties. In this example, a decrease in sidewall oxidation resulted in increased initial reaction rate for the decomposition of oxalic acid, an important reaction for the clean up of contaminated industrial waste waters. The modified MWNTs were directly applied for catalytic wet air oxidation (CWAO). No impregnated metals were used. This methodology is commonplace among other researchers for preparing Pt-based MWNT catalysts (Yang et al., 2007; Yang et al., 2008; Garcia et al., 2005). The array of functionalized MWNTs studied by Rocha et al. (2011) were as follows. Original, untreated MWNTs (CNT-O) were oxidized in nitric acid and rinsed in distilled water until a neutral pH was attained, followed by drying (CNT-N). The resulting CNT-N was then treated with urea in a high pressure reactor. The MWNTs were then rinsed, dried, and subjected to gas phase thermal treatment under N_2 flow at 600°C for 60 minutes to produce CNT-NUT. Excluding CNT-O, which was used as the starting material, the successive treatments resulted in a lowering of the density of oxygen-containing functional groups on the MWNT sidewalls in the descending order: CNT-N > CNT-NU > CNT-NUT. These catalyst surfaces were then examined for their ability to degrade oxalic acid. Fig. 9 shows the relationship between PZC values and initial reaction rate constants, as well as with the basicity (indicated by the decrease in PZC values). The decrease in reaction rates were as follows: CNT-NUT > CNT-O > CNT-NU > CNT-N. Accompanying reaction rate increase, the PZC increased with decreasing oxygen-containing moiety density. Noteworthy is the fact that the 1st-order rate constant for oxalic acid decomposition was elevated with increasing PZC while the density of oxygen-containing functional groups decreased. The data indicated that the there were fewer oxygen-containing groups in the CNT-NUT than in the original untreated CNT-Os. The CNT-NUT MWNTs was the least acidic in this series of MWNT catalysts. Catalytic performance for oxalic acid decomposition in CWAO depends mostly on the acid/base nature of MWNTs. Weak activity for CNT-N (having the second largest available surface area in this series of catalysts) can be correlated to the acidic character of the nanotube sidewall surface. The result implies that MWNTs with lower acidic character are more efficient for decomposing oxalic acid.

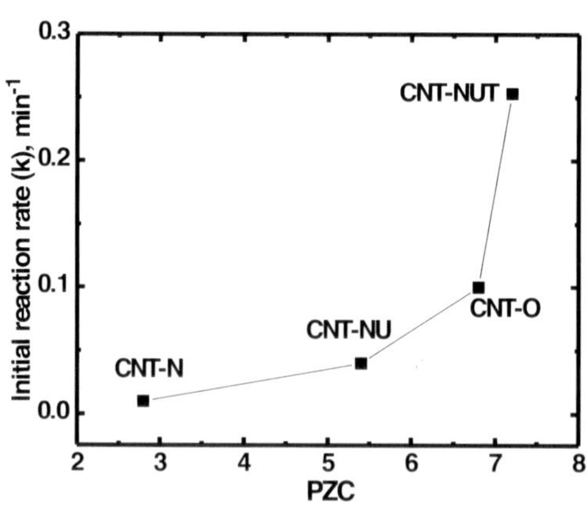

Figure 9. Apparent first-order initial reaction rate constants (k) (for the decomposition of oxalic acid) vs PZC for the original and treated MWNTs (Rocha et al., 2011).

6. Conclusions

In summary, the surface chemistry of CNT sidewalls markedly affects its properties relevant to an array of applications. Non-reversible, covalent functionalization often damages the carbon structure and/or creates defects for moiety attachment in order to make these surfaces chemically active. CNT sidewall surface structure, which can be engineered via surface functionalization in solution, can significantly affect heterogeneous catalytic properties. More recently, methods for electrochemical functionalization and manipulation of the solid surface isoelectric point have been developed to diversify our ability to engineer CNT sidewall structures. The effects of oxidizing agents on the colloidal stability of these materials and the role of the PZC have become increasingly important for engineering nanomaterials in aqueous solution environments. The direction of future research will undoubtedly involve detailed elucidation of structure-property relationships involving these parameters.

Acknowledgements

AKD and CCC gratefully acknowledge support from the Chemistry Department and the Faculty Research and Creative Activity Committee (FRCAC) of Middle Tennessee State University.

Author details

Anup K. Deb and Charles C. Chusuei*

*Address all correspondence to: Charles.Chusuei@mtsu.edu

Chemistry Department, Middle Tennessee State University, Murfreesboro, Tennessee, USA

References

[1] Ajayan, P. M. (1999). Nanotubes from Carbon. *Chem. Rev.*, 99, 1787-1800.

[2] Arias, L. R., & Yang, L. (2009). Inactivation of Bacterial Pathogens by Carbon Nanotubes in Suspensions. *Langmuir*, 25, 3003-3012.

[3] Ball, W. P., Cho, H. H., Smith, B. A., Wnuk, J. D., & Fairbrother, D. H. (2008). Influence of Surface Oxides on the Adsorption of Naphthalene onto Multiwalled Carbon Nanotubes. *Environ. Sci. Technol.*, 42, 2899-2905.

[4] Banerjee, S., Hemraj-Benny, T., & Wong, S. S. (2005). Covalent Surface Chemistry of Single-walled Carbon Nanotubes. *Adv. Mater.*, 17, 17-29.

[5] Blanchard, N. P., Hatton, R. A., & Silva, S. R. P. (2007). Tuning the Work Function of Surface Oxidised Multi-wall Carbon Nanotubes via Cation Exchange. *Chem. Phys. Lett.*, 434, 92-95.

[6] Bohem, H. P., Diehl, E., Heck, W., & Sappok, R. (1964). Surface Oxide of Carbon. *Angew. Chem. Int. Ed.*, 3, 669-677.

[7] Bottini, M., Bruckner, S., Nika, K., Buttini, N., Bellucci, S., Magrini, A., Bergamaschi, A., & Mustelin, T. (2006). Multi-walled Carbon Nanotubes Induce T Lymphocyte Apoptosis. *Toxicol. Lett.*, 160, 121-126.

[8] Brown, G. E. Jr., Henrich, V. E., Casey, W. H., Clark, D. L., Eggleston, C., Felmy, A., Goodman, D. W., Grätzel, M., Maciel, G. E., McCarthy, M. I., Nealson, K., Sverjensky, D. A., Toney, M. F., & Zachara, J. M. (1999). Chemical Interaction of Metal Oxide-Aqueous Solution Interfaces. *Chem. Rev.*, 99, 77-174.

[9] Brown, M. E., & Gallagher, P. K. (2007). *Handbook of Thermal Analysis and Calorimetry: Recent Advances, Techniques and Applications*, Vol. 5, Elsevier: Amsterdam.

[10] Brundle, C. R., Evans, C. A., & Wilson, S. (1992). *Encyclopedia of Materials Characterization*, Butterworth-Heinemann: Stoneham.

[11] Cho, H.-H., Smith, B. A., Wnuk, J. D., Fairbrother, D. H., & Ball, W. P. (2008). Influence of Surface Oxides on the Adsorption of Naphthalene onto Multiwalled Carbon Nanotubes. *Environ. Sci. Technol.*, 42, 2899-2905.

[12] Cho, H.-H., Wepasnick, K., Smith, B. A., Bangash, F. K., Fairbrother, D. H., & Ball, W. P. (2010). Sorption of Aqueous Zn(II) and Cd(II) by Multiwall Carbon Nanotubes: the Relative Roles of Oxygen-containing Functional Groups and Graphenic Carbon. *Langmuir*, 26, 967-98.

[13] Chusuei, C. C., & Wayu, M. (2011). Characterizing Functionalized Carbon Nanotubes for Improved Fabrication in Aqueous Solution Environments. *Electronic Properties of Carbon Nanotubes*, Book 5, InTech: Croatia, 55-68.

[14] Derjaguin, B. V., & Landau, L. (1941). Theory of the Stability of Strongly Charged Lyophobic Sols and the Adhesion of Strongly Charged Particles in Solutions of Electrolytes. *Acta Physicochim.*, 14, 633-662.

[15] Dresselhaus, M. S., Dresselhaus, G., & Avouris, P. (2001). *Carbon Nanotubes : Synthesis, Structure, Properties and Applications*, Springer : Berlin.

[16] Garcia, J., Gomes, H. T., Serp, P., Kalck, P., Figueiredo, J. L., & Faria, J. L. (2005). Platinum Catalysts Supported on MWNT for Catalytic Wet Air Oxidation of Nitrogen Containing Compounds. *Catal. Today*, 102/103, 101-109.

[17] Hirsch, A. (2002). Functionalization of Single-walled Carbon Nanotubes. *Angew. Chem. Int. Ed.*, 41, 1853-1859.

[18] Hu, C., Yiyi, Z., Bao, G., Zhang, Y., Liu, M., & Wang, Z. L. (2005). DNA Functionalized Single-walled Carbon Nanotubes for Electrochemical Detection. *J. Phys. Chem. B*, 109, 20072-20076.

[19] Hull, R. V., Li, L., Xing, Y., & Chusuei, C. C. (2006). Pt Nanoparticle Binding of Functionalized Multiwalled Carbon Nanotubes. *Chem. Mater.*, 18, 1780-1788.

[20] Iijima, S. (1991). Helical Microtubules of Graphitic Carbon. *Nature*, 354, 56-58.

[21] Kam, N. W. S., O'Connell, M., Wisdom, J. A., & Dai, H. (2005). Carbon Nanotubes as Multifunctional Biological Transporters and Near-infrared Agents for Selective Cancer Cell Destruction. *Proc. Natl. Acad. Sci. USA*, 102, 11600-11605.

[22] Lee, S., Zhang, Z., Wang, X., Pfefferle, L. D., & Haller, G. L. (2011). Characterization of Multi-walled Carbon Nanotubes Catalyst Supports by Point of Zero Charge. *Catal. Today*, 164, 68-73.

[23] Luo, H., Shi, Z., Li, N., Gu, Z., & Zhuang, Q. (2001). Investigation of the Electrochemical and Electrocatalytic Behavior of Single-Wall Carbon Nanotube Film on a Glassy Carbon Electrode. *Anal. Chem.*, 73, 915-920.

[24] Lu, X., Liao, T., Ding, L., Liu, X., Zhang, Y., Cheng, Y., & Du, J. (2008). Interaction of Quercetin with Supported Bilayer Lipid Membrands on Glassy Carbon Electrode. *Int. J. Electrochim. Sci.*, 3, 797-805.

[25] Magonov, S. N., & Whangbo, M. H. (1996). Surface analysis with STM and AFM: Experimental and Theoretical Aspects of Image Analysis, Wiley-VCH: Germany.

[26] McPhail, M. R., Sells, J. A., He, Z., & Chusuei, C. C. (2009). Charging Nanowalls: Adjusting the Carbon Nanotube Isoelectric Point via Surface Functionalization. *J. Phys. Chem. C.*, 113, 14102-14109.

[27] Mojica, E. R. E., Vidal, J. M., Pelegrina, A. B., & Micor, J. R. L. (2007). Voltammetric Determination of Lead(II) Ions at Carbon Paste Electrode Modified with Banana Tissue. *J. Appl. Sci.*, 7, 1286-1292.

[28] Ovejero, G., Sotelo, J. L., Romero, M. D., Rodríguez, A., Ocaña, M. A., Rodríguez, G., & García, J. (2006). Multiwalled Carbon Nanotubes for Liquid-Phase Oxidation. Functionalization, Characterization, and Catalytic Activity. *Ind. Eng. Chem. Res.*, 45, 2206-2212.

[29] Park, J., & Regalbuto, J. R. (1995). A Simple Accurate Determination of Oxide PZC and the Strong Buffering Effect of Oxide Surfaces at Incipient Wetness. *J. Colloid Interf. Sci.*, 175, 239-252.

[30] Piela, B., & Wrona, P. K. (2002). Oxidation of Nitrites on Solid Electrodes. I. Determination of the Reaction Mechanism on the Pure Electrode Surface. *J. Electrochem. Soc.*, 149, E55-E63.

[31] Rocha, R. P., Sousa, J. P. S., Silva, A. M. T., Pereira, M. F. R., & Figueiredo, J. L. (2011). Catalytic Activity and Stability of Multiwalled Carbon Nanotubes in Catalytic Wet Air Oxidation of Oxalic acid: the Role of the Basic Nature Induced by the Surface Chemistry. *Appl. Catal. B- Environ.*, 104, 330-336.

[32] Sampath, S., & Lev, O. (1998). Electrochemical Oxidation of NADH on Sol-gel Derived, Surface Renewable, Non-modified and Mediator Modified Composite Carbon Electrodes. *J. Electroanal. Chem.*, 446, 57-65.

[33] Sayes, C. M., Liang, F., Hudson, J. L., Mendez, L. J., Guo, W., Beach, J. M., Moore, V. C., Doyle, C. D., West, J. L., Billups, W. E., Ausman, K. D., & Colvin, V. L. (2006). Functionalization of Density Dependence of Single-walled Carbon Nanotubes Cytoxicity In Vitro. *Toxicol. Lett.*, 161, 135-142.

[34] Shamspur, T., & Mostafavi, A. (2009). Application of Modified Multiwalled Carbon Nanotubes as a Sorbent for Simultaneous Separation and Preconcentration Trace Amounts of Au(II) and Mn(II). *J. Hazard. Mater.*, 168, 1548-1553.

[35] Shieh, Y. T., Liu, G. L., Wu, H. H., & Lee, C. C. (2007). Effects of Polarity and pH on the Solubility of Acid-treated Carbon Nanotubes in Different Media. *Carbon*, 45(9), 1880-1890.

[36] Smith, B., Wepasnick, K., Schrote, K. E., Bertele, A. R., Ball, W. P., O'Mella, C., & Fairbrother, D. H. (2009a). Colloidal Properties of Aqueous Suspensions of Acid-treated, Multi-walled Carbon Nanotubes. *Environ. Sci. Technol.*, 43, 819-825.

[37] Smith, B., Wepasnick, K., Schrote, K. E., Cho, H.-H., Ball, W. P., & Fairbrother, D. H. (2009b). Influence of Surface Oxides on the Colloidal Stability of Multi-walled Carbon Nanotubes: a Structure Property Relationship. *Langmuir*, 25, 9767-76.

[38] Tavallai, H., Abdardideh, D., Aalaei, M., & Zahmatkesh, S. (2012). New Application of Chemically Modified Multiwalled Carbon Nanotubes with Thiosemicarbazide as a Sorbent for Separation and Preconcentration of Trace Amounts of Co(II), Cd(II), Cu(II), and Zn(II) in Environmental and Biological Samples Prior to Determination of Flame Atomic Absorption Spectrometry. *J. Chin. Chem. Soc.*, 59, 114-121.

[39] Unger, E., Graham, A., Kreupl, F., Liebau, M., & Hoenlein, W. (2002). Electrochemical Functionalization of Multi-walled Carbon Nanotubes for Solvation and Purification. *Curr. Appl. Phys.*, 2, 107-111.

[40] Verwey, E. J., & Overbeek, J. T. G. (1948). *Theory of the Stability of Lyophobic Colloids*, Elsevier: New York.

[41] Wepasnick, K. A., Smith, B. A., Schrote, K. E., Wilson, H. K., Diegelmann, S. T., & Fairbrother, D. H. (2011). Surface and Structural Characterization of Multi-walled Carbon Nanotubes Following Different Oxidative Treatments. *Carbon*, 49, 24-36.

[42] Xia, W., Wang, Y., Bergsträβer, R., Kumdu, S., & Muhler, M. (2007). Surface Characterization of Oxygen-functionalized Multi-walled Carbon Nanotubes by High-resolution X-ray Photoelectron Spectroscopy and Temperature-programmed Desorption. *Appl. Surf. Sci.*, 254, 247-250.

[43] Xing, Y. (2004). Synthesis and Electrochemical Characterization of Uniformly-Dispersed High Loading Pt Nanoparticles on Sonochemically-Treated Carbon Nanotubes. *J. Phys. Chem. B*, 108, 19255-19259.

[44] Xing, Y., Chusuei, C. C., & Hull, R. V. (2005). Sonochemical Oxidation of Multi-walled Carbon Nanotubes. *Langmuir*, 21, 4185-4190.

[45] Yang, S., Zhu, W., Li, X., Wang, J., & Zhou, Y. (2007). Multi-walled Carbon Nanotubes (MWNTs) as an Efficient Catalyst for Catalytic Wet Air Oxidation of Phenol. *Catal. Commun.*, 8, 2059-2063.

[46] Yang, S., Li, X., Zhu, W., Wang, J., & Descorme, C. (2008). Catalytic Activity, Stability and Structure of Multi-walled Carbon Nanotubes in the Wet Air Oxidation of Phenol. *Carbon*, 46, 445-452.

[47] Zhang, T., Nix, M. B., Yoo, B. Y., Zhang, T., & Myung, N. V. (2006). Electrochemically Functionalized Single-walled Carbon Nanotube Gas Sensor. *Electroanal.*, 18, 1153-1158.

[48] Zhang, Y., Shen, Y., Li, J., Niu, L., Dong, S., & Ivaska, A. (2005). Electrochemical Functionalization of Single-walled Carbon Nanotubes in Large Quantities at a Room-Temperature Ionic Liquid Supported Three-dimensional Network Electrode. *Langmuir*, 21, 4797-4800.

Chapter 11

Dispersion and Property Manipulation of Carbon Nanotubes by Self-Assemibles of Amphiphilic Molecules

Xia Xin , Guiying Xu and Hongguang Li

Additional information is available at the end of the chapter

1. Introduction

Over the past two decades, carbon nanotubes (CNTs) have grown as a novel type of nanomaterial and attracted great attention from scientists in different research fields [1-3]. Based on their unique one-dimensional nanostructure, CNTs exhibit excellent mechanical, optical and electronic properties as well as high chemical stability. As claimed by the Nobel laureate Richard Smalley [4], CNTs would be cheap, environmentally friendly, and do wonders for humankind. CNTs were discovered along with the research on fullerenes. In 1991, Iijima from Japan, who at that time was observing fullerenes produced by arc discharge method under high-resolution TEM, observed some tubular structures formed by coaxial cylinders of graphite layers [5]. These structures are nowadays well-known multi-walled carbon nanotubes (MWNTs) with diameters ranging from 2 to several hundred nanometers and lengths of microns. Two years later, single-walled carbon nanotubes (SWNTs) which contain only one cylinder of graphite were also observed and MWNTs can be equally regarded as a group of coaxial SWNTs with different diameters [6]. Theoretically, SWNTs can be regarded as the equivalent by rolling up a single layer of graphite (graphene). The way of the rolling will dominate the diameter and chirality of SWNTs and hence their electronic properties. Taking a_1 and a_2 as the basic vector of a graphite layer as shown in Figure 1, the tubes obtained by rolling up this layer along a_1 are called zigzag tubes, while those along a_2 are called armchair tubes. The tubes obtained along the vector a will have chirality ($a = \mathrm{n}a_1 + \mathrm{m}a_2$). If n - m = $3q$ (q is an integer), the tubes are metallic and can be conductive. When n - m ≠ $3q$, the tubes are semiconducting.

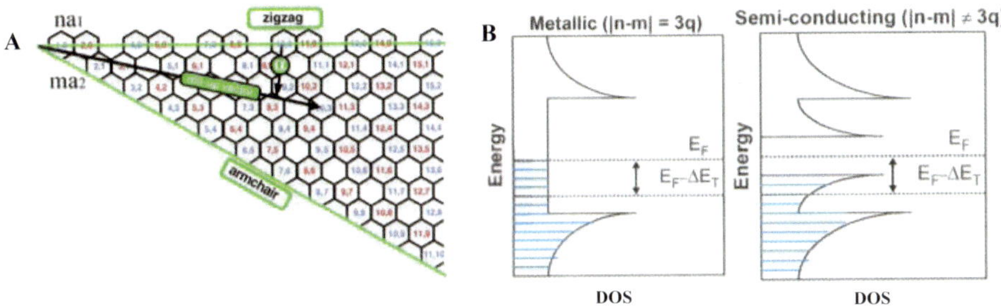

Figure 1. Physical and electronic structures of semiconducting SWNTs. (**A**) Graphene sheet segment showing indexed lattice points. Nanotubes designated (*n,m*) are obtained by rolling the sheet from (0,0) to (*n,m*) along a roll-up vector. The chiral angle α (from 0 to 30°) is measured between that vector and the zigzag axis; the tube circumference is the vector's length. (**B**) The extent of electron transfer is dependent on the density of states in that electron density near E_F leads to higher initial activity for metallic and semimetallic nanotubes [6, 9].

This unique geometry at nanometer length scale imparts CNTs many intriguing properties. However, it also leads to a problem which must be solved before many practical applications. That is, the as-produced CNTs, whatever synthetic method is used, are always a mixture of tubes with varying diameter, chirality and length. Besides the continuous effort to directly obtain the desired type of CNTs during synthesis [7], the post-sorting forms an alternative to solve this problem. In the latter case, the CNTs must be well dispersed in a medium. Moreover, the dispersion of CNTs is also a precondition of many fundamental research and practical applications because the aggregation of tubes can significantly lower the promising properties proposed to the single tube of CNTs and bring difficulties to handle these interesting nanomaterials. Unfortunately, in most cases the as-produced CNTs stay as aggregated bundles or ropes instead of single tubes driven by the van der Waals forces and π-π interactions between adjacent tubes. This aggregation trend becomes more pronounced in the case of SWNTs where the tube has a very high length to width ratio with the order of 100-1000 [8].

From a chemistry viewpoint, an idea to disperse CNTs immediately comes into mind is to covalently link suitable functional groups to the sidewalls or end-caps of the tubes to render the CNTs desired solubility or dispersability in a given solvent. Among various routes to covalently functionalize the CNTs, oxidation using mixed acids [9] has received much interest due to its simplicity and effectiveness. After oxidation, carboxylic groups can be introduced, which opens a route for further functionalizations. This covalent method, however, is proved to disturbing the π-electrons of the tubes and hence inevitably influencing the intrinsic properties of single tubes. As an alternative and improved solution, in recent years, the so-called noncovalent method using amphiphilic molecules to disperse CNTs in aqueous solutions has received much attention [10, 11]. In this chapter, we will first give a brief introduction and overview of the noncovalent method, with an emphasis on SWNTs and recent advances in this field. Then we will focus on the property manipulation of the dispersed tubes in the self-assemblies formed by amphiphilic molecules in water. Typical applications of the dispersed tubes in other research field will also be presented. These include the preparation of functional materials, fabrication of nano-devices and applications in life science. The work related to the dispersion of CNTs in organic solvents using water-insoluble conjugated polymers, which can be also included in the noncovalent method, is out of the scope of this chapter and hence will not be mentioned.

2. Dispersing CNTs by amphiphilic molecules

Amphiphilic molecule possesses both hydrophobic and hydrophilic parts in the same molecule [12]. The hydrophobic part is typically the alkyl chain while the hydrophilic part can be an ionized functional group, an ethylene oxide group or the combination of them. Amphiphilic molecule has a special name in colloid and interface science, i.e., surfactant. The surface of CNTs is intrisically hydrophobic and thus has affinity with the hydrophobic part of surfactant. In this case, the hydrophilic part of surfactant stays in water and impedes tube aggregation. This forms the base of using surfactant to disperse and stabilize CNTs. Generally, SWNTs are much more difficult to be dispersed than MWNTs due to the much stronger intertube attractions and thus attracted more attention of researchers in this field. In the following, discussions will mainly be made on SWNTs.

To facilitate the surfactant adsorption onto the tube surface, sonication is needed although example without sonication has also been reported [13]. In a typical dispersion procedure, SWNTs and surfactant are added into water and the mixture is sonicated. Recent report shows that the outermost tubes in a SWNTs bundle are treated more than the innermost tubes and the tubes tend to exfoliate from the bundle ends. Therefore, mechanical exfoliation of the bundles prior to surface treatment must occur in order to obtain individual carbon nanotubes. A mechanism of nanotube isolation from a bundle (Figure 2 i), with the combined assistance of ultrasonication and surfactant adsorption, was proposed [14]. The role of ultrasonic treatment is likely to provide high local shear, particularly to the nanotube bundle end (Figure 2 ii). Once spaces or gaps at the bundle ends are formed, they are propagated by surfactant adsorption (Figure 2 iii), ultimatelym separating the individual nanotubes from the bundle (Figure 2 iv). Since high-power sonication can potentially destroy the tubes, the power of sonication is usually low (< 10 W). On the other hand, to get a good dispersion the sonication time should be sufficiently long (several to tens of hours). After this treatment a SWNTs dispersion with single tubes as well as tube bundles can be obtained. Large tube bundles can be removed from the dispersion by simply gravity sedimentation, while small tube bundles can be stable for weeks. To remove them, ultracentrifugation is recommended. After the sonication-ultracentrifugation circle, SWNTs dispersion up to single tube level can be obtained.

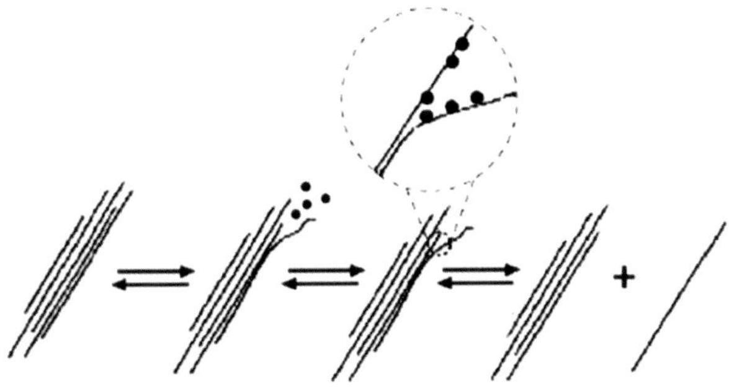

Figure 2. Mechanism of nanotube isolation from bundle obtained by ultrasonication and surfactant stabilization [14].

3. Influence of surfactant type

Up to now, various types of surfactants, both common and uncommon, have been tested to disperse SWNTs in water. Sodium dodecyl sulfate (SDS), which is an anionic surfactant, is among the earliest and most common choices [6, 9, 15], as shown in Figure 3. Later on, investigations have been expanded to cationic, nonionic and zwitterionic surfactants [16-23]. Although both ionic surfactants and nonionic ones can successfully disperse SWNTs in water, the stabilization mechanism for these two categories has subtle difference. While ionic surfactants stabilize the dispersed tubes mainly by electrostatic repulsion, the stabilization mechanism of nonionic surfactant-coated tubes is mainly achieved by steric repulsion. If surfactant molecules are in excess, they can form free micelles. When the size of micelles is above the average distance between the adjacent dispersed tubes, the micelles become difficult to arrange themselves between the tubes and, driven by the depletion attraction, phase separation occurs. Besides the surfactant concentration, the quality of an aqueous dispersion of SWNTs can be also influenced by a variety of other experimental parameters including temperature, pH and ionic strength.

Amphiphilic macromolecules such as poly (ethylene oxide)-poly (propylene oxide)-poly (ethylene oxide) (PEO-PPO-PEO) tri-block copolymers are also efficient dispersing agents for SWNTs [20, 23-25]. The unique molecular structure and property of these amphiphilic macromolecules lead to some differences in SWNTs dispersion compared to traditional low molecular weight surfactants. For example, the hydrophobic part (PPO segment) is usually longer than that of low molecular weight surfactant and can thus wrap not only on single tubes but also on tube bundles (Figure 4). This will lead to a decreased fraction of single tubes in the dispersion, which seems a disadvantage in cases where single tubes are desired. However, this feature, together with the improved steric repulsion created by the longer hydrophilic part (PEO segment), also enables these amphiphilic macromolecules to disperse and stabilize a large amount of tubes into water. This is quite important in applications where the amount of the dispersed tubes is the main concern. Consistent with the larger molecular size, the micelles formed by amphiphilic macromolecules are usually larger compared to those formed by low molecular weight surfactants, which may induce a more pronounced depletion attraction if free micelles are present in the dispersion. Temperature is also an important influencing factor. An increase in temperature can induce desorption of the macromolecule from the tube surface due to a continuous dehydration of the PPO and/or PEO segments.

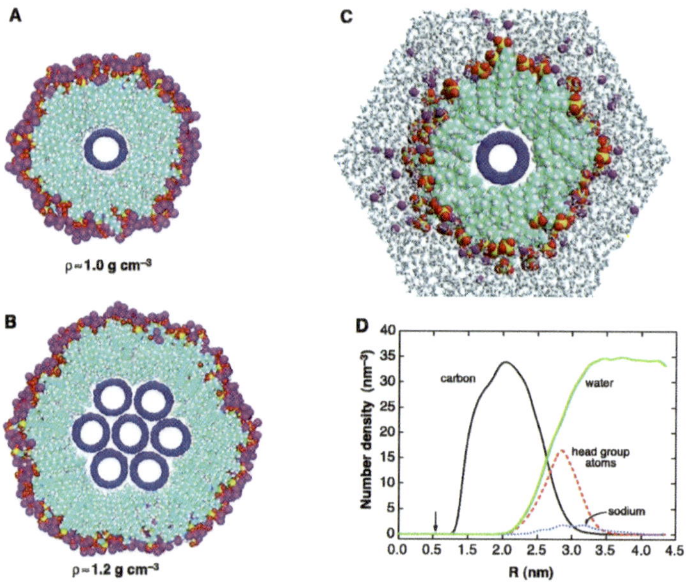

Figure 3. Cross-section model of (A) an individual and (B) a seven-tube bundle embedded in a cylindrical SDS micelle. C) A molecular dynamics simulations of water and the SDS micelle around an individual tube. D) The number density profiles for SDS carbon atoms, sulfate head group atoms, water molecules, and sodium ions [15].

Recent report reveals that the molecular architecture is also an important concern governing the quality of SWNTs dispersion. When the linear PEO-PPO-PEO is branched, an improved capability of SWNTs dispersion and stabilization is observed both experimentally and theoretically. For example, in our laboratory, we have used a starlike amphiphilic block copolymer with PPO-PEO segments (AP432) to disperse CNTs in aqueous solutions. For comparison, two commercially available linear amphiphilic block copolymers, Pluronics L64 and F127, were also selected. It was found that AP432 and F127 can get good CNT dispersions, while L64 was proved to be unable to disperse CNTs. AP432 with five branches could disperse CNTs efficiently at much lower concentrations compared with the linear F127, although it has a smaller molecular weight and shorter terminal EO groups. This indicated clearly that, once branched, copolymers would get a much better ability to disperse CNTs [26, 27]. The detailed information is shown in Figure 5.

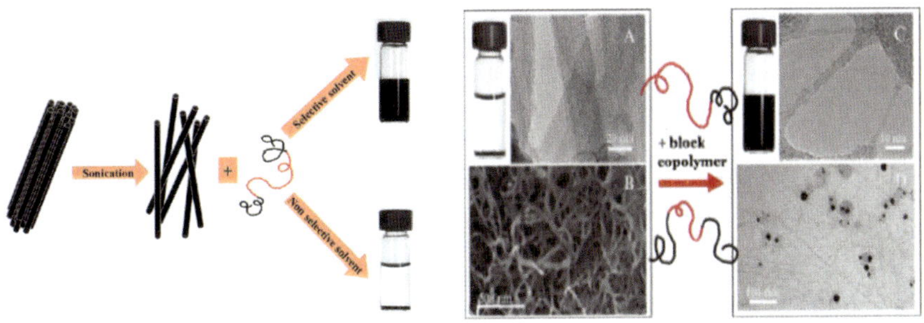

Figure 4. Block copolymers (designated A-B or A-B-A). They are comprised of covalently linked incompatible moieties and disperse SWNT in selective solvents that act as a "good solvent" for one of the blocks (i.e., A), while simultaneously acting as a "poor solvent" for the other block (B). Under these conditions polymer chains may adsorb via physical attachment of the B-block while the A blocks dangle into the solution repelling other polymer-decorated CNT and forming a long-lived dispersion. Nonselective solvents fail to disperse the SWNT [24, 25].

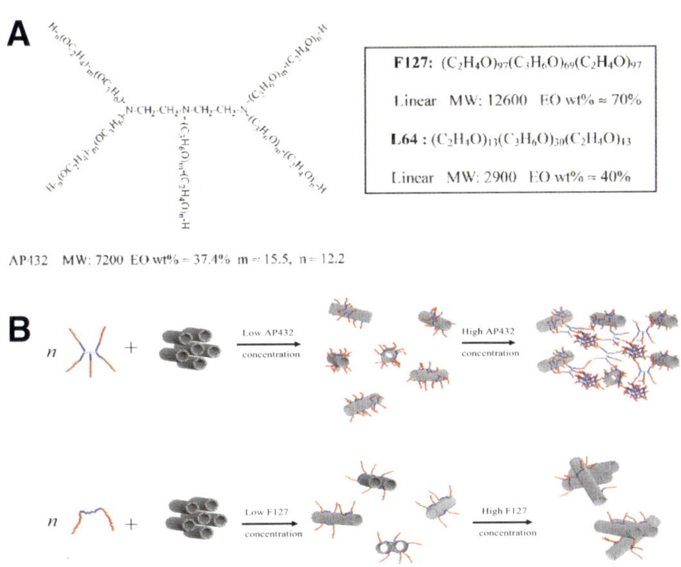

Figure 5. (A) Molecular details of linear (F127 and L64) and star-like amphiphilic block copolymers (AP432). The value of m and n for AP432 is the average value calculated according to the molecular weight and EO content. It is evident that L64 has no ability to disperse CNTs. (B) Schematic Representation of the Possible Mechanism for Nanotube Dispersion by AP432 and F127 [27].

Besides the hydrophobic/hydrophobic interaction, π–π stacking could also play a role in SWNTs dispersion (Figure 6). A typical example is the anionic surfactant sodium dodecylbenzene sulfonate (SDBS) which can greatly enhance SWNTs dispersion in water due to the existance of a benzene ring in the alkyl chain [20, 28]. Other examples include the amphiphilic pyrene [29], fluorescein [30], perylene [31] and fullerene [32, 33]. In some cases the affinity between SWNTs and the dispersant can be complicated and not so obvious. There are reports on SWNTs dispersion using polyelectrolytes such as poly (acrylic acid) (PAA) (Figure 7) [34] and naturally occurring macromolecules or biomolecules such as Gum Arabic [35], Hyaluronic Acid [36], starch [37], protein [38] and DNA [39, 40]. The surface activity of these molecules is not as high as traditional surfactants, but satisfactory SWNTs dispersion can be also obtained by optimizing the experimental parameters.

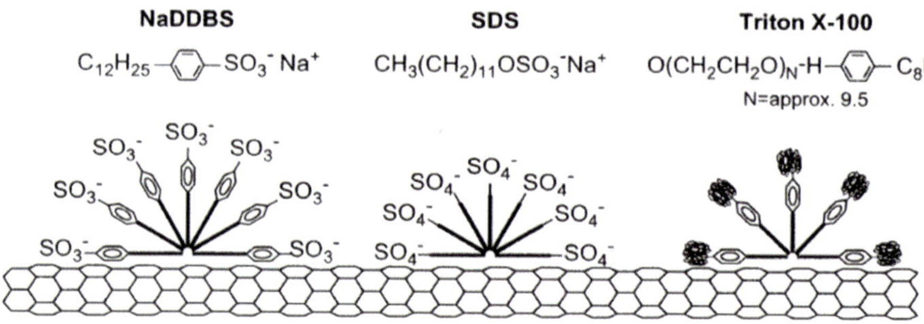

Figure 6. Schematic representation of how surfactants may adsorb onto the nanotube surface. NaDDBS and TX100 were believed to disperse the tubes better than SDS because of their benzene rings. NaDDBS disperses better than TX100 because of its headgroup and slightly longer alkyl chain. The spacing between the benzene rings on the surfactants and the tube surface is large enough to accommodate the SO_3^- charged groups [28].

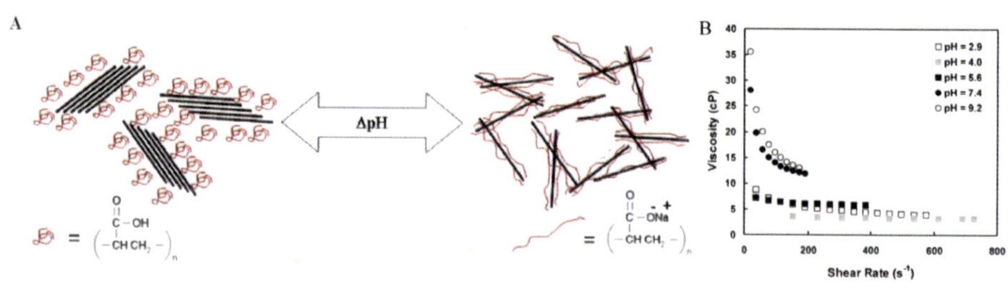

Figure 7. (A) Schematic showing the change in nanotube microstructure that occurs as the pH of poly(acrylic acid) is changed. At low pH, the polymer is uncharged and highly coiled with extensive intrachain hydrogen bonding. At high pH, the carboxylic acid groups are deprotonated and the polymer is more extended as the negatively charged side groups repel one another. **(B)** Viscosity as a function of shear rate for aqueous suspensions containing 1 wt % PAA-SWNT (SWNT is 10 wt % of the total solids) as pH is progressively increased [34].

4. Evaluation of the SWNTs dispersion

The quality of a SWNTs dispersion is evaluated mainly based on the requirement of the specific application. However, there are also some common criterions. The first one is the amount of SWNTs that can be dispersed by surfactants in water. This can be calculated by substracting the undispersed tubes from the total tubes or by simply visual inspections since the dispersion with more tubes usually has a heavier black color. The second and more important criterion is the dispersing extent of the tubes, i.e., bundles or single tubes. This can be checked either directly by imaging methods such as high resolution transmission electron microscopy (HRTEM) or atomic force microscopy observations, or indirectly by spectroscopic characterizations including UV-vis-NIR absorption, Raman spectroscopy and photoluminescence. In the latter case, researchers have shown that once SWNTs are dispersed up to a single tube level, fine structures in both absorption and emission spectra can be observed [6, 9]. While in a tube bundle, the fine spectroscopic characteristics will be significantly suppressed, highlighting the advantage of SWNTs dispersion by surfactants. The third criterion is the stability of the dispersion. The SWNTs dispersed in water is a kinetically rather than thermodynamically stable system. Thus aggregation and sedimentation can occur with time. If the dispersion is only an intermediated stage in a work and the dispersed tubes will be immediately used for the next step, short period of stability may be sufficient. In some applications, however, long term stability up to weeks or months may be required.

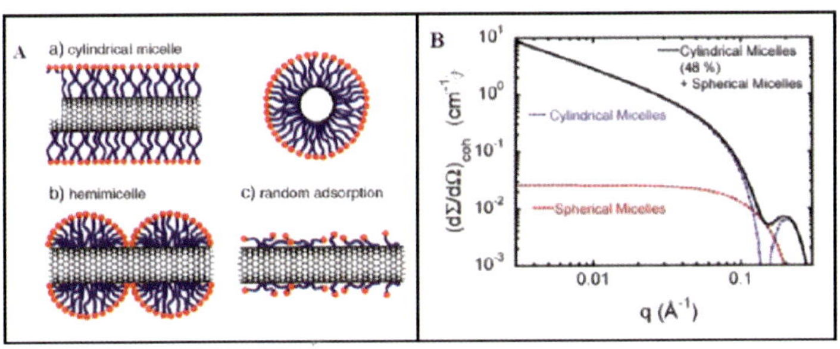

Figure 8. (A) Schematic representations of the mechanism by which surfactants help to disperse SWNT. (a) SWNT encapsulated in a cylindrical surfactant micelle (both cross section and side-view); (b) hemimicellar adsorption of surfactant molecules on a SWNT; (c) random adsorption of surfactant molecules on a SWNT [11]. **(B)** Coherent SANS intensities calculated for cylindrical (with an embedded nanotube) and spherical core-shell micelles at a concentration of 0.25 wt % SDS in D2O. The combined cylindrical and spherical micelle prediction shown is calculated on the basis of 48% of the SDS molecules participating in cylindrical micelles [41].

A concern in SWNTs dispersion is the way of surfactant adsorption on the tube surface. Understanding this can not only help to gain the insight of the interaction mechanism between surfactant molecules and SWNTs, but also provide guidelines to select or design specific surfactants to further improve the quality of the SWNTs dispersions. To fully address this issue experimentally, however, is not an easy task. Up to now, three different adsorption models have been proposed including the cylindrical micelle model [6], semi-sphere model

[28] and random adsorption model [41]. Figure 8 shows the schematic representations of the mechanism by which surfactants help to disperse SWNT. Although each of them has gained some experimental evidence, the real picture in the microscopic length scale is still to be clarified. In this context computer simulation is frequently adopted and various interaction modes between surfactant molecules and tube surfaces have been proposed [42-44]. Now it is generally accepted that for cylindrical micelle forming surfactant, SWNTs may be encapsulated in the micelles while for spherical micelle forming surfactant, random adsorption may also exist.

5. SWNTs alighment in ordered surfactant self-assemblies

In some cases, after the dispersion of SWNTs bundles the tubes are needed to be effectively aligned. SWNTs are anisotropic particles with diameters on the order of nanometers but lengths ranging from micrometers to centimeters. They display most of their expected properties along the tube axis. If the tubes are randomly oriented, the properties will be averaged which should be avoided in some applications such as nanodevice fabrication. The uniform alignment is therefore a crucial condition in these cases. Surfactant can form various self-assemblies in water above the critical micellar concentration (cmc). At medium-to-high concentrations, these self-assemblies can further organize into long range ordered phases called lyotropic liquid crystals (LLC) [45, 46]. An idea immediately comes into mind is that whether these LLC can be utilized to align SWNTs (Figure 9). A typical strategy is that SWNTs are dispersed first in a dilute surfactant aqueous solution before introduced to an LLC matrix since directly disperse SWNTs in a viscous LLC phase could be difficult. The surfactant used for SWNTs dispersion and LLC construction can be the same type or different. The incorporation of the tubes is found not to destroy the LLC matrix. This indirectly proves that the tubes are aligned along the director of the LLC phase since otherwise the system will be energetically unfavorable.

An innovative development in this research field is the fabrication of SWNTs/LLC hybrid by a spontaneous phase separation induced by hydrophilic polymer. In this method, dispersed SWNTs, surfactant and hydrophilic polymer are added into water and homogenized. Initially the concentration of surfactant is well below the critical point for LLC formation and only randomly oriented micelles exist in the solution. If the radius of gyration of the hydrophilic polymer in water is above the average distance between adjacent micelles, the polymer molecules will be driven out and phase separation will occur. The surfactant micelles together with the dispersed SWNTs will be compressed to form a new phase and the hydrophilic polymer forms another. Since the volume shrinks during phase separation, the surfactant concentration in the newly formed phase will be increased to exceed the critical point of LLC formation. This method has been successfully utilized in several surfactant and polymer combinations including nonionic surfactant/nonionic polymer system (Figure 10) [47] and ionic surfactant/polyelectrolyte system where the surfactant and polyelectrolyte have the same sign of charges [48]. The most striking advantage of this method is that the dispersed

SWNTs take part in the process of LLC formation instead of post-introduction, which is beneficial for tube alignment.

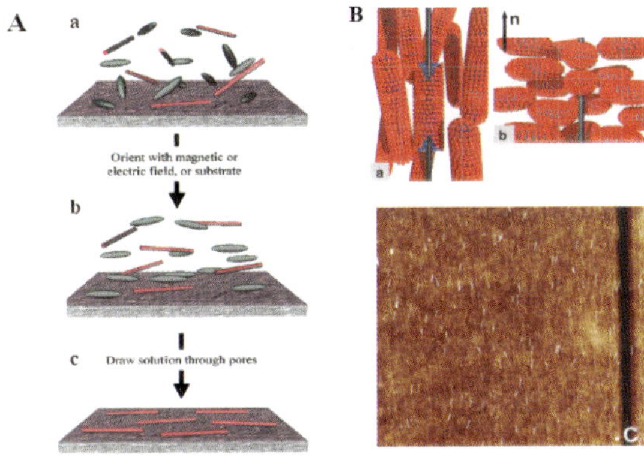

Figure 9. (A) Organizing CNTs with LC solvents. (a) A droplet of liquid crystal (ellipsoids) with suspended carbon nanotubes (rods) is applied to a porous membrane substrate. (b) The bulk LC is then aligned using a grooved surface or external field, which in turn orders the nanotubes. (c) The LC is drained through the porous membrane leaving behind an ordered nanotube film [45]. **(B)** Schematic illustrations of the mechanism of CNT alignment in a lyotropic nematic LC host, for the cases of rod- and disk-micelle-type adsorption, (a) and (b) respectively, of surfactant molecules on the CNT surface. The CNT is drawn as a vertical black rod in the center of each picture, only partially covered by surfactant molecules for clarity [46]. **(C)** AFM image of oriented CNT film of MWCNTs from 5CB (50×50 μm) [45].

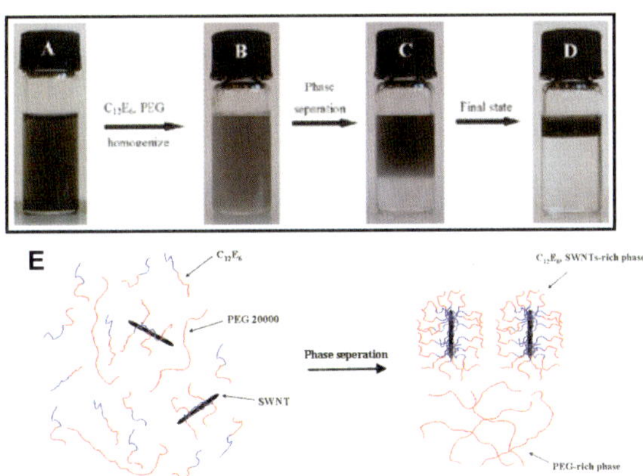

Figure 10. Schematic illustration of the phase separation process. (A) SWNTs dispersion in 0.1 wt%$C_{12}E_6$ aqueous solution; (B) mixture of SWNTs dispersion, $C_{12}E_6$ (10 wt %) and PEG(20 wt %) after being homogenized; (C) intermediate stage during phase separation; (D) final state after equilibrium. (E) Schematic Representation of the Phase Separation Process in the Four Component Mixture of the Surfactant $C_{12}E_6$, PEG 20000, SWNTs, and Water [47].

The dispersed SWNTs could also be aligned in randomly oriented surfactant micelles. For example, Tannenbaum's group probed the effects of shear flow on the alignment of dispersed single-walled carbon nanotubes in polymer solutions. Two different systems were compared: Single-walled carbon nanotubes dispersed using an anionic surfactant (sodium dodecyl benzene sulfonate, NaDDBS) and single-walled carbon nanotubes dispersed using an anionic surfactant and a weakly binding polymer (carboxyl methylcellulose, CMC). In this case, an additional force, typically shear, will be needed. It was determined that the addition of a weakly binding polymer serves two purposes: constituting the polymer matrix in which the SWNT will be dispersed and aligned and providing a secondary mechanism for the promotion of carbon nanotube dispersion [49]. The results showed that the tubes are found to be preferentially align along the direction of shear flow, as shown in Figure 11. Some reports also revealed that the presence of SWNTs can have a pronounced effect on the ordering of micelles [50]. This disadvantage of such method is, however, once the shear flow is stopped, the tubes and micelles tend to be disordered again.

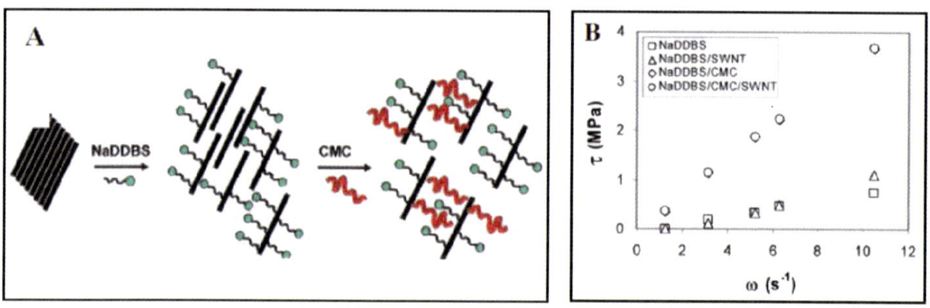

Figure 11. (A) Schematic representation of the stabilization mechanism of carbon nanotubes: Interaction between SWNT with NaDDBS, followed by the addition of CMC. It is important to note that the molecules were not drawn to scale, and the dimension of the polymer molecules is about 2 orders of magnitude larger that the dimension of the surfactant molecules (*Rg* ≈1000 Å for CMC as compared to ≈20 Å for NaDDBS); **(B)** Summary of the calculated shear stresses of the various SWNT-containing solutions: Plots of shear stresses as a function of angular velocity [49].

6. Applications of the surfactant-coated SWNTs

The formation of well-dispersed SWNTs in surfactant aqueous solutions opens the door for further investigation and practical applications of these interesting nanomaterials, as shown in Figure 12 [51]. Here we just briefly give some typical examples in different research field to elucidate the significance of the dispersion of SWNTs by surfactants. The first example comes from the preparation of advanced functional materials using SDS-coated SWNTs. When subjected to shear flow in a polymer solution, the tubes will be recondensed and aligned and finally a nanotube fiber can be obtained which has a high elastic modulus [16]. The surfactant-coated SWNTs can be also used as starting materials for SWNTs sorting i.e., separating semiconducting tubes from metallic ones. In recent years, great progress has been made in this direction by density-gradient ultracentrifugation and tubes with defined diameter and chirality can be obtained (Figure 13) [52, 53]. The sorting of SWNTs is necessary and important especially in the fabrication of high performance nanodevices such as field effect transistor arrays [54].

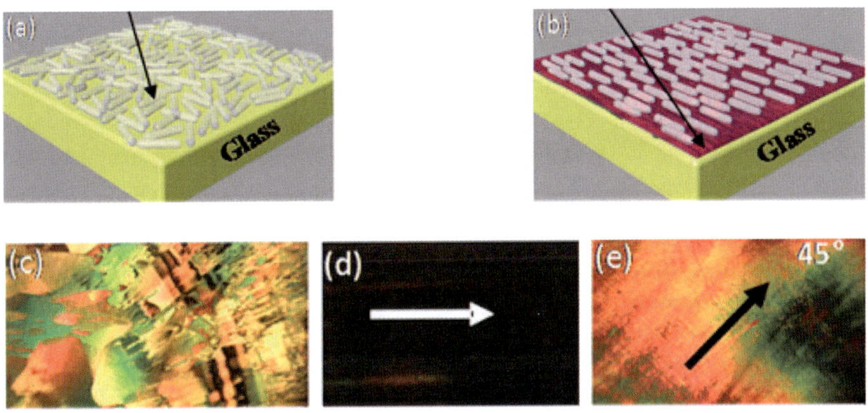

Figure 12. The aligned SWNT/polymer films can be used to orient the molecules of the liquid crystal. a) Schematic diagrams for arrays of LCs on a glass substrate showing random directional local phases. b) On a shear-aligned composite thin film (1a–SWNTarc-discharge 1:1, 2mg mL^{-1} in DCB) showing uniformly aligned LC domains. c) Transmission cross-polarized microscope image of a drop of LC 5CB on a bare glass substrate showing many small randomly aligned domains. d) A drop of LC 5CB on a glass substrate with sheared composite thin filmwith the shearing direction parallel to one of the polarizers. e) Shearing direction 45° to one of the polarizers. Arrows indicate the shearing direction [51].

Due to the unique geometry, ideal size and low cytotoxicity, SWNTs have great potential applications in life science. For this purpose the surfactant used to disperse SWNTs should be biocompatible and nonimmunogenity. One good choice is lipid derivative with a PEO segment as the hydrophilic part [55, 56]. SWNTs functionalized with single stranded-DNA can find application in gene therapy since the tubes can transport the cell membranes [55]. SWNTs are also known to absorb the near infrared light which will subsequent induce a local temperature rise, as shown in Figure 14. The surfactant-coated SWNTs can thus be used in photothermal treatments in some diseases such as cancer [55]. Besides, other functionalities including fluorescent dyes, cell receptors and drugs can be also integrated onto the surfactant-coated SWNTs to construct multifunctional materials.

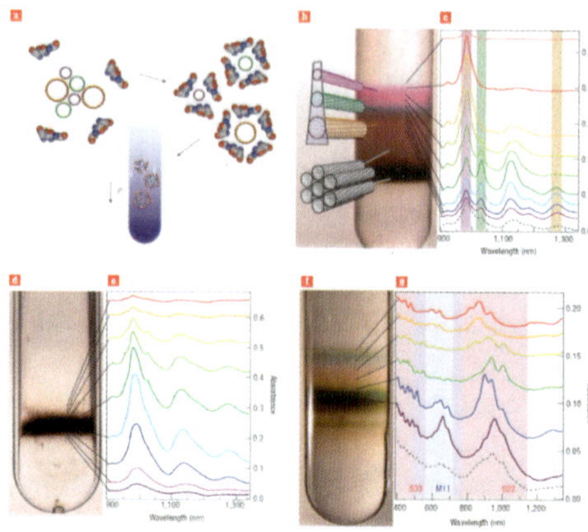

Figure 13. Sorting of SWNTs by diameter, bandgap and electronic type using density gradient ultracentrifugation. **a,** Schematic of surfactant encapsulation and sorting, where r is density. **B-g,** Photographs and optical absorbance (1 cm path length) spectra after separation using density gradient ultracentrifugation. A rich structure–density relationship is observed for SC-encapsulated SWNTs, enabling their separation by diameter, bandgap and electronic type. In contrast, no separation is observed for SDBS-encapsulated SWNTs. **b,c,** SC encapsulated, CoMoCAT-grown SWNTs (7–11 A°). **d, e,** SDBS-encapsulated CoMoCAT-grown SWNTs (7-11 A°). **f, g,** SC-encapsulated, laser-ablation-grown SWNTs (11-16 A°). pH = 7 for all parts. SWNTs before sorting are depicted as a dashed grey line in **c** and **g** [53].

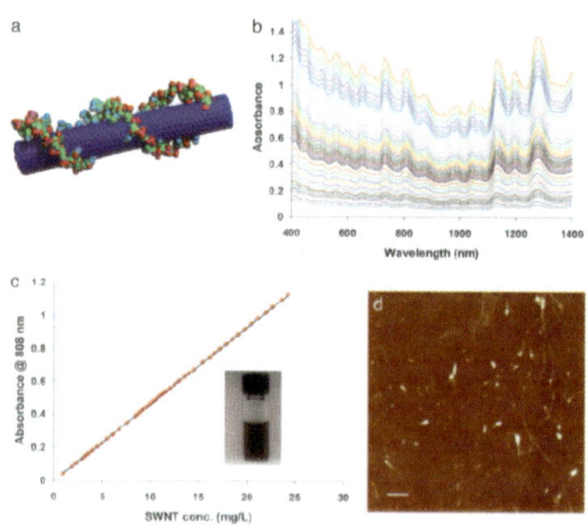

Figure 14. Carbon nanotubes with high NIR absorbance solubilized in water. (*a*) Schematic of a Cy3-DNA-functionalized SWNT. The drawing is only a graphic presentation and does not represent the precise way DNA binds on SWNTs. (*b*) UV-visible spectra of solutions of individual SWNTs functionalized noncovalently by 15-mer Cy3 labeled-DNA at various nanotube concentrations. (*c*) Absorbance at 808nmvs.SWNTconcentration (optical path = 1 cm). Solid line is Beer's law fit to obtain molar extinction coefficient of SWNT≈7.9×10^{6} M^{-1}•cm^{-1}. (*Inset*) A photo of a DNA functionalized SWNT solution. (*d*) AFM image of DNA-functionalized individual SWNTs (height of 1–10 nm) deposited on a SiO2 substrate (Scale bar: 200 nm.) [55].

7. Summary and outlook

The combined development of colloid and interface science and nanotechnology has paved an effective way to disperse CNTs, especially SWNTs, into water using surfactants. Up to now, almost all the common surfactants have been tested and quite a few newly synthesized surfactants with unique molecular structures have been tried. The effective dispersing methodology has been found and a general dispersing mechanism has been proposed. These advances enable us to get SWNTs dispersions up to single tube level, which significantly facilitates the property manipulation of the tubes and leads to a variety of important applications of SWNTs in physics, biology and life science.

At the same time, one should also keep in mind that challenges still exist. Although good dispersions with large amount of tubes or single tubes can be obtained individually, a combination of them, i.e., dispersions with large amount of single tubes, is still difficult to get. Achievement at this point will rely on the appearance of novel surfactant with improved performance and/or further optimization of the dispersion methodology. In nanodevice fabrication, the insulating surfactant layer on the tube surface may be undesirable and effective removal of them might be an issue. Despite these limitations, dispersing SWNTs using surfactants has provide an elegant way to manipulate these interesting nanomaterials and we believe the challenges mentioned above will be conquered in near future based on the continuous efforts made by scientists from related disciplines.

Acknowledgments

The authors gratefully acknowledge support of this work by the Natural Science Foundation of China (21203109) and (20873077) and the program of Hundreds of Talents of the Chinese Academy of Sciences.

Author details

Xia Xin [1], Guiying Xu [1] and Hongguang Li [2]

1 National Engineering Technology Research Center for Colloidal Materials, Shandong University, Jinan, , P. R. China

2 Laboratory of Clean Energy Chemistry and Materials, Lanzhou Institute of Chemical Physics, Chinese Academy of Sciences, Lanzhou, , P. R. China

References

[1] Kong J, Franklin NR, Zhou C, Chapline MG, Peng S, Cho K, Dai H. Nanotube molecular wires as chemical sensors. *Science* 2000; 287: 622-5.

[2] Tasis D, Tagmatarchis N, Bianco A, Prato M. Chemistry of carbon nanotubes. *Chem. Rev* 2006; 106: 1105-36.

[3] Karousis N, Tagmatarchis N. Current progress on the chemical modification of carbon nanotubes. *Chem. Rev.* 2010; 110: 5366-97.

[4] Ball P. Roll up for the revolution. *Nature* 2001; 414: 142-4.

[5] Iijima S. Helical microtubes of graphitic cabon. *Nature* 1991; *354*: 56-8.

[6] Bachilo, S. M., Strano, M. S., Kittrell. C., Hauge, R. H., Smalley, R. E., & Weisman, R. B. Structure-assigned optical spectra of single-walled carbon nanotubes. *Science* (2002)., 298, 2361-6.

[7] Wu J, Gerstandt K, Zhang H, Liu J, Hinds B. Electrophoretically induced aqueous flow through single-walled carbon nanotube membranes. *Nature Nanotechno.* 2012; 7: 133-9.

[8] Zhou W, Ding L, Yang S, Liu J. Synthesis of high-density, large-diameter, and aligned single-walled carbon nanotubes by multiple-cycle growth methods. *ACS Nano* 2011; 5: 3849-57.

[9] Strano, M. S., Dyke, C. A., Usrey, M. L., Barone, P. W., Allen, M. J., Shan. H., Kittrell, C., Hauge, R. H., Tour, J. M., & Smalley, R. E. Electronic structure control of single-walled carbon nanotube functionalization. *Science* (2003)., 301, 1519-22.

[10] Vaisman L, Wagner HD, Marom G. The role of surfactants in dispersion of carbon nanotubes, *Advances in Colloid and Interface Science* 2006; 128-130: 37-46.

[11] Wang H, Dispersing carbon nanotubes using surfactants. *Curr. Opin. Colloid Interface Sci.* 2009; 14: 364-71.

[12] Taylor DJF, Thomas RK, Penfold J. Polymer/surfactant interactions at the air/water interface, *Adv. Colloid and Interface Sci.* 2007; 132: 69-110.

[13] Wenseleers, W., Vlasov, I. I., Gppvaerts, E., Obraztsova, E. D., Lobach, A. S., & Bouwen. A. Efficient isolation and solubilization of pristine single-walled nanotubes in bile salt micelles. *Adv. Funct. Mater.* (2004)., 14, 1105-12.

[14] Strano, M. S., Moore. V. C., Miller, M. K., Allen, M. J., Haroz. E. H., Kittrell, C., Hauge, R. H., & Smalley, R. E. The role of surfactant adsorption during ultrasonication in the dispersion of single-walled carbon nanotubes. *J Nanosci Nanotech* (2003)., 3, 81-6.

[15] O'Connell, M. J., Bachilo. S. M., Huffman, C. B., Moore, V. C., Strano, Haroz. E. H., Rialon, K. L., Boul, P. J., Noon, W. H., Kittrell, C., Ma, J., Hauge, R. H., Weisman,R.

B., & Smalley, R. E. Band gap fluorescence from individual single-walled carbon nanotubes. *Science* (2002)., 297, 593-6.

[16] Pang JY, Xu GY, Tan YB, He F. Water-dispersible carbon nanotubes from a mixture of an ethoxy-modified trisiloxane and pluronic block copolymer F127. *Colloid Polym. Sci.* 2010; 288: 1665-75.

[17] Pang JY, Xu GY, Yuan SL, Tan YB, He F. Dispersing carbon nanotubes in aqueous solutions by a silicon surfactant: Experimental and molecular dynamics simulation study. *Colloids Surf. A* 2009; 350: 101-8.

[18] Panagiotis A, Andrei G, Ailsa L, Nerushev O, Bock H, Campbell EEB. Dispersing individual single-wall carbon nanotubes in aqueous surfactant solutions below the cmc. *J. Phys. Chem. B* 2010; 114: 2-9.

[19] Alvarez NT, Kittrell C, Schmidt HK, Hauge RH, Engel PS, Tour JM. Selective photochemical functionalization of surfactant-dispersed single wall carbon nanotubes in water. *J. Am. Chem. Soc.* 2008; 130: 14227-33.

[20] Moore, V. C., Strano, M. S., Haroz, E. H., Hauge, R.H., & Smalley, R. E. Individually suspended single-walled carbon nanotubes in various surfactants. *Nano Lett.* (2003)., 3, 1379-82.

[21] Nativ-Roth E, Yerushalmi-Rozen R, Regev O. Phase behavior and shear alignment in SWNT-surfactant dispersions. *Small* 2008; 4: 1459-67.

[22] Döle, S., Lechner, B. D., Park. J. H., Schymura, S., Lagerwall, J. P. F., & Scalia, G. Utilizing the krafft phenomenon to generate ideal micelle-free surfactant-stabilized nanoparticle suspensions. *Angew. Chem. Int. Ed.* (2012)., 51, 3254-57.

[23] Gong HJ, Xu GY, Liu T, Pang JY, Dou WL, Xin X. Synthesis of block polyethers with various structures and their application in dispersing single-walled carbon nanotubes. *Colloid Polym. Sci.* 2011; 289: 933-42.

[24] Shvartzman-Cohen R, Nativ-Roth E, Baskaran E, Levi-Kalisman Y, Szleifer I, Yerushalmi-Rozen R. Selective dispersion of single-walled carbon nanotubes in the presence of polymers: the role of molecular and colloidal length scales. *J. Am. Chem. Soc.* 2004; 126: 14850-7.

[25] Shvartzman-Cohen R, Levi-Kalisman Y, Nativ-Roth E, Yerushalmi-Rozen R. Generic approach for dispersing single-walled carbon nanotubes: The strength of a weak interaction. *Langmuir* 2004; 20: 6085-8.

[26] Xin X, Xu GY, Zhang ZQ, Chen YJ, Wang F. Aggregation behavior of star-like PEO-PPO-PEO block copolymer in aqueous solution. *Eur. Polym. J.* 2007; 43: 3106.

[27] Xin X, Xu G, Zhao T, Zhu Y, Shi X, Gong H, Zhang Z. Dispersing carbon nanotubes in aqueous solutions by a star-like block copolymer. *J. Phys. Chem. C* 2008; 112: 16377.

[28] Islam MF, Rojas E, Bergey DM, Johnson AT, Yodh AG. High weight fraction surfactant solubilization of single-wall carbon nanotubes in Water. *Nano Lett.* 2003; 3: 269-73.

[29] Zhou W, Islam MF, Wang H, Ho DL, Yodh AG, Winey KI, et al. Small angle neutron scattering from single-wall carbon nanotube suspensions: evidence for isolated rigid rods and rod networks. *Chem. Phys. Lett.* 2004; 384: 185-9.

[30] Chen RJ, Zhang YG, Wang DW, Dai HJ. Noncovalent sidewall functionalization of single-walled carbon nanotubes for protein immobilization. *J. Am. Chem. Soc.* 2001; 123: 3838-9.

[31] Backes C, Schmidt CD, Hauke F, Böttcher C, Hirsch A. High population of individualized SWCNTs through the adsorption of water-soluble perylenes. *J. Am. Chem. Soc.* 2009; 131: 2172-84.

[32] Takaguchi Y, Tamura M, Sako Y, Yanagimoto Y, Tsuboi S, Uchida T, Shimamura K, Kimura SI, Wakahara T, Maeda Y, Akasaka T. Fullerodendron-assisted dispersion of single-walled carbon nanotubes via noncovalent functionalization. *Chem. Lett.* 2005; 34: 1608-9.

[33] Li HG, Liu CC, Hao JC, Hartnagel U, Hirsch A. Dispersing carbon nanotubes by star-like water soluble C_{60} derivatives. *J. Nanosci. Nanotechno.* 2009; 9: 2763-7.

[34] Grunlan JC, Liu L, Kim YS. Tunable single-walled carbon nanotube microstructure in the liquid and solid states using poly(acrylic acid). *Nano Lett.* 2006; 6: 911-5.

[35] Bandyopadhyaya R, Nativ-Roth E, Regev O, Yerushalmi-Rozen R. Stabilization of individual carbon nanotubes in aqueous solutions. *Nano Lett.* 2002; 2: 25-8.

[36] Moulton SE, Maugey M, Poulin P, Wallace GG. Liquid crystal behavior of single-walled carbon nanotubes dispersed in biological hyaluronic acid solutions. *J. Am. Chem. Soc.*2007; 129: 9452-57.

[37] Star A, Steuerman DW, Heath JR, Stoddart JF. Starched carbon nanotubes. *Angew. Chem. Int. Ed.*2002; 41: 2508-12.

[38] Horn DW, Tracy K, Easley CJ, Davis VA. Lysozyme dispersed single-walled carbon nanotubes: Interaction and activity. *J. Phys. Chem. C* 2012; 116: 10341-8.

[39] Cathcart H, Quinn S, Nicolosi V, Kelly JM, Blau WJ, Coleman JN. Spontaneous debundling of single-walled carbon nanotubes in DNA-based dispersions. *J. Phys. Chem. C* 2007; 111: 66-74.

[40] Tardani F, Mesa CL, Poulin P, Maugey M. Phase behavior of DNA-based dispersions containing carbon nanotubes: Effects of added polymers and ionic strength on excluded volume. *J. Phys. Chem. C* 2012; 116: 9888-94.

[41] Yurekli, K., & Mitchell. C. A., Krishnamoorti. R. Small-angle neutron scattering from surfactant-assisted aqueous dispersions of carbon nanotubes. *J. Am. Chem. Soc.* (2004)., 126, 9902-3.

[42] Tummala NR, Striolo A. SDS surfactants on carbon nanotubes: aggregate morphology. *ACS Nano* 2009; 3: 595-602.

[43] Wallace EJ, Sansom MSP. Carbon nanotube/detergent interactions via coarse-grained molecular dynamics. *Nano. Lett.* 2007; 7:1923-8.

[44] Shvartzman-Cohen R, Nativ-Roth E, Baskaran E, Levi-Kalisman Y, Szleifer I, Yerushalmi-Rozen R. Selective dispersion of single-walled carbon nanotubes in the presence of polymers: the role of molecular and colloidal length scales. *J. Am. Chem. Soc.* 2004; 126: 14850-7.

[45] Lynch MD, Patrick DL. Organizing carbon nanotubes with liquid crystals. Nano Lett. 2002; 2: 1197-201.

[46] Lagerwall JPF, Scalia G, Haluska M, Dettlaff-Weglikowska U, Roth S, Giesselmann F. Nanotube alignment using lyotropic liquid crystals. *Adv. Mater.* 2007; 19: 359-64.

[47] Xin X, Li HG, Wieczorek SA, Szymborski T, Kalwarczyk E, Ziebacz N. Gorecka E, Pociecha D, Hołyst, R. Incorporation of carbon nanotubes into lyotropic liquid crystal by phase separation in the presence of hydrophilic polymer. *Langmuir* 2010; 26: 3562-8.

[48] Xin X, Li HG, Kalwarczyk E, Kelm A, Fialkowski M, Gorecka E. Pociecha D, Hołyst, R. Single-walled carbon nanotube/lyotropic liquid crystal hybrid materials fabricated by a phase separation method in the presence of polyelectrolyte. *Langmuir* 2010; 26: 8821-8.

[49] Camponeschi E, Florkowski B, Vance R, Garrett G, Garmestani H, Tannenbaum R. Uniform directional alignment of single-walled carbon nanotubes in viscous polymer flow. *Langmuir* 2006; 22: 1858-62.

[50] Nativ-Roth E, Regev O, Yerushalmi-Rozen R. Shear-induced ordering of micellar arrays in the presence of single-walled carbon nanotubes. *Chem. Commun.* 2008; 17: 2037-9.

[51] Lee HW, You W, Barman S, Hellstrom S, LeMieux MC, Oh JH, Liu SH, Fujiwara T, Wang WM, Chen B, Jin YW, Kim JM, Bao ZN. Lyotropic liquid-crystalline solutions of high-concentration dispersions of single-walled carbon nanotubes with conjugated polymers. *Small* 2009; 5: 1019-24.

[52] Ghosh S, Bachilo SM, Weisman RB. Advanced sorting of single-walled carbon nanotubes by nonlinear density-gradient ultracentrifugation. *Nature Nanotechno.* 2010; 5: 443-50.

[53] Arnold MS, Green AA, Hulvat JF, Stupp SI, Hersam MC. Sorting carbon nanotubes by electronic structure using density differentiation. *Nature Nanotechno.* 2006; 1: 60-5.

[54] LeMieux MC, Roberts M, Barman S, Jin YW, Kim JM, Bao ZN. Self-sorted, aligned nanotube networks for thin-film transistors. *Science* 2008; 321: 101-4.

[55] Kam NWS, O'Connell M, Wisdom JA, Dai HJ. Carbon nanotubes as multifunctional biological transporters and near-infrared agents for selective cancer cell destruction. *PNAS* 2005; 102: 11600-5.

[56] Liu Z, Davis C, Cai WB, He LN, Chen XY, Dai HJ. Circulation and long-term fate of functionalized, biocompatible single-walled carbon nanotubes in mice probed by Raman spectroscopy. *PNAS* 2008; 105: 1410-5.

Chapter 12

Mild and Nondestructive Chemical Modification of Carbon Nanotubes (CNTs): Direct Friedel-Crafts Acylation Reaction

Dong Wook Chang, In-Yup Jeon, Hyun-Jung Choi and Jong-Beom Baek

Additional information is available at the end of the chapter

1. Introduction

Since Iijima's report on carbon nanotubes (CNTs) [1], which consist of graphene sheets rolled up into a cylindrical shape, many researchers have focused on CNTs due to their superior mechanical, electrical and thermal properties. Depending on the arrangement of aromatic rings along the cylindrical surface, specifically for single-walled carbon nanotubes (SWCNTs), CNTs can possess two distinguished properties such as metallic and semiconducting. In spite of many advantages, the practical applications of CNTs have been limited by their poor processability and dispersability in solvents, polymers, ceramics and metallic matrices. Indeed, the pristine CNTs are insoluble in any solvent, due to strong van der Waals interactions between CNTs and lack of chemical affinity to organic solvents. To overcome this limitation, many chemical (covalent) and physical (noncovalent) modification methods to functionalize CNTs have been developed during last decades for improved compatibilities with both liquid and solid matrices [2-3]. Among them, chemical approaches using various chemical reactions are considered to be the most promising protocol for enhancing dispersability and processability of CNTs. However, CNTs are chemically inert for efficient chemical modifications, and thus reactions have to be carried out in harsh conditions, causing significant structural damages to CNT frameworks. As a results, a sharp decrease in their intrinsic properties is inevitable [2-3]. In this regard, physical modifications of CNTs have been considered to be more favorable methods for electronic applications, because electronic structures can be largely preserved due to the noncovalent approaches for modified CNTs [4-6]. However, homogeneous dispersion using the physical method accompanied with sonication often damages CNTs due to the effects of dose time and strength. Furthermore, they also have some disadvantages such as limited utilization of materials and insufficient modification levels for practical applications. Thus, the development of nondestructive and efficient chemical modification of CNTs is highly desirable.

Since the pioneering work from Baek *et al.*, [7], direct Friedel-Crafts acylation reaction to inherent defective sp^2C-H sites on the surface of CNTs have been widely investigated [8-13], because it has several advantages such as nondestructive reaction nature, sufficient modification level, utilization of diverse materials and suitable for mass production. Furthermore, it can be expanded to all types of carbon-based nanomaterials such as fullerenes [14], carbon nanosfibers [7, 15-17], nanodiamonds [18] and graphene [19-22]. Therefore, direct Friedel-Crafts acylation reactions could be one of ideal chemical modifications for carbon based materials, specifically CNTs. This chapter will focus on and discuss about the various aspects of direct Friedel-Crafts acylation reaction onto CNTs such as fundamental mechanisms, potential applications and perspectives. Of particular importance, this chapter is highly beneficial to general readers in research community of carbon based materials.

2. Direct Friedel-Crafts acylation of Carbon Nanotubes

2.1. Overview and mechanism

Although various chemical and physical modifications for enhancing the dispersability and processability of CNTs have been utilized for last decades, both methods have their own drawbacks depending on the platform as discussed earlier. The advantages and disadvantages of various modifications of CNTs are summarized in Table 1 [23].

Method		**Principle**	**Possible damage to CNTs**	**East to use**	**Interaction with polymer matrix**[a]	**Re-agglomeration of CNTs in matrix**
Chemical Method	Side wall	Hybridization of C atoms from sp^2 to sp^3	√	×	S	√
	Defect	Defect transformation	√	√	S	√
	Polymer wrapping	van der Waals force, -stacking	×	√	V	×
Physical Method	Surfactant adsorption	Physical adsorption	×	√	W	×
	Endohedral Method	Capillary effect	×	×	W	√

[a] S: Strong; W:Weak; V: Variable according to the miscibility between matrix and polymer on CNT

Table 1 Advantages and disadvantages of various modification methods of CNTs [23].

Additionally, the most chemical modifications are initiated by chemical oxidation of CNTs in strong acids [2-3]. Therefore, dramatic structural damages of CNTs can be easily happened during harsh oxidation reaction, which results in significant weakening of many useful intrinsic properties of CNTs. To overcome these problems, the development of alternative functionalization routes, which can not only introduce homogeneous surface functional groups with high density to enhance the compatibility of CNTs and various foreign matrixes, but also minimize the structural damages of CNTs during reactions to optimize their properties in various applications, are highly demanding.

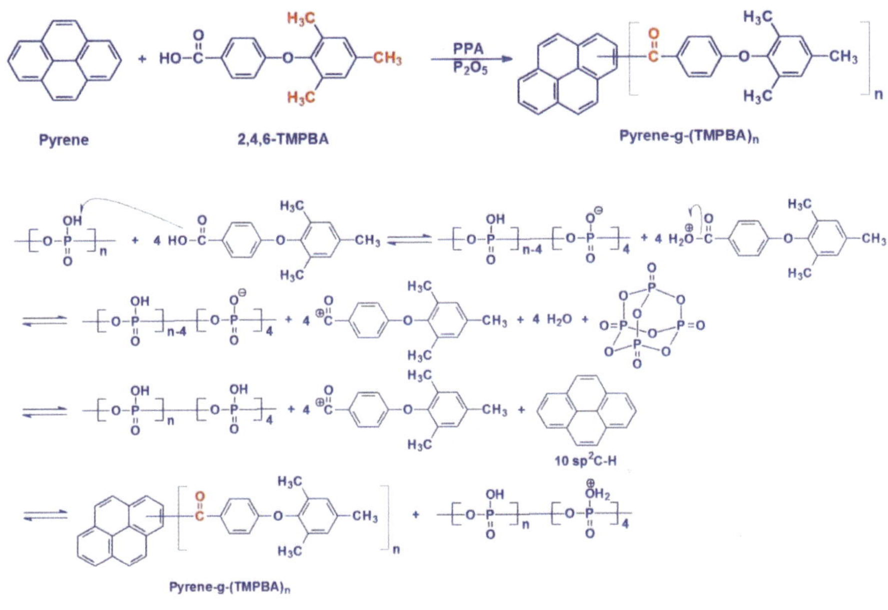

Figure 1. A summary of reaction mechanism of direct Friedel-Crafts acylation reaction using pyrene as a model compound in poly(phosphoric acid)/phosphorous pentoxide medium [24].

Recently, Baek *et al.*, [8-11, 13] have reported an efficient route to covalently functionalize CNTs *via* simple reaction called as direct Friedel-Crafts acylation. Interestingly, simple benzoic acid (-COOH) and benzamide (-CONH$_2$) groups are directly used in this newly developed synthetic strategy instead of an expensive, inconvenient and corrosive carboxylic acid chloride (COCl), which is normally utilized in Friedel-Crafts acylation. The detailed reaction mechanism of this reaction using pyrene as a model compound is shown in Figure 1 [24]. The reaction normally takes place between benzoic acid derivatives and CNTs in a mild polyphosphoric acid (PPA)/phosphorous pentoxide (P_2O_5) medium. PPA used in this study is a viscous polymeric acid and expected to play two important roles. Its mild acidic nature ($pK_a \approx 2.1$) could still be enough to protonate the surface of CNTs for deaggregation without structural damage, which is frequently observed from the oxidation reaction of CNTs using strong acids such as nitric acid ($pK_a \approx -1.5$), sulfuric acid ($pK_a \approx -3.0$) and their mixture. Thus, the outstanding properties of CNTs such as electrical, thermal and mechanical properties can be preserved. Additionally, viscous nature of PPA would help to impede reaggregation of CNTs after dispersion of CNTs with strong shear forces while mechanical stirring. Another component of reaction medium, P_2O_5, is used as a dehydrating agent to promote Friedel-Crafts reaction efficiently. In this reaction condition, defective sp^2C-H groups inherently presented on the surfaces or edges of CNTs are reactive sites for electrophilic substitution reaction with newly generated carbonium ions ($C=O^+$) from benzoic acid and benzylamide derivatives in PPA/P_2O_5 [25]. As a result, an efficient homogeneous introduction of various functional groups onto CNTs without structural damage has been obtained from the newly developed direct Friedel-Crafts acylation in PPA/P_2O_5. Furthermore, simple and scalable features of this approach could be regarded as additional advantages. In optimized reaction conditions, the fixed weight ratio of PPA/P_2O_5 (4/1) has been used as a reaction medium and the reaction takes place using high-torque mechanical stirrer at 130 °C for 48 – 72 h under dry nitrogen purge. After reaction, the solid was transferred to an extraction thimble and extracted with water for 3 days and methanol for 3 days, and finally freeze-dried for 48 h to obtain final products.

2.2. Applications

2.2.1. Functionalization of carbon nanotubes with small molecules

Recently, the functionalization of carbon nanotubes (CNTs) with small molecules containing benzoic acid [9, 13, 26] *via* direct Friedel-Crafts acylation reaction in PPA/P_2O_5 have been successfully demonstrated by Baek *et al.* For providing a fundamental concept on the relationship between structure and reactivity, a reactivity hierarchy of 4-substituted benzoic acids with multi-walled carbon nanoubes (MWCNTs) in the reaction condition has been systematically investigated [9]. Accordingly, 10 different kinds of benzoic acids with various different functionalities to 4-position of benzoic acids such as amine, hydroxyl, ethoxy, methoxy, fluoro, chloro, bromo, iodo and nitro groups were selected for the functionalization (Figure 2-left). The functionalization of MWCNTs with all benzoic acid derivatives used in this study has been efficiently occurred *via* a simple direct Friedel-Crafts acylation reaction in PPA/P_2O_5. For examples, the photograph taken of the 4-ethoxybenzoic acid and MWCNTs reaction mixture without flashlight was shiny black as shown in Figure 2-right-a. When the mixture was illuminated by flashlight, the shiny-greenish-brown color became prominent (Figure 2-right-b). The precipitated in of the mixture after reaction in distilled water was deep green as it was in the reaction mixture under the flashlight (Figure 1-right-c). The green suspension might be due to the charge complex in acidic medium. The uniformly decorated 4-ethoxybenzoyl moiety on the surface of MWCNTs and the charge complexes formed on the ether linkage could possibly display green color. These photographs provided strong visual evidence that the MWCNTs could be effectively functionalized with 4-ethoxybenzoic acid moiety. After complete purification procedures, overall yields for all cases were in the range of 53-78%. As a result, the reactivity of compounds in direct Friedel-Crafts acylation reaction in PPA/P_2O_5 could be greatly attributed to the 'electron-donating' and 'electron-accepting' natures of 4-substituted groups to the carboxylic acid [13]. The former displayed better reactivity than the latter.

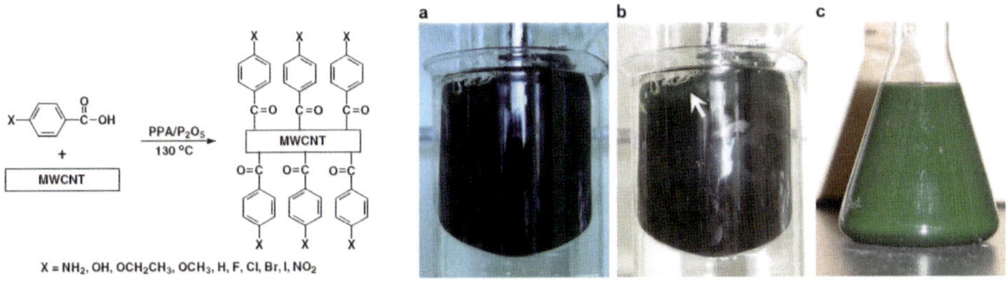

Figure 2. (left) Functionalization of MWCNTs with various 4-substituted benzoic acids using direct Friedel-Crafts acylation reaction, (right) a - reaction mixture of 4-ethoxybenzoic acids and MWCNTs without flashlight, b - reaction mixture of 4-ethoxybenzoic acids and MWCNTs with flashlight and c - precipitation of reaction mixture of 4-ethoxybenzoic acids and MWCNTs in distilled water [13].

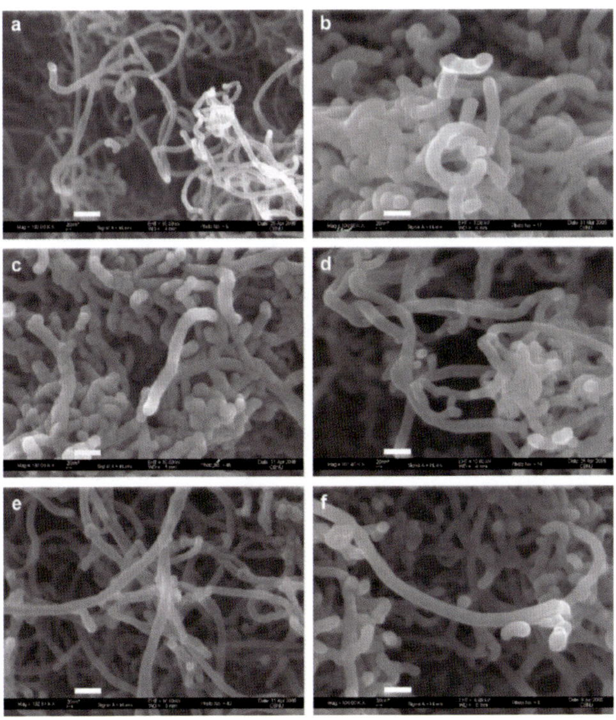

Figure 3. SEM images of MWCNTs: (a) pristine MWCNTs, (b) 4-aminobenzoic, (c) 4-ethoxybenzoic, (d) 4-hydroxybenzoic, (e) 4-bromobenzoic and (f) 4-nitorobenzoic acid functionalized MWCNTs. Scale bars are 100 nm [13].

The dispersability of MWCNTs was greatly enhanced by functionalization and debundling of MWCNTs with small molecules *via* direct Friedel-Crafts acylation reaction in PPA/P_2O_5, but the surface properties of functionalized MWCNTs could be altered significantly due to different functional groups were introduced [13]. For examples, the polar 4-hydroxybenzoyl substituted MWCNTs displayed the best solubility and they were easily dispersed in polar solvents such as tetrahydrofuran (THF), dichloromethane and *N,N*-dimethylacetamide (DMF). However, 4-bromobenzoyl functionalized MWCNTs were proved to be insoluble in all tested solvents. In addition to dispersability, the polarity of surface group on MWCNTs has also great influence on their size and morphology. The pristine MWCNTs show the clean and smooth surface with an average diameter of 10-20 nm (Figure 3a), while the surfaces of functionalized MWCNTs with 4-substituted benzoic acids reveal structurally intact with a larger diameter in the rage of 40-70 nm (Figure 2b-f). Assuming the length of 4-substituted benzoyl units to be approximately 1 nm, the diameters of functionalized MWCNTs should be within the range of 12-22 nm. However, all functionalized MWCNTs showed larger diameters, at least twice that of pristine MWCNTs. This implies that they were in the bundled state. The size of bundles was closely related to the polarity of the surface groups and the degree of functionalization. When there is enough lateral interaction among tubes to overcome axial rigidity, the larger number of tubes are aggregated to form bundle, and thus the diameters are increased. The SEM images in Figure 3 show that the average diameters of samples with polar surface groups such as amino, hydroxyl and nitro benzoic acids were larger than those of samples with non-polar surface groups such as ethoxy and bromo. Furthermore, the surface morphologies of functionalized MWCNTs with non-polar surface groups appeared to be soft and puffy (Figure 3c and d), while functionalized MWCNTs with polar surface groups showed shiny and rigidly sooth morphologies (Figure 3b, d and f).

Furthermore, this unique synthetic strategy can be applied to different types of CNTs like single- [10], few- [27-28] and multi-walled CNTs [8-13]. Recently, it has been reported that few-walled carbon nanotubes (FWCNTs), defined as nanontubes with sidewalls typically of 2 to 6 layers, diameters ranging from 3 to 8 nm, have particularly distinguished from other types of CNTs [29]. Therefore, the functionalization of FWCNTs without structural damages to generate nanocomposites hybrid materials or even thin film has attracted great attentions for their various potentials in device applications. In this purpose, Baek *et al.*, demonstrated that the functionalization of FWCNTs with two different surface groups using a direct Friedel-Crafts acylation reaction in a nondestructive PPA/P_2O_5 medium [27]. The less polar 4-ethylbenzoic acid and more polar 4-(aminomethyl)benzoic acid have been used for the surface modification of FWCNTs. Interestingly, 4-ethylbenzoic acid functionalized FWCNTs can absorb water more than 28 times its own weight, indicating that the nature of surface functional groups was significantly attributed to the sponge behavior of functionalized FWCNTs. In addition the electrical capacitance of functionalized FWCNTs was also significantly affected by the nature of surface groups [27]. Furthermore, it has been also reported that an efficient route to prepare highly conducting and flexible FWCNTs thin film by Baek *et al.*, (Figure 4) [28]. The free standing thin films were fabricated by functionalizing FWCNTs with 4-ethoxybenzoic acid *via* a direct Friedel-Crafts acylation reaction in a similar condition. The resulting 4-ethoxybenzoic acid functionalized FWCNTs (EBA-f-FWCNTs) were readily dispersed in water and the films were simply casted from the filtration of the dispersed solution. Room temperature electrical conductivity of the thin flexible film of EBA-f-FWCNTs shows a value as high as 29,400 S/cm^{-1}, while the tensile strength and modulus of it were found to be about 80 MPa and 15 Gpa, respectively. In addition cyclic voltamogram reveals a rectangular shape with superior capacitance of 133 F/g for the thin film [28]. This study demonstrated the simple and efficient preparation methods to produce highly flexible and conductive thin film of FWCNTs using a direct Friedel-Crafts acylation reaction in a mild reaction condition.

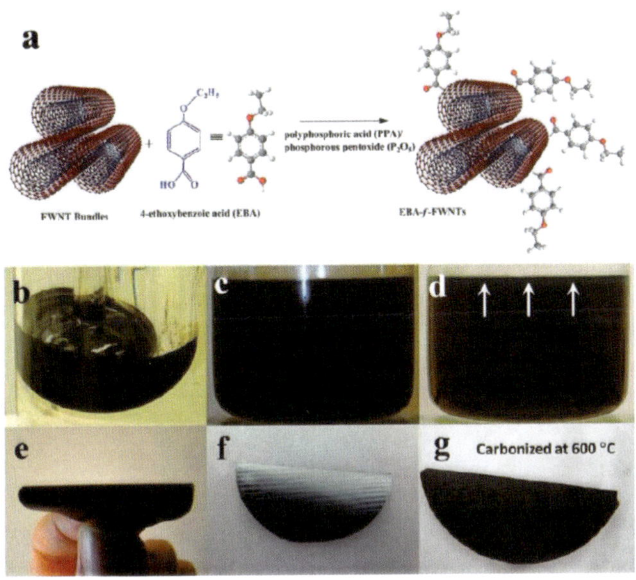

Figure 4. (a) Schematic cartoon depicting the functionalization of FWCNTs with 4-ethoxybenzoic acid. Digital photographs of (b) reaction flask, (c) EBA-f-FWCNTs dispersed in water without light, (d) EBA-f-FWCNTs dispersed in water light, (e) thin film made of EBA-f-FWCNTs (f) 180 °C folded thin film and (g) carbonized EBA-f-FWCNTs thin film at 600 °C for 2h [28].

Hitherto, a various aspects of direct Friedel-Crafts acylation reaction in PPA/P_2O_5 between 4-substituted benzoic acids and CNTs have been discussed. Interestingly, this strategy can be expanded to 4-substituted benzamides instead of 4-substituted carboxylic acids. The benzamide could also be directly attached to the surface of CNTs. As a model compound, 4-(2,4,6-trimethylphenoxy)benzamide (TMPBA) was reacted with single-walled carbon nanotubes (SWCNTs) in PPA/P_2O_5 as a mild direct Friedel-Crafts acylation reaction condition to afford TMPBA functionalized SWCNTs (Figure 5a) [10]. The covalent attachment of TMPBA onto the surface of SWCNTs was proved by elemental analysis (EA), Fourier-transform infrared spectroscopy (FT-IR), Raman spectroscopy and thermogravimatric analysis (TGA). In addition, the SEM image of TMPBA-g-SWCNT shows that the surface of SWCNTs is apparently decorated with covalently bonded moieties (Figure 5b). From the results, direct Friedel-Crafts acylation reaction in PPA/P_2O_5 could be one of powerful tools for the covalent modification of CNTs with small molecules containing various functional moieties.

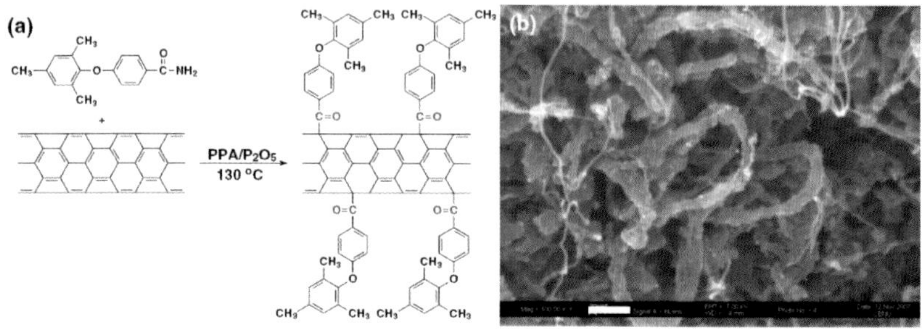

Figure 5. (a) Functionalization of SWCNTs with 4-(2,4,6-trimethylphenoxy)benzamide (TMPBA) and (b) SEM image of TMPBA-g-SWCNTs. Scale bar is 100 nm [10].

Due to the efficient modification of CNTs with a covalent attachment of small molecules with various functionalities, the functionalized CNTs are very useful for the preparation of composites *via* both solution and melt processes. For examples, 4-ethyoxybenzoic acid modified MWCNTs could be homogeneously dispersed in ethylene glycol (EG) and *in situ* polymerized with terephthalic acid. The pilot scale preparation of polyethyleneterephthalate (PET)/4-ethyoxybenzoyl modified MWCNTs composited was successfully demonstrated [30]. Various systems such as polycarbonate/4-hydroxybenzoyl modified MWCNTs, polyester thermoplastic elastomer/4-chlorobenzoyl modified MWCNTs [31], epoxy (EPON 828)/4-aminobenzoyl modified MWCNTs [32], poly(3-hexylthiophene)/4-hydroxybenzoyl modified MWCNTs [33] and Nylon 610/4-chlorobenzoyl modified MWCNTs composites were prepared *via* either *in situ* or interfacial polymerizations [26]. Furthermore, various conducting polymers such as polyaniline [11, 34-35] and polypyrrole [36] have also been successfully grafted onto 4-aminobenzoyl modified MWCNTs as an anchoring sites *via in situ* polymerization. These composite materials with conducting polymers grafted to MWCNTs show the enhanced conductivity and unique electrocatalytic activities. In addition to polymers, inorganic materials such as gold nanoparticles (GNPs) can also be immobilized onto the surface 4-mercaptobenzoyl functionalized MWCNTs as a platform (Figure 6) [37]. Firstly, the functionalization of MWCNTs with 4-mercaptobenzoic acid by a direct Friedel-Crafts acylation reaction to afford MWCNTs containing thiol groups was carried out in a nondestructive condition. Then, the separately prepared citrate stabilized GNPs were mixed with MWCNTs containing thiol moieties. Due to the strong interactions between thiol and GNPs, they can be stably immobilized onto the surface of MWCNTs covered by thiol groups without agglomeration. These hybrid inorganic-CNTs composites exhibit high electrocatalytic activity and electrochemical stability [37]. These numerous findings of CNTs based composite materials using functionalized CNTs prepared by a simple direct Friedel-Crafts acylation reaction in a mild reaction medium reveal various utilizations of them for application-specific purposes.

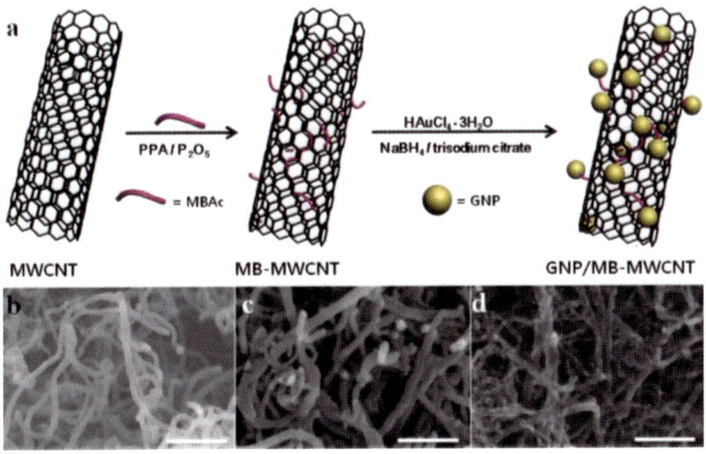

Figure 6. (a) Functionalization of MWCNTs with 4-mercaptobenzoiic acid and preparation of hybrid composites with gold nanoparticles (TMPBA) and SEM images of (b) pristine MWCNTs, (c) 4-mercaptobenzoic acid modified MWCNTs and (d) their hybrid composites with gold nanoparticles. Scale bar is 200 nm [37].

2.2.2. In situ grafting of linear or hyperbranched polymers onto carbon nanotubes

Due to the unique features of CNTs, they have been actively investigated for uses as reinforcing components to deliver outstanding properties to various matrixes such as polymers [38-39], ceramics [40] and low melting metals [41]. The resultant nanocomposites are expected to display enhanced properties providing various potential applications for light-weight and multifunctional materials. Unfortunately, CNTs usually exist in ropes and bundles due to strong lateral interactions between the tubes, causing difficulty in homogeneous dispersing them in a multi-component system. Therefore, various physical and chemical modifications to afford homogeneous dispersion of CNTs are required for the effective transfer of their outstanding properties to the matrix materials. However, chemical approach using strong acids and physical approach with sonication treatment can easily cause significant damages such as sidewall opening and tube breakage on their structures. In addition to dispersion, the strong interfacial adhesion between CNTs and matrix is also one of crucial factors in nanocomposites. It is also well known that noncovalent interactions between CNTs and matrix in nanocomposites are not expected to have any synergic effect even after homogeneous dispersions of CNTs could be achieved. Thus, it is highly desirable to covalently link desired polymers to the surface of CNTs. As a result, the development of efficient covalent polymer grafting to the surface of CNTs without structural damages is highly demanding to meet the above mentioned two important requirements for nanocomposites, *i.e.*, homogeneous dispersion and strong adhesion interaction with matrix. In this context, the chemical modification methods of CNTs with various linear and hyperbranched polymers using less destructive direct Friedel-Crafts acylation reaction in a mild PPA/P_2O_5 medium have been demonstrated by Baek, *et al.*, [8, 12, 25, 42-43]. *In situ* covalent attachments of linear and/or hyperbranched poly(ether-ketone) onto the surface of CNTs were successfully performed in a mild reaction medium. AB and AB_2 types of monomers was used for the grafting of linear and hyperbranched polymers, respectively, to the surface of CNTs. The linear poly(ether-ketone) grafted CNTs show the dramatic increase in solution viscosity due to the formation of giant molecules during polymerization and the polymer chains are uniformly coated on the surface of CNTs. The resulting nanocomposites were easily fabricated using a simple compression molding technique and the possibility of aligning the CNTs in their nanocomposites *via* solution spinning to significantly enhance the anisotropic tensile properties along the fiber axial direction was also demonstrated [42]. Compared to linear counterpart, the unique highly branched structures and available surface functionalities of hyperbanched polymers offer unusual properties such as low viscosity and enhanced solubility [44-45]. After covalent attachment of three-dimensional globular hyperbranched polymer molecules to the surface of CNTs (Figure 7) [43], the resultant hyperbranched polymer grafted (HBP-g-CNTs) nanocomposites are expected to display both enhanced dispersion and interfacial interaction. The former would be originated from impeding the lateral interaction between CNTs when hyperbranched polymers grafted to the surface of them and the latter is enhanced by the topological roughness contributed from the broad size distribution of the hyperbranched macromolecules. Furthermore, the numerous periphery surface groups and fractal molecular architecture of rigid hyperbranched polymers could provide additional chemical interactions and mechanical interlocking between HBP-g-CNTs nanocomposites and supporting matrix.

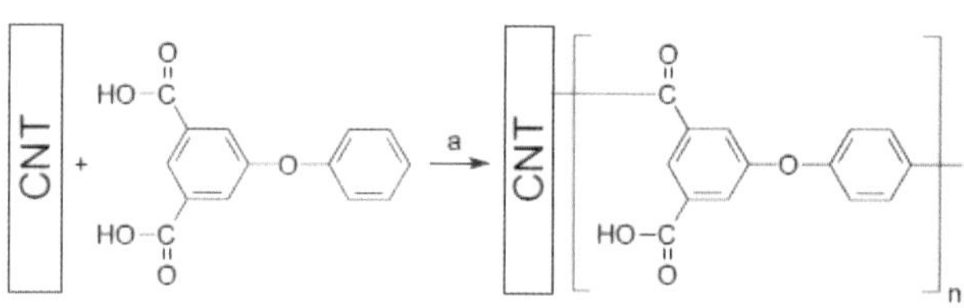

Figure 7. Grafting of hyperbranched poly(ether-ketone) onto the surface of CNTs from AB_2 monomer (a: PPA/P_2O_5) [43].

The hyperbranched poly(ether-ketone) has been attached onto the surface of both SWCNTs and MWCNTs [43]. The diameter range of pristine SWCNTs bundles is 40-60 nm (Figure 8a), while hyperbranched polymer grafted SWCNTs (HBP-g-SWCNTs) bundles show much smaller diameter range of 5-25 nm than that of pristine SWCNTs (Figure 8c). In addition, the shape of them resembles fractal structures. Some HBP-g-SWCNTs fibrils are stemmed out like tree branches and imbedded into hyperbranched matrix. The overall state of dispersion is homogeneous. Therefore, it could be hypothesized that once split is occurred at the edge of SWCNTs bundle when mechanical stirring shear force is applied, viscous polymeric reaction medium containing AB_2 monomer, which is readily react, is penetrated in between split and finally wedged by hyperbranched poly(ether-ketone). As a result, the splits are started from the tips of SWCNTs bundles and propagated further into the bundles (Figure 6c). In case of MWCNTs, the pristine MWCNTs show the seamless and smooth surfaces (Figure 8b). However, heavy amount of hyperbranched poly(ether-ketone) attached to MWCNTs could be clearly seen from the SEM images after grafting of hyperbranched poly(ether-ketone) (Figure 8d). The resultant hyperbranched polymer grafted MWCNTs (HBP-g-MWCNTs) have the diameter range of 40-150 nm, which is strong indication that the covalent attachment of hyperbranched poly(ether-ketone) to the surface of MWCNTs. Furthermore, the surfaces of nanocomposites are appeared to be puppy and bumpy compared to pristine MWCNTs.

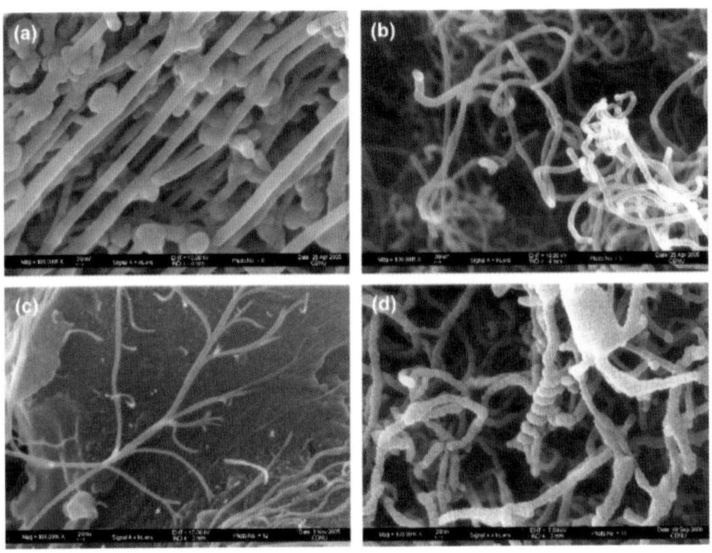

Figure 8. SEM image of (a) SWCNTs, (b) MWCNTs, (c) HBP-g-SWCNTs and (d) HBP-g-MWCNTs. All images are captured under the same magnification (100,000×) [43].

Similarly, the grafting of hyperbranched poly(ether-ketone) to the surface of MWCNTs could be realized in alternative way using unique self-controlled polycondensation methodology directly from the mixture of commercially available A_3 and B_2 monomers in the same reaction medium of PPA/P_2O_5 without gelation problem [8]. In addition, linear and hyperbranched poly(ether-ketone) containing flexible oxymethylene spacers grafted MWCNTs were also prepared by a direct Friedel-Crafts acylation reaction [25]. The resultant nanocomposites are soluble in most strong acids such as trifluoroacetic acid, methanesulfonic acid and sulfuric acid, and they are expected to display enhanced melt processability due to the flexible spacers in structural unit. It is worth to note that the semimetallic nanocomposites, linear or hyperbranched poly(phenylene sulfide) (PPS) grafted MWCNTs, could be successfully prepared by two-step reaction sequences [12]. Firstly, MWCNTs were functionalized with 4-chlorobenzoic acid using a direct Friedel-Crafts acylation reaction in PPA/P_2O_5 to afford 4-chlorobenzoyl functionalized MWCNTs (CB- MWCNTs). A subsequent nucleophilic substitution reaction between CB- MWCNTs and 4-chlrobenzenethiol as an AB monomer or 3,5-dichlrobenzenethiol as an AB_2 monomer was conducted to graft the linear PPS (LPPS) or hyperbranched PPS (HPPS) in NMP/toluene in the presence of sodium carbonate to afford LPPS grafted MWCNTs (LPPS-g-MWCNTs) or HPPS grafted MWCNTs (HPPS-g-MWCNTs), respectively (Figure 9). The covalent attachment of corresponding polymers onto the surface of MWCNTs was indirectly confirmed by a model study without MWCNTs.

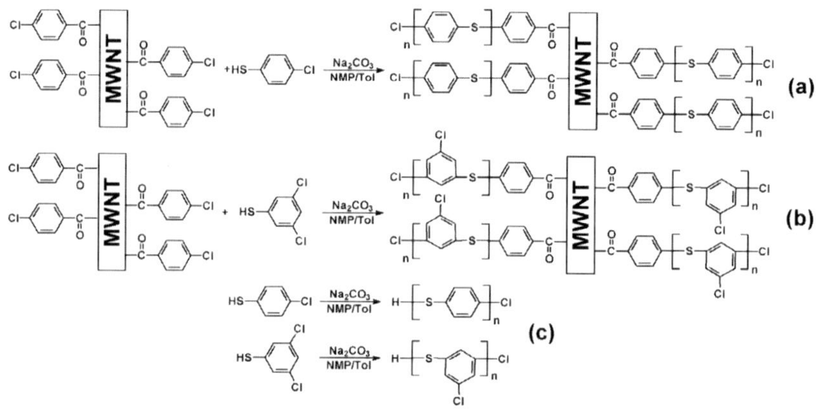

Figure 9. Grafting of (a) LPPS and (b) HPPS onto CB-g-MWCNTs, (c) Synthesis of LPPS and HPPS [12].

The SEM image of pristine MWCNTs shows that the tubes have seamless and smooth surfaces with an average diameter of 10-20 nm (Figure 10a). However, the average diameter of CB- MWCNTs is approximately 40 nm, which is 2-4 times thicker than that of pristine MWCNTs (Figure 10b). Interestingly, the shape of tube could be discerned by two parts. Opaque inner-hard core is covered by translucent outer-shadow-like part. The diameter of inner part in a rage of 10-20nm agrees well with that of the parent MWCNTs. Out-shadow-like part could be due to the 4-chlorobenzoyl moieties that have uniformly covered the surface of CB- MWCNTs. The SEM images of LPPS-g-MWCNTs reveal that the diameter approximately 100 nm, which is much larger than that of pristine MWCNTs and CB-MWCNTs (Figure 10c). Therefore, it is estimated that LPPS is heavily grafted to the CB-MWCNTs. In case of HPPS-g-MWCNTs, although the diameter dimension is close to that of CB-g-MWCNTs, the original outer-shadow-like part of CB- MWCNTs appears to be completely covered with newly attached HPPS (Figure 10d).

For the verification of structural integrity of MWCNTs during reaction sequences and the covalent attachment of the relevant polymers, transmission electron microscopic (TEM) analysis was conducted. The TEM images of LPPS-g-MWCNTs and HPPS-g-MWCNTs show that the tubes are heavily decorated with polymers (Figure 11). Furthermore, the clear wall-to-wall stripes of MWCNTs framework with its structural integrity suggest that the structural stability of MWCNTs under the two-step reaction sequence. The resultant nanocomposites show the enhanced dispersability and melt-processability, and they could be easily compression molded. Due to the synergetic effect originated from two components of MWCNTs and PPS, even without chemical doping, the surface conductivities of LPPS-g-MWCNTs and HPPS-g-MWCNTs molded samples could be reached to the semimetallic transport region at 11.76 and 3.56 S/cm, respectively [12].

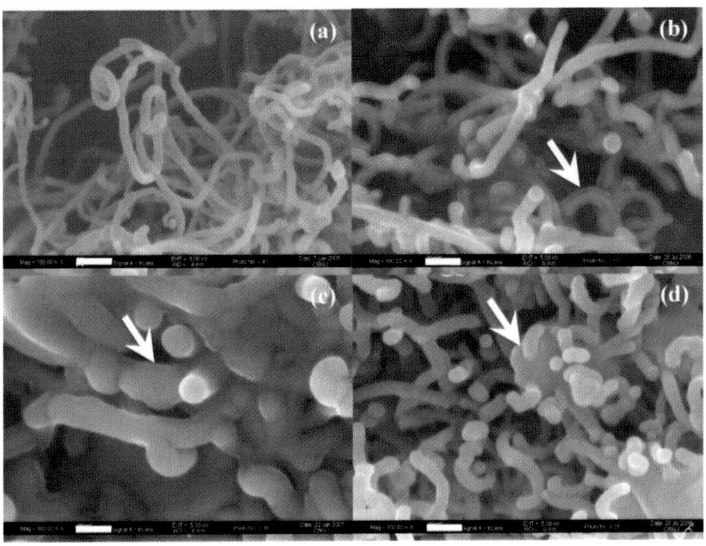

Figure 10. SEM image of (a) pristine MWCNTs, (b) CB-g-MWCNTs, (c) LPPS-g-MWCNTs and (d) HPPS-g-MWCNTs. All images are captured under the same magnification (100,000×) [12].

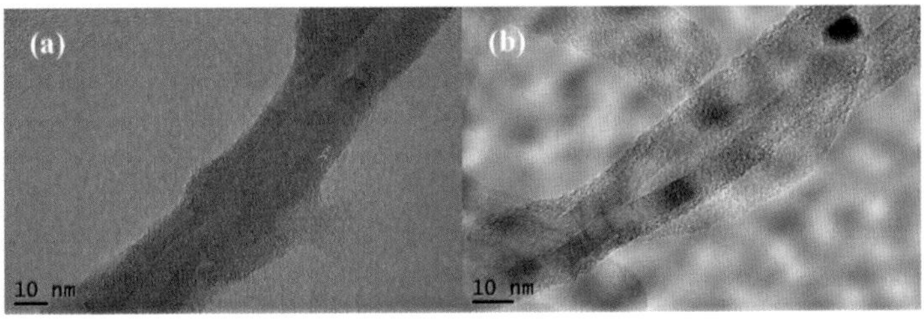

Figure 11. TEM images of (a) LPPS-g-MWCNTs and (b) HPPS-g-MWCNTs.

2.3. Other carbon-based nanomaterials: fullerene (C_{60}), carbon nanofiber, and graphene

In addition to CNTs, the covalent modification method of direct Freidel-Crafts acylation reaction in a mild PPA/P_2O_5 medium can be expanded to other carbon-based nanomaterials such as fullerene (C_{60}) [14], carbon nanofiber [7, 15-17] and graphene [19-22]. Buckminster fullerene, C_{60}, which is of the most abundant carbon sphere, is generally considered as a stable electron deficient material. Due to the electron affinity, C_{60} is considered as to be more susceptible to nucliophilic reaction than to electrophilic one. However, Baek *et al.*, firstly reported the covalent electrophilic functionalization of C_{60} *via* direct Friedel-Crafts acylation reaction in a mild PPA/P_2O_5 medium using 4-(2,4,6-trimethylphenoxyl)benzamide (TMPBA) as a substituent (Figure 12) [14]. After careful characterizations, it is suggested to that multiple destructive covalent attachments of TMPBAs onto C_{60} has successfully occurred and an average of 6.4 carbons was regioselectively detached from C_{60} framework to give $C_{53.6}$(TMPBA)$_6$.

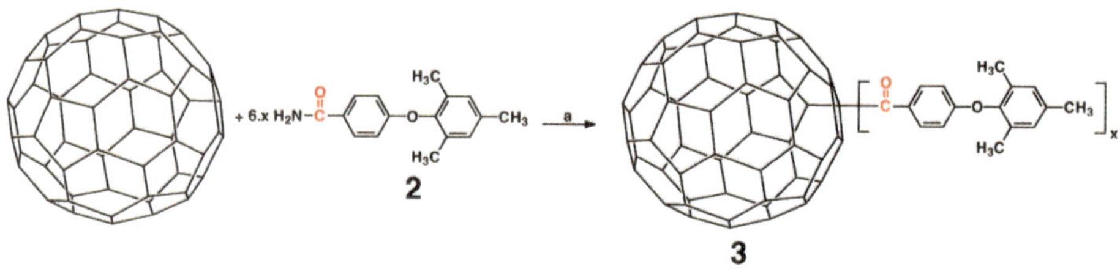

Figure 12. Synthesis of 4-(2,4,6-trimethylphenoxy) benzoyl functionalized fullerene. (a: PPA/P_2O_5) [14].

In comparison to CNTs, vapor-grown carbon nanofibers (VGCNFs), which are structurally hollow and multi-walled but several orders of magnitude larger in diameter and length than those of CNTs, are more attractive from a standpoint of practicality in terms of their relatively low cast and availability in larger quantities as a result of their more advanced stage in commercial production. These carbon nanofibers (CNFs) are typically produced by a vapor-phase catalyst process in which a carbon-containing feedstock (e.g. CH_4, C_2H_4, etc.) is pyrolyzed in the presence of small metal catalyst (e.g. ferrocene, $Fe(CO)_5$, etc.) and have an outer diameter of 60-200 nm, a hollow core of 30-90 nm, and length in the order of 50-100 μm [15-16]. Furthermore, VGCNFs have been widely used for tailoring properties in their polymer composites via cost-effective way, because of their inherent electrical and mechanical properties. To enhance compatibility and dispersability of VGCNF in polymeric matrix, various covalent grafting methods including ring-opening, atom-transfer radical and self-condensing polymerizations have been developed [17]. However, these approaches generally require multi-step synthetic procedures and limited species of materials can be utilized. To overcome these problems, Baek *et al.*, developed efficient functionalization and grafting methods onto the surface of VGCNF in a mild PPA/P_2O_5 medium, called as a direct Friedel-Crafts acylation reaction (Figure 13) [15-17]. As a result, the dispersion, interfacial adhesion and solution processabiliy of VGCNF have been greatly improved, which is quite beneficial for the development of high performance polymer-based nanocomposites.

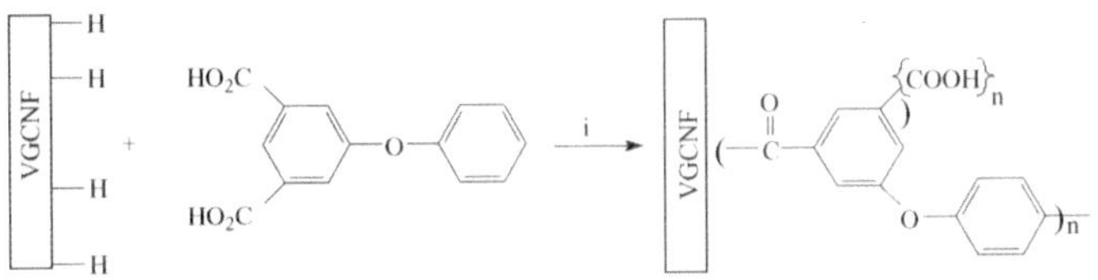

Figure 13. Synthesis of 3,5-diphenoxybenzoic acid functionalized VGCNF. (i: PPA/P_2O_5) [16].

Similarly to CNTs, the functionalization of VGCNF *via* direct Friedel-Crafts acylation reaction in PPA/P_2O_5 with various materials such as small molecules [7, 32], linear [15] and hyperbranched poly(ether-ketone) [16-17] have been successfully demonstrated. Specifically, the covalent attachment of hyperbranchedhyperbranched poly(ether-ketone) onto the surface of VGCNF has been clearly verified by TEM analysis. The TEM image of pristine VGCNF shows a smooth surface (Figure 14a). However, the nanofiber surfaces of VGCNF containing 20 wt % of grafted hyperbranchedhyperbranched poly(ether-ketone) polymers show a rough and fuzzy surface (Figure 14b). Furthermore, there is an obvious increase in the diameter due to the heavy coating by the attached hyperbranched poly(ether-ketone) with a thickness range of 10-20 nm [17]. Due to intrinsic nature of hyperbranched polymers such as a reduced viscosity, for example, the hyperbranched poly(ether-ketone) grafted VGCNF would be amenable to applications where speed and large-area coverage are required, such as spraying and painting techniques.

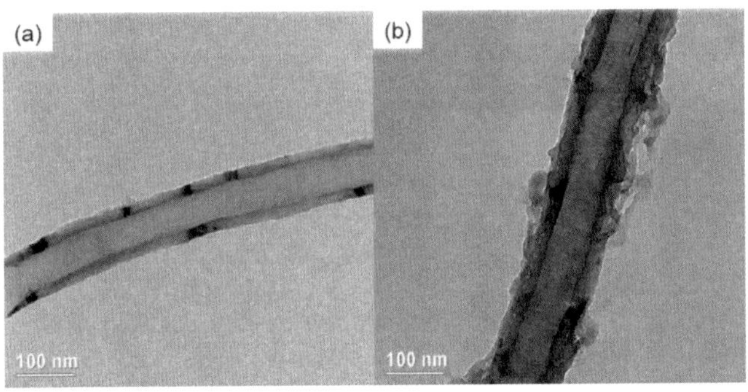

Figure 14. TEM images of (a) prisitne VGCNF and (b) hyperbranched poly(ether-ketone) polymer grafted VGCNF [17].

Graphene as a one of carbon-based nanomaterials, is currently the focal point for research into condensed matter due to its promising properties such as exceptional mechanical strength (~ 1100 GPa), high thermal conductivity (~ 5000 $Wm^{-1}K^{-1}$), large specific surface area (~ 2630 m^2g^{-1}) and ultra high electron transport properties (200,000 $cm^2V^{-1}s^{-1}$) [46]. There are two major approaches used in the preparation of graphene. The first method is the exfoliation of pristine graphite into graphene, which involves physical and chemical methods

[47-48]. The second method is where graphene can be directly grown using chemical vapor deposition (CVD) on a metal substrate [49] or from single crystal carbide [50]. For mass production, the chemical methods belong to the first approach is more preferred, but they still need to be optimized. In this regard, graphene oxide (GO) are widely investigated for the various applications of graphene, however GO has larger structural damages during the harsh preparation methods using strong acids and requires reduction, which has a limited conversion to reduced graphene oxide (rGO). Hence, the original graphitic structures cannot be efficiently restored in final graphitic structure, when GO is used as a starting material.

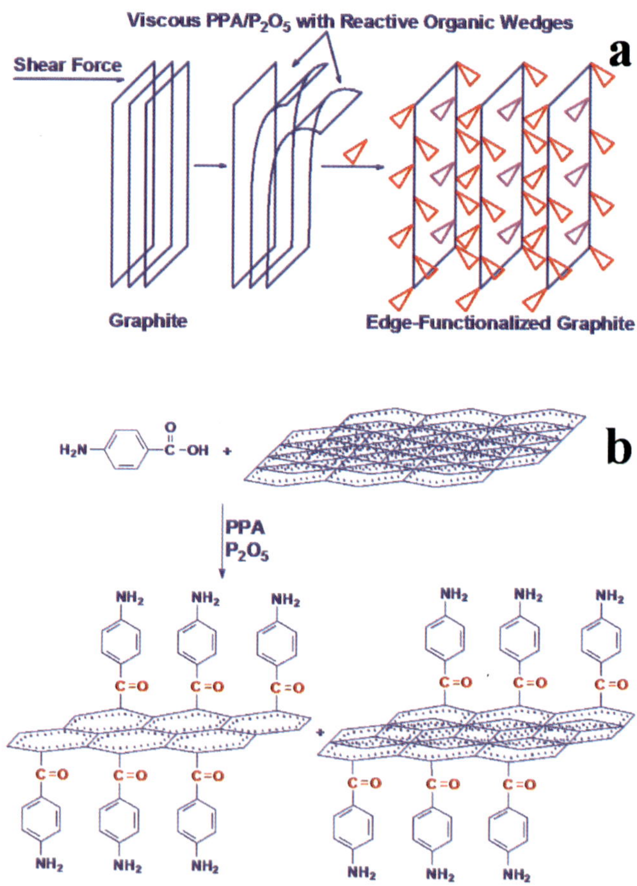

Figure 15. (a) Schematic presentation of graphite exfoliation mechanism and (b) schematic representation of the reaction between graphite and 4-aminobenzoic acid as amoelcualr wedge *via* Friedel-Crafts acyaltion in PPA/P_2O_5 medium [20].

Therefore, the development of less destructive and highly efficient method to exfoliate graphite into two-dimensional graphene and/or graphene-like sheets is highly required for the graphene research community. To meet this strong demand, Baek *et al.*, developed a new approach to chemical exfoliation of graphite by grafting organic moleculear wedges to the defect sites (mostly sp^2C-H) located mainly on the edges of graphite *via* a direct Friedel-Crafts acylation reaction in a mild PPA/P_2O_5 medium [19-22]. The reaction condition has been previously optimized for the functionalizaiton of carbon-based nanomateirals such as fullerene [14], CNTs [8-13] and VGCNF [7, 15-17]. This method is the first attempt at large-scale direct chemical exfoliation of graphite not involving strong acid and sonication that are known to damage graphitic carbon framework. The schematic presentation of graphite exfoliation mechanism and the reaction between graphite and 4-aminobenzoic acid as a molecular wedge *via* direct Friedel-Crafts acylation are shown in Figure 15 [20].

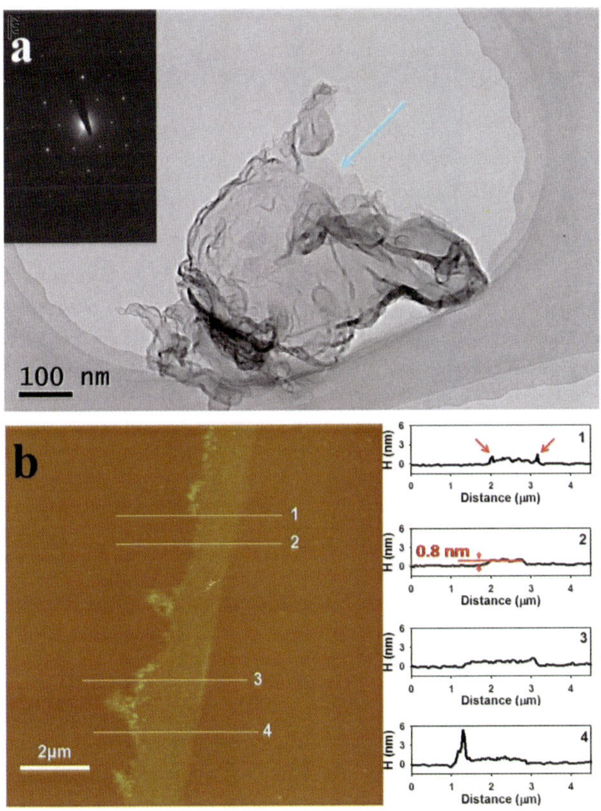

Figure 16. (a) TEM images with electron diffraction pattern (inset) of EFG and (b) AFM image with topological height profiles [20].

The resultant edge-selectively functionalized graphene (EFG) becomes dispersible without damaging the inner crystalline graphitic structure. The TEM image for EFG dispersion in NMP and dip-coated on an aperture carbon-grid, along with the corresponding selected-area electron diffraction (SAED) pattern is shown in Figure 16a. The graphene sheet is wrinkled due to its flexibility, and its surface is clean without noticeable flaws. Most of EFG consists of less than five graphene sheets. AFM images obtained from EFG on a silicon wafer clearly show EFG with approximately ~ 2 μm width and a few micron lengths (Figure 16b). Many bright spots on the edges of graphene are seen due to the covalent attachment of organic wedges. The thickness of graphene is 0.8 nm, whose value indicates single layer graphene. All topological height profiles clearly show that the interior (basal plane) are lower than the edges, implying that edge-functionalization is exclusively occurred at edges, where presumably sp^2C-H defects are located [20]. Thus the efficient exfoliation of graphite and edge-selective functionalization of graphene for improving dispersability and processabiliy have been successfully achieved by simple one-pot reaction using a direct Friedel-Crafts acylation reaction in a mild PPA/P_2O_5 medium. In addition to small molecular wedges, various macromolecular wedges using linear [51] or hyperbranched [21] polymer have also been introduced to graphene. Due to the enhanced dispersability and compatibility without structural damages, the resultant EFG has huge potentials in various applications such as polymer nanocomposites [51-52], fuel cells [22] and optoelectronic devices [19].

3. Conclusion

"Direct" Friedel-Crafts acylation reaction of electron-deficient CNTs in a mild PPA/P_2O_5 medium is a simple but less destructive functionalization method. Numerous results envision that various functional materials such as small molecules, linear and hyperbranched polymers could be covalently attached to the surface of CNTs without or with minimal damages to their carbon framework. The dispersability and compatibility of the functionalized CNTs have been greatly improve keeping their intrinsic properties, which could be regarded as a feasible approach to hybridization of CNTs and organic materials such as polymers. Furthermore, this nondestructive synthetic strategy can be expanded to other carbon-based nanomaterials such as fullerene, carbon nanofiber and graphene. Therefore, a direct Friedel-Crafts acylation reaction in a mild PPA/P_2O_5 medium possesses indeed significant potentials for the development of functional materials in various fields from all types of carbon-based nanomaterials.

Acknowledgements

This research work was supported by World Class University (WCU), US-Korea NBIT, and Basic Research (MCR) programs through the National Research Foundation (NRF) of Korea funded by the Ministry of Education, Science and Technology (MEST) and the U.S.Air Force Office of Scientific Research (AFOSR).

Author details

Dong Wook Chang[1], In-Yup Jeon[2], Hyun-Jung Choi[2] and Jong-Beom Baek[2*]

*Address all correspondence to: jbbaek@unist.ac.kr

1 Department of Chemical Systematic Engineering, Catholic University of Daegu, S. Korea

2 Interdisciplinary School of Green Energy/Low-Dimensional Carbon Materials Center, Ulsan; National University of Science and Technology (UNIST), South Korea

References

[1] Iijima, S. (1991). Helical microtubules of graphitic carbon. *Nature*, 354, 56-8.

[2] Karousis, N., Tagmatarchis, N., & Tasis, D. (2010). Current progress on the chemical modification of carbon nanotubes. *Chemcal Reviews*, 110, 5366-97.

[3] Tasis, D., Tagmatarchis, N., Bianco, A., & Prato, M. (2006). Chemistry of carbon nanotubes. *Chemical Reviews*, 106, 1105-36.

[4] Bahr, J. L., & Tour, J. M. (2002). Covalent chemistry of single-wall carbon nanotubes. *Journal of Materials Chemistry*, 12, 1952-8.

[5] Chen, R. J., Zhang, Y., Wang, D., & Dai, H. (2001). Noncovalent sidewall functionalization of single-walled carbon nanotubes for protein immobilization. *Journal of the American Chemical Society*, 123, 3838-9.

[6] Zhao, Y. L., & Stoddart, J. F. (2009). Noncovalent functionalization of single-walled carbon nanotubes. *Accounts of Chemical Research*, 42, 1161-71.

[7] Baek, J. B., Lyons, C. B., & Tan, L. S. (2004). Covalent modification of vapour-grown carbon nanofibers via direct Friedel-Crafts acylation in polyphosphoric acid. *Journal of Materials Chemistry*, 14, 2052-6.

[8] Choi, J. Y., Oh, S. J., Lee, H. J., Wang, D. H., Tan, L. S., & Baek, J. B. (2007). In-situ grafting of hyperbranched poly (ether ketone) s onto multiwalled carbon nanotubes via the A3+ B2 approach. *Macromolecules*, 40, 4474-80.

[9] Han, S. W., Oh, S. J., Tan, L. S., & Baek, J. B. (2008). One-pot purification and functionalization of single-walled carbon nanotubes in less-corrosive poly (phosphoric acid). *Carbon*, 46, 1841-9.

[10] Han, S. W., Oh, S. J., Tan, L. S., & Baek, J. B. (2009). Grafting of 4-(2, 4, 6-Trimethylphenoxy) benzoyl onto Single-Walled Carbon Nanotubes in Poly (phosphoric acid) via Amide Function. *Nanoscale Research Letters*, 4, 766-72.

[11] Jeon, I. Y., Kang, S. W., Tan, L. S., & Baek, J. B. (2010). Grafting of polyaniline onto the surface of 4-aminobenzoyl-functionalized multiwalled carbon nanotube and its electrochemical properties. *Journal of Polymer Science Part A: Polymer Chemistry*, 48, 3103-12.

[12] Jeon, I. Y., Lee, H. J., Choi, Y. S., Tan, L. S., & Baek, J. B. (2008). Semimetallic transport in nanocomposites derived from grafting of linear and hyperbranched poly (phenylene sulfide) s onto the surface of functionalized multi-walled carbon nanotubes. *Macromolecules*, 41, 7423-32.

[13] Lee, H. J., Han, S. W., Kwon, Y. D., Tan, L. S., & Baek, J. B. (2008). Functionalization of multi-walled carbon nanotubes with various 4-substituted benzoic acids in mild polyphosphoric acid/phosphorous pentoxide. *Carbon*, 46, 1850-9.

[14] Lim, D. H., Lyons, C. B., Tan, L. S., & Baek, J. B. (2008). Regioselective chemical modification of fullerene by destructive electrophilic reaction in polyphosphoric acid/ phosphorus pentoxide. *The Journal of Physical Chemistry C.*, 112, 12188-94.

[15] Baek, J. B., Lyons, C. B., & Tan, L. S. (2004). Grafting of vapor-grown carbon nanofibers via in-situ polycondensation of 3-phenoxybenzoic acid in poly (phosphoric acid). *Macromolecules*, 37, 8278-85.

[16] Wang, D. H., Baek, J. B., & Tan, L. S. (2006). Grafting of vapor-grown carbon nanofibers (VGCNF) with a hyperbranched poly (ether-ketone). *Materials Science and Engineering: B.*, 132, 103-7.

[17] Wang, D. H., Mirau, P., Li, B., Li, C. Y., Baek, J. B., & Tan, L. S. (2008). Solubilization of carbon nanofibers with a covalently attached hyperbranched poly (ether ketone). *Chemistry of Materials*, 20, 1502-15.

[18] Wang, D. H., Tan, L. S., Huang, H., Dai, L., & Osawa, E. (2008). In-Situ Nanocomposite Synthesis: Arylcarbonylation and Grafting of Primary Diamond Nanoparticles with a Poly (ether–ketone) in Polyphosphoric Acid. *Macromolecules*, 42, 114-24.

[19] Bae, S. Y., Jeon, I. Y., Yang, J., Park, N., Shin, H. S., Park, S., et al. (2011). Large-Area Graphene Films by Simple Solution Casting of Edge-Selectively Functionalized Graphite. *ACS Nano.*, 5, 4974-80.

[20] Jeon, I. Y., Bae, S. Y., Lee, H. J., Shin, H. S., Dai, L., & Baek, J. B. (2010). High-yield exfoliation of three-dimensional graphite into two-dimensional graphene-like sheets. *Chemical Communications*, 46, 6320-2.

[21] Jeon, I. Y., Choi, H. J., Bae, S. Y., Chang, D. W., & Baek, J. B. (2011). Wedging graphite into graphene and graphene-like platelets by dendritic macromolecules. *Journal of Materials Chemistry*, 21, 7820-6.

[22] Jeon, I. Y., Yu, D., Bae, S. Y., Choi, H. J., Chang, D. W., Dai, L., et al. (2011). Formation of Large-Area Nitrogen-Doped Graphene Film Prepared from Simple Solution Casting of Edge-Selectively Functionalized Graphite and Its Electrocatalytic Activity. *Chemistry of Materials*, 23, 3987-92.

[23] Siddiqui, P. C., Marom, N. A., Kim, G., & , J. K. (2010). Dispersion and functionalization of carbon nanotubes for polymer-based nanocomposites: A review. *Composites Part A: Applied Science and Manufacturing*, 41, 1345-67.

[24] Jeon, I. Y., Choi, E. K., Bae, S. Y., & Baek, J. B. (2010). Edge-Functionalization of Pyrene as a Miniature Graphene via Friedel-Crafts Acylation Reaction in Poly (Phosphoric Acid). *Nanoscale Research Letters*, 5, 1686-91.

[25] Jeon, I. Y., Tan, L. S., & Baek, J. B. (2008). Nanocomposites derived from in situ grafting of linear and hyperbranched poly (ether-ketone) s containing flexible oxyethylene spacers onto the surface of multiwalled carbon nanotubes. *Journal of Polymer Science Part A: Polymer Chemistry*, 46, 3471-81.

[26] Jeong, J. Y., Lee, H. J., Kang, S. W., Tan, L. S., & Baek, J. B. (2008). Nylon 610/functionalized multiwalled carbon nanotube composite prepared from in-situ interfacial polymerization. *Journal of Polymer Science Part A: Polymer Chemistry*, 46, 6041-50.

[27] Jang, S. S., Kumar, S., & Baek, J. B. (2010). Sponge Behaviors of Functionalized Few-Walled Carbon Nanotubes. *The Journal of Physical Chemistry C.*, 114, 14868-75.

[28] Kumar, N. A., Jeon, I. Y., Sohn, G. J., Jain, R., Kumar, S., & Baek, J. B. (2011). Highly Conducting and Flexible Few-Walled Carbon Nanotube Thin Film. *ACS Nano.*, 5, 2324-31.

[29] Hou, Y., Tang, J., Zhang, H., Qian, C., Feng, Y., & Liu, J. (2009). Functionalized few-walled carbon nanotubes for mechanical reinforcement of polymeric composites. *ACS Nano.*, 3, 1057-62.

[30] Tan, L. S., & Baek, J. B. (2005). In situ synthesis of poly (ethylene terephthalate)(PET) in ethylene glycol containing terephthalic acid and functionalized multiwalled carbon nanotubes (MWNTs) as an approach to MWNT/PET nanocomposites. *Chemistry of Materials*, 17, 5057-64.

[31] Lee, H. J. C. J., Cho, S. H., Kim, D., Tan, L. S., & Beek, J. B. (2006). Multi-walled carbon nanotubes/termoplastic polyester nanocomposites. *Polymer Preprint*, 47, 400-1.

[32] Ahn, S. N., Lee, H. J., Kim, B. J., Tan, L. S., & Baek, J. B. (2008). Epoxy/amine-functionalized short-length vapor-grown carbon nanofiber composites. *Journal of Polymer Science Part A: Polymer Chemistry*, 46, 7473-82.

[33] Lee, D. H., Baek, J. B., & Cho, G. W. (2008). Enhancement of the field-effect mobility of poly (3-hexylthiophene)/functionalized carbon nanotube hybrid transistors. *Organic Electronics*, 9, 317-22.

[34] Jeon, I. Y., Tan, L. S., & Baek, J. B. (2010). Synthesis and electrical properties of polyaniline/polyaniline grafted multiwalled carbon nanotube mixture via in situ static interfacial polymerization. *Journal of Polymer Science Part A: Polymer Chemistry*, 48, 1962-72.

[35] Choi, H. J., Jeon, I. Y., Kang, S. W., & Baek, J. B. (2011). Electrochemical activity of a polyaniline/polyaniline-grafted multiwalled carbon nanotube mixture produced by a simple suspension polymerization. *Electrochimica Acta*, 56, 10023-31.

[36] Jeon, I. Y., Choi, H. J., Tan, L. S., & Baek, J. B. (2011). Nanocomposite prepared from in situ grafting of polypyrrole to aminobenzoyl-functionalized multiwalled carbon nanotube and its electrochemical properties. *Journal of Polymer Science Part A: Polymer Chemistry*, 49, 2529-37.

[37] Tan, L. S., & Baek, J. B. (2011). Preparation and Electrocatalytic Activity of Gold Nanoparticles Immobilized on the Surface of 4-Mercaptobenzoyl-Functionalized Multiwalled Carbon Nanotubes. *The Journal of Physical Chemistry C.*, 115, 1746-51.

[38] Cai, L., Bahr, J. L., Yao, Y., & James, M. (2002). Ozonation of single-walled carbon nanotubes and their assemblies on rigid self-assembled monolayers. *Chemistry of Materials*, 14, 4235-41.

[39] Mitchell, C. A., Bahr, J. L., Arepalli, S., James, M., & Krishnamoorti, R. (2002). Dispersion of functionalized carbon nanotubes in polystyrene. *Macromolecules*, 35, 8825-30.

[40] Wang, H., & Fan, Y. (2004). Carbon-Nanotube-Reinforced Polymer-Derived Ceramic Composites. *Advanced Materials*, 16, 2036-40.

[41] Cha, S. I., Kim, K. T., Arshad, S. N., Mo, C. B., & Hong, S. H. (2005). Extraordinary Strengthening Effect of Carbon Nanotubes in Metal-Matrix Nanocomposites Processed by Molecular-Level Mixing. *Advanced Materials*, 17, 1377-81.

[42] Park, S. Y., & Baek, J. B. (2006). Multiwalled carbon nanotubes and nanofibers grafted with polyetherketones in mild and viscous polymeric acid. *Polymer*, 47, 1132-40.

[43] Choi, J. Y., Han, S. W., Huh, W. S., Tan, L. S., & Baek, J. B. (2007). In situ grafting of carboxylic acid-terminated hyperbranched poly (ether-ketone) to the surface of carbon nanotubes. *Polymer*, 48, 4034-40.

[44] Kim, Y. H. (1998). Hyperbranched polymers 10 years after. *Journal of Polymer Science Part A: Polymer Chemistry*, 36, 1685-98.

[45] Voit, B. (2000). New developments in hyperbranched polymers. *Journal of Polymer Science Part A: Polymer Chemistry*, 38, 2505-25.

[46] Dai, L., Chang, D. W., Baek, J. B., & Lu, W. (2012). Carbon Nanomaterials for Advanced Energy Conversion and Storage. *Small*, 8, 1130-66.

[47] Loh, K. P., Bao, Q., Ang, P. K., & Yang, J. (2010). The chemistry of graphene. *Journal of Materials Chemistry*, 20, 2277-89.

[48] Dreyer, D. R., Park, S., Bielawski, C. W., & Ruoff, R. S. (2009). The chemistry of graphene oxide. *Chemical Society Reviews*, 39, 228-40.

[49] Ahn, J. H., & Hong, B. H. (2009). Large-scale pattern growth of graphene films for stretchable transparent electrodes. *Nature*, 457, 706-10.

[50] Poon, S. W., Chen, W., Tok, E. S., & Wee, A. T. S. (2008). Probing epitaxial growth of graphene on silicon carbide by metal decoration. *Applied Physics Letters*, 92, 104102.

[51] Choi, E. K., Jeon, I. Y., Oh, S. J., & Baek, J. B. (2010). "Direct" grafting of linear macromolecular "wedges" to the edge of pristine graphite to prepare edge-functionalized graphene-based polymer composites. *Journal of Materials Chemistry*, 20, 10936-42.

[52] Kim, K. S., Jeon, I. Y., Ahn, S. N., Kwon, Y. D., & Baek, J. B. (2011). Edge-functionalized graphene-like platelets as a co-curing agent and a nanoscale additive to epoxy resin. *Journal of Materials Chemistry*, 21, 7337-42.

Section 4

Nanotube Devices

Chapter 13

Study of Carbon Nanotube Based Devices Using Scanning Probe Microscope

Hock Guan Ong and Junling Wang

Additional information is available at the end of the chapter

1. Introduction

Since the discovery in 1991 [1], carbon nanotube (CNT) has gained widespread attention. Many researchers have been uncovering the charaterizatics of this 1D material which possesses excellent electrical, mechanical and chemical properties. Single walled CNT has a diameter ranging from 3 Å to a few nanometer, which makes the fabrication and charaterization of CNT based devices much more difficult. There is a need for techniques that are suitable for nanometer scale charaterizations for better understanding of CNT based devices. Atomic force microscope (AFM) is powerful equipment for this purpose. In its basic mode of operation, it can reveal the morphology of CNT based devices with nanometer resolution. Moreover, various enhanced modes of operation make it possible to investigate the different properties of CNT as well as the performance of CNT based devices. In this chapter, we focus on two similiar techniques: electrostatic force microscpy (EFM) and Kelvin probe force microscpy (KPFM, alson know as scanning Kelvin probe microscopy (SKPM)). We will introduce the operation principles of these two techniques and review our recent studies on CNT using EFM. Studies conducted by other groups are also reviewed.

1.1. History of atomic force microscope

In 1972, Russell Young demonstrated surface imaging by measuring the electrical current between the sample and a scanning probe.[2] Even though the technique did not take off immediately, interest in achieving atomic resolution in surface characterization persisted in the scientific community. In 1981, Gerd Binning and Heinrich Rohr from IBM succeeded and gave birth to the first scanning tunneling microscope (STM).[3] In this system, the tunneling current between the sample and a scanning tip hovering a few angstroms above the surface is used to obtain the topography information. They later obtained image of the 7 × 7 reconstruction of silicon surface with atomic resolution in 1983.[4] In 1986, they were jointly awarded the Nobel Prize in Physics "for their design of the scanning tunneling microscope".[5]

Despite the atomic scale resolution of STM, the strict operation enviroment requirements such as high vacuum and clean surface limit its application. There is need for an easy-to-use surface characterization system. In 1986, Binning, Quate and Gerber invented the first AFM which can operate under ambient conditions.[6] The full history of STM and AFM can be found in many textbooks and reviews.[see for example, ref 7]

Atomic force microscope possesses several unique advantages over other techniques such as STM and scanning electron microscopy (SEM), including its capability to operate in different environments (vacuum, ambient and liquid), simple sample preparation and its capability to incorporate local electrical or magnetic measurements. Figure 1 (a) describes schematicaly the imaging mechanism of an AFM. For topography measurement, it can operate under contact or tapping mode. In contact mode, the tip is in direct contact with the sample surface during the scan. The position of the reflected laser spot on the photodiode changes as the tip being deflected by the surface morphology. Using this photodiode signal, the z axis piezoelectric stack will tune the height of the tip (in some AFMs, the height of the sample is changed instead) to maintain a constant deflection. The z-piezoelectric movements at different locations give rise to the topography image. In tapping mode, the tip is mechanically driven by a piezoelectric actuator to oscillate around its resonance frequency and is only tapping on the sample surface during the scan. Figure 1 (b) depicts the tapping mode operation. The oscillation of the cantilever leads to the same oscillation of the laser spot on the photodiode, which is used as the feedback signal for the z axis piezoelectric stack. When the tip scans along the surface of a sample, the z axis piezoelectric stack will move the cantilever (or the sample) up and down to maintain a constant oscillation amplitude. The movement of the z axis piezoelectric stack is used to construct the surface profile of the sample. For further discussion about the operation of an AFM, the readers are referred to ref 8.

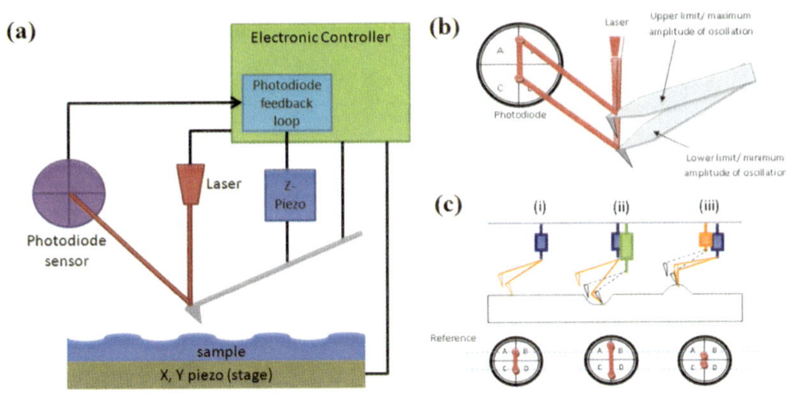

Figure 1. (a) Schematic description of an AFM. (b) Cantilever deflection and laser spot on the photodiode, (c) schematic of the tapping mode operation.

In the normal topography measurement, the AFM tip only senses the intermolecular force between the sample surface and the tip. By adjusting the scanning method, other interactions such as electrostatic and magnetostatic forces between the tip and sample can also be measured. Some of these techniques include EFM, KPFM, piezoelectric force microscopy (PFM), magnetic force microscopy (MFM), etc. In this chapter, we will forcus on EFM and KPFM and their application in the characterization of CNT based devices.

1.2. Principles of EFM and KPFM

Atomic force microscope based techniques offer unique advantages for the study of nanoelectronic devices because of their high resolution. In particular, EFM and KPFM are sensitive to local potential and space charges. Their operation principles are very similiar. They are both dual path tapping mode techniques, as shown in Figure 2 (a), and use a conductive tip for the scans. During the first scan, the system operates as per the normal tapping mode, recording the topography of the sample. During the second scan, the tip is lifted by a fixed height (usually a few tens of nanometers) from the sample surface. It then retraces the surface profile recorded in the first scan, while maintaining the lift height.

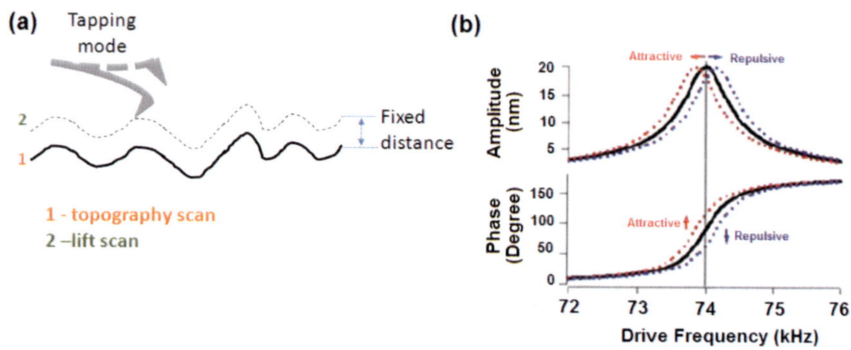

Figure 2. (a) Schematic description of the dual-pass technique. (b) Effects of attractive and repulsive forces on the cantilever oscillation amplitude and phase. [9].

For EFM, a DC bias is applied to the tip during the second scan, so that the long range electrostatic force, if any, between the sample and the tip can be detected. The electrostatic force acting on the tip will cause the resonance frequency of the cantilever to shift as indicated in Figure 2 (b). Since the cantilever is driven at its original free-standing resonance frequency, the vibration amplitude and phase will change. For example, if a positive bias is applied to the tip, it will experience a net repulsive force if positive charges exist on the surface of the sample. This will reduce the cantilever oscillation amplitude and decrease its phase shift with respect to the driving signal as shown in Figure 2 (b). On the other hand, negative charges will increase the phase shift. Thus we could tell the type of charges by looking at the phase shift of the cantilever oscillation. Mathematically, the phase changes due to the force acting on the cantilever is given by the following equation: [8]

$$\Delta\varphi = -\arcsin\left(\frac{Q}{k}\frac{dF}{dz}\right) \tag{1}$$

where k is the spring constant and Q is the quality factor of the cantilever.

Under normal condition, capacitive coupling force between the tip and sample dominates during the second scan, which is represented by

$$F(z) = \frac{1}{2}\frac{dC}{dz}V_{dc}^{2} \tag{2}$$

When net charges exist on the sample surface, electrostatic force between them and their image charges in the tip also contributes [10], so

$$F(z) = \frac{1}{2}\frac{dC}{dz}V_{dc}^{2} + \frac{q_s q_s^{'}}{4\pi\varepsilon_0 (z+z^{'})^2} + \frac{q_s C V_{dc}}{4\pi\varepsilon_0 (z+r)^2} \tag{3}$$

where r is the radius of the tip. The first term on the right hand side of the equation represents the capacitive coupling force. The second term comes from the surface charges and their image charges located at z' in the tip. The third term comes from the interaction between the surface charges and tip bias.

It has been reported that EFM can reach a resolution of ~20 nm,[11] thus is a very useful tool to study nanoelectronic devices.[12,13] Various groups have used this technique to study charge distribution,[14,15,16] defects, [17] and electrical transport. [18]

As for KPFM, it reveals the built-in potential difference between two materials when they are electrically connected. This is usually generated due to the different work functions, φ, of them. When connected, electrons will redistribute to equalize the Fermi levels and generate a built-in field across the interface. In KPFM, an external dc voltage,V_{dc}, is applied between the the tip and the sample to neutralize the built-in field. If there is no static charges involved and the reference material's work function is known, the other material's work function can be calculated as $\varphi_2 = \varphi_1 - qV_{ext}$, as shown in Figure 3.

In amplitude modulated KPFM, the setup and scan process are the same as EFM. However, during the second scan, the cantilever is not oscillated mechanically. Instead, an AC bias of frequency ω, V_{ac}, is applied to the tip, which drives the tip to oscillate at the same frequency due to capacitive coupling. This oscillation can be detected by the photodiode and is feedback to the controller. A DC bias, V_{dc}, is applied to cancel the built-in potential and the oscillation.

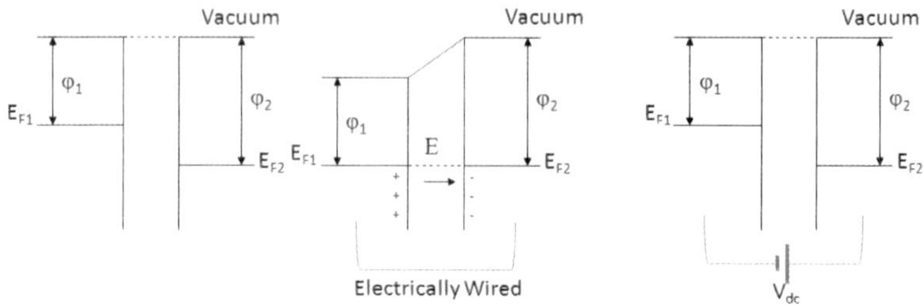

Figure 3. Build-in potential difference due to work function difference.

Approximate the tip-sample system as a parrallel plate capacitor, we can write its energy U as

$$U = \frac{1}{2} C \Delta V^2 \tag{4}$$

where C is the capacitance. The voltage between the tip and sample has three conmponents, the potential arises due to work function difference and static charges if any, the V_{dc} supplied and the oscillating V_{ac}, it can be written as:

$$\Delta V = \Delta\varphi - V_{dc} + V_{ac} \sin(\omega t) \tag{5}$$

At small oscillation amplitude, the force between the tip and sample is

$$F(z) = -\frac{\partial U}{\partial z} = -\frac{1}{2}\frac{\partial C}{\partial z} \Delta V^2 = F_{dc} + F_{\omega} + F_{2\omega} \tag{6}$$

where

$$F_{dc} = -\frac{1}{2}\frac{\partial C}{\partial z}\left[(\Delta\varphi - V_{dc})^2 + \frac{V_{ac}^2}{2}\right] \tag{7}$$

$$F_{\omega} = -\frac{\partial C}{\partial z}(\Delta\varphi - V_{dc}) V_{ac} \sin(\omega t) \tag{8}$$

$$F_{2\omega} = -\frac{1}{4}\frac{\partial C}{\partial z} V_{ac}^2 \cos(2\omega t) \tag{9}$$

When V_{dc} equals to the potential difference between the sample and tip, the oscillation at frequency ω is zero. So the V_{dc} values applied to cancel the oscillation at different locations of the sample represent the local surface potential variation across the sample.

2. Application of AFM based techniques in the study of CNT and CNT based devices

Techniques of studying electrostatic force with force microscopes have been proposed and investigated back in 1988.[19,20,21] Detection of electrostatic force of as low as 10^{-10} N has been acheived.[19] Weaver and Abraham demonstrated that using attractive-mode force potentiometry, detection of sub-millivolt signal can be acheived with spatial resolution of ~ 50 nm.[20] These early works laid the foundation of EFM. It had since improved further and reached resolution of 20 nm under ambient condition.[22] In 1991, KPFM based on volatage modulation was introduced.[23]

2.1. Seeing the CNTs more clearly

The basic capability of EFM can be demostrated in Figure 4, where both the topography and EFM signal of a SiO_2/Si substrate with CNTs on the surface are shown. Clearly, the surface roughness makes it difficult to identify the CNTs in the topography image, but they are clearly seen in the EFM image in Figure 4 (b). Furthermore, we can also identify CNTs that are connected to electrode (not shown in the figure) biased at 3 V (which are brighter) and those that are not connected. In this experiment, we have intentionally cut the CNTs by scratching the surface using a diamond cutter. Clear sharp contrast is observed at the scratch mark where the CNTs are broken. This demonstrated that EFM is an excellent tool to study CNT based nanodevices.

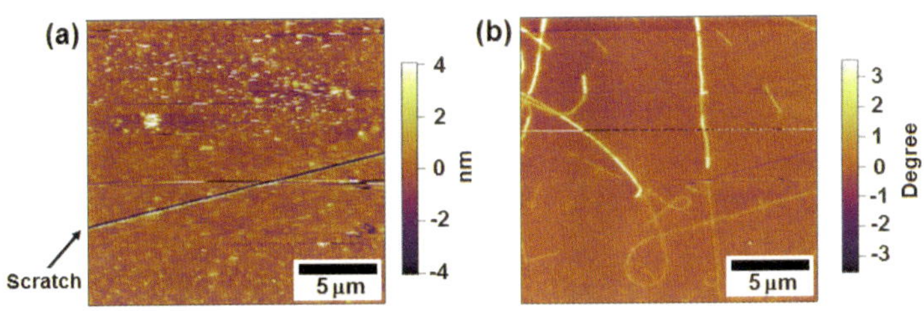

Figure 4. (a) Topography of Si substrate with CNTs on the surface. (b) EFM phase image of the same area clearly show the CNTs. The discontinuity of the CNT at the scratch mark can be observed. The electrode (not seen in the figure) is biased at 3V.

Compared with other surface characterization tools such as AFM and SEM, another advantage of EFM is that it can image CNTs that are embedded in a dielectric material noninvasively because it senses long range electrostatic force. This is especially useful for the study of CNT composite materials. For example, Jespersen *et al.* reported the mapping of individual CNT in poly-methylmethacrylate (PMMA) matrix. They have studied the EFM response vs tip-CNT distance relationship using a ~ 170 nm thick trilayer sample comprising of composite-PMMA-composite (60 nm/50 nm/60 nm) as shown in Figure 5. They used a tip bias of 7 V and lift height of 35 nm. Figure 5 (a) shows two CNTs, T_1 and T_2, locating at the top and bottom layer, respectively. Obtaining the dependence of EFM signal on tip-CNT distance allows them to estimate the the depth position of CNTs. It was reported that this subsurface characteriztion can reach a depth of 300 nm, and it is capable of 3 dimensional mapping of CNTs in the polymer matrix as shown in Figure 5 (c).[24] Subsequently, Zhao *et al.* investigated the parameters that may affect the study of polymer-CNT composite via EFM.[25] He noted that reducing humdity increases the EFM signal and improves the subsurface imaging capability. This was attributed to a reduction in the thin water layer adsorbed on the tip and sample surface, which reduces the electric field penetrating into the polymer. They also suggested that EFM subsurface imaging is useful to study high dielectric constant nanostructures in a matrix that has low dielectric constant.

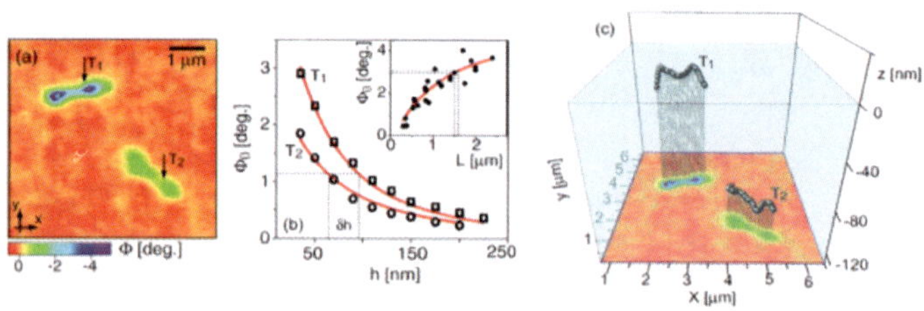

Figure 5. (a) EFM image showing two SWCNTs embedded in a ~170 nm thick film of SWCNT/PMMA composite. (b) The lift-height dependence of the length-corrected EFM signal of the two tubes in (a). The measurements were performed at the points indicated in (a). The measured amplitudes have been fitted to $\Phi(x) = A/(h+h_0)^3$. The inset shows Φ_0 vs nanotube length L for 30 isolated SWCNTs measured with known tip-tube separation h=60 nm. The red line shows a fit to the theoretical prediction $\Phi_0\ ^{-1/2}\alpha\ l^{-1}$. (c) A projection view of the three dimensional map of the two nanotubes as inferred from the data in (a) and (b). The blue region illustrates the PMMA matrix. Reprinted with permission from [24]. Copyright 2012, American Institute of Physics.

2.2. Distinguish different types of CNTs

A CNT can be either metallic or semiconducting depending on its chirality. Both types usually coexist in the as grown CNTs, which is problematic for subsequent fundamental study, device fabrication and applications. Therefore, it is very important to diferentiate the different types of CNTs without means of electrical measurements.

Lu *et al.* have used KPFM (the authors called it EFM in their paper) to seperate different types of CNTs through measuring their dielectric responses.[26] This can be acheived because metallic CNTs have a larger dielectric response than semiconducting ones. To measure the difference, a volage of $V = V_{dc} + V_{ac} \sin(\omega t)$, where $V_{ac} = 5\ V_{rms}$, is applied to the tip duing the lift scan. V_{dc} is used to nullify the contact potential difference between the tip and sample. The AC bias, V_{ac}, will create a dynamic polarization in the CNT, which interacts with the tip and gives rise to an attractive force that oscillates at the frequency of 2ω. This 2ω deflection signal is proportional to the dielectric constant of CNT, and it can be plotted against the square of the tube diameter, D_2 as shown in Figure 6. The difference between metalic and semiconducting CNTs are clearly seen in Figure 6(c).

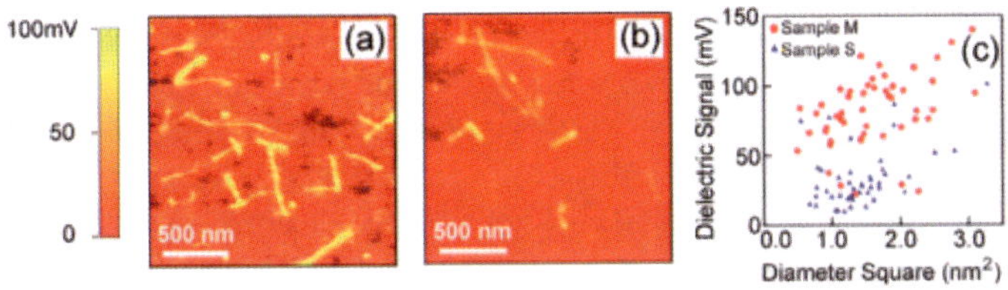

Figure 6. (a) and (b) Representative dielectric response images of semiconducting and metallic-tube-enriched samples S and M, respectively. (c) Dielectric response vs D^2 plot. Reprinted with permission from [26] Copyright 2012 American Chemical Society.

More interestingly, EFM can be used to study the dynamic tuning of CNT bandgap. It has been predicted theoretically that mechanical deformation of CNT will lead to the opening and/or closure of the bandgap.[27] And Barboza *et al.* has used EFM to study this deformation induced metal-semiconductor transition in CNT.[28]

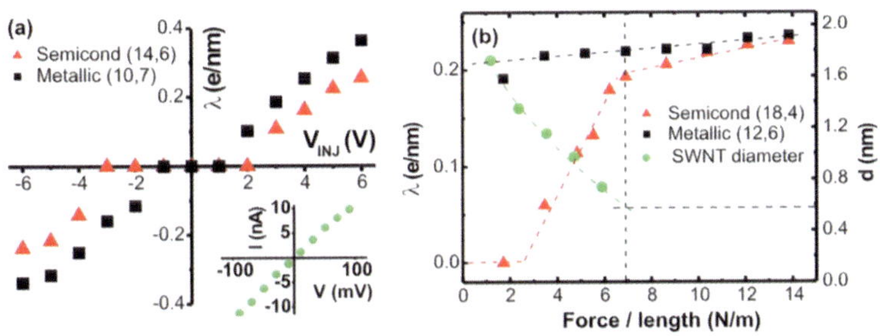

Figure 7. (a) Plot of the charge density (in electrons/nm) as a function of injection bias V_{INJ} for a (10,7) metallic nanotube (black squares) and a (14, 6) semiconducting nanotube (red triangles). The inset shows an I(V) curve acquired with the tip in contact with a thin metallic (Mo) film. (b) Plot of the charge density as function of the applied compressive force per unit length for (12,6) metallic nanotube (black squares) and (18,4) semiconducting nanotube (red triangles). The evolution of the apparent height (diameter) of the (18,4) semiconducting SWNT with applied force is also plotted in this graph (green circles). The dashed lines are guides for the eye. Reprinted figure with permission from [28]. Copyright 2012 by the American Physical Society. [1]

They first identified the nature of the CNTs using Raman spectroscopy and a pair of metallic and semiconducting CNTs with similiar diameters were chosen for the subsequent study. A compressive force is applied to the CNTs (on SiO_2) using the tip, and a bias is applied simutaneously to inject charges into the CNTs. EFM is then conducted with a tip bias of 0 V during the lift scan, this allows the authors to obtain the injected charges quantatively. From the

1 Readers may view, browse, and/or download material for temporary copying purposes only, provided these uses are for noncommercial personal purposes. Except as provided by law, this material may not be further reproduced, distributed, transmitted, modified, adapted, performed, displayed, published, or sold in whole or part, without prior written permission from the American Physical Society.

injected charge density vs bias, V_{INJ}, plot in Figure 7 (a), it is noted that metallic and semiconducting CNTs shows similiar features. But the plot is symmetric for metallic CNT with a threshold bias of ~ ±2V, while it is asymmetric for the semiconcducting CNT. Thus, there exists a bias voltage (-3 V) where metallic CNTs will show charging but semiconducting CNTs won't. Subsequently, they measured the charge density in a CNT as a function of the compressive force applied. As seen in Figure 7 (b), charge density in metallic CNT has a very weak dependance on the applied compressive force, while the semiconducting CNT shows significant changes between 2 to 7 N/m. With increasing compressive force on the semiconducting CNT, the charge density increases from zero and saturates at a similiar level as in the metallic CNT. It was concluded that compression can induce semiconducting to metallic transition in a CNT. The diameter of the CNT also decreases under compressive force until the threshold value when semiconducting-metallic transition occurs.

2.3. In-situ study of CNT based devices using EFM

2.3.1. Interface in CNT based devices.

Because of its capability to map charge and potential variations at nanometer scale, EFM is an ideal tool to study the electrical characteristics of CNT based devices. For example, Bachtold *et al.* have reported their study on the contact resistance between CNT and metal electrodes.[18] They used EFM to obtain the potential drop across the channel of a carbon nanotube field effect transistor (CNT FET) while a bias is applied to one of the electrodes as shown in Figure 8. By comparing the EFM signal drop against a known voltage, the contact resistance and the intrinsic resistance of the CNT can be obtained. It was observed that the potential drop across a multi walled CNT is uniform, indicating that it behaves as a diffusive condutor. Using this technique, they also determined that the intrinsic resistance of a two metallic CNTs bundle is 3 kΩ at the most. Combining the measured value with the four-terminal Landauer formula, they concluded that the "transport in metallic nanotubes is ballistic over a lenght of > 1 μm, even at room temperature." [18]

Besides using EFM to measure the electical resistance in CNT based devices, researchers have also used KPFM to study the band offset at metal/CNT contacts. Shiraishi *et al.* have reported that the shift of vacuum level of single walled CNT is +5.2 meV in their CNT/Au system.[29] When tetracyano-pquinodimethane (TCNQ) molecules are used as p-type dopants, the energy band of the single walled CNT will be shifted. Using KPFM, they are able to capture the corresponding shift of vacuum level from +5.2 meV to -52 meV.

Another interesting example of EFM application in the study of CNT is the charge trapping experiments conducted by Jespersen and Nygård.[30] They observed that surface static charges can be effectively trapped within CNT loops and they can be removed by touching with a ground conductive AFM tip. These static charges were suggested to cause hysteretic behaviour in CNT FETs.

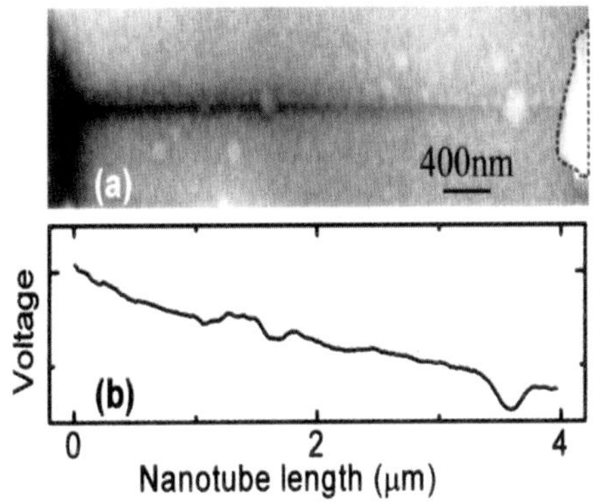

Figure 8. (a) ac-EFM image of a MWNT of diameter 9 nm. The resistance of the entire circuit is 42 kV. An ac bias of 150 mV is applied to the left electrode; the IV characteristic verified that this bias was within linear response. (b) ac-EFM signal as a function of the nanotube length. Reprinted figure with permission from [18]. Copyright 2012,the American Physical Society.[2]

2.3.2. *Origin of hysteresis in CNT FET: an in-situ EFM study*

Single walled CNT FET and prototype logic devices have been studied extensively.[31] However, hysteresis in the transfer characteristic exists in many of the CNT FETs reported so far.[32,33] It is detrimental for digital logic applications, but may be utilized in non-volatile memory devices.[34,35,36] Thus, it is important to understand the origin of the hysteresis, and to eliminate or stabilize it for different purposes.

To clarify the origin of hysteresis in CNT FET, we have conducted in-situ EFM study using the setup schematically shown in Figure 9 (a). For sample preparation and experimental details, please refer to [40]. This setup allows us to observe the charge activities around the CNT channel while the transfer charateristic of the device is being measured. We swept the gate bias, V_{gs}, from 25 V to -25 V and back. A hysteresis loop was observed as shown in Figure 9 (b). During the electrical measurements, EFM scans were performed at every 5 V intervals, with the V_{ds} and V_{gs} turned off temporarily during the scan. The EFM scans were conducted with tip bias of 3 V during its lifted scan and each image took ~ 40 s to complete. These images were displayed on the sides of the transfer loop in Figure 9 (b). In our system, the bright and dark contrasts represent negative and positive charges on the SiO_2 surface, respectively. After a gate bias of 25 V is applied, bright regions appear next to the CNT channel, indicating negative charge accumulation. It is emphasized that since the source, drain and gate were all connected to ground during the EFM scan, the contrast observed by the side of the CNT are due to residual charges on the surface of the SiO_2. Even as the gate bias was de-

2 Readers may view, browse, and/or download material for temporary copying purposes only, provided these uses are for noncommercial personal purposes. Except as provided by law, this material may not be further reproduced, distributed, transmitted, modified, adapted, performed, displayed, published, or sold in whole or part, without prior written permission from the American Physical Society.

creased to zero, the bright regions remain. As the V_{gs} polarity is reversed, dark region then first starts to appear around the CNT before extending gradually into the bright region. This suggests that the surface charges are likely injected from the CNT channel onto the SiO_2 surface. As the negative gate bias continues to increase, the bright regions disappear gradually until at V_{gs} = -25 V, where they disappears completely. The dark region, indicating positive charge accumulation, remains even when the gate bias is increased back to zero.

These injected surface charges cannot be dissipated immediately as the gate bias changes. Thus they will screen the CNT from the gate bias, causing a shift in the threshold voltage. In order to have a better understanding of the relationship between the injected charges and the hysteresis in the transfer characteristic, we performed a semi-quantitative analysis by examining the EFM phase shift distribution around CNT as shown in Figure 10 (a) and (b). Since EFM phase shift is directly related to surface charge density,[37] we can obtain qualitatively the total injected charge effect by integrating the EFM phase shift over r, the distance away from the CNT.[38] A clear hysteresis is observed (Figure 10 (c)) when the integrated value of Figure 10 (a) and (b) are plotted against gate bias, consistent with the transfer characteristic. This is expected since the screening effect should be proportional to the total amount of injected charges. Interestingly, it is observed that the amount of positive charges injected onto the surface under negative gate bias is significantly less than the negative charges under positive gate bias. This is consistent with the shift of the hysteresis loop towards positive bias side (Figure 9 (b)) and indicates that the SiO_2 surface can trap electrons more effectively than holes, consistent with other report.[39]

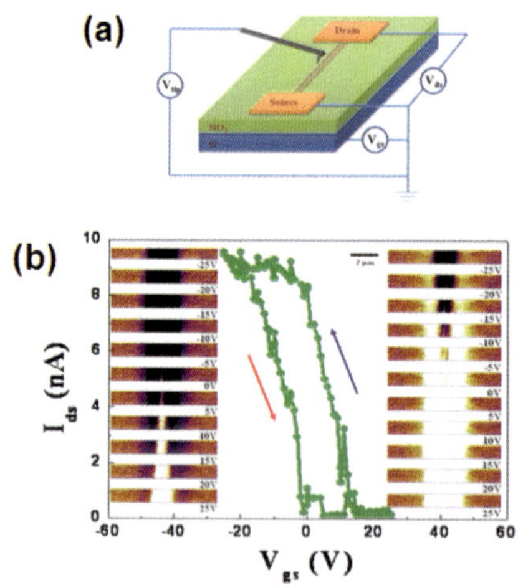

Figure 9. (a) Experiment setup for the in-situ EFM study. (b) Transfer characteristic and charge injection of a CNT FET. Transfer curve obtained by sweeping the gate bias from 25 V to – 25 V and back. The EFM images are taken at 5 V intervals. The gate bias is temporally turned off during EFM scan. Injected charges are observed around the CNT and are correlated to the hysteresis loop. Reprinted with permission from [40] Copyright 2012 IOP Publishing.

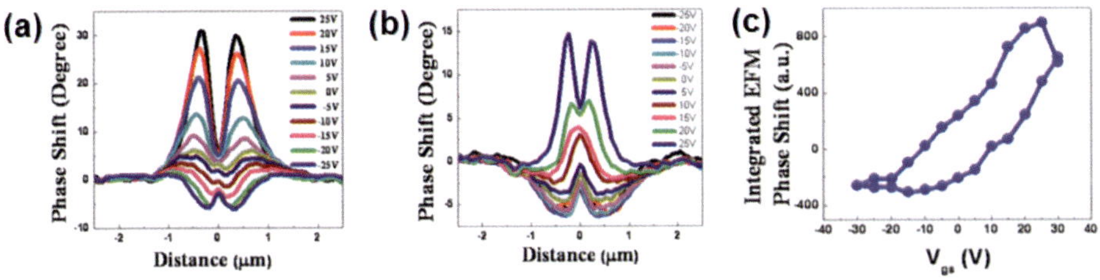

Figure 10. The cross section profiles of the EFM images, indicating the relative charge density around the CNT for (a) backward gate sweeping from 25 V to -25 V, (b) forward gate sweeping from -25 V to 25 V. (c) Integration of the phase shift, over the scan range, where r is the distance away from the CNT channel. Reprinted with permission from [40] Copyright 2012, IOP Publishing.

If the injected charges are indeed responsible for the formation of the hysteresis loop, a hysteresis-free transfer curve will be obtained if a long enough dissipation time is given before the measurement of the I_{ds}. To confirm this prediction, we repeated the same experiment with different dissipation time (Figure 11). During the measurement, the gate bias was turned off for a duration of 0 (black curve), 2 (red curve) and 15 (blue curve) minutes at every voltage step, after which it was turned back on and I_{ds} was measured immediately. It is apparent from Figure 11 (a) that longer discharging time leads to a decrease in the hysteresis width. With a discharging duration of 15 minutes, an almost hysteresis-free transfer curve was obtained. Figure 11(b) depicts the corresponding EFM images at the different discharging duration. It is obvious that the injected charges have almost disappeared completely after 15 minutes, consistent with the macroscopic transfer behavior. The remaining narrow hysteresis observed in the transfer curve is due to the little amount of residual charges (Figure 11 (c), decreased scale).

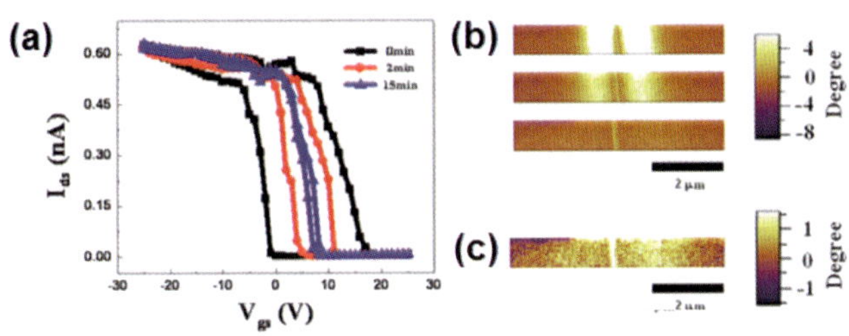

Figure 11. Hysteresis-free transfer curve under ambient condition. (a) Hysteresis width can be reduced by allowing the injected charges to dissipate at every gate bias. The experiment was conducted by turning the gate bias off at every step for 0 (black), 2 (red) and 15 (blue) minutes, and measure the Ids immediately after turning the gate bias back on. (b) The corresponding EFM images after 0 (top), 2 (middle) and 15 (bottom) minutes discharging time. (c) The bottom image in (b) replotted with reduced scale. A small amount of injected charges is observed. Reprinted with permission from [40] Copyright 2012, IOP Publishing.

Based on the above observations, a dynamic screening effect due to the injected charges can be envisioned, which is schematically described in Figure 12. When a gate bias of 25 V is applied, a layer of negative charges (purple layer) is formed on the SiO_2 surface around the

CNT (Figure 12 (a)). The CNT channel will experience a net potential from both the gate bias and the injected charges. When the gate bias is decreased, the injected charges cannot dissipate immediately, which means that the potential acting on the CNT channel from the injected charges deceases more slowly than that from the gate potential. When the gate bias decreases to a certain value, the effect from the injected charges will overwhelm that of the polarization induced by the gate bias and turns the transistor on (Figure 12 (b)). This will shift the transfer curve to the right. When $V_{gs} = 0$ V, the injected charges still exist (Figure 12 (c)) and the transistor remains on. As the gate bias decreases to the negative region, holes are injected onto the SiO_2 surface as shown in Figure 12 (d), compensating the trapped electrons. For a period of time, both holes and electrons will coexist on the SiO_2 surface side by side, as shown in Figure 12. With further decrease of the gate bias, the injected charges will be fully inverted to holes (Figure 12 (e)). This process reverts as the gate bias sweeps back from -25 V to 25 V (Figure 12 (e)-(h)).

One important conclusion that we can draw from this study is that CNT FET cannot be used for memory applications due to two reasons. First, charge dissipation leads to data retention problem. Second, the μm size of the charged area limits data storage density and generates cross talking among cells.

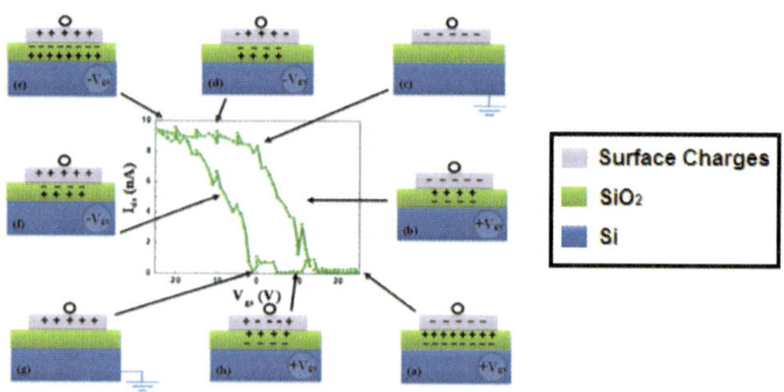

Figure 12. The dynamic screening effect of the injected charges. (a-d) Upon application of a positive gate bias, electrons are injected onto the SiO_2 surface around the CNT channel. When the gate bias is decreased, polarization charge decreases but injected charges remain. Once the gate bias turns negative, holes are injected onto the SiO_2 surface. (e-h) Opposite process takes place when the gate bias is swept from $-V_{max}$ to $+V_{max}$. Reprinted with permission from [40] Copyright 2012, IOP Publishing.

In conclusion, the in-situ EFM study helped us to establish a clear correlation between the charge injection around the CNT and the hysteresis behavior of the transistor. This technique can also be used to study other nanoelectronic devices.

2.3.3. *Surface chemistry and hysteresis in CNT FET*

We have demonstrated that injected charges around the CNT channel causes the hysteresis in the transfer characteristic of CNT FET. It is natural to ask, what are the charge traps? To answer this question, we have studied the discharging dynamics on SiO_2 surface at different temperatures.

We started with a charging process using the setup shown in Figure 13 (a) where a bias is applied to the electrode as descripted in [44]. When the bias is turned on, we can observe (Figure 13 (b)) that both the EFM phase shift and the width of the charged area increase with charging time. Since a positive tip bias is used during the lift scan, bright contrast in this context represent negative charges, and dark means positive charges.

After the charging process, the electrode is ground and EFM scan is performed on the same area at different time intervals. Figure 13 (c) shows the cross-sectional profile of the recorded EFM data with the CNT at the origin. It is observed that charges next to the CNT diffuse back to the channel immediately after the bias is turned off. This dissipation of charges results in the formation of a peak in the profile which represents the maximum surface charge density. This peak gradually moves away from the CNT as more charges diffuse back to the channel. Using Matlab 7.5, we have fitted the discharging curves at temperatures from 110 °C to 180 °C, and extracted the charge diffusion coefficients at different temperatures.[41] A transition point at 150 °C is observed as shown in Figure 13 (e). Charge dissipation at temperatures below this point experiences a barrier (trap depth) of ~ 0.46 eV, which changes to 0.91 eV at higher temperatures. According to Zhuravlev, SiO_2 surface usually terminates with silonal groups and can absorb water molecules.[42] As temperature increases, these water molecules will be released above a boundary temperature, which ranges from 120 °C to 190 °C depending the structure of silica. (Interested readers are referred to refs [42, 43].) This suggests that the change in the charge diffusion barrier is likely the result of water evaporating from the SiO_2 surface. This is consistent with the claim that water layer acts as the charge traps on SiO_2 surface at room temperature. [33]

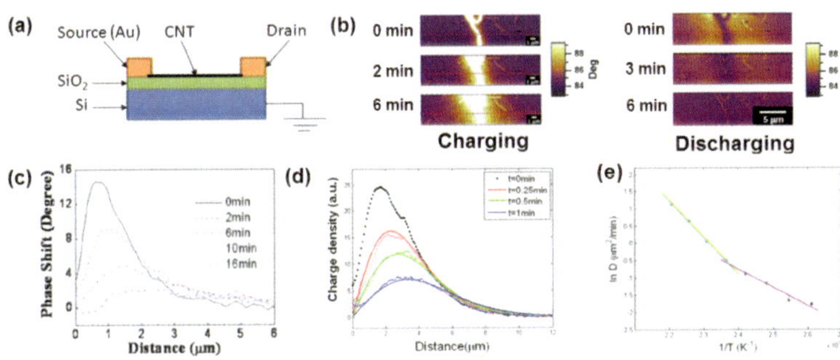

Figure 13. (a) Schematic diagram of experiment setup. (b) Images obtained during the charging and discharging processes. (c) Discharging curves obtained at room temperature. (d) Discharging curves at 180 °C (dotted) and fitting result (line). (e) Temperature dependence of diffusion coefficient reveals the activation energy change at 150 °C. Reprinted with permission from [44], Copyright 2012, American Institute of Physics. Reprinted with permission from [41], Copyright 2012, American Chemical Society.

2.3.4. *Eliminating the hysteresis*

It is clear that water layer on SiO_2 surface act as charge traps at room temperature. Thus it would be expected that hysteresis in the transfer characteristic can be eliminated simply by increasing the measurement temperature. We conducted transfer measurements at different temperatures ranging from 20 °C to 180 °C and the results are shown in Figure 14 (a, b). As temperature increases, the hysteresis width continues to decrease, though the scattering of data prevents us from obtaining accurately the activation energy of the evolution and critical temperature if any. It is interesting to note that the hysteresis disappears under negative gate bias at 180 °C and above, but remains under positive gate bias. Figure 14 (c) shows this transition more clearly from 140 °C to 180 °C. At 140 °C, a normal hysteresis (black) loop can still be observed. With an increase of 30 °C in temperature, we can observe that there is a dip (red curve) at around V_{gs} = 5 V. With further increase of temperature, there is a clear shrinkage of the hysteresis loop on the negative bias side. At 180 °C, the loop (green curve) has almost disappeared in the negative V_{gs} region. On the contrary, the loop expanded in the positive V_{gs} region. This clearly indicates that there are two types of charge traps available on the SiO_2 surface. It is likely that evaporation of water at high temperatures results in the SiO_2 surface being dominated by electron trapping defects. If this is true, at 180 °C, where there is little or no hysteresis loop in the negative V_{gs} region, there should not be any injected charges. We conducted EFM imaging at V_{gs} = -10 V at 20 °C and 180 °C as shown in insert of Figure 14 (a). At 20 °C, the dark contrast, indicating positive charges, could be observed and related to the formation of the hysteresis loop (black) in Figure 14 (a). At 180 °C, the absence of dark contrast around the CNT indicates negligible charge injection as compared with that taken at room temperature (inset of Figure 14 (a)). Upon cooling back to room temperature (20 °C), the hysteresis loop can be observed again, as shown in Figure 14 (d). Once the device reaches room temperature (0 min), main part of the hysteresis recovers immediately. However, even after 59 hours under ambient condition, the hysteresis width is still less than the original value. This observation suggests that the first few layers of water is absorbed back onto the SiO_2 surface upon cooling to room temperature, but full recover of the original surface condition takes much longer time.

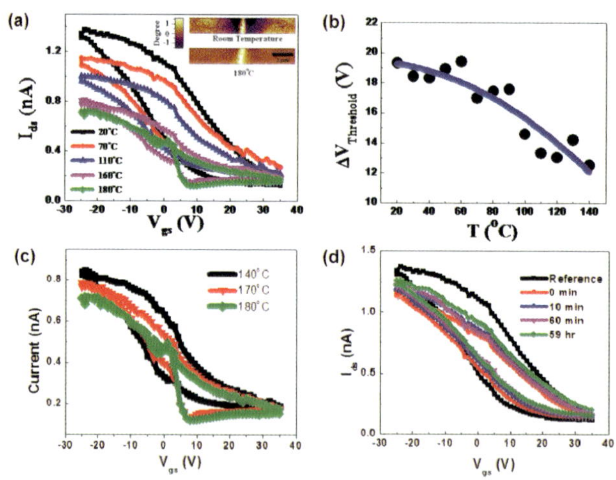

Figure 14. Temperature dependent hysteresis loops. (a) The hysteresis width decreases as temperature increases. At 180 °C, the hysteresis completely disappears under negative gate bias, but remains under positive bias, indicating the lack of hole traps on the SiO_2 surface. (b) The hysteresis width measured by threshold voltage shift at different temperatures. (c) hysteresis loops taken around the transistion temperature depicting the transformation in (a). (d) hysteresis loop time duration study upon cooled down to room temperature. [45]

Our study has demonstrated the effect of SiO_2 surface chemistry on the hysteresis behavior of the CNT FET. It is thus expected that by modifying the SiO_2 surface, we may be able to reduce or even eliminate the charge trapping and thus the hysteresis. In order to confirm this, we have prepared self assembled monolayer (SAM) of Octadecyltrichlorosilane (OTS) as a passivation layer on SiO_2. This was done by dipping the sample with CNTs into the OTS solution for a period of 72 hrs, as it is not possible for CNT to be grown on OTS treated substrates due to the high growth temperature. The sample was then tested by applying a -5 V bias to the CNT through the drain electrode while keeping the gate grounded. Figure 15 shows the results obtained.

The topography of the device has changed as seen in Figure 15 (a). The rougher surface suggests that OTS has formed on the sample surface. Figure 15 (b) shows the EFM charging images around the CNT channel before and after the OTS treatment. It is noted that after OTS treatment, the width of the charged area and the amount of surface charges decreased significantly. Unfortunately, OTS treatment is corrosive in nature and the CNT FET is easily broken after the treatment, preventing us from conducting transfer measurement. Furthermore, the increased surface roughness complicates the study. Other surface treatment techniques should be explored.

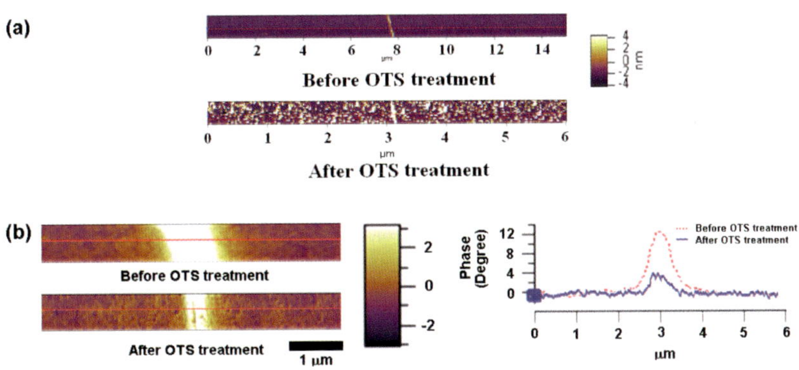

Figure 15. Surface modification with OTS (a) Topography images of sample before and after OTS treatment for 72 hrs. (b) Corresponding EFM images of (a) after charging with -5 V bias. (c) Cross section profiles of EFM images clearly shows the decrease in injected charges.

In summary, we have shown that charge injection around the CNT channel under gate bias causes the hysteresis in the transfer characteristic of the SiO_2-gated CNT FET. We suggest that CNT FET cannot be used for non-volatile memory applications because of data retention and device density issues. As for application in logic devices, surface modification may help to eliminate the hysteresis behavior.

3. Future perpectives

The unique advantages of AFM-based techniques make them ideal for nanoelectronic devices characterizations. Though much has been accomplished, there are still many areas where AFM-based techniques can be of great importance.

One area of interest where AFM-based techniques can be used is the study of sensing devices. In CNT or other nanomaterials-based sensing devices, surface interaction is important for the device functionality. As we have established in the earlier discussion, AFM-based techniques can be used for in-situ imaging with nanometer resolution. It allows users to correlate the charge activites and device performance, which will lead to a better understanding of the device operation mechnaism.

Study on graphene has exploided recently. Most of the works done on CNT that we discussed earlier can be transfered to graphene and graphene based devices. For example, the dynamics performance of graphene-based FET, the tunning of graphene band structure by external fields etc. Furthermore, graphene is structually malleable and its electronic, optical properties are strongly affected by strain. AFM-based techniques can be of great importance in this area of study. Effect of environment and surface chemistry on the properties of graphene can also be investigated using AFM in an variable environment hood. As modern technology continues to evolve into the nano era, AFM-based techniques will certainly become more important in the studies of future nanoelectronic devices.

Acknowledgements

The authors will like to acknowledge the support from Nanyang Technological University, Ministry of Education of Singapore under project number AcRF RG30/06 and National Research Foundation of Singapore under project number NRF-CRP5-2009-04. The authors will also like to thank Dr Li Bing for his help in the OTS sample preparation.

Author details

Hock Guan Ong and Junling Wang*

*Address all correspondence to: jlwang@ntu.edu.sg

School of Materials Science and Engineering
Nanyang Technological University, Singapore

References

[1] Iijima, S. (1991). Helical microtubules of graphitic carbon. *Nature*, 354, 56-58.

[2] Young, R., Ward, J., & Scire, F. (1972). The Topografiner: An Instrument for Measuring Surface Microtopography. *The Review of Scientific Instruments*, 43(7), 999-1011.

[3] Binning, G., Rohrer, H., Gerber, Ch., & Weibel, E. (1981). Tunneling through a controllable vacuum gap. *Applied Physics Letters*, 40, 178-180.

[4] Binning, G., Rohrer, H., Gerber, Ch., & Weibel, E. (1983). 7X7 Reconstruction on Si(111) Resolved in Real space. *Physical Review Letters*, 50(2), 120-123.

[5] Nobelprize.org. (1986). The official Web Site of the Nobel Prize. *The Nobel Prize in Physics Ernst Ruska, Gerd Binnig, Heinrich Rohrer*, http://www.nobelprize.org/nobel_prizes/physics/laureates/1986/, accessed 16 May 2012.

[6] Binning, G., & Quate, C. F. (1986). Atomic Force Microscope. *Physical Review Letters*, 56(6), 930-933.

[7] Mody, C. (2011). Instrumental Community: Probe Microscopy and the Path to Nanotechnology. *MIT Press*.

[8] Sarid, D. (1994). Scanning Force Microscopy. *New York : Oxford University Press*.

[9] Asylum Research Manual v04 _08. (2008). https://support.asylumresearch.com/forum/content.php?4-Software, (accessed Nov.2008).

[10] Kim, J. H., Noh, H., Khim, Z. G., Jeon, K. S., Park, Y. J., Yoo, H., Choi, E., & Om, J. (2008). Electrostatic force microscopy study about the hole trap in thin nitride/oxide/semiconductor structure. *Applied Physics Letters*, 92, 132901.

[11] Lei, C. H., Das, A., Elliott, M., & Macdonald, J. E. (2003). Conductivity of macromolecular networks measured by an electrostatic force microscopy. *Applied Physics Letters*, 83, 482-484.

[12] Mélin, T., Diesinger, H., Deresmes, D., & Stiévenard, D. (2004). Probing Nanoscale Dipole-Dipole Interactions by Electric Force Microscopy. *Physical Review Letters*, 92, 166101.

[13] Lei, C. H., Das, A., Elliott, M., & Macdonald, J. E. (2004). Quantitative electrostatic force microscopy-phase measurements. *Nanotechnology*, 15, 627-634.

[14] Paillet, M., Poncharal, P., & Zahab, A. (2005). Electrostatics of Individual Single-Walled Carbon nanotubes Investigated by Electrostatic Force Microscopy. *Physical Review Letters*, 94, 186801.

[15] Zdrojek, M., Mélin, T., Diesinger, H., Stiévenard, D., Gebicki, W., & Adamowicz, L. (2006). Charging and discharging process of carbon nanotubes probed by electrostatic force microscopy. *Journal of Applied Physics*, 100, 114326.

[16] Domanský, K., Leng, Y., Williams, C. C., Janata, J., & Petelenz, D. (1993). Mapping of mobile charges on insulator surfaces with the electrostatic force microscope. *Applied Physics Letters.*

[17] Jespersen, T. S., & Nygård, J. (2007). Probing induced defects in individual carbon nanotubes using electrostatic force microscopy. *Applied Physics A*, 88, 309-313.

[18] Bactold, A., Fuhrer, M. S., Plyasunov, S., Forero, M., Anderson, E. H., Zettl, A., & Mc Euen, P. L. (2000). Scanned Probe Microscopy of Electronic Transport in Carbon Nanotubes. *Physical Review Letters*, 84, 6082-6085.

[19] Martin, Y., Abraham, D. W., & Wickramasinghe, H. K. (1998). High-resolution capacitance measurement and potentiometry by force microscopy. *Applied Physics Letter*, 52(13), 1103-1105.

[20] Weaver, J. M. R., & Abraham, D. W. (1991). High resolution atomic force microscopy potentiometry. *Journal of Vacuum Science Technology B*, 9(3), 1559-1561.

[21] Schönenberger, C., & Alvarado, S. F. (1990). Observation of Single Charge Carriers by Force Microscopy. *Physical Review Letters*, 65(25), 3162-3164.

[22] Lei, C. H., Das, A., Elliott, M., & Macdonald, J. E. (2003). Conductivity of macromolecular networks measured by an electrostatic force microscopy. *Applied Physics Letters*, 83, 482-484.

[23] Nonnenmacher, M., O'Boyle, M. P., & Wickramasinghe, H. K. (1991). Kelvin Probe Force Microscopy. *Applied Physics Letters*, 58(25), 2921-2923.

[24] Jespersen, T. S. (2007). Nygård Mapping of individual carbon nanotubes in polymer/nanotube composites using electrostatic force microscopy. *Applied Physics Letter*, 90, 183108.

[25] Zhao, M., Gu, Z., Lowther, S. E., Park, C., Jean, Y. C., & Nguyen, T. (2010). Subsurface characterization of carbon nanotubes in polymer composites via quantitative electric force microscopy. *Nanotechnology*, 21, 225702.

[26] Lu, W., Xiong, Y., Hassanien, A., Zhao, W., Zheng, M., & Chen, L. (2009). A Scanning Probe Microscopy Based Assay for Single-Walled Carbon Nanotube Metallicity. *Nano Letters*, 9(4), 1668-1672.

[27] Park, C-J., Kim, Y-H., & Chang, K. J. (1999). Band-gap modification by radial deformation in carbon nanotubes. *Physical Review B*, 60(15), 10656-10659.

[28] Barboza, A. P. M., Gomes, A. P., Archanjo, B. S., Araujo, P. T., Jorio, A., Ferlauto, A. S., Mazzoni, M. S. C., Chacham, H., & Neves, B. R. (2008). A Deformation Induced Semiconductor-Metal Transition in Single Wall Carbon Nanotubes Probed by Electric Force Microscopy. *Physical Review Letters*, 100, 256804.

[29] Shiraishi, M., Takebe, K., Matsuoka, K., Saito, K., Toda, N., & Kataura, H. (2007). Surface potential analyses of single-walled carbon nanotube/metal interfaces. *Journal of Applied Physics*, 101, 014311.

[30] Jespersen, T.S., & Nygård, J. (2005). Charge Trapping in Carbon Nanotube Loops Demostrated by Electrostatic Force Microscopy. *Nano Letters*, 5(9), 1838-1841.

[31] Tans, S. J., Verschueren, A. R. M., & Dekker, C. (1998). Room-temperature transistor based on a single carbon nanotube. *Nature*, 393, 49-52.

[32] Wang, C., Zhang, J., Ryu, K., Badmaev, A., Arco, L. G. D., & Zhou, C. (2009). Wafer-scale fabrication of separated carbon nanotube thin-film transistors for display applications. *Nano Letters*, 9(12), 4285-4291.

[33] Kim, W., Javey, A., Vermesh, O., Wang, Q., Li, Y., & Dai, H. (2003). Hysteresis Caused by Water Molecules in Carbon Nanotube Field-Effect Transistors. *Nano Letters*, 3(2), 193-198.

[34] Radosavljevi, M., Freitag, M., Thadani, K.V, & Johnson, A.T. (2002). Nonvolatile Molecular Memory Elements Based on Ambipolar Nanotube Field Effect Transistors. *Nano Lett.*, 2(7), 761-764.

[35] Fuhrer, M. S., Kim, B. M., Dürkop, T., & Brintlinger, T. (2002). High-Mobility Nanotube TransistorMemory. *Nano Letters*, 2(7), 755-759.

[36] Cui, J. B., Sordan, R., Burghard, M., & Kern, K. (2002). Carbon nanotube memory devices of high charge storage stability. *Applied Physics Letters*, 81, 3260-3262.

[37] Kim, J. H., Noh, H., Khim, Z. G., Jeon, K. S., Park, Y. J., Yoo, H., Choi, E., & Om, J. (2008). Electrostatic force microscopy study about the hole trap in thin nitride/oxide/ semiconductor structure. *Applied Physics Letters*, 92, 132901.

[38] Chazalviel, J-N. (1999). Coulomb Screening by Mobile Charges: Applications to Material Science, Chemistry, and Boilogy. *Birkhäuser: Boston.*

[39] Chua, L-L., Zaumseil, J., Chang, J-F., Ou, E. C.-W., Ho, P. K.-H., Sirringhaus, H., & Friend, R. H. (2005). General observation of n-type field effect behavior in organic semiconductors. *Nature*, 434, 194-199.

[40] Ong, H. G., Cheah, J. W., Zou, X., Li, B., Cao, X. H., Tantang, H., Li, L-J., Zhang, H., Han, G. C., & Wang, J. (2011). Origin of hysteresis in the transfer characteristic of carbon nanotube field effect transistor. *Journal of Physics D:Applied Physics*, 44, 285301.

[41] He, Y., Ong, H. G., Zhao, Y., He, S., Li, L-J., & Wang, J. (2009). Study of Charge Diffusion at the Carbon Nanotube-SiO_2 Interface by Electrostatic Force Microscopy. *Journal of Physical Chemistry C*, 113, 15476-15479.

[42] Zhuravlev, L.T. (2000). The surface chemistry of amorphous silica. *Zhuravlev model. Colloids and Surfaces A*, 173, 1-38.

[43] Iler, R. K., editor. (1979). The Chemistry of Silica Solubility, Polymerization, Colloid and Surface Properties, and Biochemistry. New York : Wiley.

[44] Ong, H. G., Cheah, J. W., Chen, L., Tangtang, H., Xu, Y., Li, B., Zhang, H., Li, L-J., & Wang, J. (2008). Charge injection at carbonnanotube-SiO_2 interface. *Applied Physics Letters*, 93, 093509.

[45] Ong, H.G. (2012). Study of Carbon Nanotube Field Effect Transistor Using Electrostatic Force Microscopy. *PhD thesis. Nanyang Technological University.*

Chapter 14

Control of Single-Hole Transition in Carbon Nanotube Transistor with Quantum Dot in Gate Insulator at Room Temperature

Takafumi Kamimura, Yutaka Hayashi and
Kazuhiko Matsumoto

Additional information is available at the end of the chapter

1. Introduction

Flash memory has been leading player in the nonvolatile memories in recent ten years because of a small cell size and a high scalability. A gate length of a flash memory has reached to 20 nm. A number of charges also have been getting smaller. a several hundred in the 20 nm device. However, a flash memory needs high operating voltage for short write time. The high operating voltage causes low durability [1, 2], and a signal due to capacitance coupling between neighbor cells becomes larger with further miniaturization [1, 2]. Therefore, lower operating voltage and stronger capacitive coupling between a gate electrode and a charge storage compare to neighbor cells are desired.

A single walled carbon nanotube (SWNT) is a cylindrical structure with a diameter of about 1 nm [3-8]. A SWNT with a semiconductor property is very sensitive to the charge around the SWNT, because whole SWNT channel can be easily modulated by the arounded charge because of the small diameter of a SWNT. This high sensitivity can sense even a single-charge [9-14]. Therefore, many sensor applications of SWNT have been reported, e.g. bio- and gas-sensors [15-31]. Moreover, a cylindrical type memory can achieve higher electric field concentration compare to a planer type memory because of higher capacitive coupling between charge storage and a gate electrode. Therefore, the signal noise attributed to the parasitic capacitances between a neighbor cells can be reduced by this higher capacitive coupling.

In this study, we succeeded in fabricating a multifunctional quantum transistor using the particle nature and wave nature of holes in SWNT. This transistor can operate as anresonant tunneling transistor (RTT) and also as an single-hole transistor (SHT). An RTT is a device that uses the wave nature of hole and an SHT uses the particle nature of hole in the SWNT. Both devices need tunneling barriers at both sides of the quantum island. The RTT needs strong coupling while the SHT needs weak coupling between the quantum island and the electrodes. Usually, these tunneling barriers are made from thin oxide layers, etc. Therefore, the thickness of the tunneling barriers and the coupling strength cannot normally be controlled in a given device. In the present device, however, the Schottky barriers act as the tunneling barriers between the SWNT quantum island and electrodes. Therefore, the thickness of the tunneling barriers and the coupling strength between the SWNT and electrodes can be controlled by the applied gate voltage V_G.

SWNT electron devices show hysteresis characteristics in gate voltage-drain current characteristics. The hysteresis characteristics are caused by gate-voltage-dependent charge fluctuation, e.g., adsorption of water molecules around a SWNT [32], charging into insulator layer around a SWNT [33], and charging into amorphous carbon around a SWNT [34] By eliminating these origins of the hysteresis characteristics, the number of fluctuating charges becomes small and a single-charge fluctuating around the SWNT channel can be distinguished by a SWNT multi-functional quantum transistor.

Moreover, a SWNT transistor surrounded by SiN_x /Al_2O_3 double gate insulator layers with quantum dot in the insulator layers was fabricated, demonstrating discrete threshold voltage shift resulting in discrete drain current modulation at room temperature.

2. Detection of single-charge around SWNT channel

2.1. Method

We have eliminated the three origins of the hysteresis characteristics of a SWNT field effect transistor mainly pointed out in current reports [32-35]. To burn out amorphous carbon, we annealed a SWNT at low temperature in oxidizable atmosphere [33]. To reduce the number of adsorbed atmosphere molecules, we covered the channel with a silicon dioxide layer. To reduce the number of trap sites in the insulator, we reduced channel length to 73 nm. The SWNT multi-functional quantum transistor fabricated by the process mentioned above shows almost no hysteresis characteristics in the gate voltage rage from -40 to 40 V. Moreover, an abrupt discrete switching of the source-drain current is observed in the electrical measurements of the SWNT multi-functional quantum transistor at 7.3 K. These random telegraph signals (RTS) are attributed to charge fluctuating charge traps near the SWNT multi-functional quantum transistor conduction channel. The current-switching behavior associated with the occupation of individual electron traps is demonstrated and analyzed statistically.

2.2. Sample preparations

A schematic of the sample structure is shown in Fig. 1. SWNT was prepared as follows. An n^+-Si wafer with a thermally grown 300 nm thick oxide was used as a substrate. Layered Fe/Mo/Si (2 nm/20 nm/40 nm) catalysts were evaporated using an electron-beam evaporator under a vacuum of 10^{-6} Pa. These layered catalysts were patterned on the substrate using the conventional photo-lithography process. SWNT was grown by thermal chemical vapor deposition (CVD) using the mixed gases of hydrogen and argon-bubbled ethanol. After the growth of the SWNT, it was purified by burning out the amorphous carbon around the SWNT in an air atmosphere at a temperature of several hundred degrees Celsius [31]. Ti (30 nm) electrodes were deposited on the patterned catalysts as the source and drain, and on the back side of the n^+-Si substrate for the gate, using the electron-beam evaporator under a vacuum of 10^{-6} Pa. The distance (*L*) between the source and drain was 73 nm. Thus, a back gate type multi-functional quantum transistor with an SWNT channel was fabricated that had the functions of an RTT and an SHT. The single-charge measurement was carried out with the structure that silicon dioxide layer is on the SWNT channel.

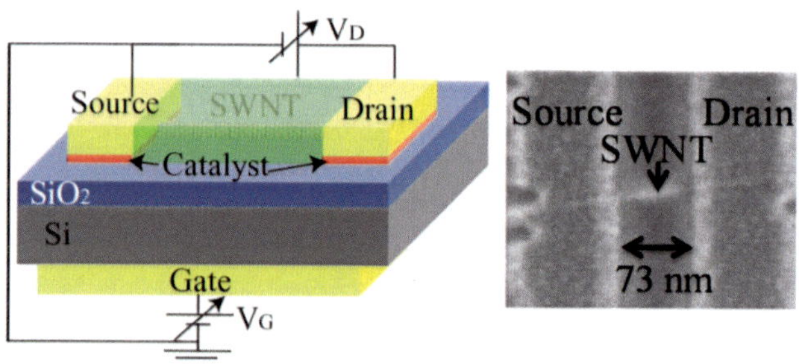

Figure 1. Schematic structure of SWNT multi-functional quantum transistor covered by silicon dioxide layer. The channel length is 73 nm. The inset shows a SEM image around the channel before silicon dioxide deposition. A few charge storages are fabricated in the SiO_2 layer.

2.3. Results and discussions

2.3.1. Single charge sensitivity of SWNT multi-functional quantum transistor

Figure 2 (a) shows the time dependence of the conductance of the SWNT multi-functional quantum transistor at 7.3 K with a gate bias of V_G=-25.36 V. A short sampling time of 10 ms was set in the dynamic characteristic measurements shown in Fig. 10.

The applied gate voltage was under the Fabry-Perot interference region. The SWNT multi-functional quantum transistor shows RTS, as shown in Fig. 2(a), The RTS showed three levels, n, n+1 and n+2, of the conductance shown in Fig. 2(a). At a lower applied gate voltage, current levels higher than n+2 such as n+3 and n+4 appeared. The multiple levels of RTS are attributed to charge fluctuating charge storages near the conduction channels of the SWNT multi-functional quantum transistor. Moreover, because there was a single-charge storage including multiple energy levels or were some charge storages being at almost the same distances from the conductance channel of the SWNT multi-functional quantum transistor, the RTS appeared. Figures 2(b) and (c) show histograms of the conductance levels of RTS at V_G= -25.36 V and V_G= -25.39 V, respectively. The three peaks of conductance of RTS expressed as

P_n, P_{n+1}, and P_{n+2} are shown in Figs. 2(b) and 2(c). The conductance levels of the three peaks directly correspond to the conductance levels of RTS. On the other hand, the relative heights of the peaks correspond to occupation probabilities at each conductance level of the RTS. The heights of the peaks depend on applied gate voltage. P_n decreases and P_{n+1} and P_{n+2} increase with slightly increasing applied gate voltage from V_G= -25.36 to -25.39 V, which means that the energy levels in the charge storage are modulated by applied gate voltage.

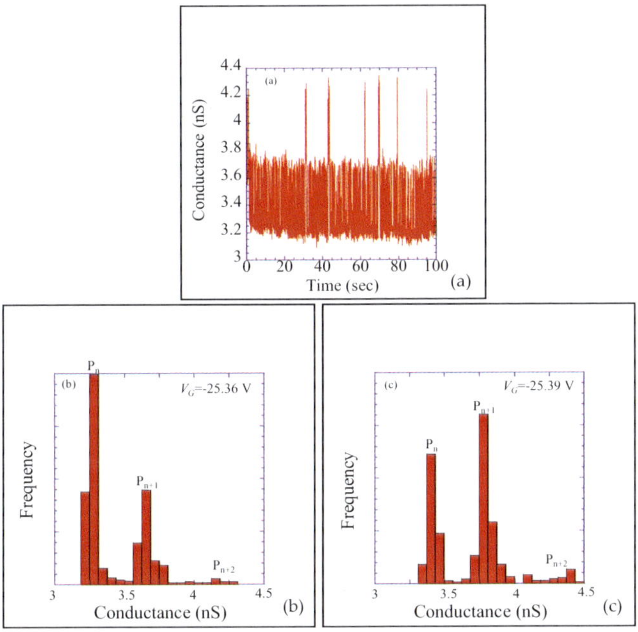

Figure 2. a) Time dependence of drain current with RTS at a gate voltage of V_G=-25.36 V. The time dependence of drain current was sampled for 100 s and the sampling time was 10 ms; (b) and (c) show histograms of the conductance levels of RTS at V_G= -25.36 and -25.39 V, respectively. $P_n$decreases and P_{n+1} and P_{n+2} increase with slightly increasing applied gate voltage from V_G= -25.36 to -25.39 V.

The gate voltage dependences of the natural log of the ratio between the m_{th} peak and the $(m+1)_{th}$ peak in the conductance histogram P_{m+1}/P_m (m= n, n+1, n+2, , n+4) are shown in Fig. 3(a). The natural log of P_{m+1}/P_m linearly depends on applied gate voltage and saturates in each V_G. Each starting point of saturation is marked by an arrow in Fig. 3(a). The charge storage energy levels are floating. Therefore, modulations of energy by V_G may be different at each charge storage. We believe that the reason why the natural log of P_{m+1}/P_m saturates at each V_G may depend on the difference in energy modulation by $V_G$at each charge storage.

The charge transition is modeled, as shown in Fig. 4(a), in which the energy barrier is between the SWNT and the charge storage.Figures 3(b)-3(e) are enlargement plots of each P_m

$_{+1}/P_m$ (m= n, n+1, n+2, , n+4).The energy differences ΔE_n between the charge storage energy level E_n and the Fermi level E_f are expressed as $\Delta E_n = E_f - E_n$, which is modulated by the applied gate voltage V_G. According to equilibrium statistical mechanics, P_{m+1}/P_m is given by

$$P_{m+1}/P_m = (g_f/g_s)e^{-\beta(E_f - E_n)} = (g_f/g_s)e^{-\beta \Delta E_n} \quad (1)$$

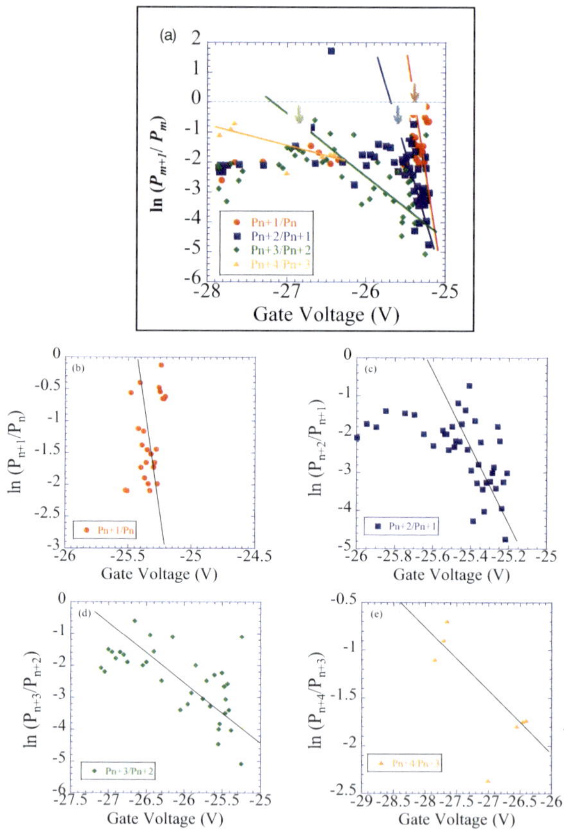

Figure 3. a) Gate voltage dependence of the natural log of the ratio between the occupancy probabilities of the *m*th current levels P_{n+1}/P_n.(b)-(e) Enlargement plots of each P_{m+1}/P_m (*m*=n, n+1, n+2, n+4). The natural log of P_{n+1}/P_n was linearly dependent on gate voltage, the slopes of which are -10.4, -4.78, -1.47, and -0.690 V^{-1}, respectively.

where g_f and g_S are the degeneracy of the top of valence band and the charge storage, respectively [32]. g_f/g_S is assumed to be 1. β is 1/kT. E_n includes the contributions of e electrostatic potential induced by V_G, intrinsic energy level in the storage, and Coulomb charging energy. The basis of eq. (1) is the Arrhenius equation. In this model, the height of the barrier is the energy difference between E_f and E_n. Assuming a linear dependence of ΔE_n on V_G, ΔE_n can be written as $\Delta E_n = \alpha e(V_0 - V_G)$, where α is the gate modulation coefficient, and is a constant value of 0.062. α is obtained from the periods of Fabry-Perot interference characteristic on V_D and V_G. V_0 is the offset voltage, and is obtained from the intersecting point of the extrapolating line of the fitting lines and the line of $\ln(P_{m+1}/P_m)=0$. Therefore, eq. (1) is transformed to

$$\ln(P_{m+1}/P_m) = -\beta e\alpha(V_0 - V_G) \tag{2}$$

Equation (2) is the transformed Arrhenius equation, in which V_G is the parameter.

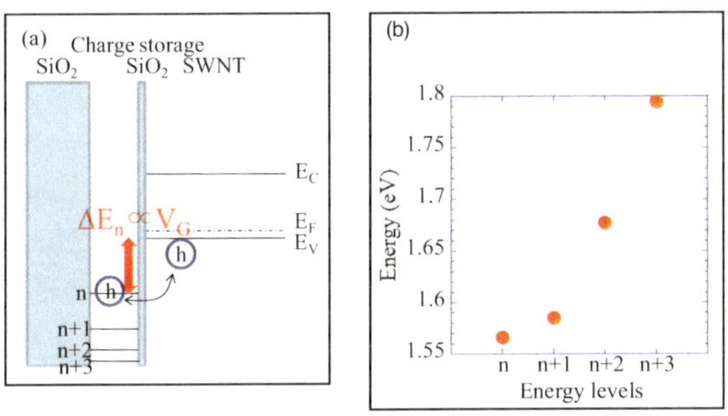

Figure 4. a) Schematic model of charge storage. When the charge storage energy level is coincident with the top of the valence band owing to applied gate voltage, the carrier goes and comes between them through the barrier with tunneling. The exiting probabilities of the carrier at the charge storage and the valence band depend on the relative height of their energy levels under equilibrium condition..(b) Estimated charge storage energy levels from the dependence of P_{n+1}/P_n on V_G shown in Figs. 3(b)-(e) and eq. (1). The energy levels increase from 0.17 to 0.28 eV with increasing number of energy levels in the region of V_G from -25 to -28 V.

From the dependence of P_{n+1}/P_n on V_G shown in Fig. 11 and eqs. (1) and (2), ΔE_n can be obtained, and is shown in Fig. 12(b). The obtained energy levels are from 1.57 to 1.79 eV.

2.4. Conclusion

In summary, we succeeded in fabricating and demonstrating a multi-functional quantum transistor using the particle nature and wave nature of holes in SWNT. This transistor can operate in the wave nature mode as an RTT and in the particle nature mode as an SHT. We were able to reveal that the principle of the characteristic transition from an SHT to an RTT is the modulation of the coupling strength between the SWNT quantum island and the electrodes by the applied V_G.

3. Control of single-hole transition at room temperature

3.1. Method

A SWNT transistor surrounded by SiNx /Al_2O_3 double gate insulator layers with quantum dot in the insulator layers was fabricated, demonstrating discrete threshold voltage shift resulting in discrete drain current modulation by single-hole transfer at room temperature.

3.2. Sample preparation

The fabrication process of the SWNT transistor surrounded by SiN_x /Al_2O_3 double gate insulator layers with quantum dot in the insulator layers is shown in Fig. 1. The fabrication process of the single-charge memory is shown in Fig. 5. A SWNT was grown by the chemical vapor deposition process on the SiO_2 substrate, and the source and drain electrodes were formed on the SWNT, where a distance between the electrodes was 70 nm. The SiO_2 under the SWNTwas etched off by chemical wet process and the SWNT bridge was formed between source and drain electrodes as shown in Fig. 5 (a). Then, the SWNT was surrounded by double-layers of SiN_x of 27 nm over Al_2O_3 of 3 nm using atomic layer deposition (ALD) process (FlexAL, Oxford Inst.) using tris(dimethylamino)silane for SiNx and trimethylaluminum for Al_2O_3 as a precursor as shown in Fig. 1(b).Figure 5 (c) – (g) shows scanning electron microscope (SEM) images of a device after an ALD process. Owing to thin insulator layers and highacceleration energy of SEM, the insulator layers can be seen through and the difference of the materials are recognized because of the difference of the contrast as shown in Fig. 5(c)-(g). Fig. 5(c) and (d) shows the top view and bird's eye view of the device near the gap. The gap of 10 nm length is realized between the source and drain electrode. Fig. 5(e) shows the cross sectional view around the drain electrode indicated by dashed square in Fig. 5(b). The drain electrode is fully covered by the insulator layers, and the space under the channel can be seen. Fig. 5(f)-(g) shows the side view and cross sectional view of the SWNT surrounded by the insulator layer of 30 nm thick in which the SWNT is seen as a light gray line at the center of the insulator. Fig. 5(h) and (i)are the schematic cross sectional view and 3D image of the device after the formation of the top gate electrode.

Using an isotropic deposition of the ALD process 10 nm length of top gate electrode is realized self-assembly as shown in Fig. 5(h). The 30 nm thick insulator layer was deposited from the source and drain electrode which narrowed a gap between the electrodes from 70 nm to 10 nm by the isotropic ALD deposition. Ti and Au were deposited through this gap to form a top gate electrode. Thus, the 10 nm long top gate electrode was self-assembly formed at a center of the source and drain electrode which is covered by the ALD insulator layers. The SiNx deposition by ALD was carried out with low deposition rate of 0.31 Å/ cycle. The low deposition rate enables to form the composite of nano particles of Si, N and C at the beginning of the deposition because of the low uniformity and the incomplete reaction. The composite of nano particles may act as a dot for the charge storage.

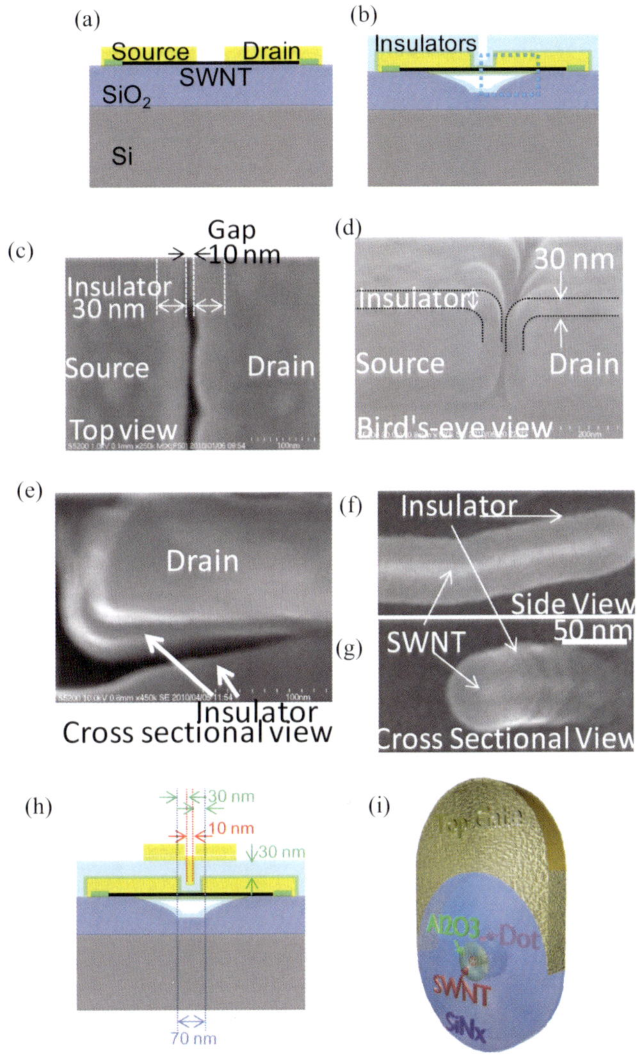

Figure 5. The device fabrication process. (a) Schematic of the device after the SiO_2etching, which is the cross sectional view along with the SWNT. (b) Schematic of the device after the ALD process, which is the cross sectional view along with the SWNT. Scanning electron micro scope images of (c) Top view, (d) Bird's eye view and (e) Cross sectional view around the drain electrode. (f) Side view and (g) Cross sectional view of the SWNT surrounded by the insulator layer. (h) Schematic of the device after the top gate electrode fabrication, which is the cross sectional view along with the SWNT. (i) 3D image around the SWNT channel.

3.3. Results and discussions

Figure 6(a) shows drain current characteristic as a function of the applied top gate voltage at room temperature. The source and drain voltage were set at -25 mV and 25 mV, respectively.

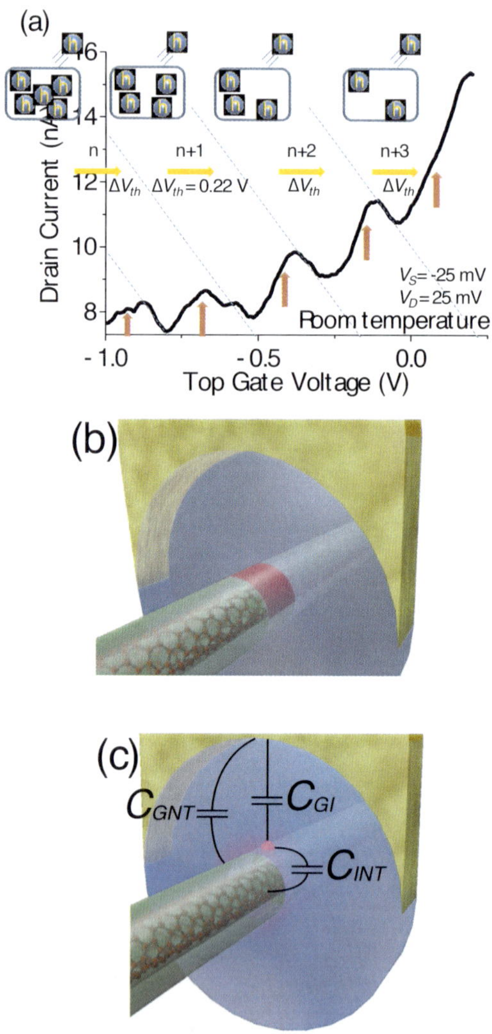

Figure 6. a) Drain current characteristic as a function of applied top gate voltage at room temperature.(b) The area under the top gate electrode, which is 10 nm length.(c) The mutual capacitances in the device.

The drain current showed the repeated threshold shift characteristic indicated by right-arrows,where the drain current repeatedly showed the drastic increases indicated by up-arrows and the gradual decreases indicated by tangential lines with increase of the applied top gate voltage as shown in Fig. 6(a).The threshold shift was observed in period of 0.22 V with increasing applied top gate voltage. The holes had been accumulated in the charge storage dot by negative gate voltage at the beginning of the measurement. By increasing the applied top gate voltage, the hole started to transfer to the SWNT channel. At this moment, the potential of the charge storage dot decreased and blocked the transfer of other holes i.e., Coulomb blockade effect. Therefore, the hole could transfer one by one, where each transfer was separated by Coulomb blockade effect. The decreased potential of the charge storage dot also impacted the SWNT channel and the drain current drastically increased because of p type channel. Therefore, the drastic increases of the drain current in Fig. 6(a) are attributed to a single-hole transfer from the accumulated charge storage dot to the SWNT channel. The gradual decreases of the drain current are attributed to the channel modulation by the applied top gate voltage as well as a conventional MOS-FET device. The charge storage dot density of $D = 1{\times}10^{12}$ cm^{-2} was estimated from the C-V measurement. The area just under the top gate electrode indicated by the red band in the Fig. 6(b) was 91.4 nm^2 from πrL, where r is the thickness of the Al_2O_3 layer, L is the length of the top gate electrode. The estimated number of charge storage dot was 0.914 from DπrL. Therefore, almost one charge storage dot exists in the area just under the top gate electrode.

Moreover, the threshold voltage shift caused by single charge transition was given by ΔV_{th}~ e/ (C_{GI} + C_{GNT}) [37-41], where C_{GI} and C_{GNT} was mutual capacitances of the top gate electrode and the charge storage dot and of the top gate electrode and the SWNT as shown in Fig. 6(c)

C_{GI} and C_{GNT} were estimated to be C_{GI} = 734 zF and C_{GNT} = 20.7 zF, from the simulation by the finite element method, where conductor sphere of 1 nm diameter at 3 nm above the SWNT was assumed in the simulation as the single charge storage dot. The estimated threshold voltage shift was estimated to be 0.201 V. This is in good agreement with the threshold voltage shift of the drain current characteristic as shown in the Fig. 6(a).

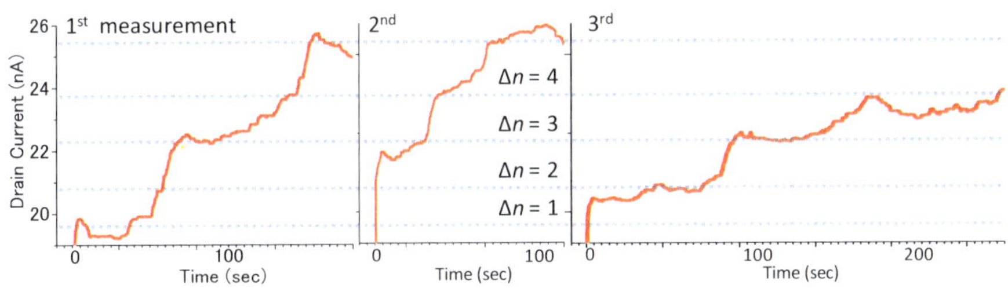

Figure 7. a) – (c) The time dependence characteristics of the drain current at room temperature. The top gate voltage of ΔV_{TG} = – 4 V had been applied before the measurements, and ΔV_{TG} = 1 V was applied in the measurements. Same measurements were repeated in three times and plotted in (a) – (c).

Figure 7(a) – (c) shows the time t dependence of the drain current characteristics at top gate voltage of V_{TG} = 1 V at room temperature, in which V_{TG} = -4 V had been applied before t = 0 sec. The source and drain voltages were set at 25 mV and -25 mV in the measurement. The measurement was carried out three times, and each measurement was plotted in Fig. 7(a) – (c), respectively.

The drain currents were plotted in the same time scale, however, time lengths of the measurement were different. The drain current increased with elapsed time and also showed five steps indicated by dotted lines as shown in Fig. 7(a) – (c). The increases of the drain current at each step were the same among Fig. 7(a) – (c). However, the time width of each steps showed the variety even if the steps were at the same drain current levels in Fig. 7(a) – (c). From the plots, average of the drain current width was t_{ave} = 68.82 s. Tunneling provability was roughly estimated to be 0.0145 from $1/t_{ave}$. By applyingV_{TG} = -4 V before the measurement, holes had been accumulated in the storage dot. After the top gate voltage was turned to 1 V at t = 0 sec, the accumulated holes started to transfer from storage dot to the SWNT channel one by one. The SWNT channel was modulated by the transferred single-hole, and showed discrete drain current levels i.e., the current steps as shown in Fig. 7(a) – (c). In other words, discrete changes of the drain current directly corresponded to the variation of single-hole Δn in the storage dot. Therefore, the transfer of single-hole could be directly counted as the discrete modulation of the drain current at room temperature in real-time [34]. Moreover, each transfer of single-hole took several tens of second because of thick tunneling barrier of Al_2O_3 of 3 nm. The variety of time widths of each step attributed to the stochastic transfers of single-hole. These characteristics also indicate the evidence of the single-hole transfer in the device.

4. Summary

In conclusion, control of single transition in carbon nanotube transistor with quantum dot in gate insulator at room temperature was demonstrated. To obtain the narrow top gate electrode of 10 nm, the isotropic deposition by ALD process for the insulators formation was used. At the same time, the concentric circle structure of insulators was formed around the SWNT channel which was on the center. The defective deposition of SiNx on Al_2O_3 may formnano size particles at the beginning of the deposition process, which worked as charge storage dots. Because of the narrow top gate electrode, only single-dot was just under the top gate electrode which stored single-hole. Though top gate electrode could surround only the upper half of insulators that realized electric field concentration all-around the SWNT channel and caused Fowler-Nordheim tunneling which make the hole transfer to the charge storage dot. The drain current affected by the stored charge showed the threshold voltage shift as a function of the applied top gate voltage at room temperature. This threshold voltage shift is attributed to the abrupt potential energy change by the transfer of single-holes from the dot to the SWNT channel. The time dependence of the drain current after changing the top gate voltage showed the step like characteristics. The time widths of the steps corresponded to the interval of stochastic transfer of single-holes, and the number of the steps corresponded to the variation of single-holes in the dot. By observing the steps, the individual transfers of single-holes could be counted in real-time at room temperature. The SWNT property with high sensitivity for the charges is suitable to the application of the single-charge memory. And the SWNT must be one of the leading candidates for the single-charge applications.

Author details

Takafumi Kamimura[1], Yutaka Hayashi[2] and Kazuhiko Matsumoto[1]

1 ISIR, Osaka University, CREST-Japan Science and Technology Agency, Japan

2 National Institute of Advanced Industrial Science and Technology, CREST-Japan Science and Technology Agency, Japan

References

[1] Bez, R., Camerlenchi, E., Modelli, A., & Visconti, A. (2003). Introduction to Flash Memory. *Proceedings of the IEEE*, 91(4), 489-502.

[2] Aritome, S., Shirota, R., Hemink, G., Endoh, T., & Masuoka, F. (1993). Reliability Issues of Flash Memory Cells. *Proceedings of the IEEE*, 81(5), 776-788.

[3] Iijima, S. (1991). Helical microtubules of graphitic carbon. *Nature*, 354-56.

[4] Iijima, S., & Ichihashi, T. (1993). Single-shell carbon nanotubes of 1 -nm diameter.

[5] Bethune, D. S., Kiang, C. H., de Vries, M. S., Gorman, G., Savoy, R., Vazquez, J., & Beyers, R. (1993). Cobalt-catalysed growth of carbon nanotubes with single-atomic-layer walls. *Nature*, 363-605.

[6] Ajyayan, P. M., Ebbesen, T. W., Ichihashi, T., Iijima, S., Tanigaki, K., & Hiura, H. Opening carbon nanotubes with oxygen and implications for filling. Nature (1993). , 362-522.

[7] Tans, S. J., Verschueren, A. R. M., & Dekker, C. (1998). Room-temperature transistor based on a single-carbon nanotube. *Nature*, 393-49.

[8] Single- and multi-wall carbon nanotube field-effect transistors. Appl. Phys. Lett.; , 73-2447.

[9] Biercuk, M. J., Reilly, D. J., Buehler, T. M., Chan, V. C., Chow, J. M., Clark, R. G., & Marcus, C. M. M.Charge sensing in carbon-nanotube quantum dots on microsecond timescales. Phys. Rev. B (2006). , 73-201402.

[10] Guo, J., Kan, E. C., Ganguly, U., & Zhang, Y. High sensitivity and nonlinearity of carbon nanotube charge-based sensors. J. Appl. Phys. (2006). , 99-084301.

[11] Peng, H. B., Hughes, M. E., & Golovchenko, J. A. Room-temperature single charge sensitivity in carbon nanotube field-effecttransistors. Appl. Phys. Lett. (2006). , 89-243502.

[12] Kamimura, T., Ohono, Y., & Matsumoto, K. Carbon Nanotube Fabry-Perot Device for Detectionof Multiple Single Charge Transitions. Jpn. J. Appl. Phys. (2009). , 48-025001.

[13] Andresen, S. E. S., Danneau, F. W., Gunnarsson, D., & Hakonen, P. J. Highly sensitive and broadband carbon nanotube radio-frequencysingle-electron transistor. J. Appl. Phys. (2008). , 104-0033715.

[14] Ohori, T., Ohno, Y., Maehashi, K., Inoue, K., Hayashi, Y., & Matsumoto, K. Quantized characteristics in carbon nanotube-based single-hole memory with a floating nanodot gate. Appl. Phys. Lett. (2011). , 98-223101.

[15] An, K. H., Jeong, S. Y., Hwang, H. R., & Lee, Y. H. Enhanced sensitivity of a gas sensor incorporating single-walled carbon nanotube-polypyrrole nanocomposites. Adv. Mater. (2004). , 16(12), 1005-1009.

[16] Ong, K. G., Zeng, K., & Grimes, C. A. A Wireless, Passive Carbon Nanotube-BasedGas Sensor. IEEE Sensor J. (2002). , 2(2), 82-88.

[17] Li, J., Lu, Y., Ye, Q., Cinke, M., Han, J., & Meyyappan, M. Carbon Nanotube Sensors for Gas andOrganic Vapor Detection. Nano Lett. (2003). , 3(7), 929-933.

[18] Suehiro, J., Zhou, G., & Hara, M. Fabrication of a carbon nanotube-based gas sensor using dielectrophoresis and itsapplication for ammonia detection by impedance spectroscopy. (2003). LL114., 109.

[19] Abraham, J. K., Philip, B., Witchurch, A., Varadan, V. K., & Redy, C. C. A compact wireless gas sensor usinga carbon nanotube/PMMA thin filmchemiresistor. Smart Mater. Struct. (2004). , 13-1045.

[20] Kojima, A., Hyon, C. K., Kamimura, T., Maeda, M., & Matsumoto, K. Protein Sensor Using Carbon Nanotube Field Effect Transistor. Jpn. J. Appl. Phys. (2005). A) , 1596-1598.

[21] Chen, R. J., Choi, H. C., Bangsaruntip, S., Yenilmez, E., Tang, X., Wang, Q., Chang, Y., & Dai, H. An Investigation of the Mechanisms of Electronic Sensing ofProtein Adsorption on Carbon Nanotube Devices. J. Am. Chem. Soc. (2004). , 126-1563.

[22] Abe, M., Ohno, Y., & Matsumoto, K. Schottky barrier control gate-type carbon nanotube field-effect transistor biosensors. J. Appl. Phys. (2012). , 111-034506.

[23] Hakamata, Y., Ohno, Y., Maehashi, K., Inoue, K., Matsumoto, K., & External-Noise, . External-Noise-Induced Small-Signal Detection with Solution-Gated Carbon Nanotube Transistor.Appl. Phys. Express (2011). , 4-045102.

[24] Ohno, Y., Maehashi, K., Inoue, K., & Matsumoto, K. Improving faint-signal sensitivity of electrolyte-gated carbon nanotube field-effect transistors using external noise. Proc. of 2010 IEEE Nanotechnology Materials and Devices Conference (2010). , 10-13.

[25] Abe, M., Ohno, Y., & Matsumoto, K. Dependence of sensitivity of biosensor for carbon nanotube field-effect transistor with top-gate structures. J. Appl. Phys. (2010). , 107-084504.

[26] Yamamoto Y., Ohno Y., Maehashi K., Inoue K., Matsumoto K.Electrical Detection of Negatively Charged Proteins Using n-Type Carbon Nanotube Field-Effect Transistor Biosensors.Jpn. J. Appl. Phys. (2010). BD10-1-02BD, 10-4.

[27] Maehashi, K., Katsura, T., Kerman, K., Takamura, Y., Matsumoto, K., Tamiya, E., & Label, . Label-Free Protein Biosensor Based on Aptamer-Modified Carbon Nanotube Field-Effect Transistors. Anal. Chem. (2007). , 79-782.

[28] Star, J., Gabriel, K., Bradley, , & , G. Gruner Electronic Detection of Specific Protein Binding Using Nanotube FET Devices Nano Lett. (2003). , 3-459.

[29] High-Performance Carbon Nanotube Field-Effect Transistors with Local Electrolyte Gates.Jpn. J. Appl. Phys.; , 47-2060.

[30] Besteman, K., Lee, J., Wiertz, F. G. M., Heering, H. A., Dekker, C., & Enzyme, . Enzyme-Coated Carbon Nanotubes as Single-Molecule Biosensors. Nano Lett. (2003). , 3-727.

[31] Maehashi, K., Matsumoto, K., Takamura, Y., Tamiya, E., & Aptamer, . (2009). Aptamer-Based Label-Free Immunosensors Using Carbon Nanotube Field-Effect Transistors. *Electroanalysis*, 21-1285.

[32] Ohno Y., Maehashi K., Inoue K., Matsumoto K.Fabrication of room-temperature-operating carbon nanotube single-charge transistors.Sensor andMaterials.(2009). , 21-393.

[33] Kim, W., Javey, A., Vermesh, O., Wang, Q., Li, Y., & Dai, H. Hysteresis Caused by Water Molecules in Carbon Nanotube Field-Effect Transistors. Nano Lett. (2003). , 3-193.

[34] Radosavljevic′, M., Freitag, M., Thadani, K. V., & Johnson, A. T. Nonvolatile Molecular Memory Elements Based on Ambipolar Nanotube Field Effect Transistors. Nano Lett. (2002). , 2-761.

[35] Martel, R., Derycke, V., Lavoie, C., Appenzeller, J., Chan, K. K., Tersoff, J. ., & Avouris, Ph. Ambipolar Electrical Transport in Semiconducting Single-Wall Carbon Nanotubes. Phys. Rev. Lett. (2001). , 256805 EOF.

[36] Kamimura, T., & Matsumoto, K. Reduction of Hysteresis Characteristics in Carbon Nanotube Field-Effect Transistors by Refining Process. IEICE Trans. Electron. (2004). E-87C , 1795-1798.

[37] Peng, H. B., Hughes, M. E. ., & Golovchenkoa, J. A. Room-temperature single charge sensitivity in carbon nanotube field-effect transistors. Appl. Phys. Lett. (2006). , 243502 EOF.

[38] Room-temperature Si single-electron memory fabricated bynanoimprint lithography. Appl. Phys. Lett.;, 83(11), 2268-2270.

[39] Yano, K., Ishii, T., Hashimoto, T., Kobayashi, T., Murai, F., & Seki, Koichi. Room-temperature single-electron memory. IEEE Trans. Elect. Dev.(1994). , 41(9), 1628-1638.

[40] Guo, L., Leobandung, E., & Chou, S. Y. Silicon Single-Electron Transistor MemoryOperating at Room Temperature. Science(1997). , 275-649.

[41] Nakajima, A., Futatsugi, T., Kosemura, K., Fukano, T., & Yokoyama, N. Room temperature operation of Si single-electron memory with selfalignedfloating dot gate. Appl. Phys. Lett. (1997). , 70-1742.

[42] Gruneis, A., Esplandiu, M. J., Gracia-Sanchez, D., & Bachtold, A. Detecting Individual Electrons Using aCarbon Nanotube Field-Effect Transistor. Nano Lett. (2007). , 7(12), 3766-3769.

Chapter 15

Carbon Nanotubes as Suitable Interface for Improving Neural Recordings

Gemma Gabriel, Xavi Illa, Anton Guimera,
Beatriz Rebollo, Javier Hernández-Ferrer,
Iñigo Martin-Fernandez, Mª Teresa Martínez,
Philippe Godignon, Maria V. Sanchez-Vives and
Rosa Villa

Additional information is available at the end of the chapter

1. Introduction

In the last decades, system neuroscientists around the world have dedicated their research to understand how neuronal networks work and how they malfunction in various diseases. Furthermore in the last years we have seen a progressively increased interaction of brain networks with external devices either for the use of brain computer interfaces or through the currently extended brain stimulation (e.g. transcranial magnetic stimulation) for therapy. Both techniques have evidenced even more the need for a better understanding of neuronal networks. These studies have resulted in the development of different strategies to understand the ongoing neuronal activity, such as fluorescence microscopy for genetic labelling and optogenetic techniques, imaging techniques, or the recording/stimulation with increasingly large numbers of electrodes in the whole brain or in both cell cultured neurons and slice preparations. It is in these last two areas where the technology developed on microelectrode arrays, commonly called multi-electrode arrays (MEAs), has become important over other technologies [1–3].

MEA devices are formed by a large number of microelectrodes arrayed on substrates with small geometry size in order to excite or register a group of neurons selectively and efficiently. There are several applications where MEA devices are crucial for nerve recording and stimulation. Some of these are limb prostheses for spinal cord injuy; bladder prostheses, cochlear and brain-stem auditory prostheses, retinal and cortical visual prostheses, cortical recordings, vagus nerve stimulation for epilepsy and depression, deep brain stimulation, Parkinson's disease, epilepsy, dystonia and depression. In such areas, advances on micro-fabrication technology have given rise to a great success in the neural interfaces field.

A MEA can be used to perform electrophysiological experiments on tissue slices or dissociated cell cultures. With acute tissue slices, the connections between the cells within the tissue slices prior to extraction and plating are more or less preserved, while the intercellular connections in dissociated cultures are destroyed prior to plating. With dissociated neuronal cultures, the neurons spontaneously form networks.

Related to the work presented here, brain slices provide more information of a realistic model where the brain architecture is maintained. Furthermore, one of our aims is to carry out developments that are as well usable for in vivo interfacing, both in acute and chronic situations. However, these emerging technologies do still face tremendous challenges mainly related with long-term experiments. Electrodes are metallic conductors (the most common ones Pt, Pt alloys, Ir oxide and TiN), however, for chronic stimulation and recording they present some drawbacks as, for example, for obtaining and maintaining good recordings. This is a consequence of the difficulty to assure both good electrochemical electrode response and good contact between the electrode and the tissue. This is mainly a consequence of the electrode material and the planarity of the substrate used to fabricate the MEAs. Recently, different technologies have been proposed to overcome these limitations such us electrochemical deposition of conductive polymers [4–6] and the use of carbon nanotube (CNT) coatings which has been extensively demonstrated to improve neuronal recordings [7–11]. Also it can be found in the literature a broad type of materials that can be deposited over the electrode to enhance the response of a recording electrodes. The most well-known are stain-less steel, tungsten, platinum, platinum-iridium alloys, iridium oxide, titanium nitride or poly(etylenedioxythiophene (PEDOT).

CNTs are high aspect ratio, exceptionally strong, tough, and show desirable chemical and electrical properties [12, 13]. Hence, they are attractive for interfacing with neural systems to develop biocompatible, durable and robust neuroprosthetic devices turning into an excellent candidate for the improvement of neural interfaces [14,15]. CNTs can be grown or assembled on a great variety of surfaces and can give rise to electrodes with different morphologies.

Based in the great experience of our research group, the aim of this work is to explain the different electrode modification methods we have developed. We have demonstrated the modification of the surface from multielectrode devices by drop casting Single Walled Carbon Nanotubes (SWNTs) [16, 17] and by selectively synthesizing arrays of Multi Walled Carbon Nanotubes (MWNTs) by chemical vapour deposition (CVD) [18,19]. The drop casting of SWNTs was demonstrated to be an easy method to perform electrode modification technique that results in a high purity CNT interface with spaghetti like morphology. The area of the electrode is one of the most important limitations of this technique as it cannot be implemented in electrodes higher than 100 μm [17]. Another option is the direct growth of MWCNTs on the metal substrate that results in a more robust electrode along the application lifetime. With this method, the electrode dimension is not a limitation. Moreover, fully biocompatible materials can be used during the fabrication process as we demonstrated platinum catalysts for the growth of CNTs [19]. In conclusion, we stated that the vertically aligned CNT array morphology has advantages regarding the voltammetric measurements over the drop casted one.

However, in our first attempts the impedance characteristics of the obtained MWNTs showed a lack of improvement compared to the bare electrode. This was mainly attributed to the presence of amorphous carbon covering the carbon nanotubes that finally inhibit the electron exchange. This issue was solved once the technological process for the integration of pure and dense arrays of vertically aligned MWNTs by using platinum catalysts on the MEAs was found [20]. This method, which is compatible with the wafer scale fabrication technology, is based on standard microelectronic fabrication processes, and only involves the use of bio-compatible materials. The wafer scale compatibility of the process is very important in order to ensure the reproducibility between devices.

In this work these two types of CNT-modified electrodes will be compared to the bare platinum electrode. In particular, arrays of 40 μm in diameter bare platinum electrodes have been used for all the experiments. The size of the electrodes has a strong influence on the impedance value, as the impedance increases with decreasing the electrode area. Thus, the 40 μm bare platinum electrodes used here have very high impedance values that make them useless for the aim of this work. Then, in order to compare the modified CNT-based microelectrodes with useful metallic based electrodes, they were alternatively prepared with electrodeposited black platinum. Electrochemical deposition of black platinum is a common approach for the modification of the surface of microelectrodes in order to reduce their interface impedance by increasing the surface roughness [21,22]. Actually, their electrode-electrolyte impedance values are 10-fold below the bare platinum electrodes.

Besides the contact problems which can be overcome by using CNT or black platinum coatings, the use of silicon or pyrex-based multisite substrates has another main limitation related to the lack of adaptability to biological tissues. This is another cause for obtaining bad measurements due to bad contact that if wanted to be improved in-situ may dramatically damage the biological tissue. The use of the technologies employed in microelectronic fabrication processes along with the development of new polymers have paved the way for the fabrication of polymer-based flexible microprobes with integrated MEAs. In addition, their simple fabrication process and biocompatibility have given to polymeric substrates even more relevance [23,24].

In particular, flexible neural microprobes have been mainly fabricated in polyimide wherein a metal layer is used for the recording sites [25–27]. Other materials like parylene [23,28], benzocyclobutene (BCB) [21] and SU-8 [29,30] have been also employed to fabricate flexible microprobes, demonstrating the interest in this field. In this work, SU-8 has been chosen due to the expertise of our group and its low-cost and versatile fabrication process.

Dealing with the surface electrode improvements, the use of flexible materials implies the use of new strategies as the above described surface modifications with carbon nanotubes are not compatible with these polymeric materials. Basically, the CNTs growth method can not be used due to the high temperatures required, while the drop casting SWNTs method can not be applied to flexible MEAs due to the fragility of these substrates. On the contrary, regarding the black platinum electrodeposition, the use of flexible substrates does not suppose a problem.

However, an alternative method to use CNTs for the modification of the microelectrodes on polymeric substrates has been lately described by using a CNTs/polypyrrole electrodeposition [31–34]. With this method, SWCNTs/polypyrrole (Ppy) films can be electrochemically grown over the electrodes on the transparent and flexible polymeric substrates [28,29,35]. The presence of SWCNTs during a slow polymerization of Ppy results in a high rough surface electrode because the polymeric coating in course entraps the SWCNTs.

In this work, validation of the CNT integrated MEAs is performed by comparing them to non-modified metal electrodes using two strategies. Firstly, the electrode-electrolyte interface has been characterized by impedance spectroscopy and by cyclic voltammetry to compare their electrode-electrolyte interfaces as along with ex-situ techniques for film characterization. Secondly, the spontaneous activity from slices of cerebral cortex has been recorded before and after the blockade inhibition in order to demonstrate its feasibility. The obtained results demonstrate the huge potential of such nanostructured materials to build an interface between the neural system and the state of the art nanoelectronics.

2. Materials and methods

2.1. Microelectrode Arrays Fabrication

2.1.1. MEA fabrication

The MEA chips are formed of 16 platinum electrodes that are connected to metal pads located on the sides of the chip not to interfere on the liquid based testing. The electrodes may be circle or square shaped with their diameter or side being 40 or 300 μm. However, all the electrochemical characterizations and the experimental section were conducted with the round 40 μm electrodes.

The electrodes are fabricated similarly as it is described in [16,19] as shown in Figure 1. This figure describes the electrode fabrication from 1a to 1d, and also describes the post process modification of carbon nanotube growth from 1e to 1h. It has been represented in a single figure because it is really implemented as a single process. The starting point is a 4 inch Si wafer (Fig. 1a). First, a 1.5 μm thick SiO2 layer is deposited by plasma enhanced (PECVD) (Fig. 1b). Then, the electrodes, the contact pads and the strips connecting them are patterned after a photolitography, a Ti/Pt deposition (30/150 nm) and a lift-off process (Fig. 1c). Next, the wafer is passivated by a SiO2 and Si3N4 bi-layer (400/700 nm) that is deposited by PECVD and windows are only opened at the electrodes and the connection pads by a second photolithography and a reactive ion etching process (Fig. 1d).

Once the wafer is fabricated the MEAs are encapsulated before the electrodes are electrochemically characterised. First, the wafer is diced by a dicing saw. Then, the MEAs are glued to previously fabricated Printed Circuit Board (PCB), the connection pads are wire-bonded and the wires are protected with an epoxy based resist. Last, a ring lid is glued to the PCB prior to the electrode characterization so that the solution is confined to it during the experiments. A picture of the encapsulated final device is shown in Figure 3A.

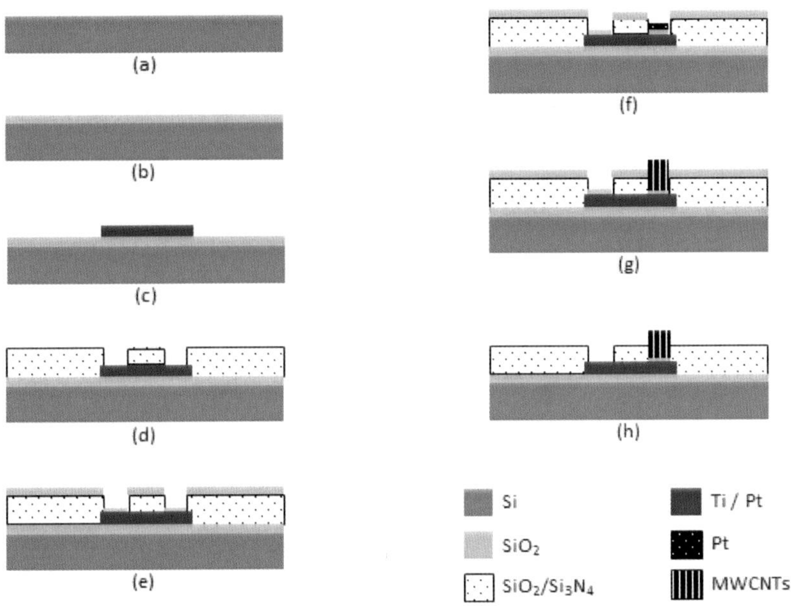

Figure 1. Schematic of the main steps of the fabrication of the electrodes modified with locally grown MWCNTs: (a) initial Si substrate; (b) deposition of a SiO2 layer; (c) patterning of the Ti-Pt electrodes, strips and connection pads; (d) passivation of the electrodes by a SiO2-Si3N4 bi-layer except for the electrodes and the connection pads; (e) deposition of a 15 nm thick layer of SiO2; (f) selective deposition of a Pt thin layer on the electrodes; (g) synthesis of the MWCNTs; (h) removal of the thin SiO2 layer

2.1.2. Flexible MEA fabrication

In this work, flexible microprobes integrating 16 platinum microelectrodes of 40 μm in diameter have been fabricated using SU-8 negative photoresist (Microchem, USA). The fabrication process has taken advantage of our recent work where SU-8-based microneedles for neural applications have been fabricated [29,30,36,37]. In brief, the fabrication process starts with the oxidation of a 4-inch silicon wafer (Fig. 2a). A 400 nm of SiO_2 will serve as a sacrificial layer for the final release of the SU-8 structures. Then a, a 25 μm thick SU-8 structural layer is deposited, baked, exposed through a mask where the shape of the microprobe is defined, and developed following the conditions defined by the SU-8 manufacturer (Fig. 2b). Afterwards, 20 nm of titanium and 200 nm of platinum are deposited by e-beam evaporation on top of the SU-8 (Fig. 2c). Subsequently, patterning of the metal layer is performed using standard photolithography steps and wet chemical etching (Fig. 2d).

In order to insulate the metal tracks a second SU-8 layer is processed on top of the wafer (Fig. 2e). This 1 μm thick passivation layer also defines the area of the microelectrodes which was designed to be 40 μm in diameter. Finally the whole wafer was immersed in a HF bath to etch the SiO_2 sacrificial layer, releasing the SU-8 microprobes with integrated MEA (Fig 2f).

To facilitate the use of the fabricated microprobes, they were connected to a printed circuit board (PCB) by means of zero insertion force (ZIF) connectors. For that, the connecting pads of the microprobe where designed to match the specifications of the desired ZIF connector. The use of these connectors to encapsulate the microprobes for both characterization and experimentation purposes provides ready-to-test microprobes, as no additional back-end fabrication process is needed. In Fig. 3b there is an image of an individual SU-8 microprobe where it can be observed the high flexibility that can be obtained with the presented fabrication process.

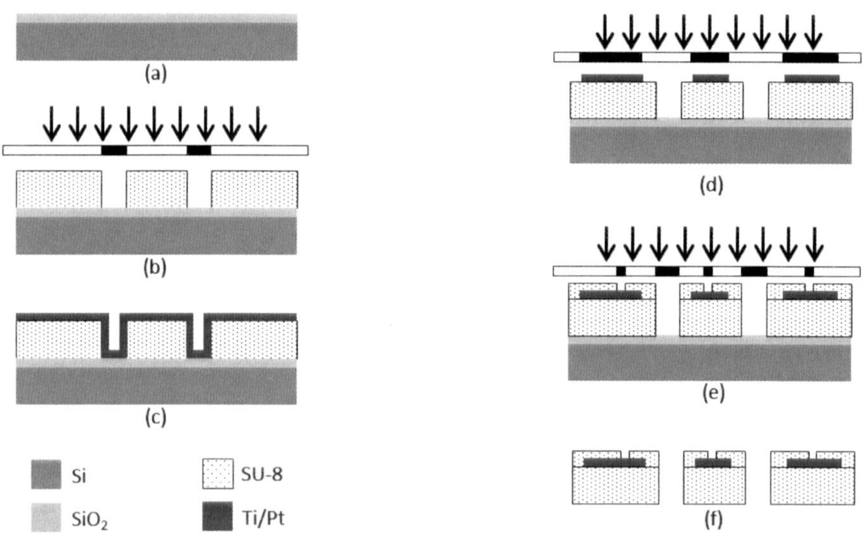

Figure 2. Schematic of the main steps of the fabrication of the electrodes in flexible SU-8 substrate: (a) deposition of a SiO2 layer; (b) deposition and patterning of the 25 μm SU-8 substrate; (c) deposition of the Ti/Pt metal layer; (d) patterning of the Ti-Pt electrodes, strips and connection pads; (e) deposition and patterning except for the electrodes and the connection pads of a 1 μm thick SU-8 layer acting as passivation; (f) etch of the sacrificial layer and release of SU-8 microprobes.

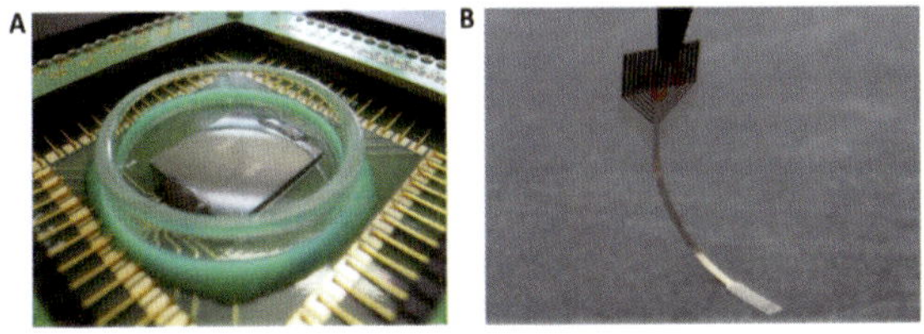

Figure 3. A, picture of standard silicon MEA of 16 electrodes provided with the lid ring and encapsulated to a Printed Circuit Board. B) Flexible SU-8 MEA of 16 electrodes.

2.2. Electrode post-processing strategies

The post-processing strategies are an enhancement of the electrode behaviour focused on the modification of the electrode surface area, so they can be described independently of the substrate of the MEA device. However, it must be taken into account the limitations that present the material substrates of the microprobes used in this work. In this way, the drop casting SWNTs methodology can not be applied to our fabricated flexible SU-8 MEAs due to the fragility of thes 20 μm thick probes. For thicker SU-8 probes this would not suppose a problem. Likewise, the CNTs growth method here described can not be applied to the SU-8 microprobes due to the high temperature requirements that the chemical vapour deposition (CVD) carbon nanotubes growth requires. The other methods can be used alike.

2.2.1. Black platinum electrodeposition

Ti/Pt electrodes on individual devices were electrochemically coated with a porous layer of black platinum to reduce their impedance through a customized process of platinization [38]. Platinization was carried out using a Pt electrode (Radiometer Analytical) in a LC20H Ultrasonic Cleaner (Elma) and involved an initial cleaning of the electrode surface for 3 min in ethanol with 35 kHz ultrasounds. Afterwards, the electrode surfaces were activated in a KCl 0.1M solution until release of H2 was apparent. Thereafter, electroplating was performed for 1 min in a solution containing platinum chloride (Hydrochloric acid 0.1M, 2.3% Platinum (IV) chloride and 0.023% Lead (IV) acetate 99 %. All reagents were analytical grade (Panreac) and used as received. The injected current was of 20mA for 40 μm Ø electrodes. Finally they were introduced again in an ultrasound cleaner, in order to blast off poorly adhered platinum from the electrode surface.

2.2.2. Drop casting SWNTs

High purity Single Walled Carbon Nanotubes (SWNTs) were purchased from Sigma Aldrich. Carbonaceous purity is about 88 % and may contain about 3 to 6 atomic % of carboxylic acid groups due to acidic purification. Thermogravimetric analyses showed a metal content of 6 % wt. Raman spectra revealed a mean diameter of 1.3-1.6 nm and confirmed a low carbonaceous content. For the modification of the Pt microelectrodes with SWNTs, 10 mg of pure SWNTs were dispersed in 10 ml of dimethyl formamide (DMF) under ultrasonic agitation resulting in a 1 mg/ml black suspension [16,17]. Once the Pt electrode was cleared with ethanol, the surface was coated by dropping the suspension of SWNTs in DMF, and dried at 90-100 ºC. Finally, the device was thoroughly rinsed with distilled water and mechanically cleaned to ensure that carbon nanotubes stayed delimited in the electrode area.

2.2.3. MWNTs growth

The integration of the MWCNT arrays can be implemented as a continuation of the described MEA device fabrication in the section 2.1.1, in this way, the CNTs growth is also compatible with large wafer scale fabrication. This facilitates to obtain an homogenous electrical response for the different modified MEAs.

The MWNT growth starts by the deposition of a 15 nm thick SiO2 layer. On the one side, this layer aims to inhibit the diffusion of the catalyst material into the electrode and, on the other side, to increase the roughness of the electrode to enhance the formation of a dense array of CNTs in the subsequent steps (Figure 1e). Afterwards, the catalyst material for the MWCNTs to grow from is selectively patterned on the electrodes by a photolithography, the deposition of a 4 nm thick layer of Pt and a lift-off process (Figure 1f). The MWCNTs are synthesized in a rapid thermal CVD system at 800ºC by H2 and CH4 as the main process gases in a 2 step process. The first step aims at dewetting the Pt layer into a dense array of ~10 nm diameter nanoparticles and, during the second step the MWCNT arrays are made to grow after the flow of the carbon containing gas (CH4 in this case)are made to grow into the chamber (Figure 1g). The last step of the fabrication is the removal of the 15 nm thick SiO2 layer by a HF based solution (Figure 1h).

2.2.4. SWNTs/polypyrrol composite electrodeposition

Sodium monohydrogenophosphate heptahydrate, sodium dihydrogenophosphate monohydrate (puriss. p.a.), sodium dodecylbenzenesulfonate (technical grade), yttrium (99.9%) and graphite powder (≥99.99%) were purchased from Aldrich. Nickel powder (99.9%) was purchased from AlfaAesar (website http://www.alfa.com). Graphite bars were obtained from CYMIT Química. Pyrrole (SAFC, ≥98% FCC) was distilled immediately before use. Ultrapure water employed in the preparation of the solutions was obtained from a Milli-Q system from Millipore. Carbon nanotubes were synthesized by the arc-discharge method using graphite electrodes and a Ni/Y 4/1 % metal catalyst mixture.

As-grown single-walled carbon nanotubes (agSWCNTs) were dispersed ultrasonically in aqueous 1% SDBS (initial nanotube concentration 4 mg/ml) and centrifuged at 13,000 rpm for 30 min (Hermle Z383) in order to increase their purity and decrease their metal content. The supernatant was decanted and the final concentration of nanotubes was estimated by UV–vis spectroscopy using absorbance at 600 nm (Shimazdu UV2401PC). For the construction of the calibration line, dilutions from the unpurified dispersions were used with a well-known concentration. The relative purity of the nanotubes in suspension was determined from near infrared (NIR) spectra (Bruker Vertex70 spectrometer). The NIR purity index was calculated by comparing the baseline-corrected peak area corresponding to the interband S22 transition for the semiconducting nanotubes with the total area under the peak, as described in [39]. The final nanotube concentration resulted to be 1.5 mg/ml and the NIR purity index was 0.080.

Electrodeposition of the composite material was carried out in galvanostatic conditions using a current value of 3 mA cm-2 during 120 s, and the obtained transient is shown in figure 1. The polymerization solution was a 0.9% NaCl, 10-2 M total phosphate concentration, pH=7 phosphate buffer solution containing 3.5 10-3 M Sodium dodecylbenzene sulfonate (SDBS) and 0.15 mg/ml agSWCNTs. An Ag/AgCl (3 M NaCl) electrode was used as a reference electrode, and a graphite bar was used as a counter electrode. The quality of the film was checked using electrochemical impedance spectroscopy (EIS) in 0.9% NaCl, 10-2 M total phosphate concentration, pH=7 phosphate buffer solution, in a two-electrode configuration at a potential of 0V versus a graphite counter electrode and a decrease in the impedance modulus and phase were observed until a constant response.

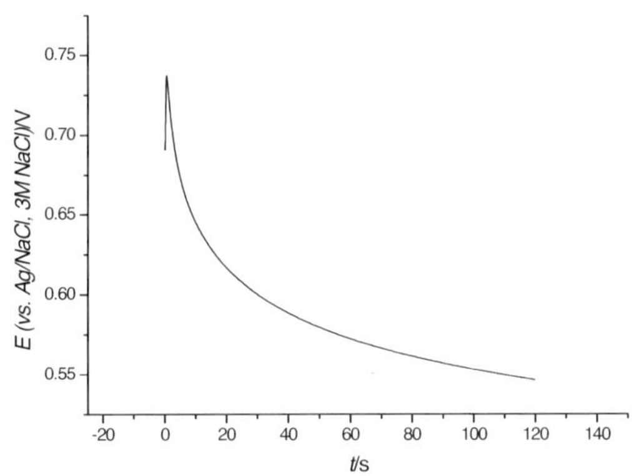

Figure 4. Galvanostatic transient obtained during the Ppy/SWCNT electrode deposition at a current of 3 mA cm^{-2}

2.3. Electrochemical impedance spectroscopy characterization

EIS was conducted by using a commercial impedance analysis system (SI 1260, Solartron Analytical) operated by Zplot software. Two-electrode impedance measurements were conducted to characterize the electrode-electrolyte interface impedance versus to a platinum reference electrode (Radiometer Analytical). The electrical properties of the electrode-electrolyte interface were evaluated by comparing the impedance and phase shifts to the frequency in physiological saline solution (0.9 wt.% NaCl, with a nominal resistivity of 71.3 Ωcm) in the 10 Hz to 1 MHz frequency range.

2.4. In vitro extracellular recordings

Coronal slices (0.4 mm thick) from occipital cortex and containing primary and secondary visual cortical areas [17, 18, and 19] were obtained from adult ferrets, as described in [40]. The MEA was inserted in the probe interface MEA1060, where the signal was pre-amplified. Further amplification (1000x) was obtained with amplifiers from Multichannel Systems. The artificial cerebrospinal fluid (ACSF) in which the slices were bathed contained (in mM): NaCl, 126; KCl, 3.5; MgSO4, 1; NaH2PO4, 1.25; CaCl2, 1.2; NaHCO3, 26; dextrose, 10, and was aerated with 95% O2, 5% CO2 to a final pH of 7.4. To induce spontaneous activity, a gabaergic blocker (5 μM) bicuculline methiodide (Sigma) was added at some point of the recording. The recording chamber where the slice was placed simulated an interface-style recording chamber, being closed on top and the air being humidified and enriched with oxygen. Bath temperature was maintained at 34.5 – 36 ºC.

3. Results and discussion

3.1. Electrochemical impedance spectroscopy electrode characterization

The neuronal activity is recorded as an extracellular potential or, as it is commonly called, an action potential. An action potential can be described as a short-lasting event in which the electrical membrane potential of a cell rapidly rises and falls. Specifically, in neurons, the action potentials play a central role in the cell-to-cell communication.

In an extracellular recording, the electrical activity detection is generated by the neurons adjacent to the electrode. Thus, the electrode area can be related with the number of neurons which activity can be detected. In general, recordings can be produced by the firing of a single neuron (single-unit activity) or can be generated by several neurons (multi-unit activity). In our case, the use of electrodes at the micro scale, gives the opportunity to detect signals of only one neuron. In living animals the single-unit recordings have provided insights into how does the brain processes information, while the multi-unit activity has usually been used to record changes during normal activity.

The use of MEAs, where the microelectrodes are closely spaced, provides the opportunity to register the activity of one neuron by several electrodes simultaneously. These recordings can be used to identify the number of neurons around each electrode as well as to locate the neurons in the space. This process is called spike sorting and is suitable in areas with well-defined spike characteristics where the type of cells is identified.

The main objective with the neuronal recording is to detect signals with a functional signal-to-noise ratio value of approximately 5:1 or greater in order to differentiate the neural activity from the background noise [41]. Therefore, the noise level represents a limit in the signals that can be detected. In general, two noise sources can be defined; the first one, known as neural noise, can be associated to the large amount of similar background action potentials produced by all the neurons surrounding the electrode. The second one, known as thermal noise, can be associated to the electrode impedance and is defined by equation 1:

$$V_{noise} = \sqrt{4 \cdot K_B \cdot T \cdot R \cdot \Delta f} \quad (1)$$

where K_B is the Boltzman Constant, T is the temperature, R is the real part of the impedance of the electrode and Δf the registered frequency range. It is important to note that only the real part of the impedance contributes to thermal noise. Consequently, for higher electrode impedances, lower signal-to-noise ratio will be obtained.

As it has been mentioned before, the use micro-electrodes with small area enables the possibility of recording activity from only one cell. However, this will increase the electrode impedance and so the associated thermal noise. Thus, it is necessary to use post-processing techniques in order to decrease the electrode impedance. As described in the previous section, the most accepted strategy is increasing the surface area of the electrode without modifying the effective area of the electrode. This can be achieved by increasing the roughness of the electrode surface. By this strategy the thermal noise will be reduced, and the electrode impedance characterization will become a powerful tool to analyze the neural behaviour and to enable the comparison between electrodes.

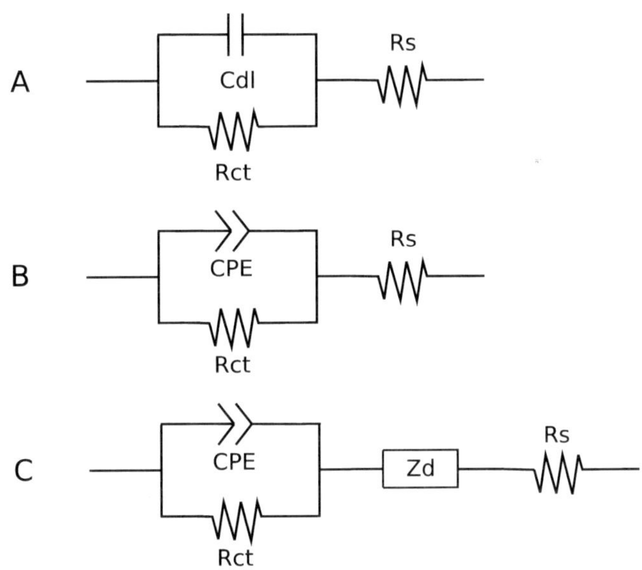

Figure 5. Equivalent circuits used to fitting the EIS characterization data. A) Simplest model where the double layer is modelled with a pure capacitance. B) Model used to fit the measurements of Pt, BkPt and dcSWNT electrodes, where the double layer behaviour is modelled by a CPE component instead of the pure capacitance. C) Model used to fit the measurements of the grMWNT and ppy/SWNT electrodes where have been added a Zd impedance to model the diffusion impedance produced by the porous thin film modifications.

In order to better understand the electrode characterization that will be hold in this section, a brief introduction of the Electrochemical Impedance Spectroscopy (EIS) technique as a tool to characterize electrode-electrolyte interface will be given. This method is based on the application of an AC potential (E(t) = Eo cos(ω + t)) of small amplitude (typically E_0 = 10 mV) that generates an AC current, I (t) = I0 cos(ωt −φ). From the relation of both signals the impedance (Z) is defined (Z= E(t)/I(t)). The obtained impedance data results in a complex number (Z= Z_{real} + j Z_{imag}), which is needed to express the signal attenuation (impedance modulus) and the delay between signals (impedance argument) in a same number. The measurements are carried out at different AC frequencies, and thus the name of impedance spectroscopy. Moreover, from the two ways to plot the impedance data, the Bode Plot, used to represent the polar notation of the complex number, has been chosen to show the data in this work. There, the impedance modulus |Z| and the phase shift angle φ are represented as a function of the frequency ω usually in a logarithmic scale (i.e. Fig 6). In this plot, the resistive processes show a phase angle close to 0 and a flat modulus behaviour, whereas the ones that are dependent on the frequency are more related to capacitive or diffusive processes (phase angles between -90º and -45º). As a result, the impedance spectra can give us a broad overview of the different processes taking place at the electrochemical interface (capacitive, resistive, diffusion effects) showing which one is dominating at a specific range of frequencies.

Figure 5 shows several equivalent circuits that can be used to understand the electrode-electrolyte processes. The equivalent circuit models can be divided in two parts, one related to the access resistance and the other to charge transfer at the electrode double layer. The components used to describe each phenomenon depend on the electrode materials properties. The first approach (model A, Fig. 5A) describes the double layer behaviour as a pure capacitance *Cdl* in parallel combination with the charge transfer resistance *Rct*. In our case this last value is infinitely large (> 10^{10} ohm) and therefore it can be omitted. The access resistance is modelled by the solution resistance *Rs* that mainly depends on the geometric area of the electrode and the conductivity of the solution. Based on this model the impedance modulus at 1 kHz is generally used by neurophysiologists as an indicator of the electrode quality.

However this approximation can only describe the electrode behaviour in all the frequency range when the electrode-electrolyte interface behaves as a pure capacitance. To achieve a better representation of the dissipative double layer behaviour it is necessary to substitute the pure capacitance by a faradaic pseudocapacitance known as the constant phase element CPE (model B, Fig. 5B). The impedance of this CPE is defined in the following equation:

$$Z_{CPE}(\omega) = \frac{1}{q(j\omega)^n} \tag{2}$$

where j = √-1 and ω is the angular frequency in rad s^{-1}. Moreover, the CPE is defined by two parameters; *q* and *n*. *q* indicates the value of the capacitance of the CPE when *n* approaches to 1, while *n* can be correlated with several factors like the surface roughness and a non-uniform current distribution as the more important ones. For *n*=1, CPE describes an ideal capacitor and for *n*=0, CPE describes an ideal resistor.

The third model (model C, Fig. 5c) is related to the electrode modification with conducting polymer films. This approximation introduces diffusion processes in the electrode surface that can be modelled by the inclusion of Z_d impedance in the described equivalent circuit. Diffusion processes in polymer coatings are usually modelled by transmission lines. These models were first proposed by de Levie [42] for porous electrodes and Bisquert et al. [23,24], who applied this theory for the thin film coatings. Thus, the diffusion impedance Zd is generally described by

$$Z_d(\omega) = \left(\frac{R}{\sqrt{\Lambda}}\right)\coth(\sqrt{\Lambda}) \qquad (3)$$

$$\Lambda = 1/R_1 + Q_1(j\omega)^{\varphi}$$

where *R* gives the resistance of the ionic pores of the film; while the R_1, Q_1 and φ, represents the charge transfer at the electrolyte polymer interface at the pores wall.

The experimental results for the different proposed electrodes; bare platinum (Pt), electrodeposited black platinum (BkPt), drop casted SWNTs (dcSWNT), grown MWNTs (grMWNT) and electrodeposited ppy/SWNT composite (ppy/SWNT), have been fitted (Figure 6) to the above described equivalent circuits. The model B was used in the case of Pt, BkPt and dcSWNTs, which behaviour do not present any diffusion process at the electrode-electrolyte interface (Fig. 6 left). On the other hand, the model C was used for the electrodes modified with grMWNT or ppy/SWNT as they present a porous thin film coating than can be associated to an impedance diffusion behaviour (Fig. 6 right). It can be observed the goodness of the fitting in both cases. In Table 1 the obtained values for each parameter of the model is shown.

As expected, the increase of the electrode surface roughness due to all the proposed post-processing techniques lead to a decrease in the initial impedance modulus approximately a 10-fold. This is as a direct consequence of the enlargement of the effective surface area which can be reflected on the equivalent capacitance value of the CPE, increasing from 1 E-9 of the Pt electrode to 2.1 E-7 for the ppy/SWNT (detailed values of *q* in Table 1). It is important to note that the *n* values of the CPE found are significantly lower than 1, demonstrating that using a CPE component is better than using a pure capacitive one. Furthermore, as can be observed in Table 1, the calculated thermal noise according to eq. (1) is reduced from 3.1 μV for the bare platinum to 1.2 -1.5 μV for the modified electrodes. This increase of the signal-to-noise ratio is one of the main reasons for the neural recording improvement.

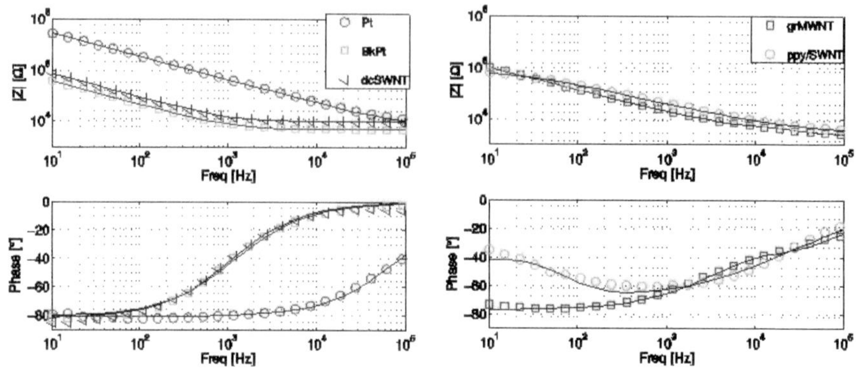

Figure 6. Bode plot (left) of electrochemical impedance of bare platinum electrodes (Pt), electrodeposited black platinum (BkPt) and drop casted SWNTs (dcSWNT). Bode plot (right) of electrochemical impedance of grown MWNTs (grMWNT) and electrodeposited ppy with SWNTs (ppy/SWNT). The fitting results are shown by the solid line, electrodes represented in the left are fitted to the proposed model B and electrodes represented in the right are fitted to the proposed model C.

From Table 1 it can be noticed, that the *Rs* value of the modified electrodes are lower than the value corresponding to the bare platinum electrode. Albeit the *Rs* of BkPt and dcSWNT (7.0 kΩ) are similar to the bare Pt electrode (7.6 kΩ), the values of the grSWNT (2.1 kΩ) and the ppy/SWNT (4.2 kΩ) present a significant reduction. Taking into account, as it has been mentioned before, that *Rs* mainly depends on the electrode area and the conductivity of the solution, which can be assumed to be the same for all cases, the observed variations on this parameter could be considered as an expansion of the real surface area. This assumption is based on a phenomenon that has been previously reported by Abidian et al. [34] and Lu et al. [43].

Parameter	Unit	Pt	BkPt	dcSWNT	grMWNT	ppy/SWNT
|Z| (at 1 kHz)	Ω	3,9E+5 ± 1,7E+5	6,1E+4 ± 5,0E+4	1,4E+4 ± 3,0E+3	3,0E+4 ± 7,1E+3	3,3E+4 ± 1,8E+4
q	μFs^{n-1}	1,0E-9 ± 5,4E-10	3,3E-8 ± 4,8E-8	3,3E-8 ± 1,0E-8	1,9E-8 ± 5,4E-9	2,1E-7 ± 4,1E-8
n	$0 \le n \ge 1$	0,900 ± 0,001	0,899 ± 0,002	0,898 ± 0,017	0,856 ± 0,003	0,627 ± 0,054
Rs	Ω	7,6E+3 ± 1,6E+3	7,0E+3 ± 1,9E+3	7,0E+3 ± 1,5E+2	2,1E+3 ± 3,0E+2	4,2E+3 ± 1,1E+3
R	Ω	—	—	—	1,9E+2 ± 1,9E+2	8,0E+2 ± 7,0E+2
R_1	Ω	—	—	—	5,3E+1 ± 3,9E+1	4,7E+2 ± 3,0E+2
Q_1	$\mu Fs^{\varphi-1}$	—	—	—	8,3E-6 ± 7,5E-6	3,1E-5 ± 4,7E-5
φ	$0 \le \varphi \ge 1$	—	—	—	0,722 ± 0,033	0,952 ± 0,026
Thermal noise	V	3,1E-6 ± 8,4E-7	1,4E-6 ± 3,7E-7	1,2E-6 ± 5,5E-8	1,3E-6 ± 1,6E-7	1,5E-6 ± 3,5E-7
Fittinig model	—	B	B	B	C	C

Table 1. Experimental results and fitting parameters for impedance measurements shown in Figure 6.

For the electrodes that present diffusion processes, grMWNT and ppy/SWNT, the contribution to the total impedance measurement of the CPE parameter and the *Zd* has been separated. This is shown in Figure 7 where it can be observed that the contribution at the low frequency range of *Zd* is significantly higher for the ppy/SWNT coating than for the grMWNT case. However, at high frequencies the *Zd* contribution of ppy/SWNT is lower. Then, it can be stated that for the case of grMWNT the low frequency range is dominated by the CPE behaviour while the high frequency range is dominated by the *Zd* behaviour. On the contrary, in the case of ppy/SWNT the low frequency range is dominated by the *Zd* behaviour and at high frequency range is dominated by the CPE behaviour. This different behaviour can probably be attributed to the small pores that the grMWNT based electrode presents. Fig. 8B shows a SEM image of the vertically aligned MWNTs from where its porous morphology can be related to two causes: the separation between nanotubes (estimated as 10-15 nm after SEM imaging) and the own carbon nanotube inner diameter (typically observed to be 2-3 nm after TEM imaging). The presence of these small diameter pores suggests that ions may not pass through them at low frequency ranges; hence, the active area is only related to the superficial area assuming a CPE behaviour in these frequencies. On the other hand, and as it can be observed in Fig 8D, the ppy/SWNT coating presents less compact porous morphologies with bigger pores. This is also reflected in the *n* value of the CPE (0.627) suggesting a high superficial roughness.

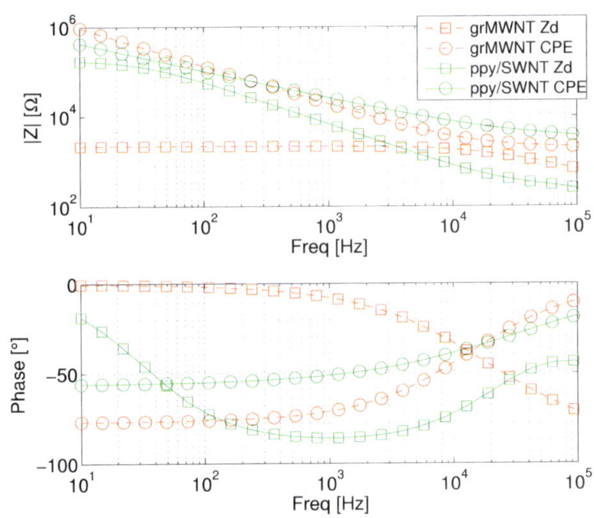

Figure 7. Detailed Bode plot where the total measured impedance of the grMWNT and the ppy/SWNT has been separated into the contribution of the CPE parameter (circles) and the Zd (squares).

The SEM images shown in Figure 8 are useful to relate the morphologies of the above proposed materials for the electrode modification with the parameters of the different impedance models that have been already discussed. The black platinum, BkPt, (Fig. 8A) acquires a fractal structure when it is electrodeposited. This structure contains numerous sub-micrometer even nanometer particles that contribute to increase the final effective surface area, which, sometimes, acquires a cauliflower-like structure. The grown multi-walled carbon nanotubes, grMWNT, (Fig. 8B) consist in a high density of vertically-aligned CNTs that remain stable thanks to Van der Waals forces. These CNTs, depending on the fabrication process, usually have an inner diameter of about 2-3 nm, and an inter-tube space of 10-15 nm. This structure confers to this type of electrode a high porosity all along the length of the CNTs. Albeit the topographical surface of the grMWNTs seems less rough than the surface obtained with the black platinum electrodeposition, the individual tips of the nanotubes at a nanometer scale, highly contribute to the surface roughness. Consequently, the effective surface area of the grMWNT can be considered as a sum of the electrolyte-CNT walls interface and the electrolyte-CNTs tips interfaces.

The drop-casted SWNTs (Fig. 8C) present the typical spaghetti-like structure where the individual nanotubes are entangled producing a compact material. It can be observed that this type of modification does not produce a porous material; only the small diameter (1-2 nm) of the individual SWNTs contribute to an increase of the effective surface area, in a similar way as the black platinum coating. In contrast, the ppy/SWNTs (Fig. 8D) exhibit a very different morphology. A three-dimensional porous microstructure formed by the individual nanotubes covered by the polypyrrole film can be discerned. Because of the diameter of the individual SWNTs (in the range of 1-2 nm), and assuming that they can be grouped in ropes, the thickness of the structures observed in the microscope is basically due to a thick polymer layer.

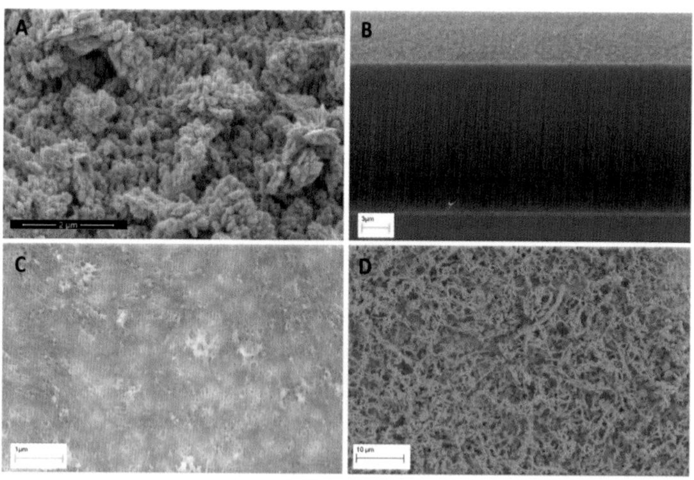

Figure 8. SEM images of the different materials used for the electrode modification surface, A) electrodeposited black platinum; B) grown MWNTs; C) drop casted SWNTs; D) electrodeposited ppy/SWNTs composite.

As a general conclusion of the electrochemical impedance spectroscopy characterization it can be stated that the different proposed post-processing strategies lead to an impedance improvement of approximately 10-fold reduction with respect to the initial bare electrode impedance values. This is especially beneficial for the neural recording electrodes as it supposes a reduction of the thermal noise value and, therefore, a better signal-to-noise ratio. Despite the similar electrochemical characteristics of the described electrodes, the main weakness of the electrodeposited black platinum is its lack of adhesion to the electrode, which compromises its mechanical stability [1]. As expected, this issue affects negatively to the recording electrodes and constitutes a limitation for the electrode reusability as the detachment of the black platinum due to the friction between the electrode and the tissue produces a progressive increase of the electrode impedance.

The motivation for using CNTs is based on the need for finding an alternative electrode improvement method that fulfils the two requirements; low-impedance electrode interface and good mechanical stability required for successful long term recordings. Here, it has been demonstrated that the CNTs modifications present the same good electrical properties as the black platinum along with good mechanical stability of this material that has been previously reported [11, 44, 45].

3.2. *In vitro* recordings with standard and flexible MEAs

The use of standard rigid MEAs poses problems for the recordings from cortical slices. It is especially problematic if the purpose is not the recording of stimulus-evoked responses but of spontaneously generated activity. The generation of specific patterns of activity in cortical slices, such as slow [40] and fast [46] rhythms requires an optimal state of the cortical slices. The generation of these patterns of activity also requires brain tissue from adult rather than juvenile rats [47], which is always more vulnerable and sensitive to factors such as low oxygenation. The main problem that we have encountered with the use of standard MEAs is the combination of ACSF (artificial cerebrospinal fluid) flow under the slice with a good and continued contact of the electrodes with the tissue. This is an easy problem to understand: the brain tissue in the form of a cortical slice is 400 micrometers thick and has to be continuously bathed and oxygenated. Dryness is a killer to the tissue. Given that the standard MEAs have flat electrodes, the key of a good electrical recording resides on a close contact between the electrode and the tissue. However, to keep the slices alive, liquid has to be flowing between the electrodes and the tissue, what not only increases the distance between the electrodes and the slice, but also compromises the mechanical stability that guarantees that the recording is always obtained from the same point.

To deal with these problems we used different strategies: one was to cut the slices thinner down to 300 micrometers, with the objective that the fluid over the slice would be enough to keep the area under bathed and oxygenated. However, even with thinner tissue this problem was not solved. Another strategy that we used in order to bath the slice while maintaining mechanical stability was to put a thin stripe of filter paper on top of the slice and thus keeping the slice in place while the fluid was circulating through the filter paper. This was also helpful to create a kind of interface chamber, which has advantages to maintain slices active and well oxygenated [47,48]. Still, under these conditions there was a loss of activity probably due to the deterioration of the tissue in area in contact with the electrodes.

An strategy used by others (Multichannel systems) to circumvent these issues has been to create a perforated base that is used with negative pressure and thus suction of the slice from below, thus achieving mechanical stability and probably maintaining a warm, humid and oxygenated environment at the bottom of the brain slice. We do not have firsthand experience with this system and we cannot say if it fulfils its purpose.

In spite of all these problems, we were able sometimes to obtain good recordings with the MEAs, and in particular with those modified with grMWNTs (Figure 9b). In Figure 9B we were able to record epileptiform activity induced by the blockade of $GABA_A$ receptors. We think that the grMWNTs makes possible a better contact of the electrode with the tissue thanks to the height of the carbon nanotubes. Still, the recordings did not last long and were not comparable to what is achieved in the same slices with conventional needle-like tungsten electrodes, where recordings can last for several hours.

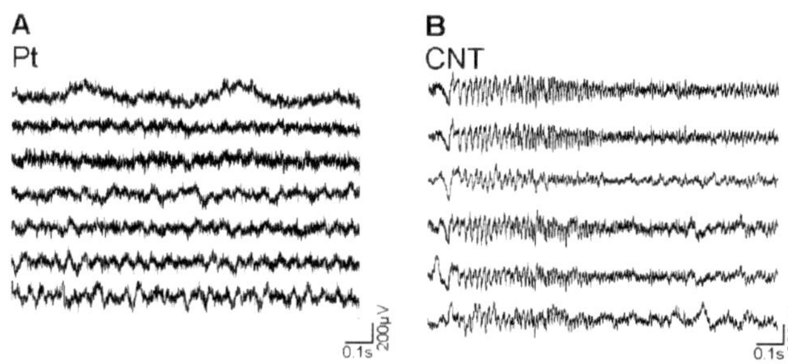

Figure 9. Recordings of spontaneous activity in the slice with Pt electrodes (a) and with grMWNT (b). An oscillation resulting from the gabaergic blockage has been recorded with the CNT-MEA. Recordings have been high passed filtered (1000 Hz) equally in (a) and (b).The same gain (x1000) and filters were used in (a) and (b), while also in both cases 50 µM bicuculline ($GABA_A$ receptor blocker) was present in the bath. Notice an epileptiform discharge in (b). Taken with permission from (20).

Given all the problems encountered with the standard rigid MEAs, we decided to try flexible MEAs that could be positioned on top of the slices in their standard interface chamber. These offers a number of advantages from the point of view of maintaining the tissue alive [47]: there is a good ACSF flow, critical for maintaining the correct ionic and glucose levels as well as temperature and oxygenation, both critical for the normal generation of cortical emergent activity [48]. Furthermore, the filter paper used on the base of interface chambers confers the slice a complete mechanical stability, making unnecessary any other kind of fixation mechanisms. A healthy brain tissue is the basis for a good electrophysiological recording. Once this is achieved, we can place the flexible MEA on top of the slice. We achieve this by means of a micromanipulator. In our experience, the flexible MEA can be held in place. Even when initially there may be problems of stability and the MEA may slip on the surface, once it settles we can obtain stable recordings for a few hours (Figures 10, 11). In Figure 10 we illustrate 5 out of 16 channels recording with our flexible MEAs. The LFP (local field potential) signal shows the ocurrence of three cycles of a spontaneous slow oscillation [40], while at the bottom a high pass filtered channel illustrates the multiunit activity corresponding to the spikes of local neurons. Because this is a flexible MEA and it is at an angle to lie on the tissue, there is a certain pressure made by the MEA on the slice. Even when we cannot measure what that pressure is, we know that it is enough to guarantee a good contact with the electrodes and to maintain the MEA in place. However, the pressure is not too much as to induce any damage on the brain tissue, indeed allowing several hours of successful recordings.

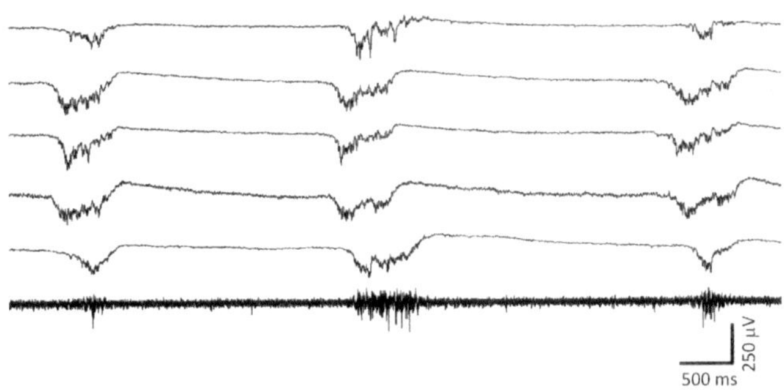

Figure 10. Recordings obtained with a flexible MEA with Pt electrodes. In five channels the LFP (local field potential) are illustrated (unfiltered). In the bottom channel the signal has been high pass filtered (500 Hz) and shows the multiunit activity. The activity of the slice corresponds to a slow oscillation that has 3 cycles in the figures. This activity is spontaneous and is a sign of the good physiological state of the cortical slice. An indication of the good quality of the physiology and the recording is the generation of high frequencies during each cycle, which are visible during the periods of activity (LFP going down).

Once we knew that the flexible MEAs had advantages over the rigid MEAs, we explored the effect of depositing CNTs on the recording points. This was the reason for trying the ppy/SWNTs electrodeposition option, as a way to fabricate electrodes with low-impedance values. As we have said above, this induces a decreased impedance of the electrodes without increasing their size. Decreasing the impedance without increasing the size of the recording point, but increasing the surface is a used strategy to obtain recordings from a small area without high electrical noise [49]. Even when we have not carried out specific measurements of signal to noise, flexible MEAs with carbon nanotubes obtained not only with low noise, but with a good detection of high frequencies (see Figure 11). During the cycles of activity of the slow oscillations, there is local synchronization in high frequencies (30-80 Hz; [46]). In Figure 11 we illustrate that the electrode with NT allowed a good view of these high frequency oscillations during the three cycles of activity displayed. Furthermore, we were able to record with this surface electrode single neurons in the multiunit channel (Figure 11, bottom trace), similarly to the recordings obtained with plated electrodes [49]. Obtaining single units with surface electrodes is unusual. Normally to obtain single units (recording axons for isolated neurons) other techniques need to be used, such as sharp glass-pulled electrodes or plated tungsten electrodes [49] that are placed inside the brain tissue rather than in the surface.

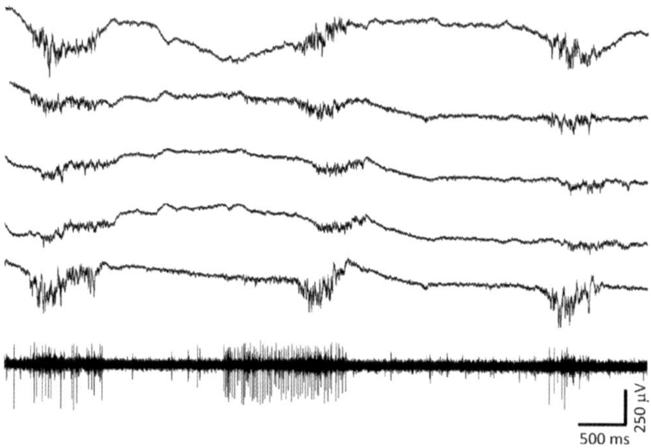

Figure 11. Recordings obtained with a flexible MEA with grMWNTs. In five channels the LFP (local field potential) are illustrated (unfiltered). In the bottom channel the signal has been high pass filtered (500 Hz) and shows multiunit activity. The activity of the slice corresponds to a slow oscillation that has 3 cycles in the figures. This activity is spontaneous and is a sign of the good physiological state of the cortical slice. An indication of the good quality of the physiology and the recording is the generation of high frequencies during each cycle, which are visible during the periods of activity (LFP going down).

We therefore find that for recording spontaneous rhythmic activity from brain slices flexible MEAs yield in our hands better results than rigid MEAs. Furthermore, we find that using CNTs as electrode interface may be a promising technique to obtain high quality electrophysiological recordings from the surface of brain slices.

4. Conclusions

It is a fact that several probe technologies have supposed a revolution on our understanding of the brain behaviour by revealing us how the network neurons work. Because the trend in the neurology field is having a large number of electrodes (MEA devices) arranged closely, that could provide local registering and stimulation, the investigation in the MEA fabrication devices have become a challenge to work in.

Related to the use of microelectrodes for neuronal recordings, one of the main objectives is to achieve low-impedance interfaces. Furthermore, the most important milestone that must be overcome, is maintaining these low-impedance properties. Here we have presented several post-processing strategies in order to decrease the microelectrode impedance. One of the most common ways to achieve it is the black platinum electrodeposition (61 kΩ at 1 kHz). However, the electrochemical response in a long term experiment is not stable for this material due to the platinum detachment. As an alternative there are techniques such as the use of hydrogel coatings that can be used to solve this problem. However we have proposed several carbon nanotube post-processes (drop casted CNTs and CNTs growth) that achieve the low impedance requirements and remain stable during acute recordings (14 kΩ and 30 kΩ at 1 kHz). This is especially beneficial to the neural recording electrodes because it supposes a reduction of the thermal noise value and as a consequence a better signal-to-noise ratio.

In this chapter we have tried to highlight that besides the impedance requirements that the microelectrodes must fulfill and that these values must be maintained for successful chronically experiments, the standard rigid MEAs deal with several other problems. In order to obtain good recordings the electrodes must be in closer contact with the brain slices and the tissue must be maintained alive. These two items have turned against the silicon standard photolithographic technologies, and has favoured the investigation on the polymer micro technologies. So, here it has been proposed the MEA fabrication in SU-8 polymer, which has the properties of being transparent, flexible and low cost. The fabrication of 25 μm thick probes has enabled the possibility to obtain acute recordings for a long period of time. However, in the case of SU-8 probes, impedance improvement strategies have been adapted to the SU-8 material properties. So, it has also been compared the common electrodeposition technique of black platinum to the electrodeposition of a composite of SWNTs and the polymer polypyrrole. The composite overcomes the usual mechanical stability problems reported before for the black platinum. The impedance properties achieved with this composite are very interesting. It presents a low impedance value at 1 kHz (33 kΩ), and a significant diffusion phenomenon consequence of its porous morphologies. This is an important characteristic in the case of microelectrodes used in stimulation, because it is responsible of increasing the safe charge injection limit (Qinj) that establishes differences between stimulating electrodes.

Acknowledgements

RV work has been funded by the project SAF2009-14724-C02-02 co-financed by the Spanish Ministry of Science and Innovation and the European Regional Development Fund. Also the GICSERV Program (6th call), Funded by MICINN has co-financed this work.

MVSV work was supported by Ministerio de Economía y Competitividad (Spain) BFU2011-27094.

MTM wish to thank Spanish Ministry of Science and Innovation (MICINN) and the European Regional Development Fund (ERDF) for financial support under project MICINN TEC2010-15736, and Mercedes Vico-Gallardo for her dedicated and helpful work. J.H.F. acknowledges the Spanish Superior Council for Scientific Research (CSIC) for his JAE-Doc contract

Author details

Gemma Gabriel[1,2*], Xavi Illa[1,2], Anton Guimera[1,2], Beatriz Rebollo[4], Javier Hernández-Ferrer[3], Iñigo Martin-Fernandez[1], Mª Teresa Martínez[3], Philippe Godignon[1,2], Maria V. Sanchez-Vives[4,5] and Rosa Villa[1,2]

1 Instituto de Microelectrónica de Barcelona (IMB-CNM), CSIC, Campus UAB, Barcelona, Spain

2 CIBER-BBN, Networking Center on Bioengineering, Biomaterials and Nanomedicine, Spain

3 Instituto de Carboquímica (CSIC), C/Miguel Luesma Castán 4, Zaragoza, Spain

4 IDIBAPS (Institute of Biomedical Research August Pi y Sunyer), Barcelona, Spain

5 ICREA (Institut Catala de Recerca i Estudis Avançats), Barcelona, Spain

References

[1] Fejtl, M., Stett, A., Nisch, W., Boven-H, K., & Möller, A. (2006). Advances in Network Electrophysiology. *Springer US*, cited 2012 Mar 22, 24-37.

[2] Pine, J. (2006). Advances in Network Electrophysiology. *Springer US*, cited 2012 Mar 22, 3-23.

[3] Whitson, J., Kubota, D., Shimono, K., Jia, Y., & Taketani, M. (2006). Advances in Network Electrophysiology. *Springer US*, cited 2012 Mar 22, 38-68.

[4] Liu, X., Yue, Z., Higgins, MJ, & Wallace, G. G. (2011). Conducting polymers with immobilised fibrillar collagen for enhanced neural interfacing. *Biomaterials*, Oct, 32(30), 7309-17.

[5] Lee, J. Y., Bashur, C. A., Goldstein, A. S., & Schmidt, C. E. (2009). Polypyrrole-coated electrospun PLGA nanofibers for neural tissue applications. *Biomaterials.*, Sep, 30(26), 4325-35.

[6] Ghasemi-Mobarakeh, L., Prabhakaran, M. P., Morshed, M., Nasr-Esfahani, M. H., & Ramakrishna, S. (2009). Electrical Stimulation of Nerve Cells Using Conductive Nanofibrous Scaffolds for Nerve Tissue Engineering. *Tissue Engineering Part A.*, Nov, 15(11), 3605-19.

[7] Gerwig, R., Stett, A., & Stelzle, M. (2012). PEDOT-CNT composite microelectrodes for recording and electrostimulation applications: fabrication, morphology, and electrical properties. *Front. Neuroeng*, 5, 8.

[8] Cellot, G., Cilia, E., Cipollone, S., Rancic, V., Sucapane, A., Giordani, S., et al. (2009). Carbon nanotubes might improve neuronal performance by favouring electrical shortcuts. *Nat Nano.*, Feb, 4(2), 126-33.

[9] Shoval, A., Adams, C., David-Pur, M., Shein, M., Hanein, Y., & Sernagor, E. Carbon Nanotube Electrodes for Effective Interfacing with Retinal Tissue. *Front Neuroengineering*, 2.

[10] Keefer, E. W., Botterman, B. R., Romero, M. I., Rossi, A. F., & Gross, G. W. (2008). Carbon nanotube coating improves neuronal recordings. *Nat Nano*, Jul, 3(7), 434-9.

[11] Voge, C. M., & Stegemann, J. P. Carbon nanotubes in neural interfacing applications. *Journal of Neural Engineering.*, 8(1), 011001.

[12] Jorio, A., Dresselhaus, G., & Dresselhaus, MS. (2008). Carbon nanotubes: advanced topics in the synthesis, structure, properties and applications. *Springer.*

[13] Reich, S., Thomsen, C., & Maultzsch, J. (2008). Carbon Nanotubes: Basic Concepts and Physical Properties. *John Wiley & Sons.*

[14] Malarkey, E. B., & Parpura, V. (2007). Applications of carbon nanotubes in neurobiology. *Neurodegener Dis*, 4(4), 292-9.

[15] Malarkey, E. B., & Parpura, V. (2010). Brain Edema XIV. *Springer Vienna*, cited 2012 Mar 22, 337-341.

[16] Gabriel, G., Gómez, R., Bongard, M., Benito, N., Fernández, E., & Villa, R. (2009). Easily made single-walled carbon nanotube surface microelectrodes for neuronal applications. *Biosensors and Bioelectronics.*, Mar 15, 24(7), 1942-8.

[17] Gabriel, G., Gomez-Martinez, R., & Villa, R. (2008). Single-walled carbon nanotubes deposited on surface electrodes to improve interface impedance. *Physiological Measurement*, S203-S212.

[18] Martin, I., Rius, G., Gabriel, G., Esplandiu, MJ, Mestres, N., Perez-Murano, F., et al. (2007). Local growth of carbon nanotubes by thermal chemical vapor deposition from iron based precursor nanoparticles. *Electron Devices, Spanish Conference on [Internet].*, cited 2010 Aug 25, 329-332.

[19] Martin-Fernandez, I., Gabriel, G., Rius, G., Villa, R., Perez-Murano, F., Lora-Tamayo, E., et al. (2009). Vertically aligned multi-walled carbon nanotube growth on platinum electrodes for bio-impedance applications. *Microelectronic Engineering*, 86(4-6), 806-808.

[20] Martin-Fernandez, I., Gabriel, G., Palomer, X., Reig, R., Sanchez-Vives, Villa. R., et al. Standardized fabrication of MWNTs-based MEA with biocompatible materials for emergent activity in the cortical network. *Biosensors & Bioelectronics.*, Submitted.

[21] Marrese, C. A. (1987). Preparation of strongly adherent platinum black coatings. *Analytical Chemistry. Enero*, 59(1), 217-8.

[22] Robinson, D. (1968). The electrical properties of metal microelectrodes. *Proceedings of the IEEE*, 56(6), 1065-1071.

[23] Rodger, D. C., Fong, A. J., Li, W., Ameri, H., Ahuja, A. K., Gutierrez, C., et al. (2008). Flexible parylene-based multielectrode array technology for high-density neural stimulation and recording. *Sensors and Actuators B: Chemical.*, Jun 16, 132(2), 449-60.

[24] Lacour, S., Benmerah, S., Tarte, E., Fitz, Gerald. J., Serra, J., Mc Mahon, S., et al. (2010). Flexible and stretchable micro-electrodes for in vitro and in vivo neural interfaces. *Medical and Biological Engineering and Computing*, 48(10), 945-54.

[25] Rousche, P. J., Pellinen, D. S., Pivin Jr, D. P., Williams, J. C., Vetter, R. J., & Kipke, D. R. (2001). Flexible polyimide-based intracortical electrode arrays with bioactive capability. *IEEE Transactions on Biomedical Engineering*, 48(3), 361-70.

[26] Metz, S., Bertsch, A., Bertrand, D., & Renaud, P. (2004). Flexible polyimide probes with microelectrodes and embedded microfluidic channels for simultaneous drug delivery and multi-channel monitoring of bioelectric activity. *Biosensors and Bioelectronics*, 19(10), 1309-18.

[27] Stieglitz, T., Beutel, H., Schuettler, M., & Meyer, J. (2000). Micromachined, Polyimide-Based Devices for Flexible Neural Interfaces. *Biomedical Microdevices*, 2, 283-94.

[28] Takeuchi, S., Ziegler, D., Yoshida, Y., Mabuchi, K., & Suzuki, T. (2005). Parylene flexible neural probes integrated with microfluidic channels. *Lab on a Chip Miniaturisation for Chemistry and Biology*, 5(5), 519-23.

[29] Tijero, M., Gabriel, G., Caro, J., Altuna, A., Hernández, R., Villa, R., et al. (2009). SU-8 microprobe with microelectrodes for monitoring electrical impedance in living tissues. *Biosensors and Bioelectronics*, 24(8), 2410-6.

[30] Altuna, A., Menendez la, Prida. L., Bellistri, E., Gabriel, G., Guimerá, A., Berganzo, J., et al. (2012). SU-8 based microprobes with integrated planar electrodes for enhanced neural depth recording. *Biosensors and Bioelectronics*, 37(1), 1-5.

[31] Green, R. A., Lovell, N. H., Wallace, G. G., & Poole-Warren, L. A. (2008). Conducting polymers for neural interfaces: challenges in developing an effective long-term implant. *Biomaterials.*, Sep, 29(24-25), 3393-9.

[32] Poole-Warren, L., Lovell, N., Baek, S., & Green, R. (2010). Development of bioactive conducting polymers for neural interfaces. *Expert Rev Med Devices.*, Jan, 7(1), 35-49.

[33] Hernández-Ferrer, J., Ansón-Casaos, A., & Martínez, M. T. (2012). Electrochemical synthesis and characterization of single-walled carbon nanotubes/polypyrrole films on transparent substrates. *Electrochimica Acta*, Mar 1, 64(0), 1-9.

[34] Abidian, M. R., & Martin, D. C. (2008). Experimental and theoretical characterization of implantable neural microelectrodes modified with conducting polymer nanotubes. *Biomaterials.*, Mar, 29(9), 1273-83.

[35] Chen-H, C., Chuang-C, S., Su-C, H., Hsu-L, W., Yew-R, T., Chang-C, Y., et al. (2011). A three-dimensional flexible microprobe array for neural recording assembled

through electrostatic actuation. *Lab on a Chip- Miniaturisation for Chemistry and Biology*, 11(9), 1647-55.

[36] Fernandez, L. J., Altuna, A., Tijero, M., Gabriel, G., Villa, R., Rodríguez, MJ, et al. (2009). Study of functional viability of SU-8-based microneedles for neural applications. *Journal of Micromechanics and Microengineering*, 19(2).

[37] Altuna, A., Gabriel, G., De La Prida, L. M., Tijero, M., Guimerá, A., Berganzo, J., et al. (2010). SU-8-based microneedles for in vitro neural applications. *Journal of Micromechanics and Microengineering*, 20(6).

[38] Ivorra, A., Gómez, R., Noguera, N., Villa, R., Sola, A., Palacios, L., et al. Minimally invasive silicon probe for electrical impedance measurements in small animals. *Biosens Bioelectron.*, 19(4), 391-9.

[39] Zhao, B., Itkis, M. E., Niyogi, S., Hu, H., Perea, D. E., & Haddon, R. C. (2004). Extinction Coefficients and Purity of Single-Walled Carbon Nanotubes. *Journal of Nanoscience and Nanotechnology*, 4(8), 995-1004.

[40] Sanchez-Vives, M. V., & Mc Cormick, D. A. (2000). Cellular and network mechanisms of rhythmic recurrent activity in neocortex. *Nat. Neurosci.*, Oct, 3(10), 1027-34.

[41] Cogan, S. F. (2008). Neural Stimulation and Recording Electrodes. *Annu. Rev. Biomed. Eng.*, Aug, 10(1), 275-309.

[42] de Levie, R. (1964). On porous electrodes in electrolyte solutions-IV. *Electrochimica Acta.*, Sep, 9(9), 1231-45.

[43] Lu, Y., Li, T., Zhao, X., Li, M., Cao, Y., Yang, H., et al. (2010). Electrodeposited polypyrrole/carbon nanotubes composite films electrodes for neural interfaces. *Biomaterials.*, Jul, 31(19), 5169-81.

[44] Ben-Jacob, E., & Hanein, Y. (2008). Carbon nanotube micro-electrodes for neuronal interfacing. *Journal of Materials Chemistry*, 18(43), 5181.

[45] Keefer, E. W., Botterman, B. R., Romero, M. I., Rossi, A. F., & Gross, G. W. (2008). Carbon nanotube coating improves neuronal recordings. *Nat Nano.*, Jul, 3(7), 434-9.

[46] Compte, A., Reig, R., Descalzo, V. F., Harvey, MA, Puccini, G. D., & Sanchez-Vives, M. V. Spontaneous High-Frequency (10-80 Hz) Oscillations during Up States in the Cerebral Cortex In Vitro. *J. Neurosci.*, 28(51), 13828-44.

[47] Sanchez-Vives, M. V. (2012). Spontaneous rhythmic activity in the adult cerebral cortex in vitro. *Isolated brain circuits. Ballanyi K. Springer.*

[48] Reig, R., Mattia, M., Compte, A., Belmonte, C., & Sanchez-Vives, M. V. (2010). Temperature modulation of slow and fast cortical rhythms. *J. Neurophysiol.*, Mar, 103(3), 1253-61.

[49] Merrill, E. G., & Ainsworth, A. (1972). Glass-coated platinum-plated tungsten microelectrodes. *Med. & biol. Engng.*, Sep, 10(5), 662-72.

Chapter 16

Phonon Scattering and Electron Transport in Single Wall Carbon Nanotube

Bo Xu, Jiang Yin and Zhiguo Liu

Additional information is available at the end of the chapter

1. Introduction

Single-walled carbon nanotube (SWCNTs) can be thought of as graphene a single graphene sheet wrapped up to form a one-atom-thick cylinders. CNTs were discovered by Iijima [1] in 1991, since then the excellent charge transport properties of CNTs have been of great interest, for its great potential applications in nanoelectronics, such as high-speed field-effect transistors (FETs) [2, 3], single-electron memories [4], and chemical sensors [5]. The CNT has an atomic and electronic structure that gives it unique advantage as an FET channel. The band gap of the semiconducting SWCNT is inversely proportional to the tube diameter, which allows such tubes to be used in various different applications. CNTs display outstanding electrical properties such as ballistic transport or diffusive transport with long mean free path, which is of the order of a micrometer. Ballistic transport in CNTs has been experimentally demonstrated for low-bias conditions at low temperatures [6, 7]. High-performance CNT transistors operating close to the ballistic limit have also been reported [8-10]. Besides, one of the most important advantages is the CNT's excellent transport properties due to the high carrier mobility. The experimentally obtained carrier mobilities are of the orders 10^4 cm^2/Vs [11, 12] so exceptional device characteristics can indeed be expected. Current transport in long metallic CNTs, however, is found to saturate at ~ 25 μA at high biases, and the saturation mechanism is attributed to phonon scattering [13]. On the other hand, for short length metallic tubes, the current is found not to saturate but to increase well beyond the above limit [14, 15].

Nevertheless, carrier transport in these shorter tubes is still influenced by phonon scattering, and warrants a detailed physical understating of the scattering mechanisms due to its implications on device characteristics for both metallic as well as semiconducting CNTs. And there have been many theoretical studies on the calculation of carrier scattering rates and mobilities in CNTs using semiclassical transport simulation based on the Boltzmann equation [16-22]. Similarly, phonon mode calculations for CNTs are also performed with varying degrees of complexity: continuum and forceconstant models [23-25] to first-principles based methods [26-28]. The determination of electron-phonon coupling strength is performed by using tight binding calculations [29-31] as well as first-principles techniques [32]. Non-equilibrium Green's function formalism also has been employed to treat the effects of phonon scattering in CNT [33-35].

In this work, we will show our physical simulation on the carrier mobilities under acoustic phonon scattering process. This work is organized as follows. In section II, we start with the basic properties of CNTs. A brief summary of the electron-phonon scattering is discussed in Section III. In this section, we will review the latest theoretical developments aimed at exploring the effect of electron-phonon interactions on carrier mobility. In the last section, we will describe the simulation approach we use. In this section, we also present the simulation results to discuss the acoustic phonon scattering effect on the charge carrier mobility.

2. Electronic strcutures of CNTs

2.1. Structure of CNTs

The SWCNT is a hollow cylinder-shaped molecule with a diameter in the order of 1 nm. SWCNT can be viewed conceptually as graphene sheets rolled up into concentric cylinders. The atomic structure of a single-walled CNT is conveniently explained in terms of two vectors C_h and T. T is called translational vector, it defines the direction of CNT axis. C_h is called chiral vector, representing the circumference of a CNT. A specific SWCNT is defined by two integers (n,m) with $n \geq m \geq 0$ related to the chiral vector $C_h = na_1 + ma_2$, where a_1 and a_2 are the basis vectors ofthe graphene lattice as shown in Fig.1a. Fig.1b shows the chiral vector for a so-called (5,5) armchair nanotube, where the SWCNT is made by joining the ends of the chiral vector, *i.e.*, dashed blue lines. Three categories of SWCNT are now defined: the armchair (n,n), the zigzag (n, 0) and the chiral nanotube (n, m) with n > m > 0 (see Figure Fig.1b, Fig.1c, Fig.1d).

2.2. Electronic structure of CNTs

The electronic structure of a SWCNT is deduced from the energy dispersion of graphene. The band structure of the SWCNT is found by imposing periodic boundary conditions around the circumference of the tube, *i.e.*, the wave function has to be single valued:

$$\Psi_k(r + C_h) = \Psi_k(r) \tag{1}$$

where k is a wave vector and r is a real space lattice vector of the graphene lattice. This leads to periodic boundary condition in momentum space

$$k \cdot C_h = 2\pi p \quad (2)$$

Where p is an integer. In other words, the k-vector projected onto the chiral vector $k_{//}$ (along the circumference) becomes quantized, while the k-vector $k\perp$ along the tube axis is continuous for an infinite nanotube. The 1D dispersion or band structure of a SWCNT is thus made of the energy bands related to different quantized values p as a function of $k\perp$. Whether or not these quantization lines cross a K-point makes the SWCNT a metal or a semiconductor.

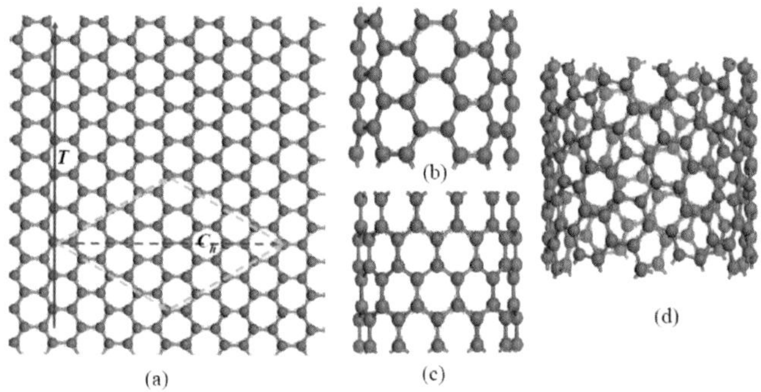

Figure 1. a) The example of folding a (4,4) armchair SWCNT from the graphene sheet. C_h and T are the chiral and the translation vectors of the SWCNT, respectively. (b), (c), and (d) are the structure of different kinds of SWCNTs. (b) A (6, 6) armchair SWCNT. (c) A (12, 0) zigzag SWCNT. (c) A chiral (12, 6) SWCNT.

For the armchair SWCNTs, these nanotubes are truly metallic and have two bands crossing at the Fermi level (Figure 2a). The bands stem from the quantization lines drawn in Figure 2d in the reciprocal lattice. The corners of the hexagons are the K-points, where the conduction and the valence band of the graphene dispersion touch. One of the quantization lines (thick dashed line) passes through two K-points making the tube metallic.

Figure 2b shows the band structure for a (15, 0) zigzag tube which is metallic judging from the degenerate band crossing the Fermi level. The bands stem from the quantization lines drawn in Figure 2e. It is seen that the bands touching at the Fermi level are two times degenerate. However, the band structure is calculated from the dispersion graphene, while the CNT has a curvature around the circumference of the tube. The curvature slightly modifies the band structure by moving the K-points.For the zigzag tube with n ≠ 3*integer, such as (16, 0) zigzag tube (Figure 2c), in the reciprocal space the quantization lines do not cross the K-points. It has a band gap in the order of ~ 1eV.

So, theoretical studies have shown that a single-walled CNT can be either metallic or semiconducting depending on its chirality and diameter. The armchair SWCNTs are a group of truly metallic conductors with two bands crossing the Fermi level. For n-m=3*integer, the nanotubes would be quasi-semiconducting with a small band gap proportional to $1/d^2$. Typical band gaps are in the order of tens of meV. Finally, a group of zigzag and chiral SWCNTs is semiconducting (n-m ≠ 3*integer) with bigger band gaps. The band gap of these tubes are in the order of ~1 eV and scales as $E_{gap} \sim 1/d$, where d is the diameter of the SWCNT [36].

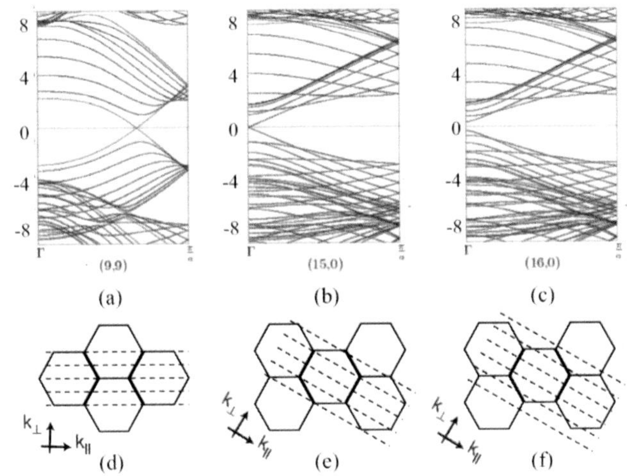

Figure 2. a) The band structure of a (9,9) armchair nanotube. One of the quantization lines bands crossing the Fermi energy stem from the quantization line in (d), which crosses the K-points and makes this SWCNT metallic. (b) The band structure of a (15,0) zigzag nanotube. It has a doubly degenerate band crossing the Fermi level in (e). When the curvature effects are taken into account the K-points move from the corners of the hexagon to the red dots, which makes metallic zigzag nanotubes to be small band gap semiconductors. (c) and (f) Some tubes are semiconducting with a bigger band gap as the (16,0) zigzag tube.

2.3. Electronic transport of CNTs

For the metallic armchair CNTs, the valence and conduction bands cross at the Fermi level just as in the case of graphene. The two crossing bands provide the tube with two conducting channels at and close to the Fermi level, where in each of these bands, two electrons of opposite spins can co-exists. By the Landauer formula, the conductance is then:

$$G = \left(4e^2/h\right)T \tag{3}$$

where e is the electron charge, h is Planck's constant, and T is the transmission coefficient for electrons through the sample. The conductance of a ballistic SWCNT with perfect contacts (T = 1) is then: $R_{CNT} = 4e^2/h \approx 150\ \mu S$, corresponding to a resistance of 6.5 kΩ. In addition, the scattering of charge carriers along the length of CNTs results in a Drude-like resistance. The presence of scatterers that gives a mean free path l for backscattering contributes an ohmic resistance to the tube, $R_d \propto L/l$, where L is the length of CNT. Thus, the total resistance of a SWCNT contacted by metal leads on both ends is sum of these two contributions: $R_{tot} = R_{CNT} + R_d$.

Different band structures are obtained for a truly metallic, a quasi-metallic and a semiconducting nanotube. The various band structures are illustrated in Figure 2 which displays a quasi-metallic (15,0) zigzag, a semiconducting (16,0) zig-zag, and an armchair (9,9) SWCNT band structure. In the case of the armchair tube, the two bands conduct the current while in the case of quasi-metallic zigzag or chiral SWCNTs, a small energy gap of few meV exists due to the nanotube curvature. This gap is important at low-temperatures and can suppress electron transport. However, at room-temperature, the thermal energy is larger than the gap and the tubes show metallic behavior. Semiconducting tubes possess an energy-gap of ≈ 0.5-1 eV, where zigzag SWCNTs have their DOS singularities at the Γ-point.

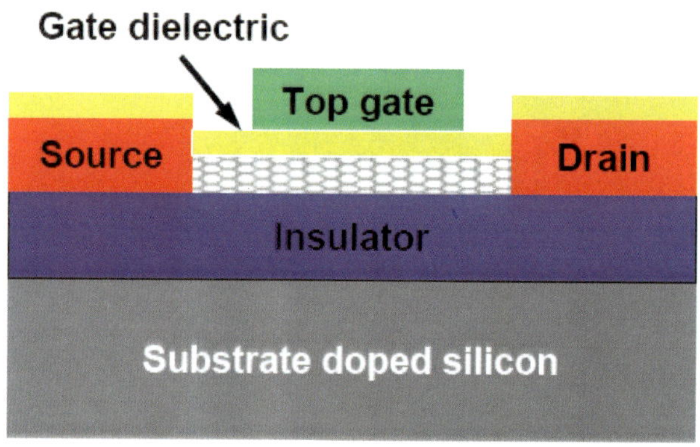

Figure 3. The model of planar-structure CNTFET

CNT filed effect transistor (CNTFET) can distinguish between the two character types. In the case of metallic tubes, the conductance is V_G independent, with the crossing bands providing conducting electrons independently on the V_G, i.e., the gate potential does not change the number of conduction channels. On the other hand, the conductance in semiconducting tubes is strongly affected by the V_G and can change by orders of magnitude. The CNT has an atomic and electronic structure that gives it unique advantage as an FET channel.

3. Electron-phonon scattering in CNTs

CNTs have been extensively explored for nanoelectronic applications due to their excellent electrical properties. Scattering plays an important role on carrier transport CNTs [13, 14]. It has demonstrated that at finite temperatures or high biases, electron-phonon scattering becomes significant. It can be divided into the low- and high-energy regimes, corresponding to acoustic-phonon scattering and optical- or zone-boundary-phonon scattering. Because of the light mass and strong bonds, the optical-phonon energy is very high in CNTs

$\hbar\omega_0 \sim 160 meV\, k_B T$ at 300 K, meaning that these phonons are not thermally populated, which is one of the reasons for the high room-temperature mobilities in CNTs.At small source-drain biases and moderate temperatures the mean free path in clean tubes is set by acoustic-phonon scattering, as shown by a number of experimental and theoretical works (15, 37, 38). A straightforward calculation shows that, when the Fermi level is in the linear part of the electron dispersion relation, the scattering rate for a tube of linear mass density ρ and sound velocity v_s is given by:

$$\frac{1}{\tau_{ac}} \cong \frac{\pi}{\hbar}(dE_g \Big/ d\varepsilon)\frac{k_B T}{\rho v_s^2}\frac{1}{h\, v_F} \tag{4}$$

The coupling is through the strain dependence of the band gap. Depending on the tube, the dominant coupling can either be through the stretching or the twisting of the tube. The linear temperature dependence comes from the thermal occupation of the (small-momentum transfer) acoustic phonon responsible for backscattering.In addition to the low-energy acoustic phonons, electron (or hole) scattering by the radial breathing mode (RBM) is important in the low bias regime. The RBM phonon energy is inversely proportional to the tube diameter8, and its energy is comparable to the thermal energy at room temperature for tubes in the diameter range of d_{CNT} = 1.5–2.0 nm, which are of interest for electronic applications. As the acoustic mean free path is very long-of the order of a micrometre at room temperature-electrons can be accelerated up to the RBM energy not only thermally, but also by an applied bias of a few Vcm^{-1}

Unlike acoustic phonon scattering, optical phonon scattering is very strong in CNTs; optical phonons contract and elongate the C–C bond length and lead to a strong modulation of the electronic structure. However, for electrons to emit an optical phonon, their energies must be larger than the optical phonon energy. This can only be achieved under high bias conditions. Such scattering processes were first observed in metallic tubes [13, 14, 15] and later in semiconducting tubes [39]. At large source-drain biases, the electrons in the tube can accelerate to energies well above the Fermi energy, and these hot electrons can scatter very efficiently by emitting optical and zone-boundary phonons. The scattering rate for this process is

$$\frac{1}{\tau_q^{\alpha}} \cong \frac{2\pi}{\hbar} \mid D_{k,q}^{\alpha} \mid^2 \frac{2\hbar}{\rho \Omega_q^{\alpha}} \frac{1}{h\, v_F}\left(\langle n \rangle + \frac{1}{2}\right) \tag{5}$$

Where $D_{k,q}^{\alpha}$ is the matrix element, Ω_q^{α} is the phonon frequency, and $<n>$ is the occupancy of the mode in the branch α with wave vector q. This process is rapid, resulting in mean free paths that are measured to be in the range of 10 nm [13, 14], a hundred times shorter than the micro-scale mean free paths at small biases.

In summary, the inelastic scattering rates determining transport properties of CNTs vary by four orders of magnitude depending on the energy of the electrons and their angular momentum (sub-band index) as shown in Figure 4 [40]. The weakest is the acoustic (primarily

RBM) phonon scattering, which has linear temperature dependence. The optical phonon scattering rate, which is two orders of magnitude stronger, is nearly temperature independent. Finally, another two orders of magnitude stronger than the optical phonon scattering is impact excitation.

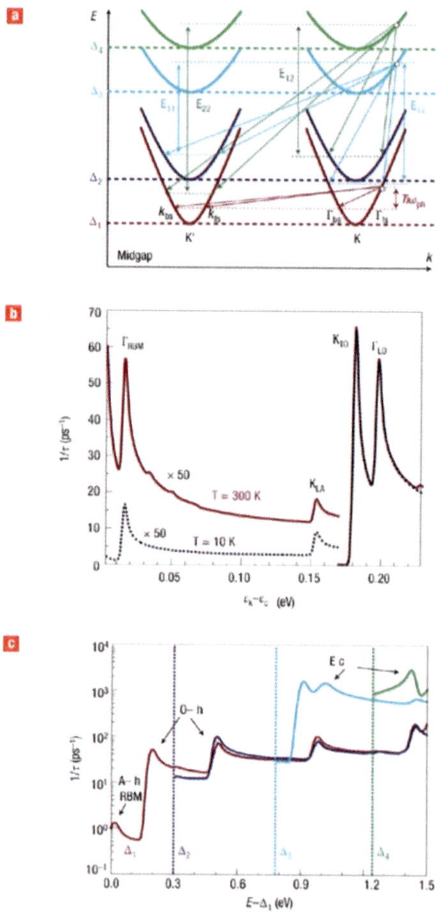

Figure 4. a), Schematic illustration of the intra-sub-band (Γ) and inter-sub-band (K) phonon scattering mechanisms (red) and electron impact excitation (blue and green curves) for the first four conduction bands.

The different conduction band edges are labelled as Δ_i and the resulting electronic excitations are denoted as E_{iij}. Subscripts bs and fs stand for the back and forward scattering. b), Calculated phonon scattering rate for a (25,0) nanotube showing weak acoustic phonon scattering and strong optical phonon scattering. c), Calculated inelastic scattering rate for a (19,0) nanotube over a wide carrier energy range. Different colours correspond to the scattering rates of electrons in bands with different circumferential angular momentum. The vertical lines show the bottoms of the conduction bands 2 (blue), 3 (cyan) and 4 (green) with respect to the fundamental band edge Δ_1. Some of the characteristic peaks in the scattering, due to the longtitudinal (LA) acoustic phonons (A-Ph), radial breathing mode (RBM), longitudinal (LO) and transverse (TO) optical phonons (O-Ph) and impact electronic excitation (I-Exc), are labelled. In b and c the electron scattering rate is shown as a function of the excess energy of the electron above the first conduction-band minimum. (Figure 2 in ref. 40)

4. Acoustic phonons scattering effect on carrier mobility of semiconducting SWCNTs

A number of groups have reported modeling and simulation studies of the carrier transport in CNTs [41-48]. Our intent in this section is not to review these works. Instead, we briefly describe the techniques we currently use to study the intrinsic carrier mobility of semicoducting SWCNTs. The semiconducting zigzag SWCNTs have large intrinsic carrier mobility due to the weak acoustic phonon scattering. Although recently much experimental progress has been achieved on improving the charge carrier mobility of semiconducting CNTs [8, 11, 49, 50], there are a lot of works on the theoretical understanding of the carrier mobility in the semiconducting zigzag SWCNTs [16, 18, 22, 51]. The carrier mobility of the semiconducting zigzag SWCNT can reach 7.9×10^4 cm^2/Vs at room temperature experimentally [8]. Even the higher mobility up to 1.2×10^5 cm^2/Vs for a 4.6 nm diameter semiconducting zigzag SWCNT at room temperature has been predicted by the zone-folding method approximation [16]. Perebeinos *et. al.* [18] studied the electron-phonon scattering in the semiconducting zigzag SWCNTs by using the tight-binding model. They found that under high fields, the dominant scattering mechanism is interband scattering by the longitudinal optical phonons, while under very low field, the scattering is entirely from the acoustic phonons. The acoustic phonon scattering process could be appropriately described by using the deformation-potential theory [52-55], which has been used extensively in carbon nanostructures and the tightly packed organic molecular crystals.

4.1. Acoustic phonons scattering based on the deformation-potential theory

The spedific conducitivity of a three-dimensional solid can be written as:

$$\sigma = q(n_e\mu_e + n_h\mu_h) \tag{6}$$

where n_e and n_h are the density of mobile electrons and holes, respectively, and μ_e and μ_h are their mobilities, respectively.

Bardeen and Shockley [52] derived an analytical expression for the intrinsic carrier mobility (μ) by assuming that the change of the energy of the electron scattered by an acoustic phonon is proportional to the deformation:

$$\mu_e = \frac{2^{3/2}\pi^{1/2}}{3} \frac{c_\perp \hbar^4 e}{\varepsilon_{1e}^2 m_e^{*5/2} (k_B T)^{3/2}} \tag{7}$$

$$\mu_h = \frac{2^{3/2}\pi^{1/2}}{3} \frac{c_\perp \hbar^4 e}{\varepsilon_{1h}^2 m_h^{*5/2} (k_B T)^{3/2}} \tag{8}$$

Here $\varepsilon_{1e}$and ε_{1h} are the deformation potentials, defined as:

$$\varepsilon_{1e} = \frac{\delta W_{c,l}}{\Delta}, \ \varepsilon_{1h} = \frac{\delta W_{v,u}}{\Delta} \tag{9}$$

Beleznay *et. al.* [53] reformulated the analytical expression of the carrier mobility in the one dimensional case to study the charge transport in the guanine stack. During the propagation of acoustic wave, the stretching vibration of the crystal lattices may be expressed as:

$$\delta\vec{r} = A_q \hat{I}_q \cos(\vec{q} \bullet \vec{r} - \omega_q t) = \frac{1}{2} A_q \hat{I}_q [e^{i(\vec{q} \bullet \vec{r} - \omega_q t)} - e^{-i(\vec{q} \bullet \vec{r} - \omega_q t)}] \tag{10}$$

where A_q is amplitude, $\vec{q}$ is wave vector, is the unit vector along the direction of propagation, $\hat{I}_q$ is the angular frequency of vibration. The displacement difference between two points with mean distance a (a is the lattice constant in one dimension):

$$\delta\vec{r}(a) - \delta\vec{r}(0) = a(\nabla_r \delta\vec{r}) \tag{11}$$

The deformation potential theory proposed by Bardeen and Schockly is based on the face that when the change of lattice is very small, the variations of the top of valance band and the bottom of conduction band are linearly related to the variation of lattice constant, therefore the energy of the top of valence band (E_v) and the energy of the bottom of conduction band (E_c) may be expressed as follows:

$$E_v = E_{0v} + E_{1v}\Delta(\vec{r}); E_c = E_{0c} + E_{1c}\Delta(\vec{r}) \tag{12}$$

Where $\Delta = \delta a / a$, E_{0c} and E_{0v} are the energies to the top of valence band and bottom of conduction band, respectively, in the undeformed crystal. Bardeen and Schockly proved that $E_1\Delta(\vec{r})$ may be considered as preturbational potential and referred to the following expression:

$$\delta U(\vec{r})=E_1\Delta(\vec{r})=E_1(\nabla_r\delta\vec{r}) \tag{13}$$

The matrix element obtained form perturbational potential is $H_{k'k}=\psi_{k'}\mid\delta U\mid\psi_k=E_1\psi_{k'}\mid\nabla_r\delta\vec{r}\mid\psi_k$We consider only the 1st Brillouin zone by expanding $u_k(\vec{r})$ and neglecting the terms of higher order and then integrate with respect to the unit cell to derive

$$\left|H_{k'k}\right|^2=\frac{1}{4}q^2A_q^2E_1^2,\ k-k'\pm q=0 \tag{14}$$

Upon considering the crystal as continuous medium, time-averaging and summing for the whole crystal, one can get the total average kinetic energy for the whole crystal :

$$E_T=\frac{1}{2}\rho L\ A_q^2\omega_q^2 \tag{15}$$

When the temperatures is higher than Debye temperature, based on classical law of equipartition energy, we have

$$E_T=\frac{1}{2}\rho L\ A_q^2\omega_q^2=\frac{1}{2}k_BT \tag{16}$$

Where $\omega_q=C_l q$, C_l is the velocity of longitudinal wave. Thus, we obtain

$$\left|H_{k'k}\right|^2=\frac{1}{4}q^2A_q^2E_1^2=k_BTE_1^2\Big/2\rho L\ C_1^2 \tag{17}$$

From quantum mechanical theory, the scattering probability from $\vec{k}$to $\vec{k}'$is:

$$\Theta(k',k)=\frac{2\pi}{\hbar}\left|H_{k'k}\right|^2\delta[E(k)-E(k')\pm\hbar\omega_q] \tag{18}$$

For the very small energy of phonon the scattering may be considered as elastic, and by summing up the probabilities of phonon absorption and emission. From quantum theory of solid, the reciprocal of relaxation time is:

$$\frac{1}{\tau}=\sum_{k'}\Theta(k',k)(1-k_z/k_z') \tag{19}$$

With the effective mass approximation, we can get:

$$\frac{1}{\tau}=\frac{2(2m^{*})^{1/2}k_{B}TE_{1}^{2}}{\hbar^{2}\rho C_{1}^{2}}(E-E_{0})^{-1/2} \tag{20}$$

In semiconductor physics the mobility is defined by

$$\mu=e\tau/m^{*} \tag{21}$$

By using Boltzmann distribution function, we can get the charge mobility in one dimensional crystal:

$$\mu=\frac{e\bar{\tau}}{m^{*}}=\frac{e\hbar^{2}C}{(2\pi k_{B}T)^{1/2}m^{*3/2}E_{1}^{2}} \tag{22}$$

where $\bar{\tau}$ is the average scattering relaxation time of the acoustic phonon, m^{*} is the effective mass of the charge, C is the stretching modulus, E_{1} is the deformation-potential constant. $\bar{\tau}$, m^{*} and C are defined as:

$$\bar{\tau}=\hbar^{2}C/(2\pi m^{*}k_{B}T)^{1/2}E_{1}^{2} \tag{23}$$

$$m^{*}=\hbar^{2}[\partial^{2}E(k)/\partial k^{2}]^{-1}|_{k=0} \tag{24}$$

$$C=a_{0}[\partial^{2}E(k_{F})/\partial a^{2}]|_{a=a_{0}} \tag{25}$$

$$E_{1}=\delta E(k_{F})a/\delta a \tag{26}$$

where $E(k)$ is the energy band and a is the lattice constant, the deformation-potential constant $E_{1}=\delta E(k_{F})a/\delta a$, where $\delta E(k_{F})$ is the conduction or valence band shift near Fermi surface that caused by the small change δa in the lattice constant. Although all of these quantities in Eq (23) are obtained from the first-principles calculations, it is a simple view of the full Boltzmann transport equation, and this method have been previously applied in the study of the graphene nanoribbons [54] and the functionalized CNTs [55].In this simple approximation, we could find that the intrinsic carrier mobility scattered by the longitudinal acoustic phonons varies with the temperature approximately as $T^{-1/2}$, not the empirical relation T^{-1} by the experimental captured, which is the combined result of other scattering mechanisms.

4.2. Calculation method and results

To calculate the carrier mobility of the semiconducting zigzag SWCNTs, there are three parameters to be determined as shown in the above formula, namely, m^{*}, C, and E_{1}. All these parameters can be calculated by the first-principles method. The density functional theory calculations were performed with *Vienna ab initio simulation pack* (VASP) code [56, 57], using Perdew-Burke-Ernzerhof exchange-correlation functional [58]. In the first principles calculations, the ion-electron interactions were treated with the projected augmented wave (PAW) approximation [59, 60]. The plane wave cutoff energy was set to 500 eV and the convergence threshold for energy was 10^{-5} eV. Brillouin Zone integration was carried out at 1×1×25 Monchorst-Pack k-grids, and 150 uniform k-points along the one-dimensional Brillouin Zone are used to obtain the band structures. The symmetric unrestricted optimizations for geometry are performed using the conjugate gradient scheme until the force acting on every atom is less than 10 meV/Å. To obtain the value of the stretching modulus C and the deformation-potential constant E_{1}, we calculated the band structures of unit cells under the uniaxial stress applied along the periodic direction, allowing a unitary deformation in the range of ±0.01%. With the changes of the energy at Fermi energy two straight lines with the correlation coefficient >0.999 are obtained. From the slope of the straight lines, the deformation-potential constants E_{1} are obtained. The stretching modulus C can be estimated from the variance obtained from the second derivative of the total energy upon unitary deformation.

Electronically, SWCNTs can behave as either metallic or semiconducting depending on the chirality of their atomic arrangements and diameter. The band structure calculations have predicted that the armchair SWCNTs with (n, n) indices are truly metallic with the finite density of states at Fermi level, whereas the zigzag SWCNTs are metallic, if n is a multiple of 3 and all others are semiconducting in the unstrained condition. So the semiconducting zigzag SWCNTs with $(n, 0)$ indices selected for the simulation correspond to $n=3q+1$ *and* $3q+2$ (q= 2, 3, 4, 5, and 6) in this paper.

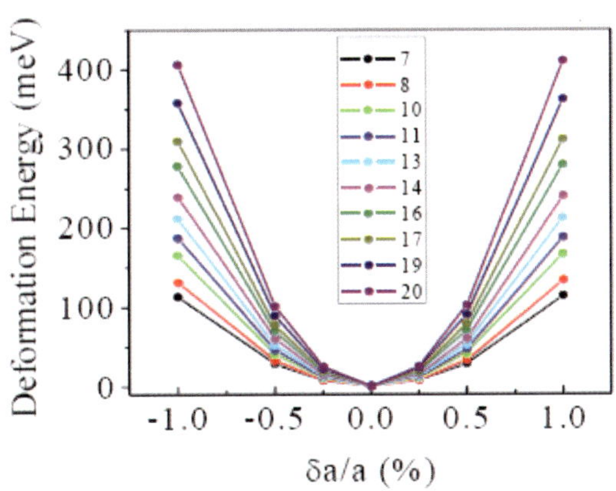

Figure 5. Deformation energy as a function of compressing and elongating the semiconducting zigzag SWCNTs along the longitudinal direction.

The stretching modulus was evaluated by compressing and elongating the semiconducting zigzag SWCNTs along the longitudinal direction. For the evaluation of the elastic properties all atomic positions were fully relaxed. Typically the unstrained configurations were calculated first and then the strain was applied in steps of 0.25% in units of strain percentage for strains less than 1%. The results of these simulations are presented in Figure 5. It clearly demonstrates the parabolic form of the strain energy as a function of the strain, reminiscent to the parabolic potential energy derived from Hook's law for the macroscopic springs. It is interesting to note that the same strain can lead to the increasingly high deformation energies in SWCNT with larger *n*, due to the additivity of the energy required to compress/elongate a larger number of carbon-carbon bonds, within the SWCNT network. The second derivative of the total energy could be obtained easily. The stretching modulus was also shown in Figure 6.

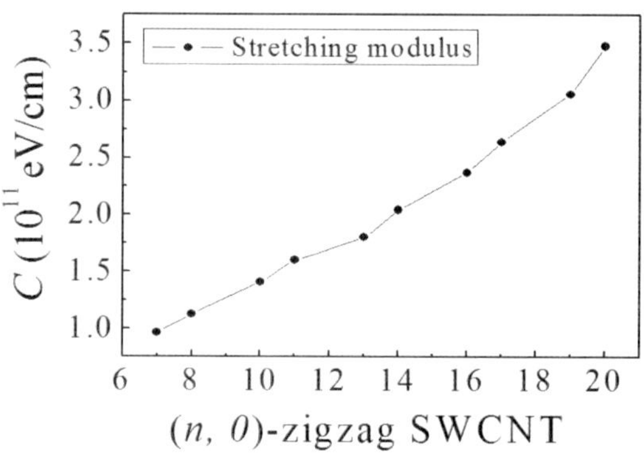

Figure 6. The calculated stretching modulus C of semiconducting zigzag SWCNTs as a function of *n*.

From the shape of the band structure, we have calculated the effective masses of the electrons and holes of the semiconducting zigzag SWCNTs. We can fit two curves the energy *E(k)* versus *k* points for the bottom of the conduction band and the top of the valence band near Γ point, so we got the effective mass m_e^* and m_h^* for electron and hole, respectively, as shown in Figure 7. We can see that the effective masses of holes of SWCNTs are smaller than those of electrons for $n=3q+1$, while it is just opposite for $n=3q+2$. This is due to the curvature effects. The obtained effective masses of the semiconducting zigzag SWCNTs, are quite well in agreement with the earlier theoretical reported values [61].

The deformation-potential constant E_{1}, which represents the scattering of an electron or hole from the acoustic phonon, was calculated from the algebraic average of the band edge shifts in the cases of the dilatation and compression. E_c and E_v are the deformation-potential constants for the conduction band and the valence band, respectively. The deformation-potential constants, E_c and E_v, calculated from the band edge shifts of the bottom of the conduction band and the top of the valence band as shown in Figure 8b and Figure 4c. The deformation-potential constants, E_c and E_v, as a function of the deformation proportion for q=2, 3, 4, 5, and 6 are displayed in Figure 8a. It is noted that E_c is larger than E_v in one order of magnitude for $n=3q+2$; while E_c is less than E_v in one order of magnitude for $n=3q+1$. Except for n=8, we find that there is always one of the deformation-potential constants about 14 eV between E_c or E_v. This is agreement with the previous result.[12]

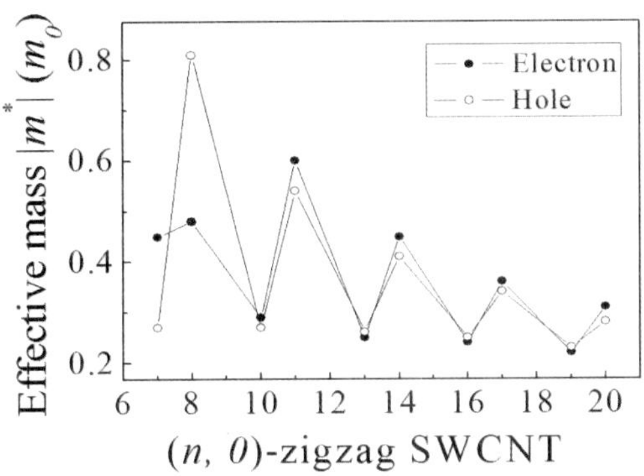

Figure 7. The electron and hole effective masses $|m^*|$ of semiconducting zigzag SWCNTs as a function of n.

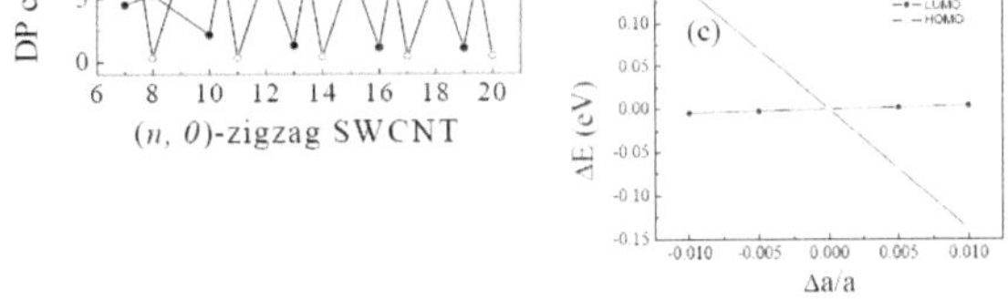

Figure 8. a): The deformation-potential (DP) constants E_c and E_v of semiconducting zigzag SWCNTs as a function of n. b) and c): Band edge shifts of the bottom of conduction band and the top of valence band as a function of deformation proportion for (13, 0) and (14, 0) SWCNT, respectively.

The calculated effective masses, the stretching modulus, and the deformation-potential constants of the semiconducting zigzag SWCNTs are summarized in Table 1. The electron and hole mobilities at room temperature can be calculated by the Eq (23) from these three parameters are also displayed in Table 1. We plotted the mobilities of electrons and holes of the semiconducting zigzag SWCNTs calculated as a function of the diameter in Figure 9.

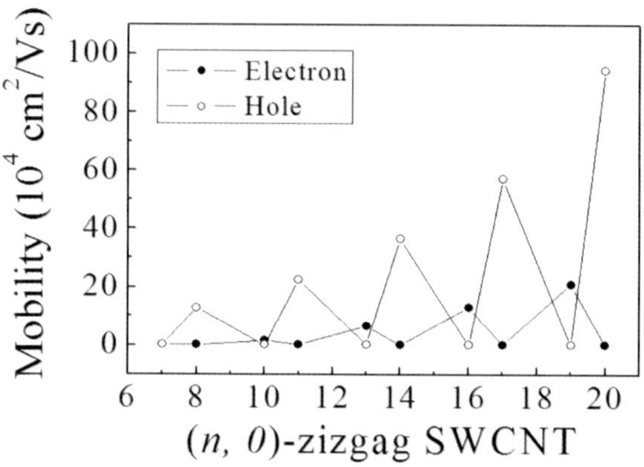

Figure 9. The mobilities of electron μ_e and hole μ_h of the semiconducting zigzag SWCNTs as a function of n.

n	*7*	*8*	*10*	*11*	*13*	*14*	*16*	*17*	*19*	*20*
$M_e^*(m_0)$	0.45	0.48	0.29	0.60	0.25	0.45	0.24	0.36	0.22	0.31
$M_h^*(m_0)$	0.27	0.81	0.27	0.54	0.26	0.41	0.25	0.34	0.23	0.28
$C(10^{11}$eV/cm)	0.97	1.13	1.41	1.60	1.80	2.04	2.37	2.64	3.06	3.48
E_c(eV)	4.52	5.35	2.15	14.3	1.32	13.9	1.12	13.7	1.06	13.5
E_v(eV)	14.8	0.31	14.2	0.37	13.8	0.41	13.5	0.43	13.2	0.44
$\mu_e(10^3$cm^2/Vs)	1.31	0.92	15.5	0.14	65.6	0.28	130	0.52	209	0.88
$\mu_h(10^3$cm^2/Vs)	0.25	128	0.41	224	0.56	367	0.89	571	1.24	945

Table 1. The calculated effective masses $|m^*|$

(m_e^*, m_h^*), the stretching modulus C, the deformation constants E_c and E_v, and the mobilities of electron μ_e and hole μ_h for the semiconducting zigzag SWCNTs for n = 7, 8, 10, 11, 13, 14, 16, 17, 19, and 20.

It is found that the intrinsic electron mobility can reach 2×10^5 cm^2/Vs at room temperature for n=19, and the hole mobility is calculated as 10^6 cm^2/Vs at room temperature for n=20. We find that the mobility exhibits a distinct alternating behavior: for n=3q+1, the intrinsic hole room-temperature mobility is about in two orders of magnitude less than that of electron;

for n=3q+2, the intrinsic hole room-temperature mobility is about in two orders of magnitude larger than that of electron. It is in consistent with DP constant, E_c or E_v, which is related to the band-edge shift induced by the scattering of an electron (conduction band edge) or hole (valence band edge) from the acoustic phonon.

To understand the alternating behavior of DP constant, we examine the frontier molecular orbitals at the Γ-point, i.e., the highest occupied molecular orbital (HOMO) for the hole and the lowest unoccupied molecular orbital (LUMO) for electron, see Figure 10, For $n = 7$, it is found that the bonding direction of HOMO is perpendicular to the longitudinal direction and it is of anti-bonding character along the transport direction. While for the LUMO, the bonding direction is along the stretching direction. The bonding state is stable and anti-bonding state is unstable, which means the site energy of anti-bonding state is more prone to change when the structure is deformed. The band-edge shift due to CNT stretching comes from the site energy change. Thus, DP constant of hole state (HOMO) is larger than that of electron state (LUMO), and hole is scattered more strongly by acoustic phonons than electron. However, for $n = 8$, the LUMO is vertically to the longitudinal direction, while the HOMO is along the longitudinal direction.

It is thus expected that for n=3q+1, the hole state (HOMO) is scattered much more strongly than the electron state (LUMO) from the acoustic phonon, while it is just opposite for n=3q+2. So the alternating behaviors of the carrier mobilities of the semiconducting zigzag SWCNT are reasonable.

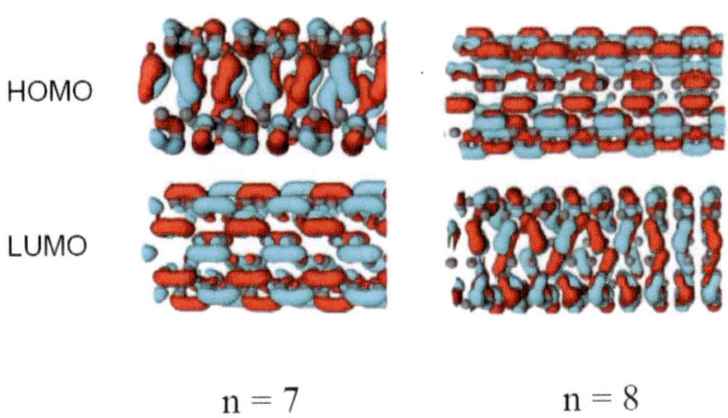

Figure 10. The Γ-point HOMO and LUMO wave functions for the zigzag SWCNT with $n = 7$ and $n = 8$.

5. Conclusion

CNTFETs are important devices with potentially important applications in nanoelectronics. In this work, we have summarized the electron-phonon scattering on the carrier transport.

The carrier mobility of the semiconducting zigzag SWCNT scattered from the acoustic phonons is investigated by using first-principles calculations. We considered only the longitudinal acoustic phonon scattering process by using the deformation-potential theory. We found that the intrinsic carrier mobility can reach 10^6 cm^2/Vs at room temperature for n=20, and the intriguing alternating behaviors of the carrier mobilities of the semiconducting zigzag SWCNTs are due to the curvature effects of the CNT. We believe that the detailed investigation of acoustic phonon scattering in CNTs [62] will also help us to study the carrier mobilities in other organic or inorganic materials by using the similar technique.

Acknowledgements

This work is supported by the Fundamental Research Funds for the Central Universities, a Project Funded by the Priority Academic Program Development of Jiangsu Higher Education Institutions (PAPD). Bo Xu thanks the support by the China Postdoctoral Science Foundation funded project (20100481119) and Jiangsu Planned Projects for Postdoctoral Research Funds (1002007B).

Author details

Bo Xu[1], Jiang Yin[1] and Zhiguo Liu[1]

1 Department of Materials Science and Engineering, Nanjing University, People's Republic of China

References

[1] Iijima, S. (1991). Helical Microtubules of Graphitic Carbon. *Nature*, 354, 56-58.

[2] Bachtold, P.H., Nakanishi, T., & Dekker, C. (2001). Logic Circuits with Carbon Nanotube Transistors. *Science*, 294, 1317-1320.

[3] Tans, S. J., Verschueren, A. R. M., & Dekker, C. (1998). Room-temperature transistor based on a single carbon nanotube. *Nature*, 393, 49-52.

[4] Radosavljevic, M., Freitag, M., Thadani, K. V., & Johnson, A. T. (2002). Nonvolatile molecular memory elements based on ambipolar nanotube field effect transistors. *Nano Lett.*, 2, 761-764.

[5] Kong, J., Franklin, N. R., Zhou, C., Chapline, M. G., Peng, S., Cho, K., & Dai, H. (2000). Nanotube Molecular Wires as Chemical Sensors. *Science*, 287, 622-625.

[6] Frank, S., Poncharal, P., Wang, Z. L., & de Heer, W. A. (1998). Carbon nanotube quantum resistor. *Science*, 280, 1744-1746.

[7] Kong, J., Yenilmez, E., Tombler, T. W., Kim, W., & Dai, H. (2001). Quantum interference and ballistic transmission in nanotube electron waveguide. *Phys. Rev. Lett*, 87, 106801.

[8] Javey, J.G., Wang, Q., Lundstrom, M., & Dai, H. J. (2003). Ballistic carbon nanotube field-effect transistors. *Nature*, 424, 654-657.

[9] Javey, J.G., Farmer, D. B., Wang, Q., Yenilmez, E., Gordon, R. G., Lundstrom, M., & Dai, H. J. (2004). Self-aligned ballistic molecular transistor and electrically parallel nanotube arrays. *Nano Lett.*, 4, 1319-1322.

[10] Lin, Y. M., Appenzeller, J., Chen, Z., Cgen, Z. G., Cheng, H. M., & Avouris, P. H. (2005). High-performance dual-gate carbon nanotube FETs with 40nm gate length. *IEEE Electron Device Lett.*, 26, 823-825.

[11] Durkop, T., Getty, S. A., Cobas, E., & Fuhrer, M. S. (2004). Extraordinary mobility in semiconducting carbon nanotubes. *Nano Lett.*, 4, 35-39.

[12] Zhou, X., Park, J. Y., Huang, S., Liu, J., & Mc Euen, P. L. (2005). Band structure, phonon scattering, and the performance limit of single-walled carbon nanotube transistors. *Phys. Rev. Lett.*, 95, 146805.

[13] Yao, Z., Kane, C. L., & Dekker, C. (2000). High-field electrical transport in single-wall carbon nanotubes. *Phys. Rev. Lett.*, 84, 2941.

[14] Javey Guo, J., Paulsson, Q., Wang, D., Mann, M., & Lundstrom, H.D. (2004). High-field quasiballistic transport in short carbon nanotubes. *Phys. Rev. Lett.*, 92, 106804.

[15] Park, J. Y., Rosenblatt, S., Yaish, Y., Sazonova, V., Üstünel, H., Braig, S., Arias, T. A., Piet, W., Brouwer, Paul. , & Mc Euen, L. (2004). Electron-phonon scattering in metallic single-walled carbon nanotubes. *Nano Lett.*, 4, 517-520.

[16] Pennington, G., & Goldsman, N. (2003). Semiclassical transport and phonon scattering of electrons in semiconducting carbon nanotubes. *Phys. Rev. B.*, 68, 045426.

[17] Pennington, G., & Goldsman, N. (2005). Low-field semiclassical carrier transport in semiconducting carbon nanotubes. *Phys. Rev. B*, 71, 205318.

[18] Perebeinos, V., Tersoff, J., & Avouris, P. H. (2005). Electron-phonon interaction and transport in semiconducting carbon nanotubes. *Phys. Rev. Lett.*, 94, 086802.

[19] Perebeinos, V., Tersoff, J., & Avouris, P. H. (2006). Mobility in semiconducting carbon nanotubes at finite carrier density. *Nano Lett.*, 6, 205-208.

[20] Verma, M. Z., Kauser, Z., & Ruden, P.P. (2005). Ensemble Monte Carlo transport simulations for semiconducting carbon nanotubes. *J. Appl. Phys.*, 97, 114319.

[21] Verma, M. Z., Kauser, , & Ruden, P. P. (2005). Effects of radial breathing mode phonons on charge transport in semiconducting zigzag carbon nanotubes. *Appl. Phys. Lett.*, 87, 123101.

[22] d'Honincthun, H. C., Retailleau Galdin, S., See, J., & Dollfus, P. (2005). Electron-phonon scattering and ballistic behavior in semiconducting carbon nanotubes. *Appl. Phys. Lett.*, 87, 172112.

[23] Saito, R., Dresselhaus, G., & Dresselhaus, M. S. (1998). Physical Property of Carbon Nanotubes. Imperial College Press London

[24] Suzuura, H., & Ando, T. (2002). Phonons and electron-phonon scattering in carbon nanotubes. *Phys. Rev. B*, 65, 235412.

[25] Goupalov, S. V. (2005). Continuum model for long-wavelength phonons in twodimensional graphite and carbon nanotubes. *Phys. Rev. B*, 71, 085420.

[26] Kurti, J., Kresse, G., & Kuzmany, H. (1998). First-principles calculation of the radial-breathing mode of single-wall carbon nanotubes. Phys. Rev. B. , 58, R8869.

[27] Sanchez, D., Portal, E., Artacho, J. , Soler, M., Rubio, A., & Ordejon, P. (1999). Ab initio structural, elastic, and vibrational properties of carbon nanotubes. *Phys. Rev. B*, 59, 12678.

[28] Dubay, O., & Kresse, G. (2003). Accurate density functional calculations for the phonon dispersion relations of graphite layer and carbon nanotubes. *Phys. Rev. B.*, 67, 035401.

[29] Mahan, G. D. (2003). Electron-optical phonon interaction in carbon nanotubes. *Phys. Rev. B*, 68, 125409.

[30] Jiang, J., Saito, R., Ge, G., Samsonidze, S. G., Chou, A., Jorio, G., & Dresselhaus, M. S. (2005). Electron-phonon matrix elements in single-wall carbon nanotubes. *Phys. Rev. B*, 72, 235408.

[31] Popov, V. N., & Lambin, P. (2006). Intraband electron-phonon scattering in single-walled carbon nanotubes. *Phys. Rev. B*, 74, 075415.

[32] Machon, M., Reich, S., Telg, H., Maultzsch, J., Ordejon, P., & Thomsen, C. (2005). Strength of radial breathing mode in single-walled carbon nanotubes. Phys. Rev. B. , 71, 035416.

[33] Guo, J. (2005). A quantum-mechanical treatment of phonon scattering in carbon nanotube transistors. *J. Appl. Phys.*, 98, 063519.

[34] Pourfath, M., Kosina, H., & Selberherr, S. (2006). Rigorous modeling of carbon nanotube transistors. *J. Phys.: Conf. Ser.*, 38, 29.

[35] Koswatta, S. O., Sayed, Hasan., Mark, S., Lundstrom, M. P., Anantram, Dmitri. E., & Nikonov, . (2007). Non-equilibrium Green's function treatment of phonon scattering in carbon nanotube transistors. *IEEE Trans. Elec. Dev.*, 54, 2339-2351.

[36] Dresselhaus, M. S., Dresselhaus, G., & Eklund, P. C. (1995). Science of Fullerenes and Carbon Nanotubes: Their Properties and Applications. *(Academic Press, Inc.)*, 0-12221-820-5.

[37] Kane, L., Mele, E. J., Lee, R. S., Fischer, J. E., Petit, P., Dai, H., Thess, A., Smalley, R. E., Verschueren, A. R. M., Tans, S. J., & Dekker, C. (1998). Temperature-dependent resistivity of single-wall carbon nanotubes. *Europhys. Lett.*, 41, 683-688.

[38] Appenzeller, J., Martel, R., Avouris, P., Stahl, H., & Lengeler, B. (2001). Optimized contact configuration for the study of transport phenomena in ropes of single-wall carbon nanotubes. *Appl. Phys. Lett.*, 78, 3313.

[39] Chen, Y. F., & Fuhrer, M. S. (2005). Electric-field-dependent charge-carrier velocity in semiconducting carbon nanotubes. *Phys. Rev. Lett.*, 95, 236803.

[40] Avouris, P., Chen, Z. H., & Perebeinos, V. (2007). Carbon-based electronics. *Nature Nanotech.*, 2, 605-615.

[41] Guo, J. (2005). A quantum-mechanical treatment of phonon scattering in carbon nanotube transistors. *J. Appl. Phys.*, 98, 063519.

[42] Guo, J., & Lundstrom, M. (2005). Role of phonon scattering in carbon nanotube field-effect transistors. *Appl. Phys. Lett.*, 86, 193103.

[43] John, L., Castro, L. C., Clifford, J. P., & Pulfrey, D. L. (2003). Electrostatics of coaxial schottky-barrier nanotube field-effect transistors. *IEEE Trans. Nanotech.*, 2, 175-180.

[44] Pennington, G., & Goldsman, N. (2003). Semi-classical transport and phonon scattering on electrons in semiconducting carbon nanotubes. *Phys. Rev. B.*, 68, 045426.

[45] Alam, K., & Lake, R. (2005). Performance of 2 nm gate length carbon nanotube field-effect transistors with source-drain underlaps. *Appl. Phys. Lett.*, 87, 073104.

[46] Verma Kauser, M. Z., & Ruden, P. P. (2005). Ensemble Monte Carlo transport simulations for semiconducting carbon nanotubes. *J. Appl. Phys.*, 97, 114319.

[47] Xia, T. S., Register, L. R., & Banerjee, S. K. (2004). Calculations and applications of the complex band structure for carbon nanotube field-effect transistors. Phys. Rev. B. , 70, 045332.

[48] Svizhenko Anantram, M. P. (2005). Effect of scattering and contacts on current and electrostatics in carbon nanotubes. *Phys. Rev. B.*, 72, 085430.

[49] Martel, R., Schmidt, T., Shea, H. R., Hertel, T., & Avouris, P. (1999). *Single- and Multi-Wall Carbon Nanotube.*

[50] Field-Effect Transistors,. (1998). *Appl. Phys. Lett.*, 73, 2447.

[51] Appenzeller, J., Knoch, J., Derycke, V., Martel, R., Wind, S., & Avouris, P. (2002). Field-Modulated Carrier Transport in Carbon Nanotube Transistors. *Phys. Rev. Lett.*, 89, 126801.

[52] Bardeen, J., & Shockley, W. (1950). Deformation Potentials and Mobilities in Non-Polar Crystals. *Phys. Rev.*, 80, 72.

[53] Beleznay, F. B., Bogar, F., & Ladik, J. (2003). Calculation of the charge-carrier mobility of polyguanylic acid: The simultaneous effect of stretching and twisting,. *J. Chem. Phys.*, 119, 5690.

[54] Long, M. Q., Tang, L., Wang, D., Wang, L. J., & Shuai, Z. G. (2009). Theoretical predictions of size dependent carrier mobility and polarity in graphene. *J. Am. Chem. Soc.*, 131, 177728-177731.

[55] Sa, N., Wang, G., Yin, B., & Huang, Y. (2008). Theoretical study on non-covalent functionalization of armchair carbon nanotube by tetrathiafulvalene molecule. Physica E. , 40, 2396-2399.

[56] Kresse, G., & Furthmüller, J. (1996). Efficiency of ab-initio total energy calculations for metals and semiconductors using a plane-wave basis set. *Comput. Mater. Sci.*, 6, 15.

[57] Kresse, G., & Furthmüller, J. (1996). Efficient iterative schemes for ab initio total-energy calculations using a plane-wave basis set, Phys. *Rev. B.*, 54, 11169.

[58] Perdew, J. P., Burke, K., & Ernzerhof, M. (1996). Generalized gradient approximation made simple. *Phys. Rev. Lett.*, 77, 3865.

[59] Blöchl, P. E. (1994). Projector augmented-wave method. Phys. Rev. B, , 50, 17953.

[60] Kresse, G., & Joubert, D. (1999). From ultrasoft pseudopotentials to the projector augmented-wave method. *Phys. Rev. B.*, 59, 1758.

[61] Sreekala, S., Peng, X. H., Ajayan, P. M., & Nayak, S. K. (2008). Effect of strain on the band gap and effective mass of zigzag single-wall carbon nanotubes: First-principles density-functional calculations. *Phys. Rev. B.*, 77, 155434.

[62] Xu, Y. D., Xia, J., Yin, X. G., Wan, K., Jiang, A. D., Li, D., Wu, Z. G., & Liu, . (2010). The effect of acoustic phonon scattering on the carrier mobility in the semiconducting zigzag single wall carbon nanotubes. *Appl. Phys. Lett.*, 96, 183108.

Permissions

The contributors of this book come from diverse backgrounds, making this book a truly international effort. This book will bring forth new frontiers with its revolutionizing research information and detailed analysis of the nascent developments around the world.

We would like to thank Dr. Satoru Suzuki, for lending his expertise to make the book truly unique. He has played a crucial role in the development of this book. Without his invaluable contribution this book wouldn't have been possible. He has made vital efforts to compile up to date information on the varied aspects of this subject to make this book a valuable addition to the collection of many professionals and students.

This book was conceptualized with the vision of imparting up-to-date information and advanced data in this field. To ensure the same, a matchless editorial board was set up. Every individual on the board went through rigorous rounds of assessment to prove their worth. After which they invested a large part of their time researching and compiling the most relevant data for our readers. Conferences and sessions were held from time to time between the editorial board and the contributing authors to present the data in the most comprehensible form. The editorial team has worked tirelessly to provide valuable and valid information to help people across the globe.

Every chapter published in this book has been scrutinized by our experts. Their significance has been extensively debated. The topics covered herein carry significant findings which will fuel the growth of the discipline. They may even be implemented as practical applications or may be referred to as a beginning point for another development. Chapters in this book were first published by InTech; hereby published with permission under the Creative Commons Attribution License or equivalent.

The editorial board has been involved in producing this book since its inception. They have spent rigorous hours researching and exploring the diverse topics which have resulted in the successful publishing of this book. They have passed on their knowledge of decades through this book. To expedite this challenging task, the publisher supported the team at every step. A small team of assistant editors was also appointed to further simplify the editing procedure and attain best results for the readers.

Our editorial team has been hand-picked from every corner of the world. Their multi-ethnicity adds dynamic inputs to the discussions which result in innovative outcomes. These outcomes are then further discussed with the researchers and contributors who give their valuable feedback and opinion regarding the same. The feedback is then collaborated with the researches and they are edited in a comprehensive manner to aid the understanding of the subject.

Apart from the editorial board, the designing team has also invested a significant amount of their time in understanding the subject and creating the most relevant covers. They scrutinized every image to scout for the most suitable representation of the subject and create an appropriate cover for the book.

The publishing team has been involved in this book since its early stages. They were actively engaged in every process, be it collecting the data, connecting with the contributors or procuring relevant information. The team has been an ardent support to the editorial, designing and production team. Their endless efforts to recruit the best for this project, has resulted in the accomplishment of this book. They are a veteran in the field of academics and their pool of knowledge is as vast as their experience in printing. Their expertise and guidance has proved useful at every step. Their uncompromising quality standards have made this book an exceptional effort. Their encouragement from time to time has been an inspiration for everyone.

The publisher and the editorial board hope that this book will prove to be a valuable piece of knowledge for researchers, students, practitioners and scholars across the globe.

List of Contributors

Maria Alexandra Fonseca, Bruno Lamas, Bruno Abreu and Hugo Calisto
TEMA - Centre for Mechanical Technology and Automation, Portugal

Maria Alexandra Fonseca, Sylvio Freitas, Bruno Lamas, Bruno Abreu, Hugo Calisto, Nelson Martins and Mónica Oliveira
Department of Mechanical Engineering, University of Aveiro, Portugal

Metin Aydin
Department of Chemistry, Faculty of Art and Sciences, Ondokuz Mayıs University, Samsun, Turkey

Daniel L. Akins
Center for Analysis of Structures and Interfaces (CASI), Department of Chemistry, The City College of The City University of New York, New York, USA

Masaru Tachibana
Department of Nanosystem Science, Yokohama City University, Japan

Takeo Oku
The University of Shiga Prefecture, Japan

Toshiaki Kato and Rikizo Hatakeyama
Department of Electronic Engineering, Tohoku University, Sendai, Japan

Rolant Eba Medjo
Department of Physics, Faculty of Science, University of Douala, Republic of Cameroon

Najib A. Kasti
Department of Mechanical Engineering, American University of Beirut, Lebanon

Zhen Xia and Ruhong Zhou
Computational Biology Center, IBM Thomas J. Watson Research Center, Yorktown Heights, NY 10598

Zhen Xia
Department of Biomedical Engineering, The University of Texas at Austin, Austin, TX 78712

Peng Xiu
Department of Engineering Mechanics, and Soft Matter Research Center, Zhejiang University, Hangzhou, 310027, China

Ruhong Zhou
Department of Chemistry, Columbia University, New York, NY 10027

Eun-Soo Park
Youngchang Silicone Co., Ltd., Korea

Anup K. Deb and Charles C. Chusuei
Chemistry Department, Middle Tennessee State University, Murfreesboro, Tennessee, USA

Xia Xin and Guiying Xu
National Engineering Technology Research Center for Colloidal Materials, Shandong University, Jinan, P. R. China

Hongguang Li
Laboratory of Clean Energy Chemistry and Materials, Lanzhou Institute of Chemical Physics, Chinese Academy of Sciences, Lanzhou, , P. R. China

Dong Wook Chang
Department of Chemical Systematic Engineering, Catholic University of Daegu, S. Korea

In-Yup Jeon, Hyun-Jung Choi and Jong-Beom Baek
Interdisciplinary School of Green Energy/Low-Dimensional Carbon Materials Center, Ulsan, National University of Science and Technology (UNIST), South Korea

Hock Guan Ong and Junling Wang
School of Materials Science and Engineering Nanyang Technological University, Singapore

Iñigo Martin-Fernandez
Instituto de Microelectrónica de Barcelona (IMB-CNM), CSIC, Campus UAB, Barcelona, Spain

Gemma Gabriel, Xavi Illa, Anton Guimera, Philippe Godignon and Rosa Villa
Instituto de Microelectrónica de Barcelona (IMB-CNM), CSIC, Campus UAB, Barcelona, Spain
CIBER-BBN, Networking Center on Bioengineering, Biomaterials and Nanomedicine, Spain

Javier Hernández-Ferrer and Mª Teresa Martínez
Instituto de Carboquímica (CSIC), C/Miguel Luesma Castán 4, Zaragoza, Spain

Beatriz Rebollo
IDIBAPS (Institute of Biomedical Research August Pi y Sunyer), Barcelona, Spain

Maria V. Sanchez-Vives
IDIBAPS (Institute of Biomedical Research August Pi y Sunyer), Barcelona, Spain
ICREA (Institut Catala de Recerca i Estudis Avançats), Barcelona, Spain

Bo Xu, Jiang Yin and Zhiguo Liu
Department of Materials Science and Engineering, Nanjing University, People's Republic of China